AF361639

Vascular Inflammation: From Target Molecules to Therapeutic Approaches

Vascular Inflammation: From Target Molecules to Therapeutic Approaches

Editor

Byeong Hwa Jeon

Basel • Beijing • Wuhan • Barcelona • Belgrade • Novi Sad • Cluj • Manchester

Editor
Byeong Hwa Jeon
Department of Physiology,
College of Medicine,
Chungnam National
University
Daejeon, Republic of Korea

Editorial Office
MDPI
St. Alban-Anlage 66
4052 Basel, Switzerland

This is a reprint of articles from the Special Issue published online in the open access journal *Biomedicines* (ISSN 2227-9059) (available at: https://www.mdpi.com/journal/biomedicines/special_issues/vasc_inflamm).

For citation purposes, cite each article independently as indicated on the article page online and as indicated below:

Lastname, A.A.; Lastname, B.B. Article Title. *Journal Name* **Year**, *Volume Number*, Page Range.

ISBN 978-3-0365-8864-3 (Hbk)
ISBN 978-3-0365-8865-0 (PDF)
doi.org/10.3390/books978-3-0365-8865-0

Contents

About the Editor . . . . **vii**

Preface . . . . **ix**

David L. Bernstein, Xinpei Jiang and Slava Rom
let-7 microRNAs: Their Role in Cerebral and Cardiovascular Diseases, Inflammation, Cancer,
and Their Regulation
Reprinted from: *Biomedicines* **2021**, *9*, 606, doi:10.3390/biomedicines9060606 . . . . **1**

Tommaso Gori
Coronary Vasculitis
Reprinted from: *Biomedicines* **2021**, *9*, 622, doi:10.3390/biomedicines9060622 . . . . **19**

Sang-Wan Chung
Vasculitis: From Target Molecules to Novel Therapeutic Approaches
Reprinted from: *Biomedicines* **2021**, *9*, 757, doi:10.3390/biomedicines9070757 . . . . **33**

Sai Sahana Sundararaman, Yvonne Döring and Emiel P. C. van der Vorst
PCSK9: A Multi-Faceted Protein That Is Involved in Cardiovascular Biology
Reprinted from: *Biomedicines* **2021**, *9*, 793, doi:10.3390/biomedicines9070793 . . . . **47**

Rajiv Sanwal, Kushal Joshi, Mihails Ditmans, Scott S. H. Tsai and Warren L. Lee
Ultrasound and Microbubbles for Targeted Drug Delivery to the Lung Endothelium in ARDS:
Cellular Mechanisms and Therapeutic Opportunities
Reprinted from: *Biomedicines* **2021**, *9*, 803, doi:10.3390/biomedicines9070803 . . . . **69**

**Ghassan Bkaily, Yanick Simon, Ashley Jazzar, Houssein Najibeddine, Alexandre Normand
and Danielle Jacques**
High Na⁺ Salt Diet and Remodeling of Vascular Smooth Muscle and Endothelial Cells
Reprinted from: *Biomedicines* **2021**, *9*, 883, doi:10.3390/biomedicines9080883 . . . . **93**

Francesco Nappi, Adelaide Iervolino and Sanjeet Singh Avtaar Singh
COVID-19 Pathogenesis: From Molecular Pathway to Vaccine Administration
Reprinted from: *Biomedicines* **2021**, *9*, 903, doi:10.3390/biomedicines9080903 . . . . **107**

Francesca Gorini, Serena Del Turco, Laura Sabatino, Melania Gaggini and Cristina Vassalle
H$_2$S as a Bridge Linking Inflammation, Oxidative Stress and Endothelial Biology: A Possible
Defense in the Fight against SARS-CoV-2 Infection?
Reprinted from: *Biomedicines* **2021**, *9*, 1107, doi:10.3390/biomedicines9091107 . . . . **135**

**Patrizia Leone, Marcella Prete, Eleonora Malerba, Antonella Bray, Nicola Susca,
Giuseppe Ingravallo and et al.**
Lupus Vasculitis: An Overview
Reprinted from: *Biomedicines* **2021**, *9*, 1626, doi:10.3390/biomedicines9111626 . . . . **149**

**Jan Menzenbach, Stilla Frede, Janine Petras, Vera Guttenthaler, Andrea Kirfel,
Claudia Neumann and et al.**
Perioperative Vascular Biomarker Profiling in Elective Surgery Patients Developing
Postoperative Delirium: A Prospective Cohort Study
Reprinted from: *Biomedicines* **2021**, *9*, 553, doi:10.3390/biomedicines9050553 . . . . **169**

Yu-Ran Lee, Hee-Kyoung Joo, Eun-Ok Lee, Sungmin Kim, Hao Jin, Yeon-Hee Choi and et al.
17β-Estradiol Increases APE1/Ref-1 Secretion in Vascular Endothelial Cells and
Ovariectomized Mice: Involvement of Calcium-Dependent Exosome Pathway
Reprinted from: *Biomedicines* **2021**, *9*, 1040, doi:10.3390/biomedicines9081040 . . . . **183**

Yu-Ran Lee, Eun Young Bae, Hong Ryang Kil, Byeong-Hwa Jeon and Geena Kim
Elevated Plasma Apurinic/Apyrimidinic Endonuclease 1/Redox Effector Factor-1 Levels in
Refractory Kawasaki Disease
Reprinted from: *Biomedicines* **2022**, *10*, 190, doi:10.3390/biomedicines10010190 . . . . **199**

Vikrant Rai and Devendra K. Agrawal
Transcriptomic Analysis Identifies Differentially Expressed Genes Associated with Vascular
Cuffing and Chronic Inflammation Mediating Early Thrombosis in Arteriovenous Fistula
Reprinted from: *Biomedicines* **2022**, *10*, 433, doi:10.3390/biomedicines10020433 . . . . **207**

Diego Caicedo, Clara V. Alvarez, Sihara Perez-Romero and Jesús Devesa
The Inflammatory Pattern of Chronic Limb-Threatening Ischemia in Muscles: The TNF-α
Hypothesis
Reprinted from: *Biomedicines* **2022**, *10*, 489, doi:10.3390/biomedicines10020489 . . . . **227**

About the Editor

Byeong Hwa Jeon

Byeong Hwa Jeon, MD, PhD is a professor at the College of Medicine, Chungnam National University, Korea, who is renowned for his extensive expertise in vascular physiology, redox biology, and medical education. He earned his medical doctoral degree from Chungnam National University in 1989, completed his PhD in Vascular Physiology at the same university in 1996, and conducted postdoctoral research into Redox Biology at Johns Hopkins University, USA, in 2001. Since 2013, he has served as the Director of the Research Institutes of Medical Sciences at Chungnam National University. His research is centered on redox biology within vascular cell systems and therapeutic approaches used to treat vascular inflammation, making him a prominent figure in these fields.

Preface

This Special Issue, titled "Vascular Inflammation: From Target Molecules to Therapeutic Approaches", will focus on new target molecules or biomarkers and novel therapeutic approaches used to identify and treat vascular inflammation. The pathophysiology of vascular inflammation is complex, with many factors being involved. This Special Issue includes nine review articles and five research papers, in which several novel biomarkers and therapeutic target proteins associated with vascular inflammation are described. We would like to thank the many scientists who participated in peer review, as well as the Managing Editors of this Special Issue of *Biomedicines*.

Byeong Hwa Jeon
Editor

Review

let-7 microRNAs: Their Role in Cerebral and Cardiovascular Diseases, Inflammation, Cancer, and Their Regulation

David L. Bernstein [1], Xinpei Jiang [1] and Slava Rom [1,2,*]

[1] Department of Pathology and Laboratory Medicine, Lewis Katz School of Medicine, Temple University, Philadelphia, PA 19140, USA; david.bernstein@temple.edu (D.L.B.); xinpei@temple.edu (X.J.)

[2] Center for Substance Abuse Research, Lewis Katz School of Medicine, Temple University, Philadelphia, PA 19140, USA

* Correspondence: srom@temple.edu; Tel.: +1-(215)-707-9412; Fax: +1-(215)-707-5255

Abstract: The *let*-7 family is among the first microRNAs found. Recent investigations have indicated that it is highly expressed in many systems, including cerebral and cardiovascular systems. Numerous studies have implicated the aberrant expression of *let*-7 members in cardiovascular diseases, such as stroke, myocardial infarction (MI), cardiac fibrosis, and atherosclerosis as well as in the inflammation related to these diseases. Furthermore, the *let*-7 microRNAs are involved in development and differentiation of embryonic stem cells in the cardiovascular system. Numerous genes have been identified as target genes of *let*-7, as well as a number of the *let*-7′ regulators. Further studies are necessary to identify the gene targets and signaling pathways of *let*-7 in cardiovascular diseases and inflammatory processes. The bulk of the *let*-7′ regulatory proteins are well studied in development, proliferation, differentiation, and cancer, but their roles in inflammation, cardiovascular diseases, and/or stroke are not well understood. Further knowledge on the regulation of *let*-7 is crucial for therapeutic advances. This review focuses on research progress regarding the roles of *let*-7 and their regulation in cerebral and cardiovascular diseases and associated inflammation.

Keywords: microRNAs; let-7; stroke; cardiovascular; inflammation

Citation: Bernstein, D.L.; Jiang, X.; Rom, S. *let*-7 microRNAs: Their Role in Cerebral and Cardiovascular Diseases, Inflammation, Cancer, and Their Regulation. *Biomedicines* **2021**, *9*, 606. https://doi.org/10.3390/biomedicines9060606

Academic Editor: Byeong Hwa Jeon

Received: 9 May 2021
Accepted: 24 May 2021
Published: 26 May 2021

Publisher's Note: MDPI stays neutral with regard to jurisdictional claims in published maps and institutional affiliations.

1. The let-7 Family and Inflammation

The *let*-7 family of microRNAs is one of the earliest originally discovered microRNAs. When several isoforms were identified in *C. elegans* in 2000 [1], the miR was named *lethal-7* (*let*-7) because its knockout was lethal during development [2]. The discovery of *let*-7, along with lin-4 [3], opened much of the current field of miR research. To date, 12 genetic loci have been identified as origination sites of *let*-7 in humans [3], while mice have 3 [4] and drosophila have 1 [5]. In humans and mice, 10 of the *let*-7 microRNAs (miRs) are present (*let*-7*a, b, c, d, e, f, g, i*, and miR-98 and miR-202) [3]. However, in spite of the multiple sites of origin, all *let*-7 miRs begin as pre-pro *let*-7 transcripts and are then processed through the Drosha pathway [6]. Throughout miRNA biogenesis, after Dicer cleavage, one of the strands is loaded into an RNA-induced silencing complex (RISC) as mature miR. The other strand, which is labelled as the "star strand", is typically degraded [7,8], though for certain miRs, both strands are preserved and are loaded into RISC as mature forms. In such an instance, the mature miR is called, for example, with *let*-7*a* miR, as *let*-7*a-5p* and *let*-7*a-3p* or *let*-7*a* and *let*-7*a**, respectively. Despite being present in many different genetic regions, all final *let*-7 miRs are similar in length, differing by only 0–3 nucleotides (Figure 1), although their functions differ significantly in protein translation and physiological function.

As one of the first miR families discovered, much has been investigated about the multiple roles that the *let*-7 family plays within the body. Numerous studies have linked *let*-7 miRs to many processes, from cell proliferation [9,10] and bone remodeling [11] to cardiac output [12]. However, across many tissues and conditions, miRs from the *let*-7 family appear to be particularly involved in the signals involved in the growth and stress characteristics, *let-7e* has been shown to be an early proinflammatory marker of hypoxic damage, and it may further propagate damage [29,30].

The anti- and pro- inflammatory pathways regulated by *let-7* miRs are not confined to the CNS. *let-7i* has been associated with wound repair across many cell types, due in part to interactions with progesterone [31]. *let-7c* has been shown to be associated with regulation of dental inflammation through extracellular matrix (ECM)-specific mechanisms [32]. *let-7d* is involved in recovery from hypoxia in cardiac tissue and offers therapeutic potential for treating the aftermath of MI [33] by stimulating proliferation and activating survival pathways in cardiac cells. *let-7g* shows similarly strong potential for stimulating cell repair within ischemic heart and vascular tissue [34] and has been shown to promote angiogenesis through activation of vascular endothelial growth factor (VEGF)-mediated signaling following insult [34,35].

Taken together, the *let-7* family appears critical for the progression of inflammation. However, its individual members can have significant and sometimes contradictory impacts on such processes. One possible explanation is the timeline in which expression occurs. Certain *let-7* family members such as *let-7a* and *let-7e* are associated with early inflammatory and pre-apoptotic pathways, while others such as *let-7i* and *let-7f* are involved in later phases of transcription and downstream elements of apoptosis (Figure 2). In the following section, we further expand on the particular regulatory elements of *let-7* miRs and why the timeline of expression of these miRs and subsequent impact on cytokine and chemokine release appear to have such high variability.

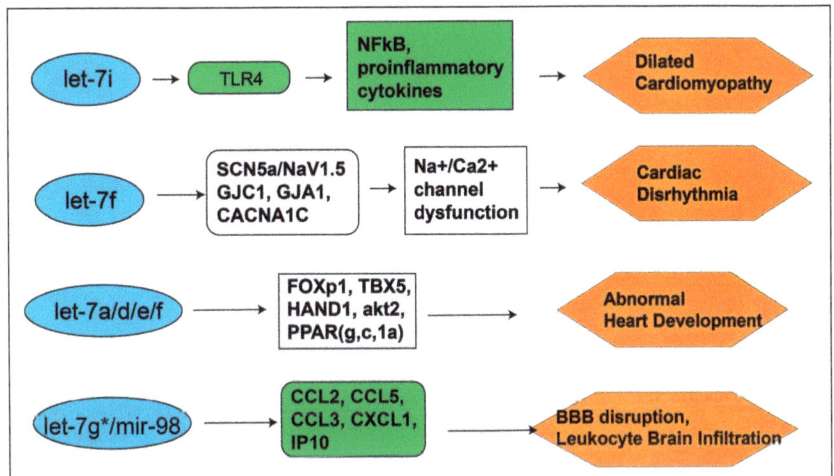

Figure 2. Targets of the *let-7* miRs in cerebral and cardiovascular disease conditions. Green boxes show confirmed targets, while white boxes show bioinformatic prediction [13,21,23–26,36,37].

1.1. let-7 miRNAs, Cell Division, and Vascular Function

let-7 is critical for prenatal development. It appears soon after fertilization, where it is responsible for mediating blastocyst attachment to uterine spiral arteries. Even from the beginning, *let-7* miRs work as clipping signals, minimizing the proliferation of non-adherent cells through suppression of transcription factors [2], thereby allowing only attached cells to grow. For this reason, *let-7* was named *"lethal-7"*, because its absence leads to uncontrolled cellular proliferation, preventing development into a viable embryo. This silencing-type role continues throughout later stages of embryonic development [36]. After the first trimester, *let-7* miRs are critical for the differentiation of different organs by arresting the proliferation of non-needed cells, functioning as a "stop" sign in many tissues, including lung [37], brain [36], heart, and vascular tissue [38]. In particular, *let-7* provides critical control of the length of the DNA replication phase in neural stem cells, thereby regulating the growth of the CNS from the first trimester through birth [39–41]. This critical regulation of cellular differentiation appears to be conserved across species; *let-7* is also necessary for differentiation in rodents [41] and invertebrates [5].

let-7's role in the vascular system begins from day 1, as it is subject to regulation by chemokines secreted by endothelial cells (EC) within the uterus [2]. *let-7* expression is critical for maintaining the integrity of endothelial cells, and its normal expression is considered critical for maintenance of the blood–brain barrier in the face of ischemic disease [26]. These processes are not confined to periods of stress; *let-7* is also critical for maintenance of endothelial cell walls. In normal functioning, *let-7* is responsible for transducing fibroblast growth factor (FGF) signaling into changes to TGF-β within endothelial cells, thereby limiting proliferation [42,43] in non-damaged blood vessels. *let-7g* has been shown to reduce EC inflammation and monocyte adhesion [21,23,24,44], diminish EC senescence, and play a role in controlling arterial stiffness and aging [45,46]. The strong effect of *let-7* on vascular health and angiogenesis underscores its role in recovery from stroke and other vascular diseases processes, such as myocardial infarction [12,36]. However, *let-7*'s critical role in regulating EC division also makes the miR family a critical regulator of disorders of cell proliferation, such as cancer.

1.2. let-7's Role in Cancer and Angiogenesis

In virtually all forms of cancer, continual tumor growth requires altered and abnormal angiogenesis [47]. Without additional vascular collaterals, tumor masses are unable to grow to larger than 1 mm in size [47]. Hence, as a critical regulator of angiogenesis, *let-7* is one of the most important elements in controlling the progression of cancer. Normal physiological levels of *let-7* effectively suppress abnormal angiogenesis and prevent tumor growth [48]. Conversely, suppression of *let-7* expression is a hallmark of most forms of cancer [49]. Moreover, tumors with lower tissue levels of *let-7* proliferate significantly faster than those with higher *let-7* expression [48,50]. Consequently, lower levels of *let-7* are associated with more aggressive growth and poor prognosis in cancer patients [51]. Perhaps most critically, restoration of *let-7* has been shown to have strong anticancer properties. This has led to it being classified as a tumor suppressor [52], as well as a promising target for future cancer therapies [53].

The inverse relationship between tissue expression of *let-7* and cell growth can be attributed in large part to actions on the vasculature. First, in patients with leukemia and lymphoma, *let-7* expression correlated inversely with the spread of the disease [54]. It has also been shown that reductions in *let-7* lead to more abnormal angiogenesis [55] and weaken vascular wall integrity [1]. Such effects may be combinatory and lead to greater inflammation, particularly IFNγ-mediated increases in cell aggregation [56], which can further exacerbate oxidative stress in the area and promote abnormal growth.

let-7 is considered a tumor suppressor gene, due in large part to its normal physiological role in arresting development. In many tissues, including vasculature, lung, and liver, proliferation is arrested by the presence of *let-7* miRNA. Currently, the *let-7* miRs have been associated with cancer. Part of this relationship is physical; *let-7* binds to coding regions and untranslated regions (UTRs) of genes critical for DNA replication such as programmed cell death ligand 1 (PD-L1) [53] and high mobility group AT-hook 2 (HMGA2) [57], as well as apoptotic genes such as caspase 3 [58], B-cell CLL/lymphoma (BCL) [59,60], and caspase 8 [61].

Recently, researchers have determined the bidirectional nature of *let-7* and cancer. miR-98 overexpression has been shown to reduce proliferation in many cells [62]. Its effect is particularly strong in endothelial cells, accounting for most forms of proliferative control [63,64]. It appears to regulate endothelial cell growth through multiple mechanisms, although a critical pathway of control is propagated through interactions with VEGF/Argonaute RISC component (AGO) pathway [36,65,66]. On the list of *let-7* targets are genes controlling cell signaling and cell cycle as well as differentiation. In some cases, *let-7s* are labelled as tumor suppressors because they reduce cancer aggressiveness. Nevertheless, in sporadic conditions, *let-7* acts as an oncogene, accelerating cancer migration, invasion, and chemoresistance due to expression of genes associated with progression and metastasis. For these reasons, *let-7s* might be considered as potential diagnostic and prognostic markers and therapeutic targets for cancer treatment [67].

2. Regulation of *let-7* Expression

let-7 miRs are evolutionarily conserved across species and play essential roles in many biological processes due to their pluripotency, such as in differentiation, growth, proliferation, self-renewal, development, and diseases. Due to the broad effects of *let-7* and any dysregulation leading to disease physiology, it is necessary to tightly control *let-7* expression. Therefore, it is critical to recognize what, where, and how *let-7* expression is regulated throughout its maturation process. *let-7*, as with any other miR, can be regulated transcriptionally and post-transcriptionally throughout the maturation process. miR (*let-7*) biogenesis and maturation are largely dependent on Drosha in the nucleus for primary microRNA (pri-miRNA), Dicer for precursor microRNA (pre-mRNA), and RISC for mature miRNA in the cytoplasm. Due to the complexity of miRNA biogenesis, its expression level can be regulated at different steps. miRNA can be regulated at transcription or post-transcription. Here, we discuss some known positive and negative protein regulators of *let-7* expression at different stages (Table 1).

Table 1. *let-7* regulators.

Regulatory Protein	*let-7* Family	*Pri-let-7* (nucleus-Drosha) or *Pre-let-7* (cytoplasm-Dicer) or Mature *let-7* (RISC)	Promote or Suppress	Mechanism	References
DAF-12	*let-7* family	Transcriptional/*pri-let-7*	Promote/Suppress	1. Unliganded DAF-12 represses *let-7* and liganded DAF-12 promotes *let-7* transcriptionally through binding to *pri-let-7* 3′-UTR 2. *Pri-let-7s* synthesis	[1,68,69]
MYC	*let-7a, 7d, 7f, 7g*	Transcriptional/*pri-let-7*	Suppress	MYC represses *let-7* at the upstream promoter region	[70,71]
LIN42	*let-7* family (*let-7a, 7b* homologs)	Transcriptional/*pri-let-7*	Suppress	Suppresses *let-7* transcriptionally by binding to the pri-*let-7* 3-UTR	[1,72,73]
LIN28A-TUTases4/7	*let-7a, 7b, 7d, 7g, 7i*	Pri-*let-7*/Pre-*let-7*	Suppress	Represses *let-7s* through TUTase-dependent uridylation of *pre-let7s*	[74–80]
LIN28B	*let-7a, 7d, 7f, 7g, 7i*	Pri-*let-7*	Suppress	Represses *let-7s* by sequestering *pri-let-7s* into the nucleolus	[79,81]
TUTases2/4/7	*let-7a, 7b, 7d, 7f, 7g, 7i, miR-98*	Pre-*let-7*	Promote	Promotes *let-7s* by mono-uridylating group II *pre-let-7s*, which enhances Dicer processing	[82]
FHIT	*let-7a, 7b, 7d, 7f, 7g*	Pri-*let-7*	Suppress	Induces LIN28B leading to suppression of *let-7s* through Lin28/Let-7 axis	[83,84]
MUC1-C	*let-7c*	Pri-*let-7*	Suppress	Translocates into the nucleus and interacts with NF-κB to activate Lin28B, leading to *let-7s* repression through Lin28/Let-7 axis	[85–87]
MSI1	*let-7b, 7g, miR-98*	Pri-*let-7*	Suppress	1. Can bind to target *pri-let-7s* 3′-UTR to repress transcription 2. Recruits LIN28 to the nucleus and represses *let-7s* through Lin28/Let-7 axis	[88]
SSB	*let-7a, 7b, 7c, 7d, 7e, 7f, 7g, 7i*	Pri-*let-7*	Suppress	Enhances LIN28B transcription and represses *let-7s* through Lin28/Let-7 axis	[89–91]
TRIM25	*let-7a*	Pre-*let-7*	Suppress	A cofactor for Lin28A/TUTase4-mediated uridylation	[77,78,92]
TRIM71	*let-7a, 7b, 7c, 7d, 7e, 7f, 7g, 7i, miR-98*	Pre-*let-7*/Mature *let-7*	Promote	1. Negatively regulates Lin28B through polyubiquitination 2. Degradation of Ago2	[93–96]
TTP	*Let-7a, 7b, 7f, 7g*	Pre-*let-7*	Promote	Downregulates LIN28A through binding to its AREs	[97,98]
YAP	*let-7a*	Pri-*let-7*	Suppress	YAP translocates into the nucleus and sequesters DDX17 and interferes with Drosha processing	[99,100]
ADAR1	*let-7a, 7d, 7e, 7f; let-7 family*	Pri-*let-7*/Pre-*let-7*	Promote	Enhances Drosha and Dicer processing through direct interactions	[101–106]
hnRNPA1	*let-7a*	Pri-*let-7*	Suppress	1. Direct binding to *pri-let-7* 2. Reduces Drosha processing	[107–110]

Table 1. *Cont.*

Regulatory Protein	*let*-7 Family	*Pri-let*-7 (nucleus-Drosha) or *Pre-let*-7 (cytoplasm-Dicer) or Mature *let*-7 (RISC)	Promote or Suppress	Mechanism	References
KSRP	*let-7a*	Pri-let-7/Pre-let-7	Promote	1. Direct binding to *pri-let-7* and *pre-let-7* 2. *Enhances Drosha processing*	[109,111,112]
TDP-43	*let-7b*	*Pri-let-7*	Promote	1. Interacts with *pri-let-7* 2. Enhances Drosha processing	[109,113–116]
TRAIL-R2	*let-7a, 7b, 7c, 7d, 7e, 7g*	*Pri-let-7*	Suppress	Interacts with Drosha complex to reduce *pri-let-7* processing	[117–119]
NF90/NF45	*let-7a*	*Pri-let-7*	Suppress	1. Directly binds to *pri-let-7s* and reduces affinity 2. Interacts with Drosha complex	[117,120]
BRCA1/SMAD/p53/DHX9	*let-7a*	*Pri-let-7*	Promote	1. Enhances *pri-let-7s* processing mediated by Drosha complex 2. Binds *pri-let-7s*	[121–126]
SNIP1	*let-7i*	*Pri-let-7*	Promote	Likely binds *pri-let-7* and enhances Drosha processing	[127]
STAUFEN	*let-7s*	*Pri-let-7*	Suppress	Likely binds to pri-let-7 3′-UTR and alters structural integrity	[128,129]
SYNCRIP	*let-7a*	*Pri-let-7*	Promote	Binds to pri-let-7 terminal loop and enhances Drosha processing	[28]
BCDIN3D	*let-7b, 7d, 7d, 7e, 7f, 7g, 7i, miR-98*	*Pre-let-7*	Promote	Methylates *pre-let-7s* and enhances Dicer processing	[130,131]
MCPIP1	*let-7g*	*Pre-let-7*	Suppress	Cleaves terminal loops on the pre-let-7s leading to degradation	[131,132]
TBM3	*let-7a, 7g, 7i*	*Pre-let-7*	Promote	Binds *pre-let-7s*/enhance Dicer	[133,134]

2.1. Negative Transcriptional Regulation of let-7s

let-7 is one of the first miRNAs to be discovered, but its transcriptional regulation is not fully understood. It has been reported that DAF-12 nuclear hormone receptor and *let-7s* have a bimodal feedback loop at the transcription level in a ligand dependent manner. Unliganded DAF-12 inhibits the transcription of *let-7* in worms through a co-repressor, DIN-1 [68,69]. DAF-12 cannot bind directly to the endogenous ligands; the ligands bind DIN-1 and modulate the activity of DAF-12. With favorable environmental and developmental cues, the ligand binds the DIN-1/DAF-12 complex, and DAF-12 is able to directly activate *let-7* transcription [68,69]. Another interesting target that shares a bimodal feedback circuit with *let-7s* is MYC. Some studies have reported that MYC binds to the conserved promoter upstream of *let-7a-1/let-7f-1/let-7d* and *let-7g* polycistronic clusters in the *pri-let-7s* and suppresses its transcription [70,71]. Interestingly, DAF-12 and MYC 3′-UTRs contain *let-7* complementary sites that are targets of *let-7* [135,136]. In addition, LIN42, a period protein homolog, has been shown to regulate a wide variety of miRNAs through transcriptional repression of *let-7* family pri-miRNA production in worms [72]. LIN42 suppresses *let-7* phenotypes through transcriptional repression at the pri-let-7s promoter region as LIN42 protein level increases [1,72,73]. *let-7* has been also shown to have a complementary sequence to the 3′-UTR regions of many genes, such as, *lin41, lin28, lin42,* and *daf-12* [1]. In mammals, these proteins are represented by TRIM71, a LIN28 homolog, period circadian regulator, PCR, and nuclear hormone receptor, NHR [137,138], respectively. Bimodal regulation is prevalent in *let-7* regulation; it is essential to find more targets sharing a bimodal regulation loop with *let-7s* to further our understanding of the complexity of *let-7* regulation and biogenesis in vivo.

2.1.1. LIN28-Dependent and -Independent Regulation of *let-7s* Biogenesis

LIN28s are considered as the master regulators of *let7s*. LIN28A and LIN28B paralogs are RNA binding proteins [74,139] and have direct roles in modulating *let-7* miRNAs. LIN28A and LIN28B can post-transcriptionally suppress both *pri-let-7s* and *pre-let-7s* biogenesis and maturation via both 3′ terminal uridylyl transferase (TUTase)-dependent (LIN28A) and -independent pathways (LIN28B) by inhibiting Drosha and Dicer activities [74–76,81]. LIN28A and LIN28B bind to both *pri-let-7* and *pre-let-7*; however, they work independently and distinctively. LIN28A is mainly in the cytoplasm; it recruits TUTases4/7 to oligo-uridylate *pre-let-7s* at its 3′ end. Uridylated *pre-let-7s* cannot undergo Dicer processing, which marks these *pre-let-7s* for degradation [74,77–80]. When LIN28A is absent, *pre-let7s* processing is upregulated, resulting in more mature *let-7s* [75]. This increase in *let-7s* is regulated by TUTases2/4/7; these proteins mono-uridylate group II *pre-let-7s*, except *pre-let-7a-2, 7c*, and *7e*, and enhance Dicer processing, resulting in increased *let-7s* [82]. Interestingly, in the cytoplasm LIN28A can selectively recruit TUTase4 to a subset of *pre-let-7s* to mediate uridylation processing and suppress pre-let-7 Dicer processing [77,78,80]. Oddly, LIN28B blocks *let-7* miRNA biogenesis via TUTase-independent pathways [79]. LIN28B is mainly located in the nucleus and sequesters *pri-let-7s* into the nucleolus and prevents Drosha/DGCR8-mediated *pri-let-7* processing [75,76,79]. Despite the similarities between the LIN28 paralogs, they work by discrete pathways at multiple steps and negatively regulate nearly all of *let-7* biogenesis. Due to the complexity of the LIN28/let-7 axis and context-dependent regulation of let-7s, it is necessary to further understand what other factors can modulate LIN28/let-7 axis. Several of the factors modulating *let-7s* biogenesis through the LIN28/Let-7 axis are discussed below.

2.1.2. LIN28-Dependent Regulation

Fragile histidine triad diadenosine triphosphotase (FHIT) was found to induce LIN28B protein expression, leading to the suppression of *let-7s* [82]. Chae and colleagues showed that FHIT expression correlated inversely with *let-7* miRs, and FHIT apparently mediates the negative feedback initiated by LIN28/Let7 at the *pri-miRNA* level in the nucleus [84]. An additional LIN28B-regulating protein is mucin 1 (Muc1), a heterodimeric protein that is subsequently autocleaved to Muc1-N and Muc1-C [85]; the later translocates to the nucleus and interacts with transcription factor NF-κB p65 [87]. Kufe's group demonstrated that Muc1-C activates LIN28B in an NF-κB-dependent manner and suppresses *let-7 biogenesis* [86]. Musashi1 (MSI1) protein either works in conjunction or compensates LIN28 to post-transcriptionally negatively regulate *miR-98, let-7b*, and *let-7g* biogenesis via Drosha processing [88]. Sjögren syndrome antigen B (SSB) protein has been shown to bind to the UUUOH element located in the 3′ end of LIN28B RNA transcripts [90,91] and subsequently to enhance LIN28B′ protein levels [89]. Whereas SSB′ silencing decreased LIN28B level, it successively resulted in an increase of mature *let-7s* (*7a, 7b, 7c, 7d, 7e, 7f, 7g*, and *7i*) through released inhibition of *pri-let-7* processing [89]. TRIM25 protein is an E3 ligase that binds to *pre-let-7s* conserved terminal loop and activates LIN28A/TUTase4-mediated uridylation [92]. It has been reported that TRIM25 is a cofactor for LIN28A/TUTase4-mediated uridylation and functions in cis to provide additional specificity and regulation of LIN28A in suppressing the maturation of pre-let-7s [77,78,92]. YAP, yes-associated protein, is a transcriptional coactivator that plays important roles in various cellular processes. YAP is downstream of the Hippo signaling pathway and has been reported to regulate miRNA biogenesis in a cell-contact-dependent manner [99,100]. At low cell density, unphosphorylated YAP translocates into the nucleus and sequesters p72, a DEAD-box helicase 17 (DDX17), which is an essential component of the miRNA processing machinery, Drosha/DGCR8, resulting in downregulation of *let-7* (*7a* and *7b*) [99,100]. When cell density and cell-to-cell contact increases, phosphorylated YAP remains in the cytoplasm and is unable to sequester DDX17; consequently, the later binds to Drosha/DGCR8 complex and increases the *pri-let-7s* processing [99,100]. YAP's nuclear-cytoplasmic dynamics provides additional regulatory control to the LIN28/let-7 axis through a novel cell-contact-dependent miRNA biogenesis.

2.1.3. LIN28-Independent Regulation

Heterogenous nuclear ribonucleoprotein A1 (HnRNPA1) has been shown to bind Drosha complex [109]. HnRNPA1 also binds the conserved terminal loop of *pri-let-7a-1* and inhibits its processing by Drosha and DGCR8 complex [108,109]. The binding of HnRNPA1 with the *pri-let7a* alters the pri-miRNA structure and inhibits Drosha processing [109,110]. HnRNPA1 depletion increases *pri-let-7a-1* processing, whereas ectopic expression of hnRNPA1 decreases *let-7a* [106]. Death receptor tumor necrosis factor-related apoptosis-inducing ligand (TRAIL)-R2 associates with p68 RNA helicase (DDX5), nuclear factor 90/45 (NF90/NF45), and hnRNPA1; together this complex is involved in RNA processing and gene regulation [117]. These binding partners of TRAIL-R2 have been shown to be involved in *let-7s* maturation and biogenesis [118,119]. Knockdown of either TRAIL-R2 or NF90/NF45 results in enhanced processing of *pri-let-7s* by the Drosha/DGCR8 complex and significant intensification of levels of different mature *let-7s* (*7a, 7b, 7c, 7d, 7e*, and *7g*) [117,120]. Several studies have reported that HnRNPA1′ binding to *let-7a* interferes with the binding of the KH-type splicing regulatory protein (KSRP), known to promote *let-7a* biogenesis [111,112]. HnRNPA1 and KSRP compete for *pri-let7* binding sties and reversibly regulate *let-7* biogenesis in vivo [109]. This antagonizing regulation of hnRNPA1 and KSRP adds an additional layer to *let-7* biogenesis and adds additional complexity to its homeostatic regulation that requires further investigation, especially under pathophysiological conditions, such as neuroinflammation, cardiovascular diseases, and stroke. STAUFEN1 protein has been shown to negatively modulate *let-7s* by binding to the *pri-let-7s* 3′-UTR and altering their structure and integrity [129]. The immune regulator, monocyte chemoattractant protein 1-induced protein 1 ribonuclease (MCPIP1), suppresses *pre-let-7s* miRNA biogenesis by inhibiting Dicer processing [132]. MCPIP1 also has an oligomerization domain for *pre-let-7g* recognition leading to its degradation [131].

2.2. Positive Regulation of let-7s Biogenesis

2.2.1. LIN28-Dependent Positive Regulation

Due to the complexity of LIN28/let-7 regulation, many layers of regulation are necessary. One study revealed that TRIM71, an E3 ubiquitin ligase, negatively modulates LIN28B through polyubiquitination, leading to the upregulation of mature *let-a* and *pre-let-7s* post-transcriptionally [93]. TRIM71 levels are also dependent on LIN28 expression; a decrease in LIN28 will reduce TRIM71 expression [94]. Additionally, TRIM71 binds to the catalytically active Ago2 protein using its NHL domain, inducing the degradation of Ago2 by interfering with mature *let-7s* [94–96]. This adds a new layer of regulatory complexity to *let-7* biogenesis and maturation post-transcriptionally. Tristetraprolin (TTP) is an AU-rich pentamer element (ARE)-binding protein that has been reported to downregulate LIN28A via binding to LIN28A' AREs in its 3′-UTR, resulting in subsequent degradation and promotion of *let-7s* maturation [97,98]. It is also interesting that AREs are often located in the 3′-UTR of various mRNA of cytokine mRNAs [98]. TTP presumably plays an important role in regulating inflammatory responses as well by directly binding to ARE-containing transcripts and downregulating these inflammatory response transcripts [140,141].

2.2.2. LIN28-Independent Positive Regulation

Adenosine deaminases acting on RNAs (ADARs) have been reported to convert adenosine residues to inosine residues in pri-miRNAs, pre-miRNAs, and mature miRNAs and to modify their structures, functions, stability, and biogenesis [101,102,142]. Loss of ADAR1 was found to significantly downregulate *let-7a, 7b, 7d*, and *7e* expression through Drosha- and Dicer-mediated processing [103]. ADAR1 modulates the expression of *pri-let-7-Complex (let-7c)* locus through a single A-to-I change at the six residues of pri-miR polycistronic transcript, leading to enhanced miRNA processing by Drosha cleavage [101]. ADAR1 mediates the differential expression of many polycistronic miRNA clusters through direct binding to Drosha/DGCR8 or Dicer complexes, such as, *pri-/pre-let7-a-1, let-7a-2, let-7a-3, pri-/pre-let-7d*, and *pri-let7f* [102–106]. Tumor suppressor breast cancer 1 (BRCA1) directly promotes the processing of pri-*let-7a* [121]. BRCA1 increases the expression of both primary transcripts and mature *let-7a*. BRCA1 was shown to directly interact with DDX5 and the Drosha complex, and studies found that BRCA1 associates with SMAD3, p53, and DEAH-box RNA helicase (DHX9) [121,122]. BRCA1 can directly bind to primary transcripts' stem root via a DNA-binding domain and can regulate *let-7a* biogenesis via the Drosha/DGCR8 complex and SMAD3/p53/DHX9 [121]. It also has been reported that SMAD3 and p53 are involved in *let-7a* maturation [125,143] and interact with BRCA1–Drosha complex [126]. SMAD3, p53, and DHX9 interactions with BRCA1 likely strengthen and stabilize BRCA1-induced Drosha processing activity. Human nuclear interacting protein 1 (SNIP1) is an RNA-binding protein that interacts with Drosha complex and has been reported to function in TGF-β and NF-κB signaling pathways. SNIPP1′ downregulation resulted in *let-7i* reduction [127], confirming its positive regulation in *let-7* biogenesis. Another protein associated with Drosha/DGCR8 complex is synaptotagmin-binding cytoplasmic RNA-interacting protein (SYNCRIP), which was shown to bind to the conserved terminal loop within *pri-let-7* [144]. Silencing SYNCRIP reduces mature *let-7a* level, while overexpressing SYNCRIP promotes *let-7a* [144]. Depletion of the BCDIN3D, a member of the Bin3 family, revealed strong downregulation of a number of mature *let-7s* (*7b, 7d, 7e, 7f, 7g, 7i and miR-98*) [130]. BCDIN3D has been reported to interact with Dicer in an RNase A-dependent manner and facilitates Dicer processing [131]. Additionally, BCDIN3D has been shown to directly interact with *pre-let-7s* and methylate them in vitro with great specificity, leading to enhanced Dicer processing [130]. RBM3, a cold-inducible, developmentally regulated RNA-binding protein regulates *let-7* biogenesis [133,134]. RBM3 level has been shown to directly correlate with miRNA generation and vice versa [133]. Pilotte and colleagues have shown that changes in *pre-let7-a, pre-let-7g*, and *pre-let-7i* are affected by the presence of RBM3 [133]. RBM3 directly binds these pre-*let-7s* and enhances these precursors' association with active Dicer complexes [133]. RBM3′ ability to directly bind to pre-miRNAs and regulate subsequent Dicer processing under hypothermia makes it an interesting target for modifying miR expression in temperature-sensitive processes. Another *let-7* biogenesis-involved protein is a TAR DNA-binding protein 43 (TDP-43). TDP043 belongs to the hnRNP family and has been shown to play a major role in many cellular processes [114,115]. Since hnRNPA1 has been described as associating with Drosha [107,113], likewise TDP-43 is a Drosha-associated protein [115,116] and is reported to downregulate *let-7b* [115]. *Pri-let-7b* binds directly to TDP-43 in different positions within the miRNA and/or the hairpin [115]. When TDP-43 is depleted, *let-7b* is downregulated [115]; this shows that TDP-43 plays a positive role in *let-7b* biogenesis. Intriguingly, another study did not find that depletion of TDP-43 lowered *let-7b* level [116]. These contradictory results remind us of the complexity of *let-7* biogenesis in a context-dependent manner, and further investigations are needed to discover TDP-43's role in regulating *let-7* biogenesis.

3. *let-7*'s Protein Regulators and Their Role in Stroke and Other Cardiovascular Disease-Related Inflammation

let-7 is involved in many cellular processes, immunity, and protective functions. Regulators of *let-7* are crucial for therapeutic advances. The majority of these regulatory proteins are well studied in development, proliferation, differentiation, and cancer, but their roles in inflammation, cardiovascular diseases, and/or stroke are not well studied. Only a few of these regulatory proteins have been described as having a link with cardiovascular diseases and stroke outcomes. hnRNPA1 has recently been reported to interact with β-arrestin1 to upregulate a miRNA processing in the heart [145]. KSRP has been shown in vitro to regulate inflammatory responses [112] through controlling inflammatory mediators, such as TNFα, IL-1β, IFNα, and IFNβ expressions [146]. Similar to TTP, KSRP is involved in direct and indirect control of cytokine synthesis and degradation, potentially through miRNA regulation [146,147]. Upregulated SMAD3/TGF-β signaling has been reported to significantly increase cell survival and exhibit neuroprotective effects after cerebral ischemic stroke [148]. Cardiomyocyte apoptosis is considered a significant event during the development of cardiomyopathy. *let-7* has been shown to target TGF-3β and regulate cardiomyocyte apoptosis after MI [149]. Bioinformatic predictions have shown several genes, such as TBX5, FOXP1, HAND1, AKT2, and PPARGC1A, which are related to cardiac development, to be targets of different *let-7s* (*let-7a/7d/7e/7f*) [150]. These findings suggest that *let-7* might contribute to heart development and/or heart diseases, potentially as a target for cardiovascular disease therapeutics [36]. A recent study showed that accumulation of hnRNPA1 and TDP-43 are associated with neurodegenerative disease and ischemic stroke [151,152]. Another study has shown that TRAIL-R2 is one of the most powerful biomarkers for predicting long-term mortality in many diseases, such as diabetes, heart failure, myocardial infarction, smoking, and hypercholesterolemia [153]. MCPIP1 has recently been shown to negatively regulate inflammatory responses after ischemic stroke, to enhance blood–brain barrier integrity, and to be neuroprotective [154]. Additionally, RBM3 has recently been shown to be neuroprotective and positively correlate with good ischemic stroke outcomes [155]. *let-7* is a major player in diverse processes; any dysfunction in *let-7* regulation can cause a disease state, and it is essential to study how these regulatory elements link with inflammatory diseases. Taking all these together, since all the above-mentioned proteins are involved in *let-7* miR regulation, it is reasonable to suggest that *let-7s* play a significant role in the aforementioned outcomes of these regulators.

In recent years, advanced bioinformatic techniques allowed characterization of many circular RNAs and long noncoding RNAs. Both of these RNAs serve as competitive endogenous RNA (ceRNA) regulators for miRs. ceRNAs work as sponges and prevent miRs from acting on their target mRNA transcripts. Hundreds of ceRNAs have been described, and some of them are involved in *let-7* miRs regulation; however, their involvement in stroke or other cardiovascular diseases still remains to be explored.

4. Conclusions

It has been shown both in vivo and in vitro that *let-7* miRs are involved in numerous cellular processes, inflammation, immunity, and protective functions. Tissue- and condition-specific *let-7* expression is tightly regulated. The majority of the *let-7'* regulatory proteins are well studied in development, proliferation, differentiation, and cancer, but their roles in inflammation, cardiovascular disease, and/or stroke are not well studied. Further knowledge of the regulation of *let-7* is crucial for therapeutic advances.

Author Contributions: Writing—original draft preparation, review and editing, D.L.B. and X.J.; conceptualization, writing—original draft preparation, review and editing, supervision, and funding acquisition, S.R. All authors have read and agreed to the published version of the manuscript.

Funding: This work was supported in part by NIH research grant R01NS101135 (S.R.).

Acknowledgments: The authors express their grateful acknowledgment of Nancy L. Reichenbach for proofreading and editing.

Conflicts of Interest: The authors declare no conflict of interest.

References

1. Reinhart, B.J.; Slack, F.J.; Basson, M.; Pasquinelli, A.E.; Bettinger, J.C.; Rougvie, A.E.; Horvitz, H.R.; Ruvkun, G. The 21-nucleotide let-7 RNA regulates developmental timing in Caenorhabditis elegans. *Nat. Cell Biol.* **2000**, *403*, 901–906. [CrossRef]
2. Ali, A.; Bouma, G.J.; Anthony, R.V.; Winger, Q.A. The Role of LIN28-let-7-ARID3B Pathway in Placental Development. *Int. J. Mol. Sci.* **2020**, *21*, 3637. [CrossRef]
3. Roush, S.; Slack, F.J. The let-7 family of microRNAs. *Trends Cell Biol.* **2008**, *18*, 505–516. [CrossRef] [PubMed]
4. Hertel, J.; Bartschat, S.; Wintsche, A.; Otto, C.; The Students of the Bioinformatics Computer Lab; Stadler, P.F. Evolution of the let-7 microRNA Family. *RNA Biol.* **2012**, *9*, 231–241. [CrossRef]
5. Wu, Y.-C.; Chen, C.-H.; Mercer, A.; Sokol, N.S. let-7-Complex MicroRNAs Regulate the Temporal Identity of Drosophila Mushroom Body Neurons via chinmo. *Dev. Cell* **2012**, *23*, 202–209. [CrossRef] [PubMed]
6. Lee, H.; Han, S.; Kwon, C.S.; Lee, D. Biogenesis and regulation of the let-7 miRNAs and their functional implications. *Protein Cell* **2016**, *7*, 100–113. [CrossRef]
7. Bartel, D.P. MicroRNAs: Genomics, Biogenesis, Mechanism, and Function. *Cell* **2004**, *116*, 281–297. [CrossRef]
8. Eder, P.S.; Devine, R.J.; Dagle, J.; Walder, J.A. Substrate Specificity and Kinetics of Degradation of Antisense Oligonucleotides by a 3' Exonuclease in Plasma. *Antisense Res. Dev.* **1991**, *1*, 141–151. [CrossRef] [PubMed]
9. Li, S.; Wang, X.; Gu, Y.; Chen, C.; Wang, Y.; Liu, J.; Hu, W.; Yu, B.; Wang, Y.; Ding, F.; et al. Let-7 microRNAs Regenerate Peripheral Nerve Regeneration by Targeting Nerve Growth Factor. *Mol. Ther.* **2015**, *23*, 423–433. [CrossRef] [PubMed]
10. Gérard, C.; Lemaigre, F.; Gonze, D. Modeling the Dynamics of Let-7-Coupled Gene Regulatory Networks Linking Cell Proliferation to Malignant Transformation. *Front. Physiol.* **2019**, *10*, 848. [CrossRef] [PubMed]
11. Wei, J.; Li, H.; Wang, S.; Li, T.; Fan, J.; Liang, X.; Li, J.; Han, Q.; Zhu, L.; Fan, L.; et al. let-7 Enhances Osteogenesis and Bone Formation While Repressing Adipogenesis of Human Stromal/Mesenchymal Stem Cells by Regulating HMGA2. *Stem Cells Dev.* **2014**, *23*, 1452–1463. [CrossRef]
12. Tolonen, A.; Magga, J.; Szabó, Z.; Viitala, P.; Gao, E.; Moilanen, A.; Ohukainen, P.; Vainio, L.; Koch, W.J.; Kerkelä, R.; et al. Inhibition of Let-7 micro RNA attenuates myocardial remodeling and improves cardiac function postinfarction in mice. *Pharmacol. Res. Perspect.* **2014**, *2*, e00056. [CrossRef] [PubMed]
13. Jiang, X.-M.; Yu, X.-F.; Wang, Z.-K.; Liu, F.-F.; Wang, Y. Let-7a gene knockdown protects against cerebral ischemia/reperfusion injury. *Neural Regen. Res.* **2016**, *11*, 262–269. [CrossRef] [PubMed]
14. Na, H.S.T.; Nuo, M.; Meng, Q.-T.; Xia, Z.-Y. The Pathway of Let-7a-1/2-3p and HMGB1 Mediated Dexmedetomidine Inhibiting Microglia Activation in Spinal Cord Ischemia-Reperfusion Injury Mice. *J. Mol. Neurosci.* **2019**, *69*, 106–114. [CrossRef]
15. Cho, K.J.; Song, J.; Oh, Y.; Lee, J.E. MicroRNA-Let-7a regulates the function of microglia in inflammation. *Mol. Cell. Neurosci.* **2015**, *68*, 167–176. [CrossRef] [PubMed]
16. Wang, S.; Zhou, H.; Wu, D.; Ni, H.; Chen, Z.; Chen, C.; Xiang, Y.; Dai, K.; Chen, X.; Li, X. MicroRNA let-7a regulates angiogenesis by targeting TGFBR3mRNA. *J. Cell. Mol. Med.* **2019**, *23*, 556–567. [CrossRef]
17. Ham, O.; Lee, S.-Y.; Lee, C.Y.; Park, J.-H.; Lee, J.; Seo, H.-H.; Cha, M.-J.; Choi, E.; Kim, S.; Hwang, K.-C. let-7b suppresses apoptosis and autophagy of human mesenchymal stem cells transplanted into ischemia/reperfusion injured heart 7by targeting caspase-3. *Stem Cell Res. Ther.* **2015**, *6*, 147. [CrossRef]
18. Long, G.; Wang, F.; Li, H.; Yin, Z.; Sandip, C.; Lou, Y.; Wang, Y.; Chen, C.; Wang, D.W. Circulating miR-30a, miR-126 and let-7b as biomarker for ischemic stroke in humans. *BMC Neurol.* **2013**, *13*, 178. [CrossRef] [PubMed]
19. Chi, N.; Chiou, H.; Chou, S.; Hu, C.; Chen, K.; Chang, C.; Hsieh, Y. Hyperglycemia-related FAS gene and hsa-let-7b-5p as markers of poor outcomes for ischaemic stroke. *Eur. J. Neurol.* **2020**, *27*, 1647–1655. [CrossRef]

20. Li, S.; Chen, L.; Zhou, X.; Li, J.; Liu, J. miRNA-223-3p and let-7b-3p as potential blood biomarkers associated with the ischemic penumbra in rats. *Acta Neurobiol. Exp.* **2019**, *79*, 205–216. [CrossRef]

21. Bernstein, D.L.; Zuluaga-Ramirez, V.; Gajghate, S.; Reichenbach, N.L.; Polyak, B.; Persidsky, Y.; Rom, S. miR-98 reduces endothelial dysfunction by protecting blood–brain barrier (BBB) and improves neurological outcomes in mouse ischemia/reperfusion stroke model. *Br. J. Pharmacol.* **2019**, *40*, 1953–1965. [CrossRef]

22. Li, H.-W.; Meng, Y.; Xie, Q.; Yi, W.-J.; Lai, X.-L.; Bian, Q.; Wang, J.; Wang, J.-F.; Yu, G. miR-98 protects endothelial cells against hypoxia/reoxygenation induced-apoptosis by targeting caspase-3. *Biochem. Biophys. Res. Commun.* **2015**, *467*, 595–601. [CrossRef]

23. Rom, S.; Dykstra, H.; Zuluaga-Ramirez, V.; Reichenbach, N.L.; Persidsky, Y. miR-98 and let-7g* Protect the Blood-Brain Barrier Under Neuroinflammatory Conditions. *Br. J. Pharmacol.* **2015**, *35*, 1957–1965. [CrossRef]

24. Bernstein, D.L.; Rom, S. Let-7g* and miR-98 Reduce Stroke-Induced Production of Proinflammatory Cytokines in Mouse Brain. *Front. Cell Dev. Biol.* **2020**, *8*, 632. [CrossRef]

25. Bernstein, D.L.; Gajghate, S.; Reichenbach, N.L.; Winfield, M.; Persidsky, Y.; Heldt, N.A.; Rom, S. let-7g counteracts endothelial dysfunction and ameliorating neurological functions in mouse ischemia/reperfusion stroke model. *Brain Behav. Immun.* **2020**, *87*, 543–555. [CrossRef] [PubMed]

26. Xiang, W.; Tian, C.; Peng, S.; Zhou, L.; Pan, S.; Deng, Z. Let-7i attenuates human brain microvascular endothelial cell damage in oxygen glucose deprivation model by decreasing toll-like receptor 4 expression. *Biochem. Biophys. Res. Commun.* **2017**, *493*, 788–793. [CrossRef]

27. Jickling, G.C.; Ander, B.P.; Shroff, N.; Orantia, M.; Stamova, B.; Dykstra-Aiello, C.; Hull, H.; Zhan, X.; Liu, D.; Sharp, F.R. Leukocyte response is regulated by microRNA let7i in patients with acute ischemic stroke. *Neurology* **2016**, *87*, 2198–2205. [CrossRef] [PubMed]

28. Chen, D.; Li, L.; Wang, Y.; Xu, R.; Peng, S.; Zhou, L.; Deng, Z. Ischemia-reperfusion injury of brain induces endothelial-mesenchymal transition and vascular fibrosis via activating let-7i/TGF-βR1 double-negative feedback loop. *FASEB J.* **2020**, *34*, 7178–7191. [CrossRef]

29. Lin, Z.; Ge, J.; Wang, Z.; Ren, J.; Wang, X.; Xiong, H.; Gao, J.; Zhang, Y.; Zhang, Q. Let-7e modulates the inflammatory response in vascular endothelial cells through ceRNA crosstalk. *Sci. Rep.* **2017**, *7*, 42498. [CrossRef]

30. Huang, S.; Lv, Z.; Guo, Y.; Li, L.; Zhang, Y.; Zhou, L.; Yang, B.; Wu, S.; Zhang, Y.; Xie, C.; et al. Identification of Blood Let-7e-5p as a Biomarker for Ischemic Stroke. *PLoS ONE* **2016**, *11*, e0163951. [CrossRef] [PubMed]

31. Nguyen, T.; Su, C.; Singh, M. Let-7i inhibition enhances progesterone-induced functional recovery in a mouse model of ischemia. *Proc. Natl. Acad. Sci. USA* **2018**, *115*, E9668–E9677. [CrossRef]

32. Yuan, H.; Zhang, H.; Hong, L.; Zhao, H.; Wang, J.; Li, H.; Che, H.; Zhang, Z. MicroRNA let-7c-5p Suppressed Lipopolysaccharide-Induced Dental Pulp Inflammation by Inhibiting Dentin Matrix Protein-1-Mediated Nuclear Factor kappa B (NF-κB) Pathway In Vitro and In Vivo. *Med Sci. Monit.* **2018**, *24*, 6656–6665. [CrossRef]

33. Wong, L.L.; Saw, E.L.; Lim, J.Y.; Zhou, Y.; Richards, A.M.; Wang, P. MicroRNA Let-7d-3p Contributes to Cardiac Protection via Targeting HMGA2. *Int. J. Mol. Sci.* **2019**, *20*, 1522. [CrossRef]

34. Hsu, P.-Y.; Hsi, E.; Wang, T.-M.; Lin, R.-T.; Liao, Y.-C.; Juo, S.-H.H. MicroRNA let-7g possesses a therapeutic potential for peripheral artery disease. *J. Cell. Mol. Med.* **2016**, *21*, 519–529. [CrossRef]

35. Zhuang, Y.; Peng, H.; Mastej, V.; Chen, W. MicroRNA Regulation of Endothelial Junction Proteins and Clinical Consequence. *Mediat. Inflamm.* **2016**, *2016*, 1–6. [CrossRef]

36. Bao, M.-H.; Feng, X.; Zhang, Y.-W.; Lou, X.-Y.; Cheng, Y.; Zhou, H.-H. Let-7 in Cardiovascular Diseases, Heart Development and Cardiovascular Differentiation from Stem Cells. *Int. J. Mol. Sci.* **2013**, *14*, 23086–23102. [CrossRef]

37. Joshi, S.; Wei, J.; Bishopric, N.H. A cardiac myocyte-restricted Lin28/let-7 regulatory axis promotes hypoxia-mediated apoptosis by inducing the AKT signaling suppressor PIK3IP1. *Biochim. Biophys. Acta (BBA) Mol. Basis Dis.* **2016**, *1862*, 240–251. [CrossRef]

38. Hennchen, M.; Stubbusch, J.; Makhfi, I.A.-E.; Kramer, M.; Deller, T.; Pierre-Eugene, C.; Janoueix-Lerosey, I.; Delattre, O.; Ernsberger, U.; Schulte, J.H.; et al. Lin28B and Let-7 in the Control of Sympathetic Neurogenesis and Neuroblastoma Development. *J. Neurosci.* **2015**, *35*, 16531–16544. [CrossRef] [PubMed]

39. Gulman, N.K.; Armon, L.; Shalit, T.; Urbach, A. Heterochronic regulation of lung development via the Lin28-Let-7 pathway. *FASEB J.* **2019**, *33*, 12008–12018. [CrossRef] [PubMed]

40. Fairchild, C.L.A.; Cheema, S.K.; Wong, J.; Hino, K.; Simó, S.; La Torre, A. Let-7 regulates cell cycle dynamics in the developing cerebral cortex and retina. *Sci. Rep.* **2019**, *9*, 15336. [CrossRef] [PubMed]

41. Morgado, A.L.; Rodrigues, C.M.P.; Solá, S. MicroRNA-145 Regulates Neural Stem Cell Differentiation Through the Sox2-Lin28/let-7 Signaling Pathway. *STEM CELLS* **2016**, *34*, 1386–1395. [CrossRef] [PubMed]

42. Deng, Z.; Wei, Y.; Yao, Y.; Gao, S.; Wang, X. Let-7f promotes the differentiation of neural stem cells in rats. *Am. J. Transl. Res.* **2020**, *12*, 5752–5761. [PubMed]

43. Chen, P.-Y.; Qin, L.; Barnes, C.; Charisse, K.; Yi, T.; Zhang, X.; Ali, R.; Medina, P.P.; Yu, J.; Slack, F.J.; et al. FGF Regulates TGF-β Signaling and Endothelial-to-Mesenchymal Transition via Control of let-7 miRNA Expression. *Cell Rep.* **2012**, *2*, 1684–1696. [CrossRef]

44. Kalomoiris, S.; Cicchetto, A.C.; Lakatos, K.; Nolta, J.A.; Fierro, F.A. Fibroblast Growth Factor 2 Regulates High Mobility Group A2 Expression in Human Bone Marrow-Derived Mesenchymal Stem Cells. *J. Cell. Biochem.* **2016**, *117*, 2128–2137. [CrossRef] [PubMed]

45. Liao, Y.-C.; Wang, Y.-S.; Guo, Y.-C.; Lin, W.-L.; Chang, M.-H.; Juo, S.-H.H. Let-7g Improves Multiple Endothelial Functions Through Targeting Transforming Growth Factor-Beta and SIRT-1 Signaling. *J. Am. Coll. Cardiol.* **2014**, *63*, 1685–1694. [CrossRef] [PubMed]

46. Mozos, I.; Malainer, C.; Horbańczuk, J.; Gug, C.; Stoian, D.; Luca, C.T.; Atanasov, A.G. Inflammatory Markers for Arterial Stiffness in Cardiovascular Diseases. *Front. Immunol.* **2017**, *8*, 1058. [CrossRef]

47. Tonini, T.; Rossi, F.; Claudio, P.P. Molecular basis of angiogenesis and cancer. *Oncogene* **2003**, *22*, 6549–6556. [CrossRef]

48. Isanejad, A.; Alizadeh, A.M.; Shalamzari, S.A.; Khodayari, H.; Khodayari, S.; Khori, V.; Khojastehnjad, N. MicroRNA-206, let-7a and microRNA-21 pathways involved in the anti-angiogenesis effects of the interval exercise training and hormone therapy in breast cancer. *Life Sci.* **2016**, *151*, 30–40. [CrossRef]

49. Wang, T.; Wang, G.; Hao, D.; Liu, X.; Wang, D.; Ning, N.; Li, X. Aberrant regulation of the LIN28A/LIN28B and let-7 loop in human malignant tumors and its effects on the hallmarks of cancer. *Mol. Cancer* **2015**, *14*, 1–13. [CrossRef]

50. Buonfiglioli, A.; Efe, I.E.; Guneykaya, D.; Ivanov, A.; Huang, Y.; Orlowski, E.; Krüger, C.; Deisz, R.A.; Markovic, D.; Flüh, C.; et al. let-7 MicroRNAs Regulate Microglial Function and Suppress Glioma Growth through Toll-Like Receptor 7. *Cell Rep.* **2019**, *29*, 3460–3471.e7. [CrossRef]

51. Wu, T.; Jia, J.; Xiong, X.; He, H.; Bu, L.; Zhao, Z.; Huang, C.; Zhang, W. Increased Expression of Lin28B Associates with Poor Prognosis in Patients with Oral Squamous Cell Carcinoma. *PLoS ONE* **2013**, *8*, e83869. [CrossRef] [PubMed]

52. Tristán-Ramos, P.; Rubio-Roldan, A.; Peris, G.; Sánchez, L.; Amador-Cubero, S.; Viollet, S.; Cristofari, G.; Heras, S.R. The tumor suppressor microRNA let-7 inhibits human LINE-1 retrotransposition. *Nat. Commun.* **2020**, *11*, 1–14. [CrossRef] [PubMed]

53. Chen, Y.; Xie, C.; Zheng, X.; Nie, X.; Wang, Z.; Liu, H.; Zhao, Y. LIN28/let-7/PD-L1 Pathway as a Target for Cancer Immunotherapy. *Cancer Immunol. Res.* **2019**, *7*, 487–497. [CrossRef] [PubMed]

54. Mardani, R.; Abadi, M.H.J.N.; Motieian, M.; Taghizadeh-Boroujeni, S.; Bayat, A.; Farsinezhad, A.; Gheibi-Hayat, S.M.; Motieian, M.; Pourghadamyari, H. MicroRNA in leukemia: Tumor suppressors and oncogenes with prognostic potential. *J. Cell. Physiol.* **2019**, *234*, 8465–8486. [CrossRef]

55. Brennan, E.; Wang, B.; McClelland, A.; Mohan, M.; Marai, M.; Beuscart, O.; Derouiche, S.; Gray, S.; Pickering, R.; Tikellis, C.; et al. Protective Effect of let-7 miRNA Family in Regulating Inflammation in Diabetes-Associated Atherosclerosis. *Diabetes* **2017**, *66*, 2266–2277. [CrossRef] [PubMed]

56. Baer, C.; Squadrito, M.L.; Laoui, D.; Thompson, D.; Hansen, S.K.; Kiialainen, A.; Hoves, S.; Ries, C.H.; Ooi, C.-H.; De Palma, M. Suppression of microRNA activity amplifies IFN-γ-induced macrophage activation and promotes anti-tumour immunity. *Nat. Cell Biol.* **2016**, *18*, 790–802. [CrossRef]

57. Li, X.-X.; Di, X.; Cong, S.; Wang, Y.; Wang, K. The role of let-7 and HMGA2 in the occurrence and development of lung cancer: A systematic review and meta-analysis. *Eur. Rev. Med. Pharmacol. Sci.* **2018**, *22*, 8353–8366.

58. Tsang, W.P.; Kwok, T.T. Let-7a microRNA suppresses therapeutics-induced cancer cell death by targeting caspase-3. *Apoptosis* **2008**, *13*, 1215–1222. [CrossRef]

59. Zha, W.; Guan, S.; Liu, N.; Li, Y.; Tian, Y.; Chen, Y.; Wang, Y.; Wu, F. Let-7a inhibits Bcl-xl and YAP1 expression to induce apoptosis of trophoblast cells in early-onset severe preeclampsia. *Sci. Total Environ.* **2020**, *745*, 139919. [CrossRef]

60. Zhang, H.; Xiong, X.; Gu, L.; Xie, W.; Zhao, H. CD4 T cell deficiency attenuates ischemic stroke, inhibits oxidative stress, and enhances Akt/mTOR survival signaling pathways in mice. *Chin. Neurosurg. J.* **2018**, *4*, 1–7. [CrossRef]

61. Wang, G.; Zhang, Z.; Ayala, C.; Dunet, D.O.; Fang, J.; George, M.G. Costs of Hospitalization for Stroke Patients Aged 18-64 Years in the United States. *J. Stroke Cerebrovasc. Dis.* **2014**, *23*, 861–868. [CrossRef]

62. Shi, X.-Y.; Wang, H.; Wang, W.; Gu, Y.-H. MiR-98-5p regulates proliferation and metastasis of MCF-7 breast cancer cells by targeting Gab2. *Eur. Rev. Med. Pharmacol. Sci.* **2020**, *24*, 10914. [PubMed]

63. Chen, Y.-L.; Qiao, Y.-C.; Xu, Y.; Ling, W.; Pan, Y.-H.; Huang, Y.-C.; Geng, L.-J.; Zhao, H.-L.; Zhang, X.-X. Serum TNF-α concentrations in type 2 diabetes mellitus patients and diabetic nephropathy patients: A systematic review and meta-analysis. *Immunol. Lett.* **2017**, *186*, 52–58. [CrossRef] [PubMed]

64. Hu, C.; Huang, S.; Wu, F.; Ding, H. miR-98 inhibits cell proliferation and induces cell apoptosis by targeting MAPK6 in HUVECs. *Exp. Ther. Med.* **2018**, *15*, 2755–2760. [CrossRef]

65. Zhao, C.; Popel, A.S. Computational Model of MicroRNA Control of HIF-VEGF Pathway: Insights into the Pathophysiology of Ischemic Vascular Disease and Cancer. *PLoS Comput. Biol.* **2015**, *11*, e1004612. [CrossRef] [PubMed]

66. Wang, L.; Lin, Z.Q.; Wong, A. COVID-Net: A tailored deep convolutional neural network design for detection of COVID-19 cases from chest X-ray images. *Sci. Rep.* **2020**, *10*, 1–12. [CrossRef] [PubMed]

67. Chirshev, E.; Oberg, K.; Ioffe, Y.J.; Unternaehrer, J.J. Let-7as biomarker, prognostic indicator, and therapy for precision medicine in cancer. *Clin. Transl. Med.* **2019**, *8*, 24. [CrossRef] [PubMed]

68. Hammell, C.M.; Karp, X.; Ambros, V. A feedback circuit involving let-7-family miRNAs and DAF-12 integrates environmental signals and developmental timing in Caenorhabditis elegans. *Proc. Natl. Acad. Sci. USA* **2009**, *106*, 18668–18673. [CrossRef]

69. Bethke, A.; Fielenbach, N.; Wang, Z.; Mangelsdorf, D.J.; Antebi, A. Nuclear Hormone Receptor Regulation of MicroRNAs Controls Developmental Progression. *Science* **2009**, *324*, 95–98. [CrossRef]

70. Chang, T.-C.; Yu, D.; Lee, Y.-S.; Wentzel, E.A.; Arking, D.E.; West, K.M.; Dang, C.V.; Thomas-Tikhonenko, A.; Mendell, J.T. Widespread microRNA repression by Myc contributes to tumorigenesis. *Nat. Genet.* **2007**, *40*, 43–50. [CrossRef] [PubMed]

71. Wang, Z.; Lin, S.; Li, J.J.; Xu, Z.; Yao, H.; Zhu, X.; Xie, D.; Shen, Z.; Sze, J.; Li, K.; et al. MYC Protein Inhibits Transcription of the MicroRNA Cluster MC-let-7a-1∼let-7d via Noncanonical E-box*. *J. Biol. Chem.* **2011**, *286*, 39703–39714. [CrossRef] [PubMed]
72. Van Wynsberghe, P.M.; Finnegan, E.F.; Stark, T.; Angelus, E.P.; Homan, K.E.; Yeo, E.; Pasquinelli, A.E. The Period protein homolog LIN-42 negatively regulates microRNA biogenesis in C. elegans. *Dev. Biol.* **2014**, *390*, 126–135. [CrossRef] [PubMed]
73. McCulloch, K.A.; Rougvie, A.E. Caenorhabditis elegans period homolog lin-42 regulates the timing of heterochronic miRNA expression. *Proc. Natl. Acad. Sci. USA* **2014**, *111*, 15450–15455. [CrossRef]
74. Heo, I.; Joo, C.; Cho, J.; Ha, M.; Han, J.; Kim, V.N. Lin28 Mediates the Terminal Uridylation of let-7 Precursor MicroRNA. *Mol. Cell* **2008**, *32*, 276–284. [CrossRef]
75. Rybak, A.; Fuchs, H.; Smirnova, L.; Brandt, C.; Pohl, E.E.; Nitsch, R.; Wulczyn, F.G. A feedback loop comprising lin-28 and let-7 controls pre-let-7 maturation during neural stem-cell commitment. *Nat. Cell Biol.* **2008**, *10*, 987–993. [CrossRef] [PubMed]
76. Viswanathan, S.; Daley, G.Q.; Gregory, R.I. Selective Blockade of MicroRNA Processing by Lin28. *Science* **2008**, *320*, 97–100. [CrossRef]
77. Hagan, J.P.; Piskounova, E.; Gregory, R.I. Lin28 recruits the TUTase Zcchc11 to inhibit let-7 maturation in mouse embryonic stem cells. *Nat. Struct. Mol. Biol.* **2009**, *16*, 1021–1025. [CrossRef]
78. Heo, I.; Joo, C.; Kim, Y.-K.; Ha, M.; Yoon, M.-J.; Cho, J.; Yeom, K.-H.; Han, J.; Kim, V.N. TUT4 in Concert with Lin28 Suppresses MicroRNA Biogenesis through Pre-MicroRNA Uridylation. *Cell* **2009**, *138*, 696–708. [CrossRef] [PubMed]
79. Piskounova, E.; Polytarchou, C.; Thornton, J.E.; Lapierre, R.J.; Pothoulakis, C.; Hagan, J.P.; Iliopoulos, D.; Gregory, R.I. Lin28A and lin28B Inhibit let-7 microRNA biogenesis by distinct mechanisms. *Cell* **2011**, *147*, 1066–1079. [CrossRef] [PubMed]
80. Thornton, J.E.; Chang, H.-M.; Piskounova, E.; Gregory, R.I. Lin28-mediated control of let-7 microRNA expression by alternative TUTases Zcchc11 (TUT4) and Zcchc6 (TUT7). *RNA* **2012**, *18*, 1875–1885. [CrossRef]
81. Newman, M.A.; Thomson, J.M.; Hammond, S.M. Lin-28 interaction with the Let-7 precursor loop mediates regulated microRNA processing. *RNA* **2008**, *14*, 1539–1549. [CrossRef] [PubMed]
82. Heo, I.; Ha, M.; Lim, J.; Yoon, M.-J.; Park, J.-E.; Kwon, S.C.; Chang, H.; Kim, V.N. Mono-Uridylation of Pre-MicroRNA as a Key Step in the Biogenesis of Group II let-7 MicroRNAs. *Cell* **2012**, *151*, 521–532. [CrossRef] [PubMed]
83. Barnes, L.D.; Garrison, P.N.; Siprashvili, Z.; Guranowski, A.; Robinson, A.K.; Ingram, S.W.; Croce, C.M.; Ohta, M.; Huebner, K. Fhit, a Putative Tumor Suppressor in Humans, Is a Dinucleoside 5′,5′″-P1,P3-Triphosphate Hydrolase†. *Biochemistry* **1996**, *35*, 11529–11535. [CrossRef]
84. Chae, H.-J.; Seo, J.B.; Kim, S.-H.; Jeon, Y.-J.; Suh, S.-S. Fhit induces the reciprocal suppressions between Lin28/Let-7 and miR-17/92miR. *Int. J. Med Sci.* **2021**, *18*, 706–714. [CrossRef] [PubMed]
85. Kufe, D.W. Mucins in cancer: Function, prognosis and therapy. *Nat. Rev. Cancer* **2009**, *9*, 874–885. [CrossRef] [PubMed]
86. Alam, M.; Ahmad, R.; Rajabi, H.; Kufe, D. MUC1-C Induces the LIN28B→LET-7→HMGA2 Axis to Regulate Self-Renewal in NSCLC. *Mol. Cancer Res.* **2015**, *13*, 449–460. [CrossRef]
87. Kufe, D.W. MUC1-C oncoprotein as a target in breast cancer: Activation of signaling pathways and therapeutic approaches. *Oncogene* **2012**, *32*, 1073–1081. [CrossRef]
88. Kawahara, H.; Okada, Y.; Imai, T.; Iwanami, A.; Mischel, P.S.; Okano, H. Musashi1 Cooperates in Abnormal Cell Lineage Protein 28 (Lin28)-mediated Let-7 Family MicroRNA Biogenesis in Early Neural Differentiation. *J. Biol. Chem.* **2011**, *286*, 16121–16130. [CrossRef]
89. Kim, S.H.; Park, B.-O.; Kim, K.; Park, B.C.; Park, S.G.; Kim, J.-H.; Kim, S. Sjögren Syndrome antigen B regulates LIN28-let-7 axis in Caenorhabditis elegans and human. *Biochim. Biophys. Acta (BBA) Bioenerg.* **2021**, *1864*, 194684. [CrossRef]
90. Teplova, M.; Yuan, Y.-R.; Phan, A.T.; Malinina, L.; Ilin, S.; Teplov, A.; Patel, D.J. Structural Basis for Recognition and Sequestration of UUUOH 3′ Temini of Nascent RNA Polymerase III Transcripts by La, a Rheumatic Disease Autoantigen. *Mol. Cell* **2006**, *21*, 75–85. [CrossRef]
91. Stefano, J.E. Purified lupus antigen la recognizes an oligouridylate stretch common to the 3′ termini of RNA polymerase III transcripts. *Cell* **1984**, *36*, 145–154. [CrossRef]
92. Choudhury, N.R.; Nowak, J.S.; Zuo, J.; Rappsilber, J.; Spoel, S.H.; Michlewski, G. Trim25 Is an RNA-Specific Activator of Lin28a/TuT4-Mediated Uridylation. *Cell Rep.* **2014**, *9*, 1265–1272. [CrossRef]
93. Lee, S.H.; Cho, S.; Kim, M.S.; Choi, K.; Cho, J.Y.; Gwak, H.-S.; Kim, Y.-J.; Yoo, H.; Lee, S.-H.; Park, J.B.; et al. The ubiquitin ligase human TRIM71 regulates let-7 microRNA biogenesis via modulation of Lin28B protein. *Biochim. Biophys. Acta (BBA) Bioenerg.* **2014**, *1839*, 374–386. [CrossRef] [PubMed]
94. Chang, H.-M.; Martinez, N.J.; Thornton, J.E.; Hagan, J.P.; Nguyen, K.D.; Gregory, R.I. Trim71 cooperates with microRNAs to repress Cdkn1a expression and promote embryonic stem cell proliferation. *Nat. Commun.* **2012**, *3*, 923. [CrossRef]
95. Rybak, A.; Fuchs, H.; Hadian, K.; Smirnova, L.; Wulczyn, E.A.; Michel, G.; Nitsch, R.; Krappmann, D.; Wulczyn, F.G. The let-7 target gene mouse lin-41 is a stem cell specific E3 ubiquitin ligase for the miRNA pathway protein Ago2. *Nat. Cell Biol.* **2009**, *11*, 1411–1420. [CrossRef]
96. Liu, Q.; Chen, X.; Novak, M.K.; Zhang, S.; Hu, W. Repressing Ago2 mRNA translation by Trim71 maintains pluripotency through inhibiting let-7 microRNAs. *eLife* **2021**, *10*, 10. [CrossRef] [PubMed]
97. Kim, C.W.; Vo, M.-T.; Kim, H.K.; Lee, H.H.; Yoon, N.A.; Lee, B.J.; Min, Y.J.; Joo, W.D.; Cha, H.J.; Park, J.W.; et al. Ectopic overexpression of tristetraprolin in human cancer cells promotes biogenesis of let-7 by down-regulation of Lin28. *Nucleic Acids Res.* **2011**, *40*, 3856–3869. [CrossRef] [PubMed]

98. Shaw, G.; Kamen, R. A conserved AU sequence from the 3′ untranslated region of GM-CSF mRNA mediates selective mRNA degradation. *Cell* **1986**, *46*, 659–667. [CrossRef]

99. Mori, M.; Triboulet, R.; Mohseni, M.; Schlegelmilch, K.; Shrestha, K.; Camargo, F.D.; Gregory, R.I. Hippo Signaling Regulates Microprocessor and Links Cell-Density-Dependent miRNA Biogenesis to Cancer. *Cell* **2014**, *156*, 893–906. [CrossRef] [PubMed]

100. Chaulk, S.G.; Lattanzi, V.J.; Hiemer, S.E.; Fahlman, R.P.; Varelas, X. The Hippo Pathway Effectors TAZ/YAP Regulate Dicer Expression and MicroRNA Biogenesis through Let-7. *J. Biol. Chem.* **2014**, *289*, 1886–1891. [CrossRef] [PubMed]

101. Chawla, G.; Sokol, N.S. ADAR mediates differential expression of polycistronic microRNAs. *Nucleic Acids Res.* **2014**, *42*, 5245–5255. [CrossRef] [PubMed]

102. Bahn, J.H.; Ahn, J.; Lin, X.; Zhang, Q.; Lee, J.-H.; Civelek, M.; Xiao, X. Genomic analysis of ADAR1 binding and its involvement in multiple RNA processing pathways. *Nat. Commun.* **2015**, *6*, 1–13. [CrossRef] [PubMed]

103. Nemlich, Y.; Greenberg, E.; Ortenberg, R.; Besser, M.J.; Barshack, I.; Jacob-Hirsch, J.; Jacoby, E.; Eyal, E.; Rivkin, L.; Prieto, V.G.; et al. MicroRNA-mediated loss of ADAR1 in metastatic melanoma promotes tumor growth. *J. Clin. Investig.* **2013**, *123*, 2703–2718. [CrossRef] [PubMed]

104. Zipeto, M.A.; Court, A.C.; Sadarangani, A.; Santos, N.P.D.; Balaian, L.; Chun, H.-J.; Pineda, G.; Morris, S.R.; Mason, C.N.; Geron, I.; et al. ADAR1 Activation Drives Leukemia Stem Cell Self-Renewal by Impairing Let-7 Biogenesis. *Cell Stem Cell* **2016**, *19*, 177–191. [CrossRef]

105. Ota, H.; Sakurai, M.; Gupta, R.; Valente, L.; Wulff, B.-E.; Ariyoshi, K.; Iizasa, H.; Davuluri, R.V.; Nishikura, K. ADAR1 Forms a Complex with Dicer to Promote MicroRNA Processing and RNA-Induced Gene Silencing. *Cell* **2013**, *153*, 575–589. [CrossRef]

106. Germanguz, I.; Lowry, W.E. RNA editing as an activator of self-renewal in cancer. *Stem Cell Investig.* **2016**, *3*, 68. [CrossRef]

107. Michlewski, G.; Guil, S.; Semple, C.; Cáceres, J.F. Posttranscriptional Regulation of miRNAs Harboring Conserved Terminal Loops. *Mol. Cell* **2008**, *32*, 383–393. [CrossRef]

108. Jain, N.; Lin, H.-C.; Morgan, C.E.; Harris, M.E.; Tolbert, B.S. Rules of RNA specificity of hnRNP A1 revealed by global and quantitative analysis of its affinity distribution. *Proc. Natl. Acad. Sci. USA* **2017**, *114*, 2206–2211. [CrossRef] [PubMed]

109. Michlewski, G.; Cáceres, J.F. Antagonistic role of hnRNP A1 and KSRP in the regulation of let-7a biogenesis. *Nat. Struct. Mol. Biol.* **2010**, *17*, 1011–1018. [CrossRef]

110. Burd, C.; Dreyfuss, G. RNA binding specificity of hnRNP A1: Significance of hnRNP A1 high-affinity binding sites in pre-mRNA splicing. *EMBO J.* **1994**, *13*, 1197–1204. [CrossRef]

111. Trabucchi, M.; Briata, P.; Garcia-Mayoral, M.; Haase, A.D.; Filipowicz, W.; Ramos, A.; Gherzi, R.; Rosenfeld, M.G. The RNA-binding protein KSRP promotes the biogenesis of a subset of microRNAs. *Nat. Cell Biol.* **2009**, *459*, 1010–1014. [CrossRef]

112. Gherzi, R.; Chen, C.-Y.; Ramos, A.; Briata, P. KSRP Controls Pleiotropic Cellular Functions. *Semin. Cell Dev. Biol.* **2014**, *34*, 2–8. [CrossRef] [PubMed]

113. Guil, S.; Caceres, J. The multifunctional RNA-binding protein hnRNP A1 is required for processing of miR-18a. *Nat. Struct. Mol. Biol.* **2007**, *14*, 591–596. [CrossRef] [PubMed]

114. Buratti, E.; Baralle, F.E. Multiple roles of TDP-43 in gene expression, splicing regulation, and human disease. *Front. Biosci.* **2008**, *13*, 867–878. [CrossRef]

115. Buratti, E.; De Conti, L.; Stuani, C.; Romano, M.; Baralle, M.; Baralle, F. Nuclear factor TDP-43 can affect selected microRNA levels. *FEBS J.* **2010**, *277*, 2268–2281. [CrossRef]

116. Kawahara, Y.; Mieda-Sato, A. TDP-43 promotes microRNA biogenesis as a component of the Drosha and Dicer complexes. *Proc. Natl. Acad. Sci. USA* **2012**, *109*, 3347–3352. [CrossRef]

117. Haselmann, V.; Kurz, A.; Bertsch, U.; Hübner, S.; Olempska–Müller, M.; Fritsch, J.; Häsler, R.; Pickl, A.; Fritsche, H.; Annewanter, F.; et al. Nuclear Death Receptor TRAIL-R2 Inhibits Maturation of Let-7 and Promotes Proliferation of Pancreatic and Other Tumor Cells. *Gastroenterology* **2014**, *146*, 278–290. [CrossRef]

118. Boyerinas, B.; Park, S.-M.; Hau, A.; Murmann, A.E.; Peter, M.E. The role of let-7 in cell differentiation and cancer. *Endocr.-Relat. Cancer* **2010**, *17*, F19–F36. [CrossRef]

119. Salzman, D.W.; Shubert-Coleman, J.; Furneaux, H. P68 RNA Helicase Unwinds the Human let-7 MicroRNA Precursor Duplex and Is Required for let-7-directed Silencing of Gene Expression. *J. Biol. Chem.* **2007**, *282*, 32773–32779. [CrossRef] [PubMed]

120. Sakamoto, S.; Aoki, K.; Higuchi, T.; Todaka, H.; Morisawa, K.; Tamaki, N.; Hatano, E.; Fukushima, A.; Taniguchi, T.; Agata, Y. The NF90-NF45 Complex Functions as a Negative Regulator in the MicroRNA Processing Pathway. *Mol. Cell. Biol.* **2009**, *29*, 3754–3769. [CrossRef]

121. Kawai, S.; Amano, A. BRCA1 regulates microRNA biogenesis via the DROSHA microprocessor complex. *J. Cell Biol.* **2012**, *197*, 201–208. [CrossRef]

122. Anderson, S.F.; Schlegel, B.P.; Nakajima, T.; Wolpin, E.S.; Parvin, J.D. BRCA1 protein is linked to the RNA polymerase II holoenzyme complex via RNA helicase A. *Nat. Genet.* **1998**, *19*, 254–256. [CrossRef] [PubMed]

123. Wilson, B.J.; Giguère, V. Identification of novel pathway partners of p68 and p72 RNA helicases through Oncomine meta-analysis. *BMC Genom.* **2007**, *8*, 419. [CrossRef] [PubMed]

124. Fuller-Pace, F.V. DExD/H box RNA helicases: Multifunctional proteins with important roles in transcriptional regulation. *Nucleic Acids Res.* **2006**, *34*, 4206–4215. [CrossRef] [PubMed]

125. Davis, B.N.; Hilyard, A.C.; Lagna, G.; Hata, A. SMAD proteins control DROSHA-mediated microRNA maturation. *Nat. Cell Biol.* **2008**, *454*, 56–61. [CrossRef] [PubMed]

126. Dubrovska, A.; Kanamoto, T.; Lomnytska, M.; Heldin, C.-H.; Volodko, N.; Souchelnytskyi, S. TGFβ1/Smad3 counteracts BRCA1-dependent repair of DNA damage. *Oncogene* **2005**, *24*, 2289–2297. [CrossRef]

127. Yu, B.; Bi, L.; Zheng, B.; Ji, L.; Chevalier, D.; Agarwal, M.; Ramachandran, V.; Li, W.; Lagrange, T.; Walker, J.C.; et al. The FHA domain proteins DAWDLE in Arabidopsis and SNIP1 in humans act in small RNA biogenesis. *Proc. Natl. Acad. Sci. USA* **2008**, *105*, 10073–10078. [CrossRef]

128. Li, P.; Yang, X.; Wasser, M.; Cai, Y.; Chia, W. Inscuteable and Staufen Mediate Asymmetric Localization and Segregation of prospero RNA during Drosophila Neuroblast Cell Divisions. *Cell* **1997**, *90*, 437–447. [CrossRef]

129. Ren, Z.; Veksler-Lublinsky, I.; Morrissey, D.; Ambros, V. Staufen Negatively Modulates MicroRNA Activity in Caenorhabditis elegans. *G3 Genes | Genomes | Genet.* **2016**, *6*, 1227–1237. [CrossRef] [PubMed]

130. Reinsborough, C.W.; Ipas, H.; Abell, N.S.; Gouws, E.B.; Williams, J.P.; Mercado, M.; Berg, C.V.D.; Xhemalçe, B. BCDIN3D RNA methyltransferase stimulates Aldolase C expression and glycolysis through let-7 microRNA in breast cancer cells. *Oncogene* **2021**, *40*, 2395–2406. [CrossRef]

131. Xhemalce, B.; Robson, S.C.; Kouzarides, T. Human RNA Methyltransferase BCDIN3D Regulates MicroRNA Processing. *Cell* **2012**, *151*, 278–288. [CrossRef] [PubMed]

132. Suzuki, H.; Arase, M.; Matsuyama, H.; Choi, Y.L.; Ueno, T.; Mano, H.; Sugimoto, K.; Miyazono, K. MCPIP1 Ribonuclease Antagonizes Dicer and Terminates MicroRNA Biogenesis through Precursor MicroRNA Degradation. *Mol. Cell* **2011**, *44*, 424–436. [CrossRef]

133. Pilotte, J.; Dupont-Versteegden, E.E.; Vanderklish, P.W. Widespread Regulation of miRNA Biogenesis at the Dicer Step by the Cold-Inducible RNA-Binding Protein, RBM3. *PLoS ONE* **2011**, *6*, e28446. [CrossRef]

134. Dannoab, S.; Nishiyamaa, H.; Higashitsujia, H.; Yokoia, H.; Xuea, J.-H.; Itoha, K.; Matsudab, T.; Fujita, J. Increased Transcript Level of RBM3, a Member of the Glycine-Rich RNA-Binding Protein Family, in Human Cells in Response to Cold Stress. *Biochem. Biophys. Res. Commun.* **1997**, *236*, 804–807. [CrossRef] [PubMed]

135. Großhans, H.; Johnson, T.; Reinert, K.L.; Gerstein, M.; Slack, F.J. The Temporal Patterning MicroRNA let-7 Regulates Several Transcription Factors at the Larval to Adult Transition in C. elegans. *Dev. Cell* **2005**, *8*, 321–330. [CrossRef] [PubMed]

136. Kumar, M.S.; Lu, J.; Mercer, K.L.; Golub, T.R.; Jacks, T. Impaired microRNA processing enhances cellular transformation and tumorigenesis. *Nat. Genet.* **2007**, *39*, 673–677. [CrossRef] [PubMed]

137. Nguyen, D.T.T.; Richter, D.; Michel, G.; Mitschka, S.; Kolanus, W.; Cuevas, E.; Wulczyn, F.G. The ubiquitin ligase LIN41/TRIM71 targets p53 to antagonize cell death and differentiation pathways during stem cell differentiation. *Cell Death Differ.* **2017**, *24*, 1063–1078. [CrossRef]

138. Mooijaart, S.; Brandt, B.; Baldal, E.; Pijpe, J.; Kuningas, M.; Beekman, M.; Zwaan, B.; Slagboom, P.; Westendorp, R.; Van Heemst, D. C. elegans DAF-12, Nuclear Hormone Receptors and human longevity and disease at old age. *Ageing Res. Rev.* **2005**, *4*, 351–371. [CrossRef]

139. Balzer, E.; Moss, E.G. Localization of the Developmental Timing Regulator Lin28 to mRNP Complexes, P-bodies and Stress Granules. *RNA Biol.* **2007**, *4*, 16–25. [CrossRef]

140. Carballo, E.; Lai, W.S.; Blackshear, P.J. Feedback Inhibition of Macrophage Tumor Necrosis Factor-α Production by Tristetraprolin. *Science* **1998**, *281*, 1001–1005. [CrossRef]

141. Lykke-Andersen, J. Recruitment and activation of mRNA decay enzymes by two ARE-mediated decay activation domains in the proteins TTP and BRF-1. *Genes Dev.* **2005**, *19*, 351–361. [CrossRef] [PubMed]

142. Yang, W.; Chendrimada, T.P.; Wang, Q.; Higuchi, M.; Seeburg, P.H.; Shiekhattar, R.; Nishikura, K. Modulation of microRNA processing and expression through RNA editing by ADAR deaminases. *Nat. Struct. Mol. Biol.* **2005**, *13*, 13–21. [CrossRef] [PubMed]

143. Suzuki, H.; Yamagata, K.; Sugimoto, K.; Iwamoto, T.; Kato, S.; Miyazono, K. Modulation of microRNA processing by p53. *Nat. Cell Biol.* **2009**, *460*, 529–533. [CrossRef]

144. Chen, Y.; Chan, J.; Chen, W.; Li, J.; Sun, M.; Kannan, G.S.; Mok, Y.-K.; Yuan, Y.A.; Jobichen, C. SYNCRIP, a new player in pri-let-7a processing. *RNA* **2020**, *26*, 290–305. [CrossRef]

145. Kim, I.-M.; Wang, Y.; Park, K.-M.; Tang, Y.; Teoh, J.-P.; Vinson, J.; Traynham, C.J.; Pironti, G.; Mao, L.; Su, H.; et al. β-Arrestin1–Biased β 1 -Adrenergic Receptor Signaling Regulates MicroRNA Processing. *Circ. Res.* **2014**, *114*, 833–844. [CrossRef] [PubMed]

146. Li, X.; Lin, W.-J.; Chen, C.-Y.; Si, Y.; Zhang, X.; Lu, L.; Suswam, E.; Zheng, L.; King, P.H. KSRP: A checkpoint for inflammatory cytokine production in astrocytes. *Glia* **2012**, *60*, 1773–1784. [CrossRef]

147. Winzen, R.; Thakur, B.K.; Dittrich-Breiholz, O.; Shah, M.; Redich, N.; Dhamija, S.; Kracht, M.; Holtmann, H. Functional Analysis of KSRP Interaction with the AU-Rich Element of Interleukin-8 and Identification of Inflammatory mRNA Targets. *Mol. Cell. Biol.* **2007**, *27*, 8388–8400. [CrossRef]

148. Zhu, H.; Gui, Q.; Hui, X.; Wang, X.; Jiang, J.; Ding, L.; Sun, X.; Wang, Y.; Chen, H. TGF-β1/Smad3 Signaling Pathway Suppresses Cell Apoptosis in Cerebral Ischemic Stroke Rats. *Med. Sci. Monit.* **2017**, *23*, 366–376. [CrossRef]

149. Chen, C.-Y.; Choong, O.K.; Liu, L.-W.; Cheng, Y.-C.; Li, S.-C.; Yen, C.Y.; Wu, M.-R.; Chiang, M.-H.; Tsang, T.-J.; Wu, Y.-W.; et al. MicroRNA let-7-TGFBR3 signalling regulates cardiomyocyte apoptosis after infarction. *EBioMedicine* **2019**, *46*, 236–247. [CrossRef]

150. Cao, L.; Kong, L.-P.; Yu, Z.-B.; Han, S.-P.; Bai, Y.-F.; Zhu, J.; Hu, X.; Zhu, C.; Zhu, S.; Guo, X.-R. microRNA expression profiling of the developing mouse heart. *Int. J. Mol. Med.* **2012**, *30*, 1095–1104. [CrossRef]

151. Kahl, A.; Blanco, I.; Jackman, K.; Baskar, J.; Milaganur Mohan, H.; Rodney-Sandy, R.; Zhang, S.; Iadecola, C.; Hochrainer, K. Cerebral ischemia induces the aggregation of proteins linked to neurodegenerative diseases. *Sci. Rep.* **2018**, *8*, 2701. [CrossRef]
152. Thammisetty, S.S.; Pedragosa, J.; Weng, Y.C.; Calon, F.; Planas, A.; Kriz, J. Age-related deregulation of TDP-43 after stroke enhances NF-κB-mediated inflammation and neuronal damage. *J. Neuroinflammation* **2018**, *15*, 1–15. [CrossRef]
153. Skau, E.; Henriksen, E.; Wagner, P.; Hedberg, P.; Siegbahn, A.; Leppert, J. GDF-15 and TRAIL-R2 are powerful predictors of long-term mortality in patients with acute myocardial infarction. *Eur. J. Prev. Cardiol.* **2017**, *24*, 1576–1583. [CrossRef] [PubMed]
154. Jin, Z.; Liang, J.; Li, J.; Kolattukudy, P.E. Absence of MCP-induced Protein 1 Enhances Blood–Brain Barrier Breakdown after Experimental Stroke in Mice. *Int. J. Mol. Sci.* **2019**, *20*, 3214. [CrossRef] [PubMed]
155. Ávila-Gómez, P.; Vieites-Prado, A.; Dopico-López, A.; Bashir, S.; Fernández-Susavila, H.; Gubern, C.; Pérez-Mato, M.; Correa-Paz, C.; Iglesias-Rey, R.; Sobrino, T.; et al. Cold stress protein RBM3 responds to hypothermia and is associated with good stroke outcome. *Brain Commun.* **2020**, *2*, fcaa078. [CrossRef] [PubMed]

biomedicines

Review

Coronary Vasculitis

Tommaso Gori

Kardiologie I and DZHK Standort Rhein-Main, Universitätsmedizin Mainz, 55131 Mainz, Germany; Tommaso.gori@unimedizin-mainz.de

Abstract: The term coronary "artery vasculitis" is used for a diverse group of diseases with a wide spectrum of manifestations and severity. Clinical manifestations may include pericarditis or myocarditis due to involvement of the coronary microvasculature, stenosis, aneurysm, or spontaneous dissection of large coronaries, or vascular thrombosis. As compared to common atherosclerosis, patients with coronary artery vasculitis are younger and often have a more rapid disease progression. Several clinical entities have been associated with coronary artery vasculitis, including Kawasaki's disease, Takayasu's arteritis, polyarteritis nodosa, ANCA-associated vasculitis, giant-cell arteritis, and more recently a Kawasaki-like syndrome associated with SARS-COV-2 infection. This review will provide a short description of these conditions, their diagnosis and therapy for use by the practicing cardiologist.

Keywords: coronary artery disease; vasculitis; inflammatory diseases

Citation: Gori, T. Coronary Vasculitis. *Biomedicines* **2021**, *9*, 622. https://doi.org/10.3390/biomedicines9060622

Academic Editor: Arnab Ghosh

Received: 28 April 2021
Accepted: 28 May 2021
Published: 31 May 2021

Publisher's Note: MDPI stays neutral with regard to jurisdictional claims in published maps and institutional affiliations.

1. Introduction

The term vasculitis refers to a group of conditions whose pathophysiology is mediated by inflammation of blood vessels. Most forms of vasculitis are systemic and may present variable clinical manifestations, requiring a multidisciplinary approach. A number of etiologies have been reported; independently of the specific organ involvement, vasculitis can be primary or secondary to another autoimmune disease or can be associated with other precipitants such as drugs, infections or malignancy [1]. Almost all cases of coronary vasculitis appear as a manifestation of systemic (primary) vasculitis, which are classified based on the type and size of the vessels affected and the cellular component responsible for the tissue infiltration. The classifications of the American College of Rheumatology, and the Chapel Hill International Consensus Conference distinguish among the following groups of vasculitis [2]: (a) large vessels vasculitis with giant cell (temporal) arteritis and Takayasu arteritis; (b) medium-sized vessels vasculitis including polyarteritis nodosa and Kawasaki disease; (c) small vessel vasculitis including allergic granulomatous angiitis, Wegener's granulomatosis, microscopic polyangiitis (microscopic polyarteritis), Henoch–Schonlein purpura, essential cryoglobulinemic vasculitis, and cutaneous leukocytoclastic angiitis [2]. Cardiac manifestations of systemic vasculitis are relatively rare but may lead to a sudden and rapid worsening of the patient's prognosis [3,4], including sudden death in young individuals with no prior cardiovascular disease. Beyond coronary artery vasculitis, clinical manifestations of cardiac involvement may include pericarditis, myocarditis, heart valve disease. Coronary vasculitis may present as stenosis, occlusion, aneurysm or rupture of coronaries. Although extremely rare, coronary vasculitis should be considered in the differential diagnostic of young patients with otherwise unexplained acute coronary syndromes or congestive cardiac failure and in patients with known primary or secondary vasculitis. Among the different forms listed above, the most frequent forms of coronary vasculitis include polyarteritis nodosa, Kawasaki's disease, Takayasu's arteritis and giant cell arteritis; most recently, an inflammatory involvement in patients infected with the SARS-COV-2 virus has also been reported. Early diagnosis of these rare forms has a drastic impact on the diagnostic methods and the choice of therapy [5], and appropriate treatment has the potential to modify the severity of the disease.

2. Most Common Vasculitis Entities with Coronary Involvement

Although a thorough discussion on the clinical features, pathophysiology and therapy of vasculitis goes beyond the scope of this paper focusing on coronary involvement, the following paragraphs will describe shortly the main features of each condition associated with coronary involvement, also summarized in Table 1.

Table 1. Typical forms of vasculitis with coronary involvement.

Group and Disease	Laboratory Findings	Frequency of Coronary Involvement	Location	Typical Lesion
Large Vessels				
Takayasu [6,7]	Inflammatory markers	10–45%	Ostial/proximal	Stenosis
Giant cell arteritis [8]	Inflammatory markers	Rare	Diffuse	Tapered smooth narrowings
Medium Vessels				
Polyarteritis nodosa [9,10]	Not consistently	10–50%	Not specific	Aneurysm or stenosis
Kawasaki [11,12]	Inflammatory markers	25–30%	Not specific	Aneurysm > stenosis
Small vessel				
Eosinophilic angiitis [10]	ANCA (not in all forms)	Rare	Not specific	Stenosis or aneurysm
Veins > arteries				
Behçet [13–16]	Not consistently	0.5–2%	Not specific	Thrombosis or pseudoaneurysm
Variable				
Erdheim–Chester [17]	Rare	25–55%	Right coronary prevalent	Periarteritis
IgG4 [18]	IgG4	1–3%	Not specific	Aneurysm or periarteritis

2.1. Large Vessel Vasculitis

2.1.1. Takayasu

Takayasu arteritis affects more commonly young females and it involves large vessels, typically the aortic arch and proximal branches of the aorta as well as the pulmonary arteries. The course of disease is usually chronic, with cutaneous, neurologic, gastrointestinal, and constitutional symptoms. Angina pectoris may occur following coronary artery ostial stenosis from aortitis or coronary arteritis in 10–45% of the cases and may have severe clinical sequelae, even though regression upon immunosuppression is possible [19]. In these cases, coronary artery stenosis is caused by the diffusion of the inflammatory process and intimal proliferation form the wall of the ascending aorta [6]. The pathogenesis is unclear, but cell-mediated mechanisms (particularly cytotoxic lymphocytes, especially gamma delta T lymphocytes releasing the cytolytic protein perforin) are thought to be involved [20]. The presence of a (yet unknown) specific antigen in the aorta might explain the selective localization of the disease [21]. At histology, mononuclear cells—predominantly lymphocytes, histiocytes, macrophages, and plasma cells—are present, and giant cells and granulomatous inflammation are found in the arterial media. The expansion of these inflammatory processes may lead on one side to destruction of the elastic lamina, causing the aneurysmal dilation of the affected vessel, or to narrowing of the lumen. Systemic inflammatory markers are often elevated. Systemic therapy consists of corticosteroids and immunosuppression, but anti-IL6 therapy with Tocilizumab has also been associated with a reduction in the markers of inflammation and, importantly, also the number of coronary artery lesions [22]. Takayasu is a progressive chronic disease with remissions and relapses and an 80–90% survival at five years [23].

2.1.2. Giant Cell Arteritis

This condition shares several similarities with Takaysu's arteritis, and the differential diagnostic is based on a rather artificial criterium of age (older than 50 in giant cell arteritis and younger than 40 in Takayasu) and vascular district involvement (more frequent involvement of supraaortic vessels in giant cell arteritis) [24]. Both giant cell arteritis and Takayasu involve the aorta and its major branches and are indistinguishable histopathologically. Although the prevalence of coronary artery involvement in giant cell arteritis is thought to be lower [8], large cohort studies report that the incidence of acute myocardial infarctions is twice as high as that in patients without giant cell arteritis, particularly early after diagnosis (Hazard Ratio for AMI: 11.9 [2.4–59.0]) [25]. Therapy with tocilizumab, an anti-IL6 antibody, is recommended in addition to corticosteroids, but its efficacy specifically on coronary vasculitis has not been investigated. The disease does not compromise life expectancy except for those patients in whom aortic dissection is present [8].

2.2. Medium Vessel Vasculitis

2.2.1. Polyarteritis Nodosa

Polyarteritis nodosa is a systemic necrotizing vasculitis which predominantly involves medium-sized arteries, more rarely small muscular arteries in middle-aged adults. Unlike microscopic polyarteritis, granulomatosis with polyangiitis, polyarteritis nodosa is antineutrophil cytoplasmic antibodies (ANCA)-negative [2]. It may affect virtually any organ in the body (kidneys, skin, joints, muscles, nerves, and gastrointestinal tract), usually in combination, but usually does not involve the lungs. Coronary involvement is considered to be rare but may be severe, including aggressive three-vessel disease with infaust outcome (Figure 1) [26]. Clinical sequelae of this involvement include congestive heart failure, hypertension, pericarditis, and arrhythmias.

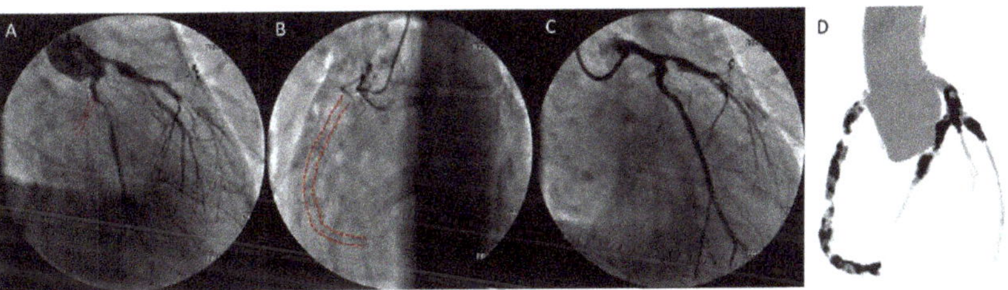

Figure 1. Coronary angiography (**A–C**) and computed tomography image (**D**) of three-vessel disease in a 22-years old patient with polyarteritis nodosa. Angiography showed chronic total occlusion of the right and circumflex coronaries. Reproduced with permission from [26] under the creative common license (http://creativecommons.org/publicdomain/zero/1.0/) (accessed on 3 March 2021).

Coronary involvement may manifest in the form of stenosis, occlusion, aneurysm, or dissection. In a recent review of cases, a total of 34 patients with an average age of 41 years were identified from 32 publications. Male sex is more frequent, coronary disease was the first manifestation of polyarteritis nodosa in 3/4. The clinical course of the disease was in general very severe, with cases of death from cardiac arrest, pulmonary edema with alveolar hemorrhage or multiple intracranial hemorrhages after thrombolytic therapy [27]. The formation of immune complexes as the result of a virus infection (hepatitis B or C virus) and hairy cell leukemia is thought to mediate the inflammatory reaction, which most often lead to media thickening and stenosis rather than aneurysm formation. Medical therapy includes cyclophosphamide in addition to corticosteroids and/or azathiprine, even though the long-term effectiveness of this approach appears to be limited. Five-year survival reaches 80%.

2.2.2. Kawasaki

Also known as mucocutaneous lymph node syndrome, Kawasaki disease is an acute, self-limited small and medium vessel systemic vasculitis with a frequent involvement of coronary arteries (right more often than left) affecting children <5 years of age with rare cases in older children [28]. Although Kawasaki disease generally has a self-limiting febrile course, given the decreasing incidence of rheumatic disease, it represents the most common cause of acquired heart disease in childhood in developed countries [29] and it accounts for as many as 5% of the cases of acute coronary syndrome in patients younger than 40 years. Clinically, it presents as polymorphous rash, mucosal changes (including dry, cracked lips and strawberry tongue), extremity changes (including palmar and/or plantar erythema, swelling, and desquamation), bilateral nonpurulent conjunctivitis, and cervical lymphadenopathy ($\geq$1.5 cm diameter), usually unilateral. Although a genetic influence has also been hypothesized, like for Behçet, the pathogenesis of Kawasaki is not clear and may be related to a wind-borne or water-borne pathogen. The most commonly affected age group are children under five years of age, but cases in adults are also common. For instance, a case of adult-onset Kawasaki Disease Shock Syndrome complicated by coronary aneurysms in a 20-year old man of East Asian ancestry has been recently reported [30]. Like for Behçet, the prevalence is higher, and the prognosis worse, in males affected by Kawasaki [12]. In the current hypothesis, it is believed that the activation of an immune reaction involving in the first phase neutrophils and then lymphocytes, cytokines, and proteinases; tumor necrosis factor alpha (TNF-a); Interleukin 1, 4, and 6 and matrix metalloproteinases (MMP3 and MMP9) triggered by exposure to an airborne virus may subtend Kawasaki. CD8+T cells, plasma cells, and monocytes cause release of pro-inflammatory cytokines IL-1β and TNFα. These processes may evolve for months to years resulting in a chronic arteritis. Oligoclonal IgA plasma cells appear to be central in the cascade leading to coronary arteritis. Clinical manifestations may include myocarditis and arteritis resulting in fibrinoid necrosis of the internal elastic lamina and subsequent formation of coronary aneurysms in up to one-third of untreated patients. Cerebral aneurysms are less frequent at 1–2% of patients. Monocytes, neutrophils and macrophages appear to be involved in the pathogenesis of these vascular lesions. Resulting from these inflammatory processes, an inappropriate healing response may also cause coronary stenosis. Otherwise, typical complications associated with the presence of the aneurysms include thrombus formation causing embolism and peripheral occlusion and rupture. Cases of regression of small coronary aneurysms upon inflammatory therapy have also been reported [31]. In the acute phase of Kawasaki disease, therapy with intravenous Immunoglobulin, corticosteroids, and aspirin monotherapy is encouraged [32]. Therapy with immunoglobulins and corticosteroids is associated with a reduction in the incidence of coronary events (odds ratio: 0.3 [0.2–0.5]) [33], but the presence of signs compatible with coronary vasculitis is a negative predictor for the responsiveness to immunoglobulins. The prognosis is variable and depends on the size of the aneurysms. Regular follow-up is therefore important.

2.3. Small Vessel Vasculitis

Eosinophilic Angiitis

Anti-neutrophil cytoplasmic antibody (ANCA)-associated vasculitides compose a family of conditions involving severe, systemic, small-vessel vasculitis featuring autoantibodies directed against the neutrophil proteins leukocyte proteinase 3 or myeloperoxidase. Their clinical presentation and therapies have been recently reviewed [34]. Within this group of diseases, eosinophilic granulomatosis with polyangiitis most commonly involves coronary vessels. This condition (which is not necessarily mediated by ANCA) has an incidence of ~0.5–2 cases per million per year, its time of onset is typically in the middle or older age and is equally frequent in males and females. Histologically, it is characterized by eosinophil-rich and necrotizing granulomatous inflammation predominantly affecting small-to-medium vessels; eosinophilia is the typical marker and pulmonary involvement and asthma are often associated. Connected, but not overlapping, with this condition, the first case of a novel medium-sized arteritis with eosinophilic inflammation limited to the adventitia and periadventitial soft tissue of epicardial coronary arteries was reported in 1989. Eosinophilic coronary periarteritis is not systemic and differs from poliarteritis nodosa and Churg–Strauss Syndrome in that it does not involve other arterial layers and it does not show fibrinoid necrosis. This form of coronary artery vasculitis has an adult onset and it has been associated with cases of spontaneous coronary artery dissection or coronary spasm [35,36]. Eosinophilic arteritis without dissection is more frequent in men than women, while a spontaneous dissection—prevalently of the left anterior descending coronary—is more frequent in females. Although the etiology and pathogenesis of this condition remain uncertain, it is possible that this disease may play a role in several cases of spontaneous dissection that remain etiologically undiagnosed. Churg–Strauss syndrome is a related disease that is diagnosed in the presence of the following six criteria: severe asthma, fleeting pulmonary infiltrates on chest radiographs, eosinophilia, paranasal sinus abnormalities, eosinophilic infiltration on biopsy, and neurologic manifestations. The highest incidence is in the fourth-fifth decade of life and is higher in females. Cardiac involvement may manifest as pericarditis, restrictive or dilated cardiomyopathy, myocarditis, arrhythmias, and sudden cardiac death. The prognosis of these diseases is markedly improved by the use of corticosteroids, and the mortality is in the range of 30% at 5 years [37].

2.4. Other Forms

2.4.1. Behçet

Behçet disease is a severe disease most frequent in the eastern Mediterranean and the east, where it reaches an incidence of 0.03% of the population. The incidence in males, most commonly in the fourth and fifth decade of life, is three times higher than in females. The most common manifestations include signs of inflammatory activation such as fever and constitutional symptoms. Clinical features of this disease include oral and/or genital ulcerations, ocular disease, and arthritis. Behçet´s disease has an unknown cause even though a genetic predisposition is suspected; its pathophysiology is based on autoinflammatory mechanisms, endothelial damage/dysfunction, and impaired fibrinolysis involving the activation of T-lymphocytes, including T helper 17 cells, immune complex formation, neutrophil activation, and the secretion of inflammatory cytokines. From the histologic perspective, Behçet is associated with perivasculitis with neutrophil infiltration, fibrinoid necrosis and endothelial swelling [15,16]. Anti-lymphocyte and anti-cardiolipin antibodies are a feature of this condition and allow its diagnosis. Arterial involvement in Behçet disease manifest itself as aneurysms of the medium- and large-sized arteries. The vascular manifestations of Behçet vary between 10 and 30% of the cases and may involve both arteries and, approximately 4 times more frequently, veins [14]. Vascular involvement portends a negative prognosis, particularly in younger males without traditional coronary artery disease risk factors (except for smoking). Although native coronary disease is most commonly described, cases of in-stent restenosis have also been reported [14]. Therapy may include colchicine and general immunosuppressive therapy with corticosteroids, azathioprine, cyclosporine A, and cyclophosphamide). The prognosis is good, although the incidence of relapses is very high.

2.4.2. Erdheim–Chester Disease

Erdheim–Chester disease is a rare non-Langerhans histiocytosis and is essentially a malignancy of myeloid progenitor cells caused by a somatic mutation of signaling molecule genes, which explains the increased expression of inflammatory cytokines (IL-6, interferon alpha, MCP-1). Fewer than 1000 cases of this disease have been reported, most commonly under the form of multifocal sclerotic lesions of the bones. Bone pain, neurologic symptoms, and cardiac symptoms are the most common manifestations. It may result in multisystem involvement including vasculitis-like phenomena of large and medium-sized arteries, particularly of the aorta (around 60% in a recent case series) [17]. Coronary artery involvement is rather common (55%), leading to about one-third of deaths [38]. The most common presentation includes infiltration, particularly at the level of the right atrium; periarterial fibrosis, thickening and pleural effusion are also common. Its histologic hallmark is perinephric fibrosis. The treatment may include a BRAF inhibitor (vemurafenib), a MEK inhibitor, interferon alfa, or immunosuprression with corticosteroids or cytotoxic therapies. Five-year survival reaches 70%.

2.4.3. IgG4-Related Disease

Immunoglobulin G4-related disease is a rare fibroinflammatory condition with multi-organ lymphoplasmacytic infiltration that affects more comonly males. Every organ may be affected and extracardiac involvement may manifest as sialadenitis, thyroiditis, nephritis, lymphadenopathy, and lung disease. The pathogenesis of this disease appears to depend on an immune response mediated by Th1 and Th2 cells, while the role of IgG4 antibodies remains unclear. Cardiovascular manifestations of IgG4-RD may include aortitis, medium-sized vessel arteritis, pulmonary vascular disease, phlebitis, valvulopathy, pericarditis, and antineutrophilic cytoplasmic antibody-associated vasculitis and more frequently involve the abdominal vasculature, even though cardiac cases have also been reported [39,40]. Therapy with glucocorticoids leads to a remission in 98% of the cases, but a spontaneous remission without therapy is also often observed. Cases that are refractory to steroids and require immunosuppression, however, also exist.

3. Epidemiology

The relative rarity of these conditions as well as a lack of standardized prospective imaging studies systematically evaluating the coronary vasculature complicates the diagnosis of coronary artery vasculitis in patients with known autoimmune diseases. The involvement of the coronary microvasculature (not visible at computed tomography or angiography) and the fact that coronary artery vasculitis may manifest under the form of coronary (micro)vascular spasm rather than fixed stenosis further complicate the diagnosis and may result in many false negatives. Finally, the co-existence of accelerated atherosclerosis caused by immunosuppressive therapies and the lack of availability of methods for the differential diagnosis of these two forms of disease is an additional hurdle to the diagnosis. Particularly in cases of isolated coronary artery vasculitis, which is considered to be extremely rare, the true incidence is probably dramatically underestimated [10]. A coronary artery vasculitis is diagnosed (Table 1) in 50% of the patients with polyarteritis nodosa (with a reported incidence of 4–10 per million per year) and in about 20% of patients with Kawasaki disease (incidence 2 per million per year [5]). A similar incidence of coronary artery vasculitis is reported for patients with large vessel vasculitides, such as Takayatsu arteritis and giant cell arteritis (incidence rates of 1–3 per million each). Other forms of large vessel vasculitides like the Erdheim–Chester disease are rarely associated with coronary artery vasculitis (about 5% of the cases). The incidence of vascular involvement in Behçet has also been reported to be in the range of 50% [41].

4. Pathogenesis and Pathology

The pathogenesis of coronary artery vasculitis is complex, multifaceted, depends on the specific disease, and it is essentially mediated by both extrinsic and host factors such as immuno-mediated inflammation and auto-antibody dependent reactions [9]. Inflammatory cytokines, in part specific for each condition, including interferon-gamma (IFN-γ), tumor necrosis factor-alpha (TNF-α), and T-1 interleukins have all been reported. Histopathological findings on autopsy are also specific for each condition. The coronary artery vasculitis caused by polyarteritis nodosa is characterized by intramural, perivascular lymphocyte, and macrophage infiltration as well as fibrinoid necrosis which provoke the destruction of the arterial wall. Takayatsu and giant cell arteritis are associated with intimal hyperplasia, granulomatous arteritis, and coronary atherosclerosis. In contrast, Kawasaki disease is characterized by multi-cellular infiltration of the arterial wall causing necrosis of the internal elastic lamina. As a result of these different mechanisms, findings of chronic inflammation, scar tissue, necrosis, and stenosis may be present in all forms of coronary artery vasculitis. Coronary artery aneurysm are a rarer entity caused, as mentioned above, by weakening of the arterial wall mediated by an overactivity of metalloproteinases and metalloelastases. The same mechanisms, coupled with the release of pro-angiogenetic substances from the eosinophils and the resulting formation of weak capillaries that may rupture in the arterial media, may lead to spontaneous dissection in eosinophilic arteritis [42]. Arterial thromboses may result from the blood stasis in these aneurysms and/or in stenosis leading to myocardial ischemia [43]. As in all inflammatory conditions, the incidence of typical coronary atherosclerosis is also higher in patients with vasculitis. Rarely, embolisms of Wegener granulomatosis of heart valve lesions causing coronary artery occlusion have also been described [44].

5. Clinical Manifestations and Diagnosis

The symptoms and signs associated with system involvement differ significantly among different disease entities reflecting the different organs involved. Because of the relative rarity of these disease, a diagnosis can easily be missed. The clinical presentation may not be specific, but clinical features that might suggest an inflammatory etiology include elevated inflammatory markers that cannot be explained otherwise, signs of inflammation—such as fever, chills, night sweats, weight loss, subclavian or aortic bruits—suggestive of atypical stenosis in other districts, a history of multiple system involvement such as abdominal ischemic events, early age particularly in the absence of genetic predisposition or severe risk factors. Patients with suspected vasculitis, particularly Takayasu, should undergo imaging of the arterial tree by magnetic resonance angiography (which has the advantage of avoiding contrast medium and radiation exposure) or computed tomography to evaluate the presence of aneurysms, stenoses, arterial wall thickening, and/or masslike lesions surrounding the coronary arteries as recently reviewed in [45,46]. In particular, computed tomography provides accurate non-invasive information on both luminal and mural lesions in the aorta and its branches, which is important in both the diagnosis and the follow-up of disease progression. Computed tomography allows detection and quantification of stenoses and aneurysms which may be present in virtually every form of vasculitis [46]; findings that are typical of specific conditions include "skip lesions" (focal stenoses in Takayasu disease) and "pigs-in-a-blanket" lesions (rings of soft-tissue attenuation surrounding the coronary arteries) in coronary periarteritis [18]. Ultrasound imaging of other districts, including the chest, abdomen, head, and neck may be useful, particularly to detect wall thickenings which cannot be diagnosed by angiography. Positron emission tomography, often in combination with computed tomography or magnetic resonance, has been used for the diagnosis of large-vessel vasculitis. Arterial segments featuring increased standardized uptake values may be suggestive of disease [47]; these abnormalities are however not present in coronaries.

At angiography, coronary artery vasculitis may manifest under the form of coronary stenosis, aneurysm (Figure 2), dissection (Figure 3, most frequent in eosinophilic periarteritis), spasm (Figure 4), or coronary rupture (Table 1). Sudden death, typical angina, acute myocardial infarction, atrial and ventricular arrhythmias, conduction disturbances, or cardiac failure have all been described [48]. Although no specific finding at coronary angiography allows a safe diagnosis, features associated with early, advanced, or atypical coronary artery disease may suggest an inflammatory etiology. For example, Kawasaki disease and polyarteritis nodosa are often associated with large aneurysms [11]. Polyarteritis nodosa often features multifocal aneurysms with a "beads on a string" or nodular appearance [49], which are also present in patients with anti-neutrophil cytoplasmic antibody (ANCA)-associated vasculitis and Behcet's disease [13]. The arterial lesions in Behçet disease may be occlusive or aneurysmal [15]. Giant cell arteritis has been associated with long coronary lesions [50], and in Takayasu's arteritis, coronary lesions have been classified in three main types: 1, stenosis or occlusion of the coronary ostia (60–80%);

2, diffuse disease that may involve all epicardial branches or only focal segments (10–20%); and 3, coronary aneurysms (0–5%) [23].

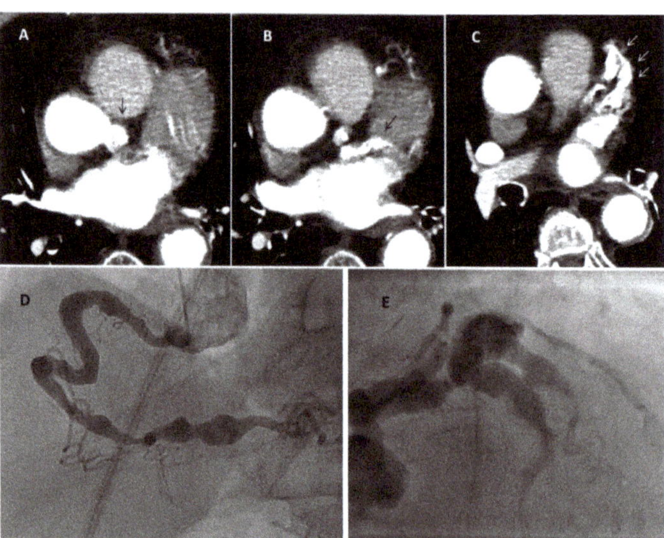

Figure 2. Coronary aneurysms in an 80-year-old female patient with stroke and rheumatoid arthritis. Panel a: right coronary; Panel b: left coronary (arrows mark the aneurysms). The patient had been admitted for sudden loss of consciousness. Chest computed tomography (panels **A–C**, Video S1) was performed for the suspect of lung embolism. The exam showed severe coronary aneurysms. At angiography (panel **D,E**), the aneurysms were so large that they could not be imaged in one single run despite large contrast volume.

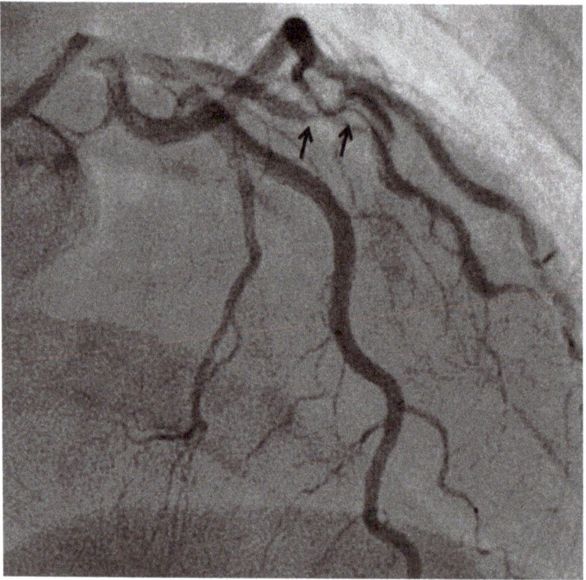

Figure 3. Spontaneous dissection (arrows) in a 45-years old woman causing non-ST elevation myocardial infarction.

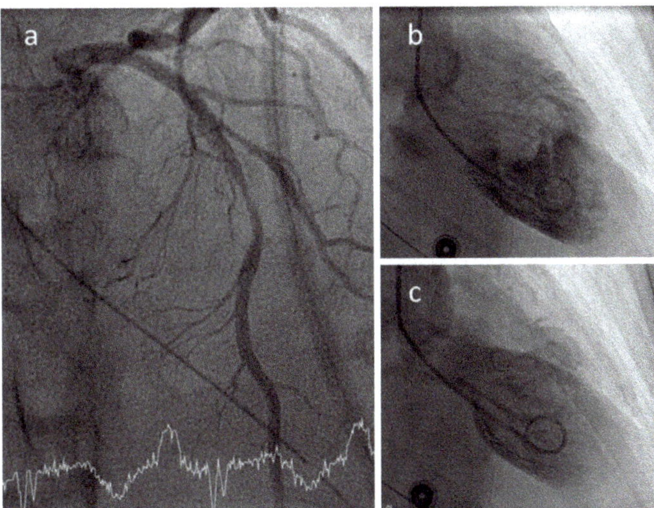

Figure 4. Coronary spasm in a 51-year-old female patient with ANCA-associated vasculitis and spasm of the proximal left anterior descending (panel **a**) causing a Tako-Tsubo-like contractile dysfunction (**b**,**c**).

Coronary angiography does not detect inflammation, necrosis or reactive thickening of the arterial wall. Non-invasive assessment of the cardiac and extracardiac circulation using computed tomography or magnetic resonance imaging is therefore recommended in patients with large (giant cell arteritis and Takayasu's arteritis), medium (poliarteritis nodosa) or variable (Behçet's disease) vessel vasculitis. Patients with small vessel vasculitis (i.e., ANCA-associated) often have associated myo- or pericarditis, which should also be investigated using magnetic resonance imaging. Periarterial soft tissue thickening or extrinsic compression are also features of more rare conditions such as Immunoglobulin G4(IgG4)-related disease or the Erdheim–Chester disease [18,51].

6. COVID-19 and Coronary Inflammation

The impact of COVID-19 on vascular biology is a topic of recent great attention as it appears to be responsible for a significant percentage of the negative outcomes [52]. Macro- and microvascular thrombosis involving arteries, veins, arterioles, capillaries, and venules in all major organs has been reported, along with evidence of endotheliitis across different vascular beds [53,54]. The lesions featured diffuse lymphocytic endotheliitis and apoptotic bodies likely following virus adhesion to the cell and proinfammatory and apoptotic pathway signaling [55]. Oxidative stress, increased production and release of chemokines, cytokines, and byproducts of damage-associated molecular patterns appear to be responsible for the endotheliitis [56]. Neutrophil recruitment by products of vascular injury is mediated by several pathways including the PI3K/AKt/eNOS/NF-Kβ and ERK1/2/ P38 MAPK-activated protein kinases and leads to further inflammation via release of TNF-α, IL-1, and IL-8, and neutrophil-extracellular traps NETs [57]. The mechanisms leading to necrosis and apoptosis have been recently summarized in dedicated reviews [58]. COVID-19 associated cytotoxicity primarily find place within the microcirculatory system, leading to ultrastructural changes and vascular dysfunction. Furthermore, increased bioavailability of the vasoconstrictor Angiotensin II due to COVID-19-mediated depletion of its receptor ACE-2, is an acknowledged mechanism of endothelial cell dysfunction. The release of von Willebrand factors which follows endothelial damage and activation can in turn recruit and activate circulating platelets contributing to the enhanced production of pro-coagulants, inflammatory cytokines, and NETs. Type I and II myocardial ischemia may result. Along with thrombotic events, vasculitis thus occurs in COVID-19 but its role in the several reported cases of acute coronary syndrome on COVID-19 patients remains unclear [57,59,60].

COVID and Kawasaki-Like Disease

As described above, Kawasaki disease is believed to result from an excess innate immune response to viral pathogens. Although the mechanism remains to be elucidated, the involvement of the stimulator of interferon genes (STING, a cytosolic DNA sensor and adaptor protein in type I IFN and nuclear factor(NF)-κB pathway, leading to hypercoagulopathy via macrophage production of tissue factor) pathway has been proposed [61]. A similar situation may occur in COVID-19 infections, where SARS–COV-2 binding to ACE2 increases STING pathway activation. The resulting immune hyper-responses, decreased lymphocyte counts, increased monocyte populations that secrete cytotoxic cytokines and heightened B and T cell responses configure a Kawasaki-like disease that may result in toxic shock syndrome or multi-systemic inflammatory disease [62]. In as many as 25% of the children affected by COVID-19, coronary dilations have been reported, and several cases of COVD-19 myocarditis have been published [63]. In a recent survey of 149 patients, children with "Kawacovid syndrome" were significantly older and presented more frequently gastrointestinal and respiratory involvement. Cardiac involvement in the form of myocarditis was more common (60%) as in traditional Kawasaki, but coronary artery abnormalities were rare. About 40% of patients with Kawacovid presented hypotension/non-cardiogenic shock [52]. At cardiac magnetic resonance imaging, evidence of diffuse myocardial edema has been reported [64]. The clinical presentation of the resulting multisystem inflammatory syndrome in children (MIS-C) reflects that of Kawasaki, but differences between the two exist with regards to age (from early childhood to late adolescence as compared to early childhood in Kawasaki); more frequent lymphopenia and thrombocytopenia, cardiac ventricular stress including myocarditis, and coagulopathy in MIS-C, more frequent gastrointestinal involvement, myocarditis leading to cardiogenic shock, and a different cytokine profile (IL-6 and IL-8 in MIS-C, IL-1 in Kawasaki).

7. Revascularization Procedures in Coronary Artery Vasculitis

7.1. Chronic Coronary Syndromes

All patients with known vasculitis should be treated with acetylsalicilic acid. With regards to the indication to revascularization, traditional methods for the assessment of vitality (low-dose dobutamine stress-echocardiography, scintigraphy, magnetic resonance imaging) and ischemia (exercise-electrocardiography, stress echocardiography or scintigraphy, etc.) maintain their validity. There is little data, anecdotal in nature, regarding the outcomes of interventions in patients with coronary arteritis; current American Heart Association guidelines recommend percutaneous interventions in patients with either a single vessel involvement or focal multivessel disease [65]. Percutaneous coronary interventions might however not provide adequate revascularization in cases of diffuse disease. Given the severity of these diseases and their rarity, no prospective randomized trial to determine the results of different revascularization methods (interventional versus by-pass surgery) has been conducted. In patients with Takayasu's arteritis, a higher rate of target lesion failure has been reported after percutaneous coronary intervention as compared to surgery (odds ratio 7.4 [2.4–23.1], $p = 0.01$) [66], such that surgery is recommended in these patients after induction of immunosuppression [19]. The involvement of the ascending aorta is a clear factor that complicates venous and free arterial by-pass graft surgery. As well, the internal mammary artery may also be a target of disease in patients with Takayasu's arteritis [7], which complicates by-pass grafting [67]. Scar healing might be compromised or slower in patients with active disease. Finally, coronary artery aneurysms can be occluded by coiling or implantation of covered graft stents, even though the risk of thrombosis and restenosis remains high. Aneurysm resection/thrombectomy and by-pass surgery also remain options [68,69].

7.2. Acute Coronary Syndromes

Patients with coronary aneurysms have an elevated risk of myocardial infarction. Given the young age, interventional therapy in this setting is often not feasible. In case of severe thrombosis occurring within the aneurysm, therapy with heparin, IIbIIIa antagonists and fibrinolytics has been reported also in pediatric age [70,71]. However, the outcomes of these procedures are not known. Regardless of the type of revascularization, antiplatelet therapy should be applied, and life-long antiplatelet therapy is to be recommended in all patients with coronary aneurysms [65]. Data on anticoagulation in patients with aneurysms are also very scarce, and this type of approach is discouraged by some authors for fear of progression of the aneurysms [72] while others report a reduced incidence of acute myocardial infarctions in patients treated with a combination of aspirin and warfarin [73]. As well, despite anticoagulation, thrombosis in giant aneurysms can occur owing to blood stasis and decreased wall shear stress [74]. Finally, immunosuppressive therapy is a mainstay of therapy, and cases of "spontaneous" regression of coronary artery stenoses have been reported [75]. Cardiac transplantation may be considered in patients deemed not suitable for revascularization.

8. Conclusions

Coronary artery vasculitis is rare, but still represent one of the most frequent causes of coronary artery disease in young patients. Its anatomical manifestations may include coronary artery stenosis, aneurysms, thrombosis, and spontaneous dissection; and its consequences may be severe. Even though prognosis of coronary vasculitis is poor, early diagnosis and therapy improve survival rates. Both non-invasive and invasive methods provide essential information in the diagnosis. Large-scale studies are now necessary to further investigate the incidence, diagnostic yield, and therapy of this rare and heterogeneous group of diseases.

Supplementary Materials: The following are available online at https://www.mdpi.com/article/10.3390/biomedicines9060622/s1, Video S1: computed tomography scan of a patient with diffuse coronary aneurysms.

Funding: This research received no external funding.

Institutional Review Board Statement: Not applicable.

Informed Consent Statement: Not applicable.

Data Availability Statement: Not applicable.

Conflicts of Interest: The author declares no conflict of interest.

References

1. Chakravarty, K.; Ong, V.H.; Denton, C.P. Secondary vasculitis in autoimmune connective tissue diseases. *Curr. Opin. Rheumatol.* **2016**, *28*, 60–65. [CrossRef]
2. Jennette, J.C.; Falk, R.J.; Bacon, P.A.; Basu, N.; Cid, M.C.; Ferrario, F.; Flores-Suarez, L.F.; Gross, W.L.; Guillevin, L.; Hagen, E.C.; et al. 2012 revised International Chapel Hill Consensus Conference Nomenclature of Vasculitides. *Arthritis Rheum.* **2013**, *65*, 1–11. [CrossRef]
3. Sakai, K.; Asakura, K.; Saito, K.; Fukunaga, T. Sudden unexpected death due to coronary thrombosis associated with isolated necrotizing vasculitis in the coronary arteries of a young adult. *Forensic Sci. Med. Pathol.* **2019**, *15*, 252–257. [CrossRef]
4. Markham, R.; Rahman, A.; Tai, S.; Hamilton-Craig, I.; Hamilton-Craig, C. Myocardial infarction from isolated coronary artery vasculitis in a young patient: A rare case. *Int. J. Cardiol.* **2015**, *180*, 40–41. [CrossRef]
5. Mavrogeni, S.; Cantini, F.; Pohost, G.M. Systemic vasculitis: An underestimated cause of heart failure-assessment by cardiovascular magnetic resonance. *Rev. Cardiovasc. Med.* **2013**, *14*, 49–55.
6. Amano, J.; Suzuki, A. Coronary artery involvement in Takayasu's arteritis. Collective review and guideline for surgical treatment. *J. Thorac. Cardiovasc. Surg.* **1991**, *102*, 554–560. [CrossRef]
7. Kuijer, A.; van Oosterhout, M.F.; Kloppenburg, G.T.; Morshuis, W.J. Coronary artery bypass grafting in Takayasu's disease-importance of the proximal anastomosis: A case report. *J. Med. Case Rep.* **2015**, *9*, 283. [CrossRef]
8. Mednick, Z.; Farmer, J.; Khan, Z.; Warder, D.; Hove, M.T. Coronary arteritis: An entity to be considered in giant cell arteritis. *Can. J. Ophthalmol.* **2016**, *51*, e6–e8. [CrossRef] [PubMed]

9. Dye, J.R.; Kaul, M.S.; St Clair, E.W. Infammatory diseases of the coronary arteries. In *PanVascular Medicine*; Springer: Berlin/Heidelberg, Germany, 2013; pp. 1–40.

10. Koster, M.J.; Warrington, K.J. Vasculitis of the Coronary Arteries. *Am. Coll. Cardiol. Online* **2019**. Available online: https://www.acc.org/latest-in-cardiology/articles/2019/03/13/06/50/vasculitis-of-the-coronary-arteries (accessed on 31 May 2021).

11. Friedman, K.G.; Gauvreau, K.; Hamaoka-Okamoto, A.; Tang, A.; Berry, E.; Tremoulet, A.H.; Mahavadi, V.S.; Baker, A.; de Ferranti, S.D.; Fulton, D.R.; et al. Coronary Artery Aneurysms in Kawasaki Disease: Risk Factors for Progressive Disease and Adverse Cardiac Events in the US Population. *J. Am. Heart Assoc.* **2016**, *5*. [CrossRef] [PubMed]

12. Saguil, A.; Fargo, M.; Grogan, S. Diagnosis and management of kawasaki disease. *Am. Fam. Physician* **2015**, *91*, 365–371. [PubMed]

13. Demirelli, S.; Degirmenci, H.; Inci, S.; Arisoy, A. Cardiac manifestations in Behcet's disease. *Intractable Rare Dis. Res.* **2015**, *4*, 70–75. [CrossRef] [PubMed]

14. Kariyanna, P.T.; Shah, P.; Jayarangaiah, A.; Chowdhury, Y.S.; Lazaro, D. Acute coronary syndrome in Behcet's syndrome: A systematic review. *Eur. J. Rheumatol.* **2021**, *8*, 31–35. [CrossRef]

15. Yazici, H.; Seyahi, E.; Hatemi, G.; Yazici, Y. Behcet syndrome: A contemporary view. *Nat. Rev. Rheumatol.* **2018**, *14*, 119. [CrossRef]

16. Zeidan, M.J.; Saadoun, D.; Garrido, M.; Klatzmann, D.; Six, A.; Cacoub, P. Behcet's disease physiopathology: A contemporary review. *Auto Immun. Highlights* **2016**, *7*, 4. [CrossRef] [PubMed]

17. Villatoro-Villar, M.; Bold, M.S.; Warrington, K.J.; Crowson, C.S.; Goyal, G.; Shah, M.; Go, R.S.; Koster, M.J. Arterial involvement in Erdheim–Chester disease: A retrospective cohort study. *Medicine* **2018**, *97*, e13452. [CrossRef]

18. Guo, Y.; Ansdell, D.; Brouha, S.; Yen, A. Coronary periarteritis in a patient with multi-organ IgG4-related disease. *J. Radiol. Case Rep.* **2015**, *9*, 1–17. [CrossRef]

19. Yokokawa, T.; Kunii, H.; Kaneshiro, T.; Ichimura, S.; Yoshihisa, A.; Furuya, M.Y.; Asano, T.; Nakazato, K.; Ishida, T.; Migita, K.; et al. Regressed coronary ostial stenosis in a young female with Takayasu arteritis: A case report. *BMC Cardiovasc. Disord.* **2019**, *19*, 79. [CrossRef]

20. Seko, Y.; Minota, S.; Kawasaki, A.; Shinkai, Y.; Maeda, K.; Yagita, H.; Okumura, K.; Sato, O.; Takagi, A.; Tada, Y.; et al. Perforin-secreting killer cell infiltration and expression of a 65-kD heat-shock protein in aortic tissue of patients with Takayasu's arteritis. *J. Clin. Investig.* **1994**, *93*, 750–758. [CrossRef]

21. Seko, Y.; Sato, O.; Takagi, A.; Tada, Y.; Matsuo, H.; Yagita, H.; Okumura, K.; Yazaki, Y. Restricted usage of T-cell receptor Valpha-Vbeta genes in infiltrating cells in aortic tissue of patients with Takayasu's arteritis. *Circulation* **1996**, *93*, 1788–1790. [CrossRef]

22. Pan, L.; Du, J.; Liu, J.; Liao, H.; Liu, X.; Guo, X.; Liang, J.; Han, H.; Yang, L.; Zhou, Y. Tocilizumab treatment effectively improves coronary artery involvement in patients with Takayasu arteritis. *Clin. Rheumatol.* **2020**, *39*, 2369–2378. [CrossRef]

23. Sun, T.; Zhang, H.; Ma, W.; Yang, L.; Jiang, X.; Wu, H.; Hui, R.; Zheng, D. Coronary artery involvement in takayasu arteritis in 45 Chinese patients. *J. Rheumatol.* **2013**, *40*, 493–497. [CrossRef] [PubMed]

24. Michel, B.A.; Arend, W.P.; Hunder, G.G. Clinical differentiation between giant cell (temporal) arteritis and Takayasu's arteritis. *J. Rheumatol.* **1996**, *23*, 106–111.

25. Tomasson, G.; Peloquin, C.; Mohammad, A.; Love, T.J.; Zhang, Y.; Choi, H.K.; Merkel, P.A. Risk for cardiovascular disease early and late after a diagnosis of giant-cell arteritis: A cohort study. *Ann. Intern. Med.* **2014**, *160*, 73–80. [CrossRef] [PubMed]

26. Huang, H.; Gong, Y.; Guo, L.; Zhang, Z. Insidious coronary artery disease in a young patient with polyarteritis nodosa: A case report and literature review. *BMC Cardiovasc. Disord.* **2021**, *21*, 115. [CrossRef] [PubMed]

27. Samson, M.; Puechal, X.; Mouthon, L.; Devilliers, H.; Cohen, P.; Bienvenu, B.; Ly, K.H.; Bruet, A.; Gilson, B.; Ruivard, M.; et al. Microscopic polyangiitis and non-HBV polyarteritis nodosa with poor-prognosis factors: 10-year results of the prospective CHUSPAN trial. *Clin. Exp. Rheumatol.* **2017**, *35* (Suppl. 103), 176–184. [PubMed]

28. Tanti, S.K.; Mishra, S. Atypical Kawasaki Disease Complicated With Coronary Artery Aneurysm: A Case Report and Review of Literature. *Cureus* **2021**, *13*, e13589. [CrossRef] [PubMed]

29. Gkoutzourelas, A.; Bogdanos, D.P.; Sakkas, L.I. Kawasaki Disease and COVID-19. *Mediterr. J. Rheumatol.* **2020**, *31*, 268–274. [CrossRef] [PubMed]

30. Zazoulina, J.; Cheung, C.C.; Hurlburt, A.; Chahal, D.; Wong, S. A Case of Adult-Onset Kawasaki Disease Shock Syndrome Complicated by Coronary Aneurysms. *CJC Open* **2019**, *1*, 206–208. [CrossRef]

31. Matsubara, T. Infliximab for the treatment of Kawasaki disease. *Pediatric Int.* **2018**, *60*, 775. [CrossRef]

32. Royal Children's Hospital Melbourne. Kawasaki Disease Clinical Practice Guidelines. Available online: https://www.rch.org.au/clinicalguide/guideline_index/Kawasaki_disease/ (accessed on 19 May 2021).

33. Chen, S.; Dong, Y.; Yin, Y.; Krucoff, M.W. Intravenous immunoglobulin plus corticosteroid to prevent coronary artery abnormalities in Kawasaki disease: A meta-analysis. *Heart* **2013**, *99*, 76–82. [CrossRef]

34. Kitching, A.R.; Anders, H.J.; Basu, N.; Brouwer, E.; Gordon, J.; Jayne, D.R.; Kullman, J.; Lyons, P.A.; Merkel, P.A.; Savage, C.O.S.; et al. ANCA-associated vasculitis. *Nat. Rev. Dis. Primers* **2020**, *6*, 71. [CrossRef] [PubMed]

35. Omalu, B.; Hammers, J.; DiAngelo, C.; Moore, S.; Luckasevic, T. Autopsy features of sudden death due to isolated eosinophilic coronary arteritis: Report of two cases. *J. Forensic Nurses* **2011**, *7*, 153–156. [CrossRef]

36. Bitar, A.Y.; Thompson, C.D.; Tan, C.W.; Allem, K.; Khachatrian, A.; Weis, P.J.; Garg, V.; Srivastava, A. Coronary artery vasospasm and cardiogenic shock as the initial presentation for eosinophilic granulomatosis with polyangiitis. *J. Cardiol. Cases* **2016**, *13*, 105–108. [CrossRef]

37. Saku, A.; Furuta, S.; Hiraguri, M.; Ikeda, K.; Kobayashi, Y.; Kagami, S.I.; Kurasawa, K.; Matsumura, R.; Nakagomi, D.; Sugiyama, T.; et al. Longterm Outcomes of 188 Japanese Patients with Eosinophilic Granulomatosis with Polyangiitis. *J. Rheumatol.* **2018**, *45*, 1159–1166. [CrossRef]
38. Haroche, J.; Amoura, Z.; Dion, E.; Wechsler, B.; Costedoat-Chalumeau, N.; Cacoub, P.; Isnard, R.; Genereau, T.; Wechsler, J.; Weber, N.; et al. Cardiovascular involvement, an overlooked feature of Erdheim–Chester disease: Report of 6 new cases and a literature review. *Medicine* **2004**, *83*, 371–392. [CrossRef]
39. Shakir, A.; Wheeler, Y.; Krishnaswamy, G. The enigmatic immunoglobulin G4-related disease and its varied cardiovascular manifestations. *Heart* **2021**, *107*, 790–798. [CrossRef]
40. Okuyama, T.; Tanaka, T.D.; Nagoshi, T.; Yoshimura, M. Coronary artery disease concomitant with immunoglobulin G4-related disease: A case report and literature review. *Eur. Heart J. Case Rep.* **2019**, *3*, ytz013. [CrossRef]
41. Geri, G.; Wechsler, B.; Huong, D.L.T.; Isnard, R.; Piette, J.C.; Amoura, Z.; Resche-Rigon, M.; Cacoub, P.; Saadoun, D. Spectrum of cardiac lesions in Behcet disease: A series of 52 patients and review of the literature. *Medicine* **2012**, *91*, 25–34. [CrossRef]
42. Aceves, S.S. Remodeling and fibrosis in chronic eosinophil inflammation. *Dig. Dis.* **2014**, *32*, 15–21. [CrossRef]
43. Emmi, G.; Silvestri, E.; Squatrito, D.; Amedei, A.; Niccolai, E.; D'Elios, M.M.; Della Bella, C.; Grassi, A.; Becatti, M.; Fiorillo, C.; et al. Thrombosis in vasculitis: From pathogenesis to treatment. *Thromb. J.* **2015**, *13*, 15. [CrossRef]
44. Parry, S.D.; Clark, D.M.; Campbell, J. Coronary arteritis in Wegener's granulomatosis causing fatal myocardial infarction. *Hosp. Med.* **2000**, *61*, 284–285. [CrossRef]
45. Broncano, J.; Vargas, D.; Bhalla, S.; Cummings, K.W.; Raptis, C.A.; Luna, A. CT and MR Imaging of Cardiothoracic Vasculitis. *Radiographics* **2018**, *38*, 997–1021. [CrossRef] [PubMed]
46. Zhou, Z.; Xu, L.; Zhang, N.; Wang, H.; Liu, W.; Sun, Z.; Fan, Z. CT coronary angiography findings in non-atherosclerotic coronary artery diseases. *Clin. Radiol.* **2018**, *73*, 205–213. [CrossRef] [PubMed]
47. Grayson, P.C.; Alehashemi, S.; Bagheri, A.A.; Civelek, A.C.; Cupps, T.R.; Kaplan, M.J.; Malayeri, A.A.; Merkel, P.A.; Novakovich, E.; Bluemke, D.A.; et al. (18) F-Fluorodeoxyglucose-Positron Emission Tomography As an Imaging Biomarker in a Prospective, Longitudinal Cohort of Patients With Large Vessel Vasculitis. *Arthritis Rheumatol.* **2018**, *70*, 439–449. [CrossRef]
48. Miloslavsky, E.; Unizony, S. The heart in vasculitis. *Rheum. Dis. Clin. N. Am.* **2014**, *40*, 11–26. [CrossRef]
49. Pagnoux, C.; Guillevin, L. Cardiac involvement in small and medium-sized vessel vasculitides. *Lupus* **2005**, *14*, 718–722. [CrossRef]
50. Jang, J.J.; Gorevic, P.D.; Olin, J.W. Images in vascular medicine. Giant cell arteritis presenting with acute myocardial infarction. *Vasc. Med.* **2007**, *12*, 379. [CrossRef]
51. Nicolazzi, M.A.; Carnicelli, A.; Fuorlo, M.; Favuzzi, A.M.R.; Landolfi, R. Cardiovascular Involvement in Erdheim–Chester Disease: A Case Report and Review of the Literature. *Medicine* **2015**, *94*, e1365. [CrossRef]
52. Maiuolo, J.; Mollace, R.; Gliozzi, M.; Musolino, V.; Carresi, C.; Paone, S.; Scicchitano, M.; Macri, R.; Nucera, S.; Bosco, F.; et al. The Contribution of Endothelial Dysfunction in Systemic Injury Subsequent to SARS-Cov-2 Infection. *Int. J. Mol. Sci.* **2020**, *21*, 9309. [CrossRef]
53. Maccio, U.; Zinkernagel, A.S.; Shambat, S.M.; Zeng, X.; Cathomas, G.; Ruschitzka, F.; Schuepbach, R.A.; Moch, H.; Varga, Z. SARS-CoV-2 leads to a small vessel endotheliitis in the heart. *EBioMedicine* **2021**, *63*, 103182. [CrossRef] [PubMed]
54. Vrints, C.J.M.; Krychtiuk, K.A.; Van Craenenbroeck, E.M.; Segers, V.F.; Price, S.; Heidbuchel, H. Endothelialitis plays a central role in the pathophysiology of severe COVID-19 and its cardiovascular complications. *Acta Cardiol.* **2021**, *76*, 109–124. [CrossRef] [PubMed]
55. Sharma, A.; Garcia, G.; Arumugaswami, V.; Svendsen, C.N. Human iPSC-Derived Cardiomyocytes are Susceptible to SARS-CoV-2 Infection. *BioRxiv* **2020**. [CrossRef]
56. Berezin, A. Neutrophil extracellular traps: The core player in vascular complications of diabetes mellitus. *Diabetes Metab. Syndr.* **2019**, *13*, 3017–3023. [CrossRef]
57. Becker, R.C. COVID-19-associated vasculitis and vasculopathy. *J. Thromb. Thrombolysis* **2020**, *50*, 499–511. [CrossRef]
58. Roberts, K.A.; Colley, L.; Agbaedeng, T.A.; Ellison-Hughes, G.M.; Ross, M.D. Vascular Manifestations of COVID-19-Thromboembolism and Microvascular Dysfunction. *Front. Cardiovasc. Med.* **2020**, *7*, 598400. [CrossRef]
59. Bangalore, S.; Sharma, A.; Slotwiner, A.; Yatskar, L.; Harari, R.; Shah, B.; Ibrahim, H.; Friedman, G.H.; Thompson, C.; Alviar, C.L.; et al. ST-Segment Elevation in Patients with Covid-19-A Case Series. *N. Engl. J. Med.* **2020**, *382*, 2478–2480. [CrossRef]
60. Saririan, M.; Armstrong, R.; George, J.C.; Olechowski, B.; O'Connor, S.; Byrd, J.B.; Chapman, A.R. ST-segment elevation in patients presenting with COVID-19: Case series. *Eur. Heart J. Case Rep.* **2021**, *5*, ytaa553. [CrossRef]
61. Berthelot, J.M.; Drouet, L.; Liote, F. Kawasaki-like diseases and thrombotic coagulopathy in COVID-19: Delayed over-activation of the STING pathway? *Emerg. Microbes Infect.* **2020**, *9*, 1514–1522. [CrossRef]
62. Toubiana, J.; Poirault, C.; Corsia, A.; Bajolle, F.; Fourgeaud, J.; Angoulvant, F.; Debray, A.; Basmaci, R.; Salvador, E.; Biscardi, S.; et al. Kawasaki-like multisystem inflammatory syndrome in children during the covid-19 pandemic in Paris, France: Prospective observational study. *BMJ* **2020**, *369*, m2094. [CrossRef]
63. Escher, F.; Pietsch, H.; Aleshcheva, G.; Bock, T.; Baumeier, C.; Elsaesser, A.; Wenzel, P.; Hamm, C.; Westenfeld, R.; Schultheiss, M.; et al. Detection of viral SARS-CoV-2 genomes and histopathological changes in endomyocardial biopsies. *ESC Heart Fail.* **2020**, *7*, 2440–2447. [CrossRef]

64. Blondiaux, E.; Parisot, P.; Redheuil, A.; Tzaroukian, L.; Levy, Y.; Sileo, C.; Schnuriger, A.; Lorrot, M.; Guedj, R.; Ducou le Pointe, H. Cardiac MRI in Children with Multisystem Inflammatory Syndrome Associated with COVID-19. *Radiology* **2020**, *297*, E283–E288. [CrossRef]
65. McCrindle, B.W.; Rowley, A.H.; Newburger, J.W.; Burns, J.C.; Bolger, A.F.; Gewitz, M.; Baker, A.L.; Jackson, M.A.; Takahashi, M.; Shah, P.B.; et al. Diagnosis, Treatment, and Long-Term Management of Kawasaki Disease: A Scientific Statement for Health Professionals From the American Heart Association. *Circulation* **2017**, *135*, e927–e999. [CrossRef]
66. Jung, J.H.; Lee, Y.H.; Song, G.G.; Jeong, H.S.; Kim, J.H.; Choi, S.J. Endovascular Versus Open Surgical Intervention in Patients with Takayasu's Arteritis: A Meta-analysis. *Eur. J. Vasc. Endovasc. Surg.* **2018**, *55*, 888–899. [CrossRef] [PubMed]
67. Yamamoto, Y.; Iino, K.; Ueda, H.; No, H.; Nishida, Y.; Takago, S.; Shintani, Y.; Kato, H.; Kimura, K.; Takemura, H. Coronary Artery Bypass Grafting in a Patient With Polyarteritis Nodosa. *Ann. Thorac. Surg.* **2017**, *103*, e431–e433. [CrossRef] [PubMed]
68. Fineschi, M.; Gori, T.; Sinicropi, G.; Bravi, A. Polytetrafluoroethylene (PTFE) covered stents for the treatment of coronary artery aneurysms. *Heart* **2004**, *90*, 490. [CrossRef] [PubMed]
69. Harandi, S.; Johnston, S.B.; Wood, R.E.; Roberts, W.C. Operative therapy of coronary arterial aneurysm. *Am. J. Cardiol.* **1999**, *83*, 1290–1293. [CrossRef]
70. Burns, J.C.; El-Said, H.; Tremoulet, A.H.; Friedman, K.; Gordon, J.B.; Newburger, J.W. Management of Myocardial Infarction in Children with Giant Coronary Artery Aneurysms after Kawasaki Disease. *J. Pediatric* **2020**, *221*, 230–234. [CrossRef]
71. Jone, P.N.; Tapia, D.; Davidson, J.; Fagan, T.E.; Browne, L.; Ing, R.J.; Kay, J. Successful Treatment of Myocardial Infarction in an Infant With Kawasaki Disease. *Semin. Cardiothorac. Vasc. Anesth.* **2015**, *19*, 255–259. [CrossRef]
72. Espinosa, G.; Font, J.; Tassies, D.; Vidaller, A.; Deulofeu, R.; Lopez-Soto, A.; Cervera, R.; Ordinas, A.; Ingelmo, M.; Reverter, J.C. Vascular involvement in Behcet's disease: Relation with thrombophilic factors, coagulation activation, and thrombomodulin. *Am. J. Med.* **2002**, *112*, 37–43. [CrossRef]
73. Sugahara, Y.; Ishii, M.; Muta, H.; Iemura, M.; Matsuishi, T.; Kato, H. Warfarin therapy for giant aneurysm prevents myocardial infarction in Kawasaki disease. *Pediatric Cardiol.* **2008**, *29*, 398–401. [CrossRef] [PubMed]
74. Sengupta, D.; Kahn, A.M.; Kung, E.; Moghadam, M.E.; Shirinsky, O.; Lyskina, G.A.; Burns, J.C.; Marsden, A.L. Thrombotic risk stratification using computational modeling in patients with coronary artery aneurysms following Kawasaki disease. *Biomech. Model. Mechanobiol.* **2014**, *13*, 1261–1276. [CrossRef] [PubMed]
75. Mohan, S.; Poff, S.; Torok, K.S. Coronary artery involvement in pediatric Takayasu's arteritis: Case report and literature review. *Pediatric Rheumatol. Online J.* **2013**, *11*, 4. [CrossRef] [PubMed]

 *biomedicines*

Review

Vasculitis: From Target Molecules to Novel Therapeutic Approaches

Sang-Wan Chung

Division of Rheumatology, Department of Internal Medicine, School of Medicine, Kyung Hee University, 26 Kyungheedae-ro, Dongdaemun-gu, Seoul 02447, Korea; wanyworld83@gmail.com; Tel.: +82-2-958-8200

Abstract: Systemic vasculitis is a group of diverse diseases characterized by immune-mediated inflammation of blood vessels. Current treatments for vasculitis, such as glucocorticoids and alkylating agents, are associated with significant side effects. In addition, the management of both small and large vessel vasculitis is challenging due to a lack of robust markers of disease activity. Recent research has advanced our understanding of the pathogenesis of both small and large vessel vasculitis, and this has led to the development of novel biologic therapies capable of targeting key cytokine and cellular effectors of the inflammatory cascade. It is anticipated that these novel treatments will lead to more effective and less toxic treatment regimens for patients with systemic vasculitis.

Keywords: systemic vasculitis; molecular target; novel treatment

Citation: Chung, S.-W. Vasculitis: From Target Molecules to Novel Therapeutic Approaches. *Biomedicines* **2021**, *9*, 757. https://doi.org/10.3390/biomedicines9070757

Academic Editor: Stefano Bellosta

Received: 27 May 2021
Accepted: 25 June 2021
Published: 30 June 2021

Publisher's Note: MDPI stays neutral with regard to jurisdictional claims in published maps and institutional affiliations.

1. Introduction

Systemic vasculitis pathologically denotes inflammation of a blood vessel, which is characterized by the presence of an inflammatory infiltrate and destruction of the vessel wall, causing stenosis and thrombosis. Vasculitis is a group of diverse disorders that demonstrate various organ involvement and clinical severity. Vasculitis can virtually affect any vessel in the organ system, and, depending on which vessel it invades, the manifestations can be very diverse. It is very common for highly variable conditions to lead to delays in diagnosis. Therefore, identifying vasculitis early, assessing response to therapy, and detecting disease relapse remain important clinical challenges [1,2].

As with many rheumatic diseases, there are no disease-specific clinical features or laboratory tests for making a definite diagnosis. Instead, vasculitis is classified according to classification criteria, of which the most widely used is the nomenclature, published by Chapel Hill Consensus Conference (CHCC) in 2012. Here, vasculitis is classified according to the size of the affected vessel: small vessel vasculitis (SVV), medium vessel vasculitis (MVV), and large vessel vasculitis (LVV) [3]. The epidemiology of systemic vasculitides varies greatly according to the type of vasculitis and the patient's age, sex, and geographic location [4].

The pathogenesis of vasculitis remains unclear. One explanation is that exposure to an unidentified antigen, such as a virus, toxin, or cryptic epitope, leads to activation of the immune response. In some people, this immune response is not down-regulated, leading to the production of immune complexes that deposit in blood vessel walls and lead to vasculitis [5]. Pauci-immune vasculitides are not immune-complex-mediated and typically associate with antineutrophil cytoplasmic autoantibodies (ANCA), which are hypothesized to cause vascular damage indirectly by priming neutrophils to degranulate and to produce oxygen-free radicals.

The last decade has seen major advances in our understanding of the pathogenesis of vasculitis. These discoveries have led to the development of novel treatments, which seek to provide greater efficacy and a more acceptable side effect profile. In this review, we discuss the recent advances in understanding the pathogenesis of primary systemic vasculitides and the development of novel treatments.

2. Systemic Vasculitis Classification

Classification criteria are intended to create homogeneous patient groups for research. The classification systems for vasculitis are limited by overlapping features of subgroups and unrecognized pathogenic mechanisms. The most used classification criteria are defined by the size of the vessel they predominantly affect, namely, small, medium, or large, or variable vessel size (Table 1). The Chapel Hill International Consensus Conference (CHCC) of 2012 defined and standardized the nomenclature of systemic vasculitides [3].

Table 1. Nomenclature of the systemic vasculitides defined during the 2012 International Chapel Hill Consensus Conference (Adapted from [3]).

Systemic Vasculitis
Large-vessel vasculitis (LVV)
Giant cell arteritis (GCA)
Takayasu arteritis (TA)
Medium-vessel vasculitis (MVV)
Polyarteritis nodosa (PAN)
Kawasaki disease (KD)
Small-vessel vasculitis (SVV)
Anti-neutrophil cytoplasmic antibody (ANCA) associated vasculitis (AAV)
Microscopic polyangiitis (MPA)
Granulomatosis with polyangitis (GPA)
Eosinophilic granulomatosis with polyangitis (EGPA)
Immune complex vasculitis
Anti-glomerular basement membrane (anti-GBM) disease
Cryoglobulinemic vasculitis (CV)
IgA vasculitis (Henoch-Schonlein) (IgAV)
Hypocomplementemic urticarial vasculitis
Variable vessel vasculitis (VVV)
Behçet's disease (BD)
Cogan's syndrome (CS)

LVV involves the aorta and its major branches, and includes giant cell arteritis (GCA) and Takayasu arteritis (TA). MVV involves the main visceral arteries and veins and their initial branches and includes polyarteritis nodosa (PAN) in adults. Kawasaki disease—the other major form of MVV and an acute arteritis of childhood—is not covered in this review. SVV involves arterioles, capillaries, intraparenchymal arteries, venules, and some veins and includes ANCA-associated vasculitis (AAV), the most common SVV in adults. There is, however, some overlap, and arteries of any size can potentially be involved in any case of the three main categories of dominant vessel pattern involvement [3]. In addition to the multi-organ systemic vasculitides, other forms of vasculitis have also been defined, such as single-organ vasculitis, including cutaneous arteritis, primary central nervous system vasculitis, and isolated aortitis; vasculitis associated with systemic disease, including rheumatoid vasculitis, lupus vasculitis, and sarcoid vasculitis; and vasculitis associated with an underlying cause: disease-related (Hepatitis B, Hepatitis C-associated cryoglobulinaemia, and cancer), or drug-related vasculitis [3].

The central feature of LVV is granulomatous arteritis. GCA exclusively affects individuals aged >50 years with a female-to-male predominance of 3:1. Additionally, GCA is more common in patients of Northern European descent than in Asian ethnic groups. GCA typically affects the branches of carotid, vertebral, and temporal arteries resulting in the classic symptoms of headache, jaw claudication, and loss of vision [6]. In contrast, TA usually affects females during the second and third decades of life. It is rare in Northern Europe but is more common in southeast Asia [7]. TA typically involves the aorta and its primary branches leading to vascular occlusion with claudication, aneurysm formation, aortic insufficiency, and cardiac failure. Current treatment of LVV is glucocorticoids [8]. Although methotrexate and azathioprine have been used as steroid-sparing agents, their effectiveness has not been proven in randomized controlled trials (RCT).

PAN is uncommon, with an estimated incidence of 1 to 10 per million. Both sexes are affected equally, and the peak age range of onset is between 40 and 60 years. The etiopathogenesis of PAN is strongly linked to viral hepatitis infection, particularly hepatitis B virus, which comprised over one-third of 348 PAN cases in the largest case series to date [9,10]. About 35% of polyarteritis nodosa (PAN) cases are associated with hepatitis B [11]. The incidence of hepatitis B virus-related PAN has declined substantially over the last four decades after improvements in immunization, transfusion practice, and hepatitis B virus therapy. PAN is characterized by a transmural necrotizing arteritis of muscular arteries [12]. The most commonly affected sites are the skin (causing livedo reticularis and ulceration) and peripheral nerves (leading to a mononeuritis multiplex). Involvement of visceral vessels is also common with multiple irregular arterial stenoses and microaneurysms demonstrable on contrast angiography in up to 90% of patients [13] with long-term immunosuppression with glucocorticoids alongside other agents such as cyclophosphamide, methotrexate, or azathioprine, improves patient outcomes, and supports an autoimmune component to pathogenesis [14].

Antineutrophil cytoplasmic antibody (ANCA)-associated vasculitides (AAV) are a group of systemic autoimmune disorders that predominately affects the small vessels. AAV are necrotizing vasculitides that are differentiated from other small vessel vasculitis by the lack of significant immune deposition in the vessel walls. AAV includes microscopic polyangiitis (MPA), granulomatosis with polyangiitis (GPA), and eosinophilic granulomatosis with polyangiitis (EGPA). The autoantibodies that define AAV are myeloperoxidase (MPO)-ANCA and proteinase 3 (PR3)-ANCA [15]. AAV are rare autoimmune conditions with a combined estimated prevalence of 46–184 per million [3]. However, they are associated with significant mortality. GPA mortality is reported to be greater than 90% at two years if left untreated [16,17]. Fortunately, the introduction of effective therapeutics has dramatically decreased the two-year mortality rate to 15% [18]. The treatment of AAV consists of remission induction followed by a maintenance phase. The treatment recommendations for induction or maintenance AAV vary based on the severity of the disease [16]. There are different definitions for determining what constitutes severe disease, but, generally, any AAV that is life- or organ-threatening is considered severe [19]. Current treatment options are effective; however, they are associated with significant patient morbidity due to treatment-related adverse effects.

Behçet disease (BD) is a rare relapsing, multisystem vasculitis, characterized by recurrent attacks of oral-genital ulcers and ocular, musculoskeletal, vascular, central nervous system (CNS), and gastrointestinal (GI) involvement. The prevalence of BD varies widely by geographic area, but, according to a recent meta-analysis, it is approximately 10.3 per 100,000 inhabitants [20]. The vascular involvement is the most frequent cause of mortality, and ocular involvement is the most important factor of morbidity in BD as it can cause blindness [21]. Treatment of BD is based on clinical manifestations. While colchicine, nonsteroidal anti-inflammatory agents, and topical treatments are often sufficient for mucocutaneous and joint involvement, immunosuppressive agents are required for major organ involvement [22].

As the understanding of the pathophysiology of systemic vasculitis increases, new therapies with fewer toxic effects are being proposed. This article will provide a review of current treatment options and an expert opinion on the future of AAV treatment.

3. Drug Discovery and Potential Targets in Vasculitis

Treatment of the various type of vasculitis mainly relies on corticosteroids and conventional immunosuppressive drugs, such as methotrexate or azathioprines. Since vasculitis is a complex, chronic inflammatory disease, treatment may be needed for many different inflammatory molecules and targets. At present, research on these molecules and targets is mainly based on a few antibodies or inhibitors. Recent advances in the era of biologic agents have improved the management of difficult-to-treat cases dramatically. (Table 2, Figure 1).

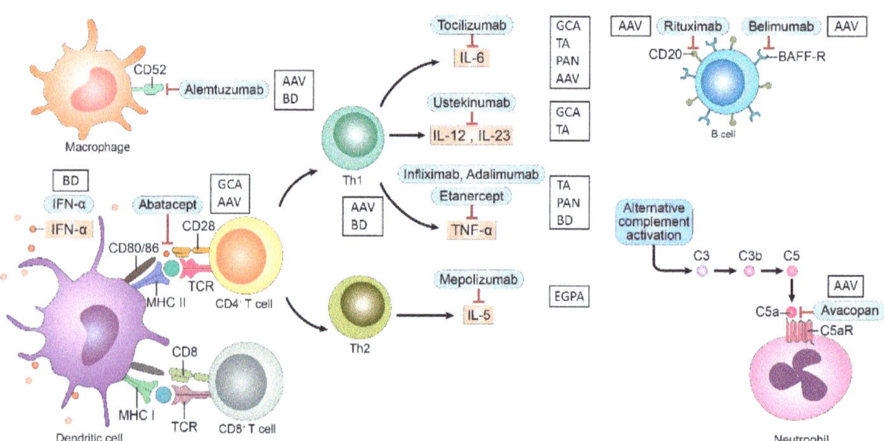

Figure 1. Targets and relative agent in vasculitis. IL, interleukin; mAb, monoclonal antibody; GCA, giant cell arteritis; TA, Takayasu's Arteritis; PAN, polyarteritis nodosa; TNF, tumor necoris factor; EGPA, Eosinophilic granulomatosis with polyangiitis, AAV, ANCA associated vasculitis; BAFF, B-cell activating factor; PDE, phosphodiesterase; IFN, interferon.

3.1. Th1 Cytokines and Relative Drug Discovery

3.1.1. IL-6

IL-6 plays a pathological effect on the inflammatory response in both the vessel wall and the systemic circulation. Tocilizumab is a humanized monoclonal antibody that competitively inhibits IL-6 by binding to circulating and membrane-bound IL-6 receptors. The first reported randomized controlled trial on the efficacy of tocilizumab in GCA randomized 20 patients to either 8 mg/kg tocilizumab delivered intravenously each month or placebo infusions in addition to glucocorticoids and found a higher relapse-free survival in the tocilizumab group (85 versus 20%, $p = 0.001$) at week 52 [23]. The effects of tocilizumab on glucocorticoid-sparing were observed in both relapsing and newly diagnosed GCA. The phase 3 Giant Cell Arteritis Actemra (GiACTA) trial enrolled 251 patients with new-onset GCA, randomized to one of four arms: tocilizumab 162 mg weekly or every other week (combined with a 26-week prednisone taper), or a prednisone taper alone (either 26 or 52 weeks). This study reported that Tocilizumab is an effective glucocorticoid-sparing therapy, demonstrating sustained glucocorticoid-free remission in 56% of patients receiving weekly tocilizumab compared with 18% of patients receiving a 52-week prednisone taper [24]. Tocilizumab is Food and Drug Administration (FDA)-approved for treatment of GCA.

In TA, a phase 3 trial about the effect of tocilizumab, Takayasu arteritis treated with tocilizumab (TAKT) was reported in 2017 [25]. Here, 36 relapsing TA patients were randomized to either tocilizumab, 162 mg weekly or placebo given weekly alongside a tapering glucocorticoid dose. Analyzed by an intention-to-treat method, tocilizumab failed to show difference in time to relapse as compared to placebo (hazard ratio [HR] 0.41, 95% confidence interval [CI] 0.15–1.10, $p = 0.0596$). However, the per-protocol analysis showed a significant difference for tocilizumab ($n = 16$) versus placebo ($n = 17$) (HR 0.34, 95% CI 0.11–1.00,

$p = 0.03$). In 2020, the long-term efficacy and safety of tocilizumab in TA was reported. In that study, 28 patients received tocilizumab for 96 weeks. 46.4% of these 28 patients treated with tocilizumab reduced their dose to <0.1 mg/kg/day, thus showing evidence of a steroid-sparing effect of Tocilizumab in TA in long-term treatment [26].

There is no RCT for the effect of tocilizumab in PAN yet. In a recent case report, tocilizumab was effective for hepatitis B virus related PAN without Hepatitis B virus reactivation [27]. In a literature review based on 11 case reports, tocilizumab is effective in cases of refractory or relapsing polyarteritis nodosa and showed its glucocorticoid-sparing effect [28].

There are several case reports describing patients with AAV treated with tocilizumab showing that complete and sustained remission was achieved in many of the patients with refractory disease [29,30]. RCTs may be warranted in the future.

3.1.2. IL-12 and IL-23

IL-23 is a pro-inflammatory cytokine composed of two subunits, IL-23A (p19) and IL-23B (p40), the latter shared with IL-12. The IL-23/IL-17 axis mainly plays a protective role against bacterial infections; its dysregulation plays a role in in immune-mediated inflammatory disorders [31–33]. As it has been reported that the IL-12/Th1 cell/IFN-γ pathway is involved in granulomatous inflammation in the pathogenesis of GCA, treatments targeting IL12 have been attempted, and the use of ustekinumab to treat LVV has been reported [34].

Ustekinumab is a monoclonal antibody that targets the p40 subunit of IL-12/23. One open-label study of 25 patients with refractory GCA treated with ustekinumab in addition to glucocorticoids demonstrated that no patients relapsed over 52 weeks. The median prednisolone dose decreased from 20 to 5 mg, and about 25% of patients were able to stop glucocorticoids. In addition, CT angiography showed an improvement in mural thickness with complete resolution in eight patients who underwent CT angiography before and after treatment [35]. However, in a recently reported prospective study, 10 out of 13 (77%) patients who failed to achieve the primary endpoint with ustekinumab in prednisone taper, and seven experienced disease flares after a mean period of 23 weeks [36]. Further research on the effect of Ustekinumab in GCA seems warranted.

Ustekinumab treatment in TA has been reported sporadically, and only in a few case series. One series of three patients with refractory TA treated with ustekinumab reported stabilization of clinical disease activity and normalization of inflammatory markers [37]. Recently, the results of a long-term follow-up on the same three patients reported that ustekinumab showed marginal effects on reducing prednisolone dose, and 2 of 3 patients discontinued ustekinumab treatment because of relapse and secondary failure [38].

3.1.3. Tumor Necrosis Factor (TNF) α Inhibitor

TNF α inhibitors were the first biologic agents tried in various vasculitides. TNF α is an important cytokine for the formation of granuloma [39], and also for activation of endothelial cells [40].

After a few cases showing successful anti-TNF-α treatment in GCA patients had been reported, a comparative double-blind study was attempted using infliximab but was subsequently stopped due to the lack of efficacy on the prevention of relapse [41]. Using etanercept, a randomized controlled study showed a significant steroid sparing effect after one year in 17 patients, however not for a longer period [42]. Adding adalimumab to a standard prednisone regimen showed no steroid sparing effect in 70 patients in a 10-week prospective randomized controlled study [43]. As a result of these studies, anti-TNF-α therapy is not recommended in GCA.

Several retrospective studies and case series reported that TNF α inhibitors were effective in most patients with refractory Takayasu's arteritis [44–46]. A two-year follow-up cohort study from Norway reported higher rates of sustained remission as well as lesser progression of angiographic lesions in patients receiving anti-TNF-α agents, when compared with conventional treatments in Takayasu's arteritis [47].

In PAN, infliximab has been used in refractory forms of the disease or because of intolerance to conventional drugs and seems to be effective [48]. In a small case series, nine refractory PAN patients were treated with infliximab and 8 of 9 patients (89%) achieved significant improvement and prednisone dose reduction of 50% [49].

A number of open-label studies and case series have reported the usefulness of anti-TNF-α therapies in AAV, although these results have not been confirmed in RCTs. In the Wegener's Granulomatosis Etanercept Trial (WGET), which recruited 174 patients with GPA, there was no benefit from etanercept on the sustained remission rate [50]. With little evidence for its effectiveness, the use of anti-TNF α treatment in AAV may be significantly limited in the future. Similarly, for EGPA, the only available information is derived from five case reports with conflicting findings that do not support anti-TNF-α use to treat EGPA [51–53].

In small RCT of 40 patients with BD, etanercept was significantly more effective in suppressing most of the mucocutaneous manifestations, such as oral ulcers and erythema nodosum, than placebo [54]. Several observational studies and case series also confirmed the beneficial effects of infliximab and adalimumab on mucocutaneous lesions of BD [55]. Most of the studies on the effects of TNF-α inhibitors in BD are in ocular manifestations and mainly reported in case series. Infliximab significantly decreases in relapse rate and glucocorticoid dosage in BD patients with ocular involvement [56–58]. In the first prospective study in 63 patients with BD uveitis, uveoretinitis improved with infliximab treatment in 92% and maintained for up to 12 months [59]. A 1-year observational multicenter study reported the results of infliximab and adalimumab use in 124 patients with refractory BD uveitis, and complete remission was achieved in 84/124 (68%) [60]. A recent retrospective observational study also reported that adalimumab was highly effective and safe for treatment of BD related uveitis [61]. An open-label study of 177 patients with BD related uveitis compared the efficacy of infliximab (103 patients) versus adalimumab (74 patients) as a first-line biologic agent. In this study, an improvement in all ocular parameters were observed in both groups after 1-year treatment; however, adalimumab had significantly better ocular outcomes in some parameters [62]. In BD related vascular manifestation, such as deep vein thrombosis, superficial thrombophlebitis, a retrospective study reported that adalimumab achieved significantly higher vascular response (34/35, 97%) compared with conventional immunosupressants (23/35, 66%) during a mean follow-up of 25.7 ± 23.2 months. Significantly lower vascular relapse was also observed in adalimumab group [63]. In recent two multicenter observational studies, clinical remission was achieved in 89% and 80% of patients with BD, respectively, with vascular involvement refractory to conventional ISs treatment [64,65].

3.2. Th2 Cytokines and Relative Drug Discovery
IL-5

IL-5 is the major cytokine responsible for eosinophil activation, chemoattraction, and survival. Several studies have reported elevated serum IL-5 levels in EGPA [66,67]. Recently, IL-5 antagonists have been studied as an EGPA-specific treatment. Mepolizumab, a humanized anti-IL-5 monoclonal antibody, selectively inhibits eosinophilic inflammation and is approved to treat severe eosinophilic asthma [68]. A benefit of mepolizumab treatment in EGPA patients was observed in previous small open-label pilot studies [69–71]. In the large scale randomized controlled trial, 136 EGPA patients with an uncontrolled disease injected 300 mg of mepolizumab once a month subcutaneously. The remission rate in patients treated with mepolizumab was significantly higher than that of patients treated with placebo (28% vs. 3%, of the participants had ≥ 24 weeks of accrued remission; odds ratio, 5.91; 95% confidence interval [CI], 2.68 to 13.03; $p < 0.001$) [72]. This study led to FDA authorization of mepolizumab as the first drug specifically approved for EGPA.

3.3. Targets and Drug Discovery of B Cells

3.3.1. CD20

B cells are clearly central to the pathogenesis of AAV, as they produce ANCAs. Rituximab is a chimeric monoclonal antibody that induces B-cell depletion by binding CD20 expressing B cells. Its development has led to advances in AAV therapy. The Rituximab in ANCA-Associated Vasculitis (RAVE) [73] and Rituximab Versus Cyclophosphamide in ANCA-Associated Vasculitis (RITUXVAS) [74] trials established noninferiority of rituximab to cyclophosphamide for AAV remission induction. Several second-generation anti-CD20 drugs have been developed, one of which, ofatumumab, has been tested in one small case series of patients with AAV, with results showing its therapeutic benefit. However, there has been no RCT yet [75].

3.3.2. BAFF

B-cell activating factor (BAFF), also known as B-lymphocyte stimulator (BlyS), plays an important role in B cell maturation and is increasingly recognized as important in the pathogenesis of relapsing AAV. Increased BAFF expression is evident in patients with active vasculitis, and preclinical data suggest that high BAFF concentrations can promote the survival of autoreactive B cells that, under normal conditions, would be degraded [76,77]. Belimumab is a fully humanized monoclonal antibody that binds to BAFF receptors on B cells. It is licensed for the treatment of systemic lupus erythematosus [78,79].

In AAV, the Belimumab in Remission of Vasculitis (BREVAS) trial examined the addition of belimumab to azathioprine and glucocorticoids for maintenance of remission in patients with GPA and MPA [80]. However, the trial was stopped early due to suboptimal recruitment, and no improvement in the relapse rate was observed. The combination of rituximab and belimumab is being investigated further in an ongoing randomized double-blind placebo-controlled trial, the Rituximab and Belimumab Combination Therapy in PR3-AAV trial (COMBIVAS) (ClinicalTrials.gov identifier: NCT03967925).

3.4. B-Cell and T-Cell Co-Stimulation and Depletion

3.4.1. CD28–CD80/CD86

Co-stimulatory molecules predominantly modulate the immune responses by activating T- and B-cell functions, but also affect dendritic cell and macrophage functions and play a crucial role in inflammation. Abatacept, a fusion protein of the extracellular domain of CTLA-4 and the Fc fragment of human IgG1 (CTLA-4–Ig), is an inhibitor of T lymphocyte activation by means of co-stimulatory blockade by binding to CD80 and CD86 receptors on APC that is needed for antigen-presenting activation of T cells [81].

In GCA, 41 patients were randomized and relapse-free survival was 48% with abatacept as maintenance therapy compared with 31% with placebo ($p = 0.049$) [82]. The duration of remission is significantly longer in the abatacept group than that of the placebo group. In a double-blind randomized controlled multicenter study, 34 patients with TA were treated with abatacept at a dose of 10 mg/kg on days 1, 15, 29, and at 8 weeks. Patients attaining remission at 12 weeks were randomized to either receive placebo ($n = 15$) or monthly abatacept ($n = 11$) and followed up until 12 months. However, there was no difference in the duration of remission and relapse-free survival at 12 months between the two groups [83].

In AAV, infiltrations of granulomatous T-cells were observed in lungs and kidneys, suggesting a pathogenic role of T cells. In an open-label trial of 20 patients with a non-severe relapsing GPA who were treated with abatacept, the remission rate was about 80% and steroid discontinuation rate 75% [84]. The ongoing phase III randomized placebo-controlled Abatacept for the Treatment of Relapsing, Non-Severe, Granulomatosis with Polyangiitis (ABROGATE) (ClinicalTrials.gov identifier: NCT02108860) trial is currently recruiting patients.

3.4.2. CD52

CD52 is expressed on monocytes, macrophages, and eosinophils. Depletion of B cells and T cells can be achieved by alemtuzumab [85]. This humanized anti-CD52 monoclonal antibody selectively depletes lymphocytes and has been shown to be effective in other systemic vasculitides such as Behçet's disease [86]. Alemtuzumab as remission induction therapy was effective in 84% of 32 BD patients, and sustained remission was achieved in 69% at 12 months [86]. Walsh et al. reported results of a retrospective long-term study of 71 patients with refractory or relapsing AAV treated with alemtuzumab and found it useful to achieve remission with a lower relapse rate [87]. A randomized, prospective, open-label study of alemtuzumab for remission induction in refractory AAV (ALEVIATE trial) was presented in the form of an abstract in 2019. In this study, remission was achieved in 65% of AAV patients at 6 months and 35% sustained remission at one year [88] However, adverse events, such as infection, are high compared to standard treatment.

3.5. Targeting Complement

C5a Receptors

The complement system is a central mediator of antibody-mediated immune responses. C5 is a potent effector protein in this pathway, exerting its effects through its cleavage products: C5a, a potent anaphylatoxin and chemoattractant, and C5b, part of the a membrane attack complex that lyses target cells [89]. In patients with AAV, complement deposition is evident at sites of tissue inflammation, such as kidneys, and high plasma levels of complements correlate with disease severity [90,91].

Avacopan contains a small molecule that binds to C5a preventing it from binding to its receptor. Clinical trial results of C5a receptor inhibition with avacopan have shown promising results [92]. Sixty-seven patients with AAV were randomized to either high-dose glucocorticoids, avacopan plus low-dose glucocorticoids, or avacopan alone alongside cyclophosphamide or rituximab induction. At 12 weeks, a 50% reduction from baseline in the Birmingham Vasculitis Activity Score (BVAS) occurred in 86% of the avacopan/glucocorticoid and 81% of the avacopan-alone groups, compared with 70% in the glucocorticoid group ($p = 0.002$ and $p = 0.01$, respectively). However, this study included only non-severe disease. The results of a phase III trial, the Avacopan in Patients With ANCA-Associated Vasculitis (ADVOCATE) trial, were reported in 2021 [93]. This study enrolled 331 patients, randomized to receive either avacopan or glucocorticoids during remission induction with either cyclophosphamide or rituximab. At 26 weeks, the number of patients in remission, as assessed by a score 0 on the BVAS and withdrawal of steroid therapy, was not inferior in both the avacopan and prednisone groups (72.3%, 70.1% respectively). Additional data showed that avacopan was superior over glucocorticoids in sustained remission at 52 weeks, with an acceptable safety profile.

An anti-C5a monoclonal Ab, IFX-1 is also being evaluated in phase II studies (INFLARX trial, NCT03895801 and NCT03712345). Recruitment is ongoing and completion is estimated by July 2021.

3.6. Other Targets

Interferon-α

Interferons (IFN), a large family of glycoproteins, produce a cellular response to the microbes, tumors, and antigens [94]. IFN-α was demonstrated to modulate the Th1/Th2 balance toward Th1 by increased IFN-γ production and inhibiting IL-5 and IL-13 production in Th2 cells [95,96].

The efficacy of IFN-α has been well established in BD, with the data coming from case series, especially in ocular manifestations [97–99]. A retrospective study reported that there was no difference between azathioprine plus colchicine and IFN-α2a treatment in BD uveitis regarding remission and relapse rates [100]. Some case reports also reported the efficacy of INF-α in neuro BD [101,102]. In an RCT of 44 patients of BD, IFN-α treatment significantly improved mucocutaneous manifestations, such as orogenital ulcers, and papulopustular lesions [103]. In a recent prospective study of 33 patients with deep venous thrombosis, one of the serious complications of BD, the relapse rate was lower and recanalization rate was higher in patients treated with IFN-α compared with AZA (12% vs. 45% and 86% vs. 45%) [104].

Table 2. Targets and relative agent in vasculitis.

Target		Agent	Vasculitis	References
Th1 cytokines	IL-6	Tocilizumab (anti-IL-6R mAb)	GCA, TA PAN AAV	[23,24] [25,26] [27,28] [29,30]
	IL-12 and IL-23	Ustekinumab (p40 subunit of IL-12/IL-23 mAb)	GCA TA	[35] [37,38]
	TNF-α	Infliximab, Adalimumab (anti-TNF-α mAb) Etanercept (TNF-α receptor fusion protein)	TA PAN BD	[44–47] [48,49] [54–59] [60–65]
Th2 cytokines	IL-5	Mepolizumab (anti-IL-5 mAb)	EGPA	[69–72]
B cells	CD20	Rituximab (anti-CD20 mAb)	AAV	[73,74]
	BAFF-R	Belimumab (BAFF-receptor mAb)	AAV	[80]
Co-stimulatory molecules	CD28–CD80/CD86	Abatacept (CTLA4Ig fusion protein)	GCA AAV	[82] [84]
	CD52	Alemtuzumab (anti-CD52 mAb)	AAV BD	[87,88] [86]
Complement	C5a	Avacopan (C5a receptor inhibitor)	AAV	[92,93]
Other targets	IFN-α	IFN-α	BD	[97–100] [101–104]

4. Conclusions

As the understanding of the pathogenesis of systemic vasculitis advances, novel target molecules and therapeutic approaches are being proposed. Treatment outcomes of vasculitis have improved with several new evidence-based treatments.

In spite of the success of blocking IL-6 in large vessel vasculitis, relapse rates remain high, suggesting that further study is needed. In AAV, recent trials of therapies that target B-cell activation, complements, and IL-5 provide encouraging evidence of better outcomes for these patients. Future clinical trials of these novel therapeutic agents will need to establish their efficacy and, as an increasing number of potential treatments become available, will need to indicate how they can be used to complement or replace existing approaches.

Funding: This research received no external funding.

Conflicts of Interest: The authors declare no conflict of interest.

References

1. Deshazo, R.D. The spectrum of systemic vasculitis: A classification to aid diagnosis. *Postgrad. Med.* **1975**, *58*, 78–82. [CrossRef]
2. Fauci, A.S.; Haynes, B.F.; Katz, P. The spectrum of vasculitis: Clinical, pathologic, immunologic, and therapeutic considerations. *Ann. Intern. Med.* **1978**, *89*, 660–676. [CrossRef]
3. Jennette, J.C.; Falk, R.; Bacon, P.; Basu, N.; Cid, M.; Ferrario, F.; Flores-Suarez, L.; Gross, W.; Guillevin, L.; Hagen, E.; et al. 2012 revised international chapel hill consensus conference nomenclature of vasculitides. *Arthritis Rheum.* **2013**. [CrossRef] [PubMed]
4. Berti, A.; Dejaco, C. Update on the epidemiology, risk factors, and outcomes of systemic vasculitides. *Best Pract. Res. Clin. Rheumatol.* **2018**, *32*, 271–294. [CrossRef]
5. Weyand, C.M.; Goronzy, J.J. Medium-and large-vessel vasculitis. *N. Engl. J. Med.* **2003**, *349*, 160–169. [CrossRef]
6. Chandran, A.K.; Udayakumar, P.D.; Crowson, C.S.; Warrington, K.J.; Matteson, E.L. The incidence of giant cell arteritis in Olmsted County, Minnesota, over a 60-year period 1950–2009. *Scand. J. Rheumatol.* **2015**, *44*, 215–218. [CrossRef]

7.   Gudbrandsson, B.; Molberg, Ø.; Garen, T.; Palm, Ø. Prevalence, incidence, and disease characteristics of Takayasu arteritis by ethnic background: Data from a large, population-based cohort resident in southern Norway. *Arthritis Care Res.* **2017**, *69*, 278–285. [CrossRef] [PubMed]

8.   Buttgereit, F.; Matteson, E.L.; Dejaco, C.; Dasgupta, B. Prevention of glucocorticoid morbidity in giant cell arteritis. *Rheumatology* **2018**, *57*, ii11–ii21. [CrossRef] [PubMed]

9.   Watts, R.; Lane, S.; Scott, D.; Koldingsnes, W.; Nossent, H.; Gonzalez-Gay, M.; Garcia-Porrua, C.; Bentham, G. Epidemiology of vasculitis in Europe. *Ann. Rheum. Dis.* **2001**, *60*, 1156–1157. [CrossRef]

10.   Pagnoux, C.; Seror, R.; Henegar, C.; Mahr, A.; Cohen, P.; Le Guern, V.; Bienvenu, B.; Mouthon, L.; Guillevin, L. Clinical features and outcomes in 348 patients with polyarteritis nodosa: A systematic retrospective study of patients diagnosed between 1963 and 2005 and entered into the French Vasculitis Study Group Database. *Arthritis Rheum. Off. J. Am. Coll. Rheumatol.* **2010**, *62*, 616–626. [CrossRef]

11.   Maslennikov, R.; Ivashkin, V.; Efremova, I.; Shirokova, E. Immune disorders and rheumatologic manifestations of viral hepatitis. *World J. Gastroenterol.* **2021**, *27*, 2073–2089. [CrossRef] [PubMed]

12.   Guillevin, L.; Mahr, A.; Callard, P.; Godmer, P.; Pagnoux, C.; Leray, E.; Cohen, P.; Group, F.V.S. Hepatitis B virus-associated polyarteritis nodosa: Clinical characteristics, outcome, and impact of treatment in 115 patients. *Medicine* **2005**, *84*, 313–322. [CrossRef] [PubMed]

13.   Stanson, A.W.; Friese, J.L.; Johnson, C.M.; McKusick, M.A.; Breen, J.F.; Sabater, E.A.; Andrews, J.C. Polyarteritis nodosa: Spectrum of angiographic findings. *Radiographics* **2001**, *21*, 151–159. [CrossRef] [PubMed]

14.   Mukhtyar, C.; Guillevin, L.; Cid, M.C.; Dasgupta, B.; de Groot, K.; Gross, W.; Hauser, T.; Hellmich, B.; Jayne, D.; Kallenberg, C.G. EULAR recommendations for the management of primary small and medium vessel vasculitis. *Ann. Rheum. Dis.* **2009**, *68*, 310–317. [CrossRef]

15.   Hilhorst, M.; van Paassen, P.; Tervaert, J.W.C. Proteinase 3-ANCA vasculitis versus myeloperoxidase-ANCA vasculitis. *J. Am. Soc. Nephrol.* **2015**, *26*, 2314–2327. [CrossRef]

16.   Carpenter, S.; Cohen Tervaert, J.W.; Yacyshyn, E. Advances in therapeutic treatment options for ANCA-associated vasculitis. *Expert Opin. Orphan Drugs* **2020**, *8*, 127–136. [CrossRef]

17.   Yates, M.; Watts, R.A.; Bajema, I.; Cid, M.; Crestani, B.; Hauser, T.; Hellmich, B.; Holle, J.; Laudien, M.; Little, M. EULAR/ERA-EDTA recommendations for the management of ANCA-associated vasculitis. *Ann. Rheum. Dis.* **2016**, *75*, 1583–1594. [CrossRef]

18.   Hilhorst, M.; Wilde, B.; van Paassen, P.; Winkens, B.; van Breda Vriesman, P.; Cohen Tervaert, J.W. Improved outcome in anti-neutrophil cytoplasmic antibody (ANCA)-associated glomerulonephritis: A 30-year follow-up study. *Nephrol. Dial. Transplant.* **2013**, *28*, 373–379. [CrossRef]

19.   McGeoch, L.; Twilt, M.; Famorca, L.; Bakowsky, V.; Barra, L.; Benseler, S.M.; Cabral, D.A.; Carette, S.; Cox, G.P.; Dhindsa, N. CanVasc recommendations for the management of antineutrophil cytoplasm antibody-associated vasculitides. *J. Rheumatol.* **2016**, *43*, 97–120. [CrossRef] [PubMed]

20.   Maldini, C.; Druce, K.; Basu, N.; LaValley, M.P.; Mahr, A. Exploring the variability in Behçet's disease prevalence: A meta-analytical approach. *Rheumatology* **2018**, *57*, 185–195. [CrossRef]

21.   Saadoun, D.; Wechsler, B.; Desseaux, K.; Huong, D.L.T.; Amoura, Z.; Resche-Rigon, M.; Cacoub, P. Mortality in Behçet's disease. *Arthritis Rheum.* **2010**, *62*, 2806–2812. [CrossRef]

22.   Hatemi, G.; Christensen, R.; Bang, D.; Bodaghi, B.; Celik, A.F.; Fortune, F.; Gaudric, J.; Gul, A.; Kötter, I.; Leccese, P. 2018 update of the EULAR recommendations for the management of Behçet's syndrome. *Ann. Rheum. Dis.* **2018**, *77*, 808–818. [CrossRef] [PubMed]

23.   Villiger, P.M.; Adler, S.; Kuchen, S.; Wermelinger, F.; Dan, D.; Fiege, V.; Bütikofer, L.; Seitz, M.; Reichenbach, S. Tocilizumab for induction and maintenance of remission in giant cell arteritis: A phase 2, randomised, double-blind, placebo-controlled trial. *Lancet* **2016**, *387*, 1921–1927. [CrossRef]

24.   Stone, J.H.; Tuckwell, K.; Dimonaco, S.; Klearman, M.; Aringer, M.; Blockmans, D.; Brouwer, E.; Cid, M.C.; Dasgupta, B.; Rech, J. Trial of tocilizumab in giant-cell arteritis. *N. Engl. J. Med.* **2017**, *377*, 317–328. [CrossRef]

25.   Nakaoka, Y.; Isobe, M.; Takei, S.; Tanaka, Y.; Ishii, T.; Yokota, S.; Nomura, A.; Yoshida, S.; Nishimoto, N. Efficacy and safety of tocilizumab in patients with refractory Takayasu arteritis: Results from a randomised, double-blind, placebo-controlled, phase 3 trial in Japan (the TAKT study). *Ann. Rheum. Dis.* **2018**, *77*, 348–354. [CrossRef] [PubMed]

26.   Nakaoka, Y.; Isobe, M.; Tanaka, Y.; Ishii, T.; Ooka, S.; Niiro, H.; Tamura, N.; Banno, S.; Yoshifuji, H.; Sakata, Y. Long-term efficacy and safety of tocilizumab in refractory Takayasu arteritis: Final results of the randomized controlled phase 3 TAKT study. *Rheumatology* **2020**, *59*, 2427–2434. [CrossRef]

27.   Carrión-Barberà, I.; Pros, A.; Salman-Monte, T.; Vílchez-Oya, F.; Sánchez-Schmidt, J.; Pérez-García, C.; Monfort, J. Safe and successful treatment of refractory polyarteritis nodosa with tocilizumab in a patient with past hepatitis B virus infection: A case-based review. *Clin. Rheumatol.* **2020**, *40*, 1–6. [CrossRef]

28.   Akiyama, M.; Kaneko, Y.; Takeuchi, T. Tocilizumab for the treatment of polyarteritis nodosa: A systematic literature review. *Ann. Rheum. Dis.* **2020**. [CrossRef]

29.   Berti, A.; Cavalli, G.; Campochiaro, C.; Guglielmi, B.; Baldissera, E.; Cappio, S.; Sabbadini, M.G.; Doglioni, C.; Dagna, L. Interleukin-6 in ANCA-associated vasculitis: Rationale for successful treatment with tocilizumab. *Semin. Arthritis Rheum.* **2015**, *45*, 48–54. [CrossRef] [PubMed]

30. Sakai, R.; Kondo, T.; Kurasawa, T.; Nishi, E.; Okuyama, A.; Chino, K.; Shibata, A.; Okada, Y.; Takei, H.; Nagasawa, H. Current clinical evidence of tocilizumab for the treatment of ANCA-associated vasculitis: A prospective case series for microscopic polyangiitis in a combination with corticosteroids and literature review. *Clin. Rheumatol.* **2017**, *36*, 2383–2392. [CrossRef]
31. Tang, C.; Chen, S.; Qian, H.; Huang, W. Interleukin-23: As a drug target for autoimmune inflammatory diseases. *Immunology* **2012**, *135*, 112–124. [CrossRef]
32. Kleinschek, M.A.; Muller, U.; Brodie, S.J.; Stenzel, W.; Kohler, G.; Blumenschein, W.M.; Straubinger, R.K.; McClanahan, T.; Kastelein, R.A.; Alber, G. IL-23 enhances the inflammatory cell response in Cryptococcus neoformans infection and induces a cytokine pattern distinct from IL-12. *J. Immunol.* **2006**, *176*, 1098–1106. [CrossRef]
33. Oppmann, B.; Lesley, R.; Blom, B.; Timans, J.C.; Xu, Y.; Hunte, B.; Vega, F.; Yu, N.; Wang, J.; Singh, K.; et al. Novel p19 protein engages IL-12p40 to form a cytokine, IL-23, with biological activities similar as well as distinct from IL-12. *Immunity* **2000**, *13*, 715–725. [CrossRef]
34. Weyand, C.M.; Goronzy, J.J. Immune mechanisms in medium and large-vessel vasculitis. *Nat. Rev. Rheumatol.* **2013**, *9*, 731–740. [CrossRef]
35. Conway, R.; O'Neill, L.; Gallagher, P.; McCarthy, G.M.; Murphy, C.C.; Veale, D.J.; Fearon, U.; Molloy, E.S. Ustekinumab for refractory giant cell arteritis: A prospective 52-week trial. *Semin. Arthritis Rheum.* **2018**, *48*, 523–528. [CrossRef]
36. Matza, M.A.; Fernandes, A.D.; Stone, J.H.; Unizony, S.H. Ustekinumab for the treatment of giant cell arteritis. *Arthritis Care Res.* **2020**. [CrossRef]
37. Terao, C.; Yoshifuji, H.; Nakajima, T.; Yukawa, N.; Matsuda, F.; Mimori, T. Ustekinumab as a therapeutic option for Takayasu arteritis: From genetic findings to clinical application. *Scand. J. Rheumatol.* **2016**, *45*, 80–82. [CrossRef]
38. Gon, Y.; Yoshifuji, H.; Nakajima, T.; Murakami, K.; Nakashima, R.; Ohmura, K.; Mimori, T.; Terao, C. Long-term outcomes of refractory Takayasu arteritis patients treated with biologics including ustekinumab. *Mod. Rheumatol.* **2021**, *31*, 678–683. [CrossRef] [PubMed]
39. Arend, W.P.; Michel, B.A.; Bloch, D.A.; Hunder, G.G.; Calabrese, L.H.; Edworthy, S.M.; Fauci, A.S.; Leavitt, R.Y.; Lie, J.T.; Lightfoot, R.W., Jr.; et al. The American College of Rheumatology 1990 criteria for the classification of Takayasu arteritis. *Arthritis Rheum.* **1990**, *33*, 1129–1134. [CrossRef] [PubMed]
40. Wallis, R.S.; Ehlers, S. Tumor necrosis factor and granuloma biology: Explaining the differential infection risk of etanercept and infliximab. *Semin. Arthritis Rheum.* **2005**, *34*, 34–38. [CrossRef]
41. Hoffman, G.S.; Cid, M.C.; Rendt-Zagar, K.E.; Merkel, P.A.; Weyand, C.M.; Stone, J.H.; Salvarani, C.; Xu, W.; Visvanathan, S.; Rahman, M.U.; et al. Infliximab for maintenance of glucocorticosteroid-induced remission of giant cell arteritis: A randomized trial. *Ann. Intern. Med.* **2007**, *146*, 621–630. [CrossRef] [PubMed]
42. Martinez-Taboada, V.M.; Rodriguez-Valverde, V.; Carreno, L.; Lopez-Longo, J.; Figueroa, M.; Belzunegui, J.; Mola, E.M.; Bonilla, G. A double-blind placebo controlled trial of etanercept in patients with giant cell arteritis and corticosteroid side effects. *Ann. Rheum. Dis.* **2008**, *67*, 625–630. [CrossRef] [PubMed]
43. Seror, R.; Baron, G.; Hachulla, E.; Debandt, M.; Larroche, C.; Puechal, X.; Maurier, F.; de Wazieres, B.; Quemeneur, T.; Ravaud, P.; et al. Adalimumab for steroid sparing in patients with giant-cell arteritis: Results of a multicentre randomised controlled trial. *Ann. Rheum. Dis.* **2014**, *73*, 2074–2081. [CrossRef]
44. Schmidt, J.; Kermani, T.A.; Bacani, A.K.; Crowson, C.S.; Matteson, E.L.; Warrington, K.J. Tumor necrosis factor inhibitors in patients with Takayasu arteritis: Experience from a referral center with long-term followup. *Arthritis Care Res.* **2012**, *64*, 1079–1083. [CrossRef] [PubMed]
45. Mekinian, A.; Neel, A.; Sibilia, J.; Cohen, P.; Connault, J.; Lambert, M.; Federici, L.; Berthier, S.; Fiessinger, J.N.; Godeau, B.; et al. Efficacy and tolerance of infliximab in refractory Takayasu arteritis: French multicentre study. *Rheumatology* **2012**, *51*, 882–886. [CrossRef]
46. Clifford, A.; Hoffman, G.S. Recent advances in the medical management of Takayasu arteritis: An update on use of biologic therapies. *Curr. Opin. Rheumatol.* **2014**, *26*, 7–15. [CrossRef]
47. Gudbrandsson, B.; Molberg, O.; Palm, O. TNF inhibitors appear to inhibit disease progression and improve outcome in Takayasu arteritis; an observational, population-based time trend study. *Arthritis Res. Ther.* **2017**, *19*, 99. [CrossRef]
48. Wu, K.; Throssell, D. A new treatment for polyarteritis nodosa. *Nephrol. Dial. Transplant.* **2006**, *21*, 1710–1712. [CrossRef]
49. Ginsberg, S.; Rosner, I.; Slobodin, G.; Rozenbaum, M.; Kaly, L.; Jiries, N.; Boulman, N.; Awisat, A.; Hussein, H.; Novofastovski, I.; et al. Infliximab for the treatment of refractory polyarteritis nodosa. *Clin. Rheumatol.* **2019**, *38*, 2825–2833. [CrossRef]
50. Wegener's Granulomatosis Etanercept Trial (WGET) Research Group. Etanercept plus standard therapy for Wegener's granulomatosis. *N. Engl. J. Med.* **2005**, *352*, 351–361. [CrossRef] [PubMed]
51. Ciledag, A.; Deniz, H.; Eledag, S.; Ozkal, C.; Duzgun, N.; Erekul, S.; Karnak, D. An aggressive and lethal course of Churg-Strauss syndrome with alveolar hemorrhage, intestinal perforation, cardiac failure and peripheral neuropathy. *Rheumatol. Int.* **2012**, *32*, 451–455. [CrossRef]
52. Tiliakos, A.T.; Shaia, S.; Hostoffer, R.; Kent, L. The use of infliximab in a patient with steroid-dependent Churg-Strauss syndrome. *J. Clin. Rheumatol.* **2004**, *10*, 96–97. [CrossRef]
53. Arbach, O.; Gross, W.L.; Gause, A. Treatment of refractory Churg-Strauss-Syndrome (CSS) by TNF-alpha blockade. *Immunobiology* **2002**, *206*, 496–501. [CrossRef] [PubMed]

54. Melikoglu, M.; Fresko, I.; Mat, C.; Ozyazgan, Y.; Gogus, F.; Yurdakul, S.; Hamuryudan, V.; Yazici, H. Short-term trial of etanercept in Behcet's disease: A double blind, placebo controlled study. *J. Rheumatol.* **2005**, *32*, 98–105. [PubMed]

55. Arida, A.; Fragiadaki, K.; Giavri, E.; Sfikakis, P.P. Anti-TNF agents for Behcet's disease: Analysis of published data on 369 patients. *Semin. Arthritis Rheum.* **2011**, *41*, 61–70. [CrossRef] [PubMed]

56. Accorinti, M.; Pirraglia, M.P.; Paroli, M.P.; Priori, R.; Conti, F.; Pivetti-Pezzi, P. Infliximab treatment for ocular and extraocular manifestations of Behcet's disease. *Jpn. J. Ophthalmol.* **2007**, *51*, 191–196. [CrossRef] [PubMed]

57. Ohno, S.; Nakamura, S.; Hori, S.; Shimakawa, M.; Kawashima, H.; Mochizuki, M.; Sugita, S.; Ueno, S.; Yoshizaki, K.; Inaba, G. Efficacy, safety, and pharmacokinetics of multiple administration of infliximab in Behcet's disease with refractory uveoretinitis. *J. Rheumatol.* **2004**, *31*, 1362–1368. [PubMed]

58. Tugal-Tutkun, I.; Mudun, A.; Urgancioglu, M.; Kamali, S.; Kasapoglu, E.; Inanc, M.; Gül, A. Efficacy of infliximab in the treatment of uveitis that is resistant to treatment with the combination of azathioprine, cyclosporine, and corticosteroids in Behçet's disease: An open-label trial. *Arthritis Rheum. Off. J. Am. Coll. Rheumatol.* **2005**, *52*, 2478–2484. [CrossRef]

59. Okada, A.A.; Goto, H.; Ohno, S.; Mochizuki, M.; Ocular Behcet's Disease Research Group of Japan. Multicenter study of infliximab for refractory uveoretinitis in Behcet disease. *Arch. Ophthalmol.* **2012**, *130*, 592–598.

60. Calvo-Río, V.; Blanco, R.; Beltrán, E.; Sánchez-Bursón, J.; Mesquida, M.; Adán, A.; Hernandez, M.V.; Hernandez Garfella, M.; Valls Pascual, E.; Martínez-Costa, L. Anti-TNF-α therapy in patients with refractory uveitis due to Behçet's disease: A 1-year follow-up study of 124 patients. *Rheumatology* **2014**, *53*, 2223–2231. [CrossRef] [PubMed]

61. Fabiani, C.; Vitale, A.; Emmi, G.; Vannozzi, L.; Lopalco, G.; Guerriero, S.; Orlando, I.; Franceschini, R.; Bacherini, D.; Cimino, L. Efficacy and safety of adalimumab in Behçet's disease-related uveitis: A multicenter retrospective observational study. *Clin. Rheumatol.* **2017**, *36*, 183–189. [CrossRef]

62. Atienza-Mateo, B.; Martín-Varillas, J.L.; Calvo-Río, V.; Demetrio-Pablo, R.; Beltrán, E.; Sánchez-Bursón, J.; Mesquida, M.; Adan, A.; Hernández, M.V.; Hernández-Garfella, M. Comparative study of infliximab versus adalimumab in refractory uveitis due to Behçet's Disease: National multicenter study of 177 cases. *Arthritis Rheumatol.* **2019**, *71*, 2081–2089. [CrossRef]

63. Emmi, G.; Vitale, A.; Silvestri, E.; Boddi, M.; Becatti, M.; Fiorillo, C.; Fabiani, C.; Frediani, B.; Emmi, L.; Di Scala, G. Adalimumab-Based treatment versus disease-modifying antirheumatic drugs for venous thrombosis in Behçet's Syndrome: A retrospective study of seventy patients with vascular involvement. *Arthritis Rheumatol.* **2018**, *70*, 1500–1507. [CrossRef] [PubMed]

64. Desbois, A.; Biard, L.; Addimanda, O.; Lambert, M.; Hachulla, E.; Launay, D.; Ackermann, F.; Perard, L.; Hot, A.; Maurier, F. Efficacy of anti-TNF alpha in severe and refractory major vessel involvement of Behcet's disease: A multicenter observational study of 18 patients. *Clin. Immunol.* **2018**, *197*, 54–59. [CrossRef]

65. Aksoy, A.; Yazici, A.; Omma, A.; Cefle, A.; Onen, F.; Tasdemir, U.; Ergun, T.; Direskeneli, H.; Alibaz-Oner, F. Efficacy of TNFα inhibitors for refractory vascular Behçet's disease: A multicenter observational study of 27 patients and a review of the literature. *Int. J. Rheum. Dis.* **2020**, *23*, 256–261. [CrossRef]

66. Hellmich, B.; Csernok, E.; Gross, W.L. Proinflammatory cytokines and autoimmunity in Churg-Strauss syndrome. *Ann. N. Y. Acad. Sci.* **2005**, *1051*, 121–131. [CrossRef] [PubMed]

67. Jakiela, B.; Szczeklik, W.; Plutecka, H.; Sokolowska, B.; Mastalerz, L.; Sanak, M.; Bazan-Socha, S.; Szczeklik, A.; Musial, J. Increased production of IL-5 and dominant Th2-type response in airways of Churg-Strauss syndrome patients. *Rheumatology* **2012**, *51*, 1887–1893. [CrossRef] [PubMed]

68. Yanagibashi, T.; Satoh, M.; Nagai, Y.; Koike, M.; Takatsu, K. Allergic diseases: From bench to clinic—Contribution of the discovery of interleukin-5. *Cytokine* **2017**, *98*, 59–70. [CrossRef] [PubMed]

69. Kahn, J.E.; Grandpeix-Guyodo, C.; Marroun, I.; Catherinot, E.; Mellot, F.; Roufosse, F.; Bletry, O. Sustained response to mepolizumab in refractory Churg-Strauss syndrome. *J. Allergy Clin. Immunol.* **2010**, *125*, 267–270. [CrossRef]

70. Kim, S.; Marigowda, G.; Oren, E.; Israel, E.; Wechsler, M.E. Mepolizumab as a steroid-sparing treatment option in patients with Churg-Strauss syndrome. *J. Allergy Clin. Immunol.* **2010**, *125*, 1336–1343. [CrossRef] [PubMed]

71. Herrmann, K.; Gross, W.L.; Moosig, F. Extended follow-up after stopping mepolizumab in relapsing/refractory Churg-Strauss syndrome. *Clin. Exp. Rheumatol.* **2012**, *30*, S62–S65.

72. Wechsler, M.E.; Akuthota, P.; Jayne, D.; Khoury, P.; Klion, A.; Langford, C.A.; Merkel, P.A.; Moosig, F.; Specks, U.; Cid, M.C.; et al. Mepolizumab or placebo for eosinophilic granulomatosis with polyangiitis. *N. Engl. J. Med.* **2017**, *376*, 1921–1932. [CrossRef]

73. Stone, J.H.; Merkel, P.A.; Spiera, R.; Seo, P.; Langford, C.A.; Hoffman, G.S.; Kallenberg, C.G.; St Clair, E.W.; Turkiewicz, A.; Tchao, N.K.; et al. Rituximab versus cyclophosphamide for ANCA-associated vasculitis. *N. Engl. J. Med.* **2010**, *363*, 221–232. [CrossRef]

74. Jones, R.B.; Tervaert, J.W.; Hauser, T.; Luqmani, R.; Morgan, M.D.; Peh, C.A.; Savage, C.O.; Segelmark, M.; Tesar, V.; van Paassen, P.; et al. Rituximab versus cyclophosphamide in ANCA-associated renal vasculitis. *N. Engl. J. Med.* **2010**, *363*, 211–220. [CrossRef] [PubMed]

75. McAdoo, S.P.; Bedi, R.; Tarzi, R.; Griffith, M.; Pusey, C.D.; Cairns, T.D. Ofatumumab for B cell depletion therapy in ANCA-associated vasculitis: A single-centre case series. *Rheumatology* **2016**, *55*, 1437–1442. [CrossRef] [PubMed]

76. Nagai, M.; Hirayama, K.; Ebihara, I.; Shimohata, H.; Kobayashi, M.; Koyama, A. Serum levels of BAFF and APRIL in myeloperoxidase anti-neutrophil cytoplasmic autoantibody-associated renal vasculitis: Association with disease activity. *Nephron. Clin. Pract.* **2011**, *118*, c339–c345. [CrossRef] [PubMed]

77. Liu, Z.; Davidson, A. BAFF and selection of autoreactive B cells. *Trends Immunol.* **2011**, *32*, 388–394. [CrossRef]

78. Furie, R.; Petri, M.; Zamani, O.; Cervera, R.; Wallace, D.J.; Tegzova, D.; Sanchez-Guerrero, J.; Schwarting, A.; Merrill, J.T.; Chatham, W.W.; et al. A phase III, randomized, placebo-controlled study of belimumab, a monoclonal antibody that inhibits B lymphocyte stimulator, in patients with systemic lupus erythematosus. *Arthritis Rheum.* **2011**, *63*, 3918–3930. [CrossRef]

79. Furie, R.; Rovin, B.H.; Houssiau, F.; Malvar, A.; Teng, Y.K.O.; Contreras, G.; Amoura, Z.; Yu, X.; Mok, C.C.; Santiago, M.B.; et al. Two-Year, Randomized, Controlled Trial of Belimumab in Lupus Nephritis. *N. Engl. J. Med.* **2020**, *383*, 1117–1128. [CrossRef]

80. Jayne, D.; Blockmans, D.; Luqmani, R.; Moiseev, S.; Ji, B.; Green, Y.; Hall, L.; Roth, D.; Henderson, R.B.; Merkel, P.A.; et al. Efficacy and Safety of Belimumab and Azathioprine for Maintenance of Remission in Antineutrophil Cytoplasmic Antibody-Associated Vasculitis: A Randomized Controlled Study. *Arthritis Rheumatol.* **2019**, *71*, 952–963. [CrossRef]

81. Mayer, E.; Holzl, M.; Ahmadi, S.; Dillinger, B.; Pilat, N.; Fuchs, D.; Wekerle, T.; Heitger, A. CTLA4-Ig immunosuppressive activity at the level of dendritic cell/T cell crosstalk. *Int. Immunopharmacol.* **2013**, *15*, 638–645. [CrossRef] [PubMed]

82. Langford, C.A.; Cuthbertson, D.; Ytterberg, S.R.; Khalidi, N.; Monach, P.A.; Carette, S.; Seo, P.; Moreland, L.W.; Weisman, M.; Koening, C.L.; et al. A Randomized, Double-Blind Trial of Abatacept (CTLA-4Ig) for the Treatment of Giant Cell Arteritis. *Arthritis Rheumatol.* **2017**, *69*, 837–845. [CrossRef]

83. Langford, C.A.; Cuthbertson, D.; Ytterberg, S.R.; Khalidi, N.; Monach, P.A.; Carette, S.; Seo, P.; Moreland, L.W.; Weisman, M.; Koening, C.L.; et al. A Randomized, Double-Blind Trial of Abatacept (CTLA-4Ig) for the Treatment of Takayasu Arteritis. *Arthritis Rheumatol.* **2017**, *69*, 846–853. [CrossRef] [PubMed]

84. Langford, C.A.; Monach, P.A.; Specks, U.; Seo, P.; Cuthbertson, D.; McAlear, C.A.; Ytterberg, S.R.; Hoffman, G.S.; Krischer, J.P.; Merkel, P.A.; et al. An open-label trial of abatacept (CTLA4-IG) in non-severe relapsing granulomatosis with polyangiitis (Wegener's). *Ann. Rheum. Dis.* **2014**, *73*, 1376–1379. [CrossRef] [PubMed]

85. Hu, Y.; Turner, M.J.; Shields, J.; Gale, M.S.; Hutto, E.; Roberts, B.L.; Siders, W.M.; Kaplan, J.M. Investigation of the mechanism of action of alemtuzumab in a human CD52 transgenic mouse model. *Immunology* **2009**, *128*, 260–270. [CrossRef]

86. Mohammad, A.J.; Smith, R.M.; Chow, Y.W.; Chaudhry, A.N.; Jayne, D.R. Alemtuzumab as remission induction therapy in Behcet Disease: A 20-year experience. *J. Rheumatol.* **2015**, *42*, 1906–1913. [CrossRef]

87. Walsh, M.; Chaudhry, A.; Jayne, D. Long-term follow-up of relapsing/refractory anti-neutrophil cytoplasm antibody associated vasculitis treated with the lymphocyte depleting antibody alemtuzumab (CAMPATH-1H). *Ann. Rheum. Dis.* **2008**, *67*, 1322–1327. [CrossRef]

88. Gopaluni, S.; Smith, R.; Goymer, D.; Broadhurst, E.; Mcclure, M.; Cahill, H. Alemtuzumab for relapsing and refractory primary systemic vasculitis: A trial of efficacy and safety (ALEVIATE)[abstract no: TH-PO1160]. *J. Am. Soc. Nephrol.* **2018**, *29*, B8.

89. Walport, M.J. Complement. Second of two parts. *N. Engl. J. Med.* **2001**, *344*, 1140–1144. [CrossRef] [PubMed]

90. Chen, M.; Xing, G.Q.; Yu, F.; Liu, G.; Zhao, M.H. Complement deposition in renal histopathology of patients with ANCA-associated pauci-immune glomerulonephritis. *Nephrol. Dial. Transplant.* **2009**, *24*, 1247–1252. [CrossRef] [PubMed]

91. Gou, S.J.; Yuan, J.; Chen, M.; Yu, F.; Zhao, M.H. Circulating complement activation in patients with anti-neutrophil cytoplasmic antibody-associated vasculitis. *Kidney Int.* **2013**, *83*, 129–137. [CrossRef]

92. Jayne, D.R.W.; Bruchfeld, A.N.; Harper, L.; Schaier, M.; Venning, M.C.; Hamilton, P.; Burst, V.; Grundmann, F.; Jadoul, M.; Szombati, I.; et al. Randomized trial of C5a receptor inhibitor avacopan in ANCA-Associated vasculitis. *J. Am. Soc. Nephrol.* **2017**, *28*, 2756–2767. [CrossRef]

93. Jayne, D.R.W.; Merkel, P.A.; Schall, T.J.; Bekker, P.; Group, A.S. Avacopan for the treatment of ANCA-Associated vasculitis. *N. Engl. J. Med.* **2021**, *384*, 599–609. [CrossRef]

94. Isaacs, A.; Lindenmann, J. Virus interference. I. The interferon. *Proc. R. Soc. Lond. Ser. B Biol. Sci.* **1957**, *147*, 258–267.

95. Shibuya, H.; Hirohata, S. Differential effects of IFN-α on the expression of various TH2 cytokines in human CD4+ T cells. *J. Allergy Clin. Immunol.* **2005**, *116*, 205–212. [CrossRef]

96. Shibuya, H.; Nagai, T.; Ishii, A.; Yamamoto, K.; Hirohata, S. Differential regulation of Th1 responses and CD154 expression in human CD4+ T cells by IFN-α. *Clin. Exp. Immunol.* **2003**, *132*, 216–224. [CrossRef]

97. Yang, P.; Huang, G.; Du, L.; Ye, Z.; Hu, K.; Wang, C.; Qi, J.; Liang, L.; Wu, L.; Cao, Q. Long-term efficacy and safety of interferon alpha-2a in the treatment of Chinese patients with Behçet's uveitis not responding to conventional therapy. *Ocul. Immunol. Inflamm.* **2019**, *27*, 7–14. [CrossRef] [PubMed]

98. Celiker, H.; Kazokoglu, H.; Direskeneli, H. Factors affecting relapse and remission in Behçet's uveitis treated with interferon Alpha2a. *J. Ocul. Pharmacol. Ther.* **2019**, *35*, 58–65. [CrossRef] [PubMed]

99. Gueudry, J.; Wechsler, B.; Terrada, C.; Gendron, G.; Cassoux, N.; Fardeau, C.; Lehoang, P.; Piette, J.-C.; Bodaghi, B. Long-term efficacy and safety of low-dose interferon alpha2a therapy in severe uveitis associated with Behçet disease. *Am. J. Ophthalmol.* **2008**, *146*, 837–844.e831. [CrossRef]

100. Hasanreisoglu, M.; Cubuk, M.O.; Ozdek, S.; Gurelik, G.; Aktas, Z.; Hasanreisoglu, B. Interferon alpha-2a therapy in patients with refractory Behçet uveitis. *Ocul. Immunol. Inflamm.* **2017**, *25*, 71–75. [CrossRef]

101. Çalgüneri, M.; Onat, A.M.; Öztürk, M.A.; Özçakar, L.; Ureten, K.; Akdogan, A.; Ertenli, İ.; Kiraz, S. Transverse myelitis in a patient with Behcet's disease: Favorable outcome with a combination of interferon-α. *Clin. Rheumatol.* **2005**, *24*, 64–66. [CrossRef] [PubMed]

102. Monastirli, A.; Chroni, E.; Georgiou, S.; Ellul, J.; Pasmatzi, E.; Papathanasopoulos, P.; Tsambaos, D. Interferon-α treatment for acute myelitis and intestinal involvement in severe Behçet's disease. *QJM Int. J. Med.* **2010**, *103*, 787–790. [CrossRef] [PubMed]

103. Alpsoy, E.; Durusoy, C.; Yilmaz, E.; Ozgurel, Y.; Ermis, O.; Yazar, S.; Basaran, E. Interferon alfa-2a in the treatment of Behçet disease: A randomized placebo-controlled and double-blind study. *Arch. Dermatol.* **2002**, *138*, 467–471. [CrossRef] [PubMed]
104. Ozguler, Y.; Hatemi, G.; Cetinkaya, F.; Tascilar, K.; Hamuryudan, V.; Ugurlu, S.; Seyahi, E.; Yazici, H.; Melikoglu, M. Clinical course of acute deep vein thrombosis of the legs in Behçet's syndrome. *Rheumatology* **2019**, *59*, 799–806. [CrossRef] [PubMed]

 *biomedicines*

Review

PCSK9: A Multi-Faceted Protein That Is Involved in Cardiovascular Biology

Sai Sahana Sundararaman [1,2,3], **Yvonne Döring** [4,5,6,*] and **Emiel P. C. van der Vorst** [1,2,3,4,5,*]

1. Interdisciplinary Centre for Clinical Research (IZKF), RWTH Aachen University, 52074 Aachen, Germany; ssahanasunda@ukaachen.de
2. Institute for Molecular Cardiovascular Research (IMCAR), RWTH Aachen University, 52074 Aachen, Germany
3. Department of Pathology, Cardiovascular Research Institute Maastricht (CARIM), Maastricht University Medical Centre, 6229 ER Maastricht, The Netherlands
4. Institute for Cardiovascular Prevention (IPEK), Ludwig-Maximilians-University Munich, 80336 Munich, Germany
5. German Centre for Cardiovascular Research (DZHK), Partner Site Munich Heart Alliance, 80336 Munich, Germany
6. Department of Angiology, Swiss Cardiovascular Center, Inselspital, Bern University Hospital, University of Bern, 3010 Bern, Switzerland
* Correspondence: Yvonne.doering@insel.ch (Y.D.); evandervorst@ukaachen.de (E.P.C.v.d.V.); Tel.: +41-31-632-5441 (Y.D.); Tel.: +49-(0)241-80-36914 (E.P.C.v.d.V.)

Abstract: Pro-protein convertase subtilisin/kexin type 9 (PCSK9) is secreted mostly by hepatocytes and to a lesser extent by the intestine, pancreas, kidney, adipose tissue, and vascular cells. PCSK9 has been known to interact with the low-density lipoprotein receptor (LDLR) and chaperones the receptor to its degradation. In this manner, targeting PCSK9 is a novel attractive approach to reduce hyperlipidaemia and the risk for cardiovascular diseases. Recently, it has been recognised that the effects of PCSK9 in relation to cardiovascular complications are not only LDLR related, but that various LDLR-independent pathways and processes are also influenced. In this review, the various LDLR dependent and especially independent effects of PCSK9 on the cardiovascular system are discussed, followed by an overview of related PCSK9-polymorphisms and currently available and future therapeutic approaches to manipulate PCSK9 expression.

Keywords: PCSK9; cardiovascular disorders; low density lipoprotein receptor; cholesterol; polymorphisms; monoclonal antibodies

Citation: Sundararaman, S.S.; Döring, Y.; van der Vorst, E.P.C. PCSK9: A Multi-Faceted Protein That Is Involved in Cardiovascular Biology. *Biomedicines* **2021**, *9*, 793. https://doi.org/10.3390/biomedicines9070793

Academic Editor: Byeong Hwa Jeon

Received: 14 June 2021
Accepted: 5 July 2021
Published: 8 July 2021

Publisher's Note: MDPI stays neutral with regard to jurisdictional claims in published maps and institutional affiliations.

1. Introduction

Pro-protein convertase subtilisin/kexin type 9 (PCSK9) is a soluble protease that has been widely studied in the field of cholesterol homeostasis and in cardiovascular biology after its discovery in 2003 [1]. Originally named Neural Apoptosis Regulated Convertase 1 (NARC1), PCSK9 is part of the secretory serine proteinase family called Pro-protein Convertases (PCs) [2]. Hepatocytes are the primary source of PCSK9 that is secreted into the circulation. However, also other cells can produce and secrete PCSK9, like cells in the intestines [3,4], pancreas [5], adipose tissue [6], kidneys [7] and brain [8]. Interestingly, the circulating levels of PCSK9 biologically increases in the late night, and decreases in the late afternoon, following a diurnal rhythm [9]. Furthermore, the total circulating PCSK9 levels are influenced by sex as females have higher levels compared to men, suggesting that hormones like oestrogens are involved in the expression and secretion of PCSK9 [10–12]. Besides sex, also age, body mass index (BMI), plasma cholesterol and triglyceride levels and blood pressure have been shown to modulate the total concentration of PCSK9 [10]. The best-known function of PCSK9 is its effect on low density lipoprotein receptor (LDLR), to which it can bind and thereby facilitate its lysosomal degradation. Lately, PCSK9 is the focus of various studies in different clinical contexts as a marker for cardiovascular risks [13–17] independent of other known traditional risk factors and independent of its influence on LDLR. PCSK9 is thought to play an important role in cardiovascular diseases (CVDs) via different mechanisms, either through binding to Epidermal Growth Factor (EGF) domains on receptors or through binding with receptors present in lipid rafts and associated cell membrane micro-domains or by influencing the gene expression and protein response of various factors involved in the development of cardiac complications [18]. Further supporting its role in CVDs, is the fact that various gain-of-function (GOF) mutations of PCSK9 are associated with hypercholesterolaemia and thereby an elevated risk for cardiac events [19–23]. In addition to these GOF mutations, also hundreds of loss-of-function (LOF) mutations of PCSK9 have already been discovered, that hinder its secretion into circulation providing protection against cardiovascular complications [24–27]. Targeting PCSK9 using monoclonal antibodies (mAbs) has emerged as an additional and additive therapy to treat hyperlipidaemia due to its regulation of LDLR [28–31]. However, several recent studies highlighted that PCSK9 can also have various additional effects that are independent of the LDLR, that provide more information on the involvement of PCSK9 in cancer, type 2 diabetes, obesity and several cardiovascular disorders [32–34].

Therefore, in this review the biological and physiological characterisation of PCSK9 will first be outlined in the context of cardiovascular biology, followed by an overview of the LDLR dependent and especially the LDLR independent effects of PCSK9 in this context. Furthermore, we will focus on state-of-the-art PCSK9 inhibitory techniques and their clinical potential.

2. PCSK9 Biology

The 22kb human PCSK9 gene is located on the chromosome 1p32 [2]. It has 12 exons and 11 introns that encode a 692 amino acid proteinase [2]. The protease is manufactured within the endoplasmic reticulum (ER) and has a molecular mass of 120 kDa [2]. It is secreted as an inactive protein that later undergoes post-translational modifications to form a 62 kDa mature protein. PreProPCSK9 consists of five segments: a signal peptide (amino acids 1–30), a N-terminal prodomain (amino acids 31–152), a catalytic domain containing the active sites (amino acids 153–404), a C-terminal cys-his-rich domain (amino acids 455–692) and a small peptide that serves as a link between the catalytic and C-terminal domain also known as the hinge region (amino acids 405–454) [35]. The C-terminal domain can further be divided into three modules, namely M1, M2 and M3. It has been demonstrated that the M2 module plays a crucial role in the extracellular PCSK9-LDLR binding [36]. The maturation or the formation of active PCSK9 requires three steps (Figure 1): first, the signal sequences are cleaved; pro-PCSK9 then undergoes proteolysis; the last step is the transportation of the protein into the circulation via the Golgi complex [37]. PCSK9 is secreted into the circulation as an enzymatically active protein. The secreted PCSK9 has only three domains and is a heterodimer protein [38]. The first is the pro-domain that is cleaved auto-catalytically, nevertheless it remains attached to the mature protein non-covalently to act as an escort for its intracellular movement. The pro-domain region contains acidic residues that have an auto-inhibitory role by interacting with the basic residues present in the catalytic domain [39]. The catalytic domain contains the proteolytic active site, which is necessary for the autocatalytic cleavage, but is not related to LDLR-PCSK9 binding [38]. Lastly, the C-terminal domain consisting of the three modules is attached to the catalytic domain [38]. While PCSK9 is a serine protease, the degradation of LDLR does not require proteolytic activity of PCSK9, only the ability of PCSK9 to chaperone LDLR towards lysosomes [40].

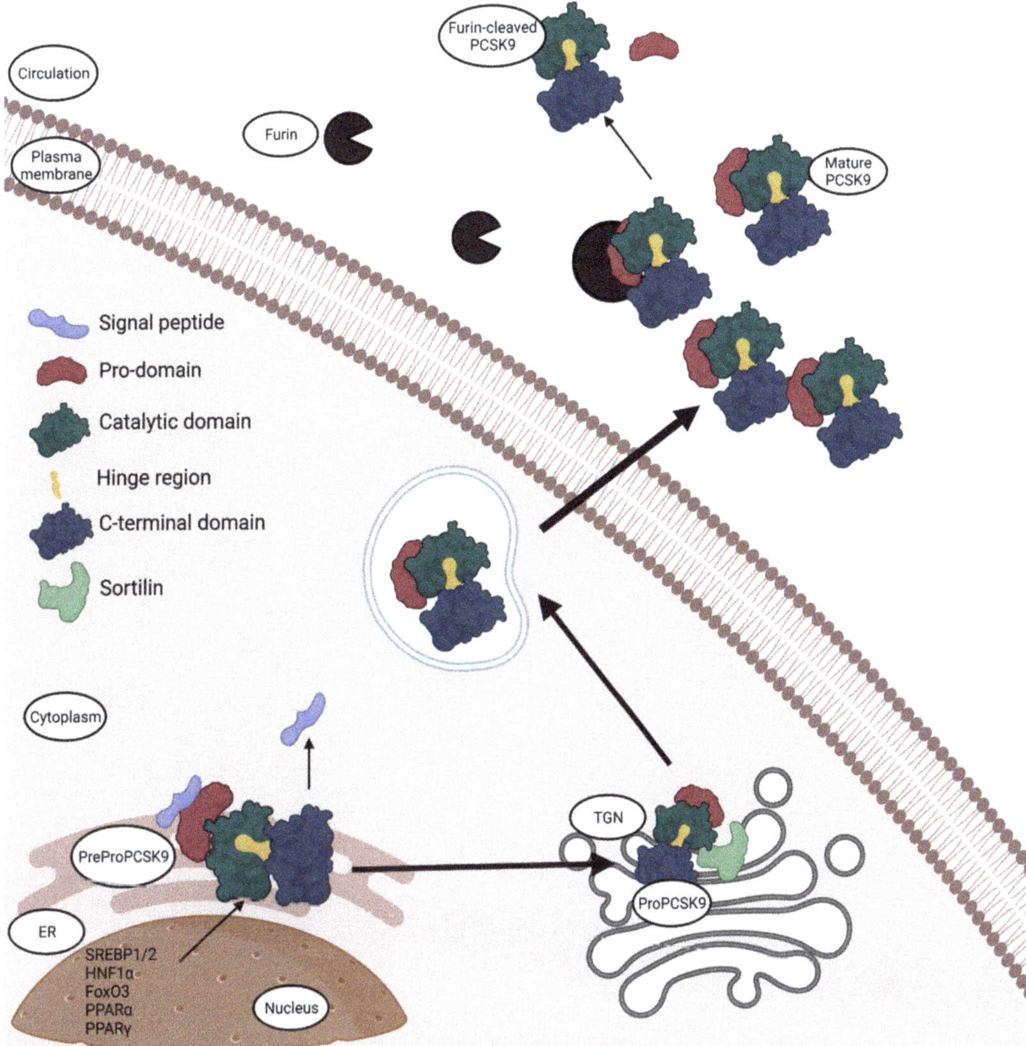

Figure 1. Synthesis and secretion of PCSK9. PCSK9 is synthesised in the endoplasmic reticulum (ER) with the help of transcription factors sterol-response element binding protein (SREBP-1/2), hepatocyte nuclear factor 1α (HNF1α), forkhead box O3 (FoxO3), Peroxisome proliferator-activated receptor α and γ (PPARα and PPARγ). The synthesised PCSK9 is in the form of inactive zymogen, called PreProPCSK9. PreProPCSK9 has five segments: a signal peptide, the pro-domain, catalytic domain, hinge region and the C-terminal domain; the protein then undergoes autocatalytic cleavage in the ER to lose the signal peptide and becomes ProPCSK9. Pro-PCSK9 is then transported to Trans-Golgi Network (TGN), where it interacts with Sortilin, undergoes proteolysis and the mature hetero-dimer PCSK9 is then transported in endosomes and secreted into the circulation. Normally, mature PCSK9 has its pro-domain non-covalently attached, although in circulation it can encounter furin, which then cleaves the pro-domain and releases a smaller peptide into circulation that is less active than the hetero-dimer PCSK9. The figure was created with Biorender.com (accessed on 19 May 2021).

PCSK9 also exists in a furin-cleaved form that is 55 kDa in size [41] (Figure 1). This form is slightly less active and has half the affinity to LDLR as well as a shorter half-life compared to the intact heterodimer form. In addition, the furin-cleaved form is in general not associated with ApoB containing lipoproteins [41], establishing that the furin-cleaved form is the less active one. PCSK9 is cleaved by furin at the N-terminal to exclude the pro-domain and releases the protein into circulation [42]. This cleavage can occur only in circulation [42], because there is no evidence that furin can cleave PCSK9 intracellularly although they co-exist in the Golgi complex. The fact that the furin-cleaved form does not contain the pro-domain to act as a chaperone for PCSK9 to get secreted extracellularly validates the hypothesis that furin only cleaves PCSK9 in circulation [42].

Various transcription factors and cofactors regulate the expression of the PCSK9 gene. The PCSK9 promoter contains an important region that is essential for transcription, called the Sterol regulatory Element (SRE) [43]. Sterol-response Element binding protein (SREBP-1/2) is a predominant transcription factor that connects to this SRE promoter in PCSK9 [35]. Low dietary cholesterol concentrations upregulate SREBP-1/2 expression that in turn regulates PCSK9 levels in circulation [44]. Additionally, hepatocyte nuclear factor 1 α (HNF1α) and forkhead box O3 (FoxO3) are also predominantly involved in the transcription of PCSK9 [43]. HNF1α acts by binding to a site that is located next to the SRE site in the promoter region [43], while FoxO3 suppresses the transcriptional activity of HNF1α by recruiting sirtuins 6 instead of HNF1α [45]. Furthermore, sirtuins 1 and 6 and histone deacetylases are known to supress SREBP2 expression in the liver resulting in suppression of PCSK9 expression and thereby decreasing the circulating PCSK9 levels [44]. Insulin is also involved in the regulation of PCSK9 expression as it could be shown that excess insulin in the blood decreases the transcription of the PCSK9 gene [46] in post-menopausal obese women, but it does not amend the mRNA levels in healthy men and type-2 diabetic patients [47]. Peroxisome proliferator-activated receptor α and γ (PPARα and PPARγ) regulate the gene expression of PCSK9 as well [2], PPARγ increases the gene expression while PPARα decreases it. Furthermore, PCSK9 interacts with sortilin in the Trans-Golgi complex, which is a transmembrane protein involved in the development of atherosclerosis and other CVDs [48], aiding in the extracellular secretion of PCSK9. It could be shown that mice that overexpress sortilin have increased circulating PCSK9 levels, while sortilin deficient mice have low circulating PCSK9 levels, suggesting that this protein–protein interaction is highly essential for the regulation of secretion of PCSK9 by hepatocytes [49]. The binding of PCSK9 to sortilin occurs at the pH 6.5 and any modification of the pH deteriorates the binding [49].

Although a lot of progress has been made over the last years to fully understand the regulation of PCSK9, it is still not completely understood as this regulation turns out to be highly dynamic. The following sections will discuss how PCSK9 not just interacts with LDLR, but also with several other receptors to influence cardiovascular biology.

3. PCSK9-LDLR Binding

In 2007, PCSK9 emerged as a novel target to treat dyslipidaemia. Numerous studies have identified a clear link between low density lipoprotein cholesterol (LDL-C) and PCSK9. Normally, LDL-C binds and gets internalised by binding to the LDLR, thereby mediating its clearance by endocytosis. In the presence of PCSK9, the LDLR undergoes lysosomal degradation, leading to an inhibition of the recycling of the receptor and therefore an increase in the plasma levels of LDL-C [50]. This degradation of the LDLR by PCSK9 occurs through two different pathways, an extracellular and intracellular one (Figure 2). The commonly known and widely studied process of LDLR degradation is by extracellular binding of PCSK9 to the cell surface of the LDLR. PCSK9 does not degrade LDLR itself, it just acts as a tag to promote the degradation process. The catalytic domain of the mature circulating PCSK9 binds to the epidermal growth factor A (EGF-A) domain of the LDLR. The PCSK9-LDLR compound then undergoes endocytosis through clathrin-coated pits and is taken up by the endosomes or lysosomes in the cells, leading to the deprivation of LDLR as well as PCSK9 [2,50]. This PCSK9-LDLR binding is dependent on calcium concentrations and is affected by pH changes as acidic pH of the endosome increases the affinity of PCSK9 to LDLR [51]. This increased affinity is caused by an interaction between the ligand-binding domain of the LDLR and the C-terminal domain of the mature PCSK9 [50]. This binding complex occurs in the endosome and the binding sustains, causing LDLR to not unfold and therefore not getting recycled back to the cell surface. The C-terminal domain of PCSK9 is quite important for its binding to LDLR, but alone it has no influence on the LDLR [39]. The interaction between the catalytic domain of PCSK9 and the EGF-A domain of LDLR contributes less to the LDLR degradation, while the positive charge at acidic pH of C-terminal domain and its size is rather responsible for the PCSK9-LDLR binding [52]. Although a review published in 2017 concluded that only hepatic LDLR is affected by circulating PCSK9 [53], studies have revealed that PCSK9 is involved in regulating the expression of LDLR in pancreas and other organs as well [5]. PCSK9 also facilitates intracellular degradation of mature LDLR present in the Golgi or Trans-Golgi complex, before it reaches the cell surface [44]. In this pathway, the immature or the budding PCSK9 binds to the LDLR found in the Golgi apparatus and redirects it to the lysosomes for degradation. Unlike the extracellular degradation of LDLR, this intracellular degradation compels the involvement of catalytic activity of PCSK9 [44].

On the other hand, PCSK9 also interacts with LDLR in non-destructive ways (Figure 2). The catalytic domain of the pre-pro-PCSK9 binds to the EGF-A repeat of the precursor of LDLR in the ER that is essential for the transport of LDLR to the Golgi apparatus wherein LDLR matures by obtaining various carbohydrates [44,54]. This binding is not just advantageous for the LDLR, but it also helps pre-pro-PCSK9 to undergo auto-catalytic cleavage and thus undergo maturation. The mature PCSK9 then tethers with an additional salt bridge to LDLR and this complex is released by the cell to increase the number of cell-surface receptors [54]. Nonetheless, this chaperone activity is not critical for the transportation of LDLR to cell surfaces and happens much less often than the degrading activity, considering that even in the absence of PCSK9 the LDLR is transported to the cell membrane without any hindrance [54]. Although, the reason why PCSK9 is able to chaperone LDLR to the surface without leading to lysosomal degradation is not known yet.

The binding of PCSK9 to LDLR resulting in its degradation plays an important role in the cholesterol homeostasis, as it could be shown that a full-body knockout of PCSK9 in mice leads to an extreme reduction of cholesterol levels in the serum. More particularly, liver-specific PCSK9 knockout also leads to a considerable reduction in circulating cholesterol levels concluding that the hepatic PCSK9 contributes most to the cholesterol phenotype [2]. The destruction of LDLR by PCSK9 leads to hyperlipidaemia, which is associated with numerous cardiovascular complications. In particular, lipid uptake and accumulation by macrophages in the vessel wall and formation of 'foamy' macrophages leads to the development of atheromas. PCSK9 inhibition using mAbs reduces the accumulation of lipid particles inside monocytes and thereby also inhibits monocyte chemotaxis [55]. Apart from the effects on lipid uptake, silencing of PCSK9 using siRNAs also increases the expression of the chemokine receptor CCR2 on monocytes, thereby increasing their potential to migrate towards and infiltrate the arterial wall, triggering arterial inflammation and promoting atherogenesis [56]. The degradation of hepatic LDLR by PCSK9 also causes hypertriglyceridaemia, due to increased amounts of circulating ApoB containing lipoproteins that could not be degraded by LDLR [53]. This has been confirmed by several murine models where overexpression of PCSK9 leads to an increase in the plasma ApoB particles and VLDL-triglycerides, whereas deficiency of PCSK9 results in a decline of triglyceride levels in the plasma of these mice [53]. Therefore, the degradation of LDLR influences not just the cholesterol homeostasis, but also the triglyceride homeostasis, that eventually leads to diversified pathologies in the vasculature and beyond.

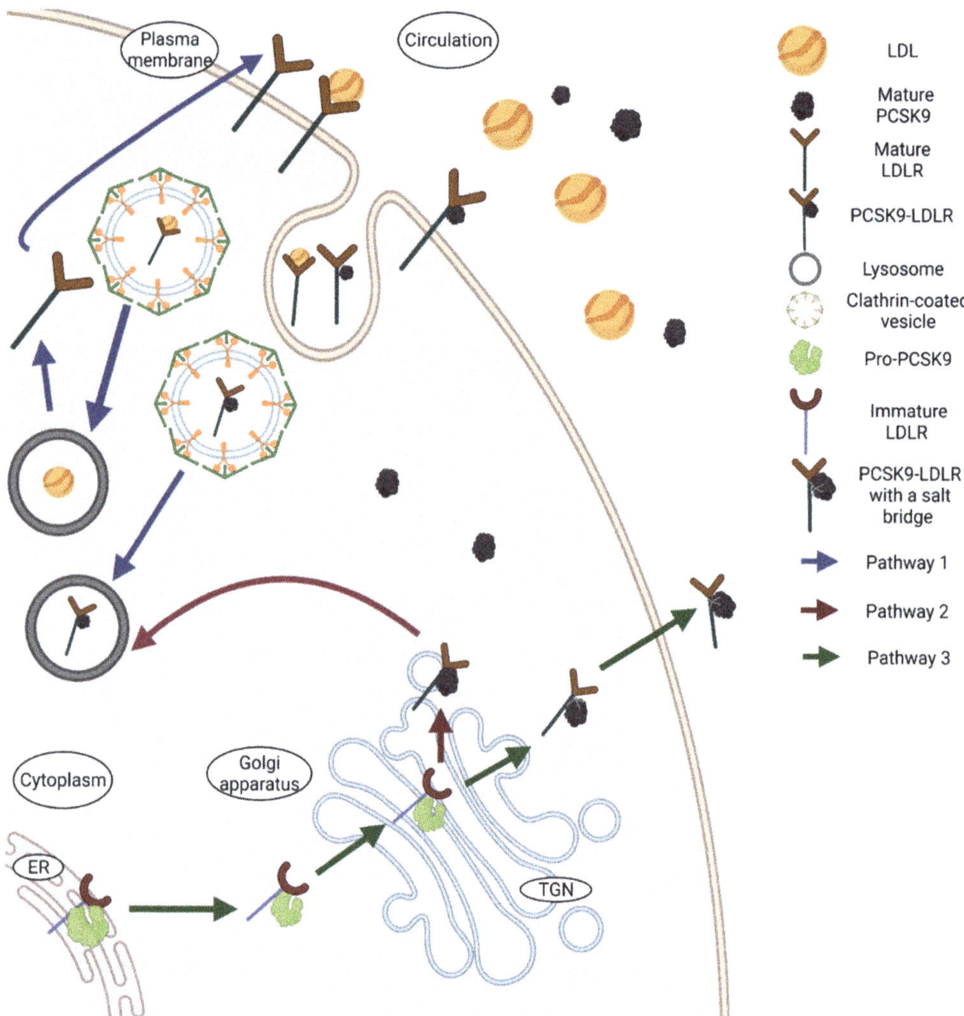

Figure 2. Interactions between PCSK9 and LDLR. PCSK9 interacts with LDLR through three different approaches. Pathway 1 is represented in the figure with blue arrows, while pathway 2 with red arrows and pathway 3 with green arrows. Typically, LDLR on the surface binds to the circulating LDL and transports the cholesterol particle through clathrin coated pits for degradation by lysosomes. The receptor then gets recycled back to the surface. In the presence of PCSK9, a bond between the epidermal growth factor-A (EGF-A) domain of LDLR and the catalytic domain of PCSK9 is formed. PCSK9 transports LDLR via clathrin-coated vesicles to lysosomes for degradation. In this process, PCSK9 gets degraded as well (pathway 1). Intracellularly, PCSK9 binds in the same manner with LDLR in the Trans-Golgi network (TGN) and directs the compound again towards lysosomes (pathway 2). Lastly, Pro-PCSK9 also binds to LDLR in the endoplasmic reticulum (ER) and transports the complex to Golgi apparatus, where both PCSK9 and LDLR undergo maturation. The matured PCSK9 binds to LDLR with an additional salt bridge that is released from the cells into circulation (pathway 3). The figure was created with Biorender.com (accessed on 19 May 2021).

4. PCSK9's Activity Independent of LDLR

The function of PCSK9 is not just limited to the regulation of the LDLR, it also influences the lysosomal degradation of various receptors that structurally relate to the LDLR. Receptors such as very low density lipoprotein receptor (VLDLR), LDLR-related protein 1 (LRP1) [53] and apolipoprotein E receptor 2 (ApoER2) [57] also contain EGF-A domains enabling the interaction with PCSK9. The expression of PCSK9 by vascular endothelial cells (ECs), smooth muscle cells (SMCs), cardiomyocytes along with various immune cells also sparks our interest to substantially acknowledge the role of PCSK9 in pathologies of the cardiovascular system. In this section, we review the myriad of recently published studies that relate the LDLR-independent effects of PCSK9 on vascular diseases that include atherosclerosis, myocardial infarction and calcification and discuss the assorted mechanisms by which PCSK9 targets the cardiovascular system (summarised in Figure 3).

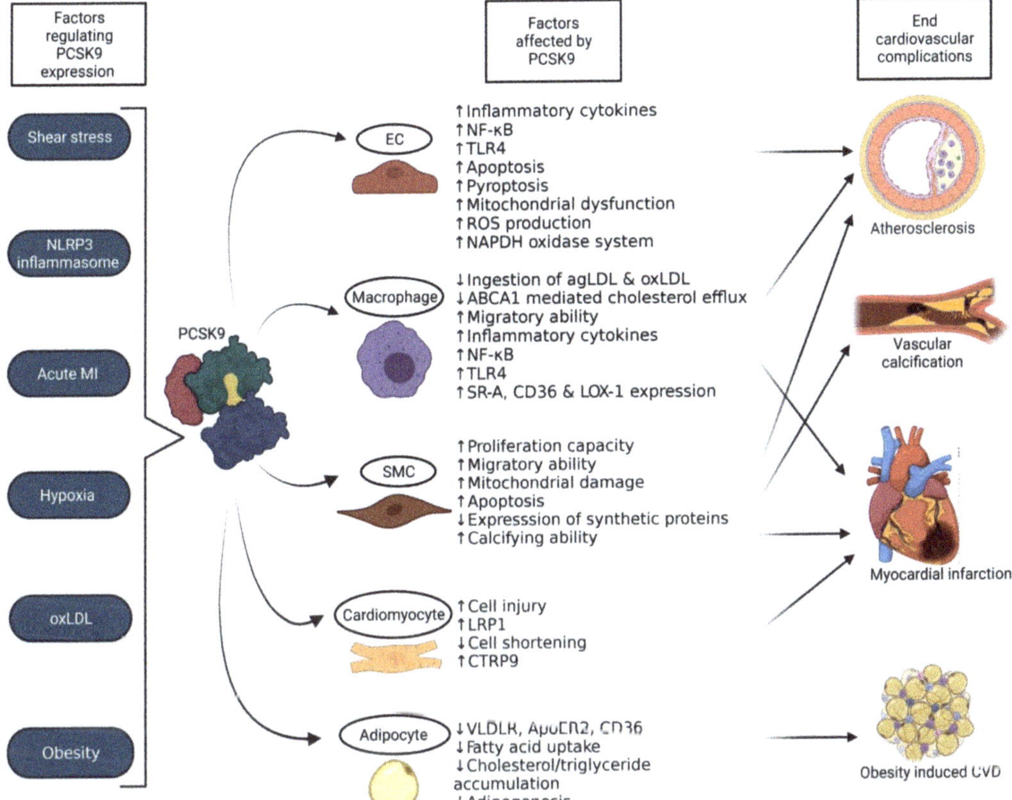

Figure 3. LDLR-independent mechanisms that regulate cardiovascular complications. The secretion of PCSK9 is affected by several factors, including shear stress, inflammasome, acute MI, hypoxia, oxLDL and obesity. The secreted PCSK9 then influences the expression of various factors in endothelial cells, macrophages, smooth muscle cells, cardiomyocytes and adipocytes to eventually influence the development and progression of atherosclerosis, vascular calcification, myocardial infarction and obesity induced cardiovascular disorders. ABCA1—ATP-binding cassette transporter 1; agLDL—aggregated low-density lipoprotein; ApoER2—apolipoprotein E receptor 2; CD36—Cluster of differentiation 36; CTRP9—C1q/TNF-related protein 9; CVD—cardiovascular disorder; EC—endothelial cells; LOX-1—Lectin-like oxidised low-density lipoprotein receptor-1; LRP1—low-density lipoprotein receptor-related protein 1; NADPH—Nicotinamide adenine dinucleotide phosphate; NF-κB—nuclear factor kappa-light-chain-enhancer of activated B-cells; oxLDL—oxidised low-density lipoprotein; ROS—reactive oxygen species; SMC—smooth muscle cells; SR-A—scavenger receptor-A; TLR4—Toll-like receptor 4; VLDLR—very low-density lipoprotein receptor. The figure was created with Biorender.com (accessed on 19 May 2021).

4.1. Inflammation and Atherosclerosis

Atherosclerosis is characterised by chronic inflammation of the vasculature, accompanied by lipid retention in arteries and the formation of plaques. Activation of the endothelial cell layer by atherogenic and pro-inflammatory stimuli initiates atherogenesis, which is followed by the infiltration of monocytes, lymphocytes, and various other immune cells [58]. The recruited monocytes differentiate into macrophages and internalise atherogenic lipoproteins and transform into foam cells. This lipoprotein and thus cholesterol accumulation continues which will eventually result in apoptosis and secondary necrosis, forming a lipid-rich and necrotic core in the plaques. Meanwhile, during disease progression medial SMCs proliferate and migrate to form a fibrous cap that stabilises the plaques and shields the necrotic core [2]. If this cap is broken down in later stages of the disease, plaques can rupture leading to thrombus formation, one of the primary causes of cardiovascular complications and related deaths. Albeit the steps involved in atherogenesis have already been clearly outlined by various studies, the exact role of PCSK9 in atherosclerosis has only recently been (partly) unravelled.

PCSK9 is expressed on aortic ECs, SMCs, macrophages, dendritic cells and epithelial cells [57], suggesting that PCSK9 not only regulates atherosclerosis by influencing serum LDL-C levels, but also by interacting and influencing cellular processes in the vessel wall to aggravate atherosclerosis [33,34]. The severity of atherosclerosis is positively correlated with circulating PCSK9 levels [14]. Furthermore, it could be demonstrated that low shear stress, such as in the aortic arch branch points and aorta-iliac bifurcations, increases the expression of PCSK9 by ECs, while high shear stress does the opposite [59–61]. Studies observed that silencing of PCSK9 reduces the levels of inflammatory cytokines in aortic tissues, resulting in an attenuated plaque inflammation without affecting LDLR levels [62,63]. This effect is caused by the fact that PCSK9 influences the inflammatory pathways to activate NF-κB and upregulates Toll-like receptor 4 (TLR4) expression [62]. The same study also identified that PCSK9 is overexpressed in atherosclerotic plaques of $Apoe^{-/-}$ mice [62]. Further confirming an important role of PCSK9 in inflammation and apoptosis, it could be shown that silencing of PCSK9 in ECs in vitro reduces the ability of the cells to go into apoptosis when exposed to oxLDL by reducing pro-apoptotic factors Bcl2-associated protein (Bax), Caspase 3 and 9, while increasing the anti-apoptotic factor Bcl-2 as well as activating p38/JNK/MAPK pathways [2]. Additionally, PCSK9 induces pyroptosis, mitochondrial dysfunction and reactive oxygen species (ROS) production in human umbilical vein endothelial cells (HUVECs) after an exposure to oxLDL, suggesting that PCSK9 also plays a valuable role in the antioxidant response in the context of atherosclerosis [64]. The increased expression of PCSK9 by low shear stress also induces ROS generation via the nicotinamide adenine dinucleotide phosphate (NADPH) oxidase system [60], clearly demonstrating an important role of PCSK9 in ECs.

Besides ECs, vascular SMCs are also affected by shear stress when the EC layer is disrupted, as it could be demonstrated that low shear stress upregulates their proliferation and migration capability while increasing the secretion of PCSK9 by the SMCs. Several studies have demonstrated that SMCs secrete functional PCSK9 into the atheroma that exerts effects on monocytes migration in the intima. The overexpression of PCSK9 by SMCs in atherosclerotic plaques also reduces the ability of macrophages to ingest aggregated LDL (agLDL) and oxLDL molecules through scavenger receptors and LDLR related proteins [4,65]. PCSK9 secreted by SMCs not just plays a paracrine effect, PCSK9 also regulates the metabolism in SMCs. This could be perceived by several studies: for instance, treating SMCs in vitro with recombinant PCSK9 stimulates mitochondrial damage that in turn activates apoptosis pathways [66]. Studies were performed in vitro to validate this, and it was seen that mice that are deficient in PCSK9 show less mitochondrial damage in SMCs compared to wild type mice when injected with LPS [66]. Mediated by mitochondrial ROS generation, PCSK9 and mitochondrial DNA damage influence each other in a positive feedback loop to facilitate cell injury and thereby advance atherosclerosis [56]. Contrarily, PCSK9 might provide a protective effect against atherosclerosis progression by regulating SMCs. Deficiency of PCSK9 in mice has been shown to reduce the ability of the SMCs to proliferate and migrate, with the cells expressing more than usual levels of contractile, such as alpha-actin and myosin proteins [2,67]. These SMCs also express very low levels of synthetic proteins, such as extracellular matrix components and collagen that are involved in the formation of fibrous cap [68]. Combined, the lack of PCSK9 therefore seems to reduce the fibrous cap formation and thereby destabilises the lesions. Altogether, it could be shown that SMCs do not only express PCSK9, but that PCSK9 can also influence cellular processes in SMCs to influence plaque stability.

Besides its influence on vascular cells, PCSK9 has also been shown to exert pro-inflammatory and pro-atherogenic effects on macrophages in vitro even in the absence of LDLR [2]. For example, PCSK9 has been shown to inhibit ATP-binding cassette transporter (ABCA1) mediated cholesterol efflux in macrophages and thereby disturbs the cholesterol homeostasis [69]. Furthermore, PCSK9 increases the infiltration of Ly6c^{hi} monocytes into the atherosclerotic plaques [70]. Inhibition of PCSK9 also supresses the expression of inflammatory cytokines IL-1α, IL-6, IL-1β, MCP-1 and TNFα and the activation of NF-κB pathway when macrophages are exposed to oxLDL and inflammation [56,62]. In line with this, macrophages that are stimulated with recombinant PCSK9 express pro-inflammatory cytokines in a dose-dependent fashion [71]. These pro-inflammatory effects are LDLR-independent as it could be shown that PCSK9 has similar effects on macrophages from LDLR$^{-/-}$ mice [72]. Macrophages can also secrete PCSK9 themselves and in vitro and in vitro experiments have discovered that the NLR family pyrin domain containing 3 (NLRP3) inflammasome triggers the expression of PCSK9 in macrophages by IL-1β release [73]. PCSK9 secreted from lipid-loaded macrophages can also regulate several LDLR like receptors. For example, members of the LRP family, such as LRP5 form a complex with PCSK9, that is highly essential for further lipid uptake by macrophages. When this LRP5-PCSK9 bond is formed, the complex triggers the TLR4/NF-κB pathway resulting in increased inflammation [74]. Similarly, PCSK9 upregulates the expression of various scavenger receptors such as SR-A, CD36 and LOX-1 to boost the uptake of oxLDL by macrophages, to further facilitate inflammation [56,75]. There is also a feed-forward loop in which the activation of LOX-1 triggers PCSK9 expression as well [67]. Although this process can play an import role in atherogenesis, the exact effects of inhibition of PCSK9 on LOX-1 remain elusive [53].

Considering all the above-mentioned studies, it can be summarised that PCSK9 has a direct effect on both vascular and immune cells to directly influence atherogenesis not only by modulating the LDLR but also by LDLR-independent mechanisms.

4.2. Myocardial Infarction

Complete cessation or lowering of blood flow to parts of the heart, due to atherosclerosis formation, causes the myocardium to be deprived of oxygen, resulting in a myocardial infarction (MI) which is associated with myocardial cell death and necrosis [76]. PCSK9 levels were observed to be upregulated in rats after acute MI [77], which is also validated in humans as serum levels of PCSK9 are also elevated in patients with acute MI [11,60]. This could be explained by the fact that acute MI leads to an increase in the expression of SREBP-2, hepatocyte nuclear factor 1 α (HNF1α) and NLRP3 resulting in an elevated expression of PCSK9 [78]. Furthermore, MI leads to hypoxia of cardiomyocytes and it was demonstrated that hypoxia induces the expression of PCSK9 in cultured cardiomyocytes [60,79]. The PCSK9 produced by hypoxic cardiomyocytes promotes injury even in healthy cardiomyocytes [79]. Moreover, the elevated PCSK9 levels also stimulate the secretion of pro-inflammatory cytokines and activate NF-κB signalling in the recruited macrophages at the sites of injury [79]. This can be further confirmed by inhibiting or knocking out PCSK9 in murine models as PCSK9 deficiency improves the cardiac function and reduces the infarct size in mice [80]. Consistent with this, when mice are subjected to coronary artery occlusion, the mice develop infarcts and the area surrounding the infarct zones have more than normal, non-infarcted, levels of PCSK9 protein expression [60]. These zones with elevated PCSK9 expression also show elevated autophagy. Inhibiting PCSK9 reduces the size of the infarct and rescues the phenotype, suggesting that PCSK9 plays a causal role in the development of these infarcts.

MI is very closely associated with coronary artery disease (CAD), as MI causes occlusion of coronary arteries [81]. A study observed that PCSK9 levels in patients with a history of CAD are associated with high levels of circulating cholesterol, triglyceride, and inflammation which did not correspond to the severity and progression of the disease [82], although contradicting studies also exist [83]. PCSK9 is involved in the regulation of diverse lipoproteins involved in the development and progression of CAD. For example, oxLDL induces the expression of PCSK9 in cardiomyocytes which, in turn, reduces the cardiomyocyte cell shortening [84]. Serum lipoprotein (a) (Lp(a)) is yet another high-risk factor for CADs, but the relation between Lp(a) and PCSK9 remains debateable. Different studies revealed a positive association between serum Lp(a) and PCSK9 [85], however, other studies did not find associations between these two [86].

To summarise, in situations of MI, the expression of PCSK9 is elevated to not just alter the lipid metabolism but also to promote the cytotoxic capacity of oxLDL and increase the expression of apoptotic markers and promotes autophagy [78]. Inhibition of PCSK9 is believed to play a role in MI by limiting vascular remodelling and by tampering autophagy and inflammatory markers on top of influencing the lipid profile and cardiac function, although studies are yet to find all the mechanistic insights [87].

4.3. Obesity Induced CVD

Obesity being a major health problem at the moment, leads to the development of metabolic comorbidities in the form of cardiovascular complications. Obesity is characterised by the enlargement of adipocytes and adipose tissue, and secretion of inflammatory cytokines to instigate pathological complications to the cardiovascular system [88]. At first, it was believed that adipocytes do not express PCSK9, but only express targets of PCSK9 like VLDLRs and LDLRs on their surface [89]. On the contrary, recent studies have discovered the presence of PCSK9 in human adipocytes, that is easily detectable on gene and protein levels [16] although these levels are relatively low compared to the levels of hepatic PCSK9. Studies have identified that obesity upregulates the expression of PCSK9 and high levels of PCSK9 are again associated with a progression of the disease, therefore making it a vicious feedback cycle between adiposity and PCSK9. The levels of PCSK9 expression in adipose tissue positively correlates with the body mass index (BMI) of the individual, suggesting that obesity and adiposity induces the expression of PCSK9 [16]. Likewise, excessive dietary fat consumption leads to an upregulation of hepatic PCSK9 [90]. The function of PCSK9 in relation to obesity seems controversial. Validating the hypothesis that PCSK9 provides a defence mechanism against obesity, studies showed that visceral adipose tissue content increases in PCSK9 knockout mice due to increased expression of VLDLR [44]. PCSK9 targets VLDLR and ApoER2, receptors that are responsible for the hydrolysis of triglyceride-rich lipoproteins (TRLs). This hydrolysis is essential for fat storage in adipose tissues as well as utilisation of fat by the vascular tissues [41]. Through such VLDLR regulation, PCSK9 limits adipogenesis in visceral adipose tissue and protects against adiposity [89]. On the contrary, PCSK9 also influences the expression of receptors and molecules other than VLDLR to promote obesity. For example, PCSK9 is also involved in the degradation of CD36, reducing fatty acid uptake and cholesterol/triglyceride accumulation in adipose tissue and liver [91]. Furthermore, the PCSK9 secreted from adipocytes is involved in the modification of myocardial LRP1 levels and the glucose metabolism in the cardiomyocytes by influencing C1q/tumour necrosis factor-related protein-9 (CTRP9) [6], which is involved in various cardiac complications [92]. In conclusion, adipose tissue expresses and secretes PCSK9 which via LDLR-independent mechanisms can contribute to obesity and thus CVDs.

4.4. Calcification

Ectopic calcification is associated with old age and is a very relevant comorbidity of CVDs. Calcific aortic valve stenosis (CAVS) is a form of ectopic calcification. Studies have indicated that there is an association between the levels of plasma PCSK9 and the severity of aortic valve stenosis [93]. The expression of PCSK9 is higher in the calcified valves of patients suffering from CAVS, and the valvular interstitial cells isolated from these calcified valves show elevated expression of markers of calcification with elevated PCSK9 expression [94]. In line with this, aortic valves isolated from PCSK9$^{-/-}$ mice show lower markers for calcification compared to mice with functional PCSK9 [93]. Histological examination of aortic samples from patients with coronary artery calcification further validated the abnormal expression of PCSK9 in calcified area [95]. As a validation of the causal role of PCSK9 in vascular calcification, LDLR$^{-/-}$ mice overexpressing PCSK9 using an adenovirus containing PCSK9 develop more vascular calcification in response to being fed with Western diet compared to mice with normal PCSK9 levels [96]. All these studies suggest that PCSK9 is a marker and perhaps even a causal factor of vascular calcification that deserves more attention in future research.

5. PCSK9 Polymorphisms

Several polymorphisms occur in the PCSK9 gene that tend to cause an effect on CVDs [19,97–104]. For example, the PCSK9 gene undergoes loss-of-function (LOF) mutations that disturb the secretory pathways of PCSK9 into circulation. Certain LOF mutations of PCSK9 result in the hindrance of the transport of PCSK9 from the ER complex to the Golgi apparatus [44,105], leading to reduced circulating levels of mature or furin-cleaved PCSK9. Additionally, nonsense mutations to PCSK9 also give protection against CAD by reducing the LDL-C levels [106]. LOF mutations are also related to an increase in the proteolysis [39,107,108] and lowering of the levels of lp(a), LDL-C and reduce the risk of cardiac complications such as MI and aortic valve stenosis [109,110]. Furthermore, LOF mutations attenuate the cytokine response in healthy as well as septic patients when they are administered LPS [60]. Arg46Leu and Asp301Gly are two LOF mutations, that can be inherited from the parents to the children, and they have been associated with lowering LDL-C levels and providing protection against cardiovascular pathologies [27]. Mutations in the hinge region of the protein can be caused by variants W428X, A443T and R434W, resulting in misfolded and non-functional PCSK9 [111]. In conclusion, LOF mutations result in low circulating LDL-C levels and reduce the risk of developing CVDs [53].

The PCSK9 gene also undergoes gain-of-function mutations (GOF), that usually result in hypercholesterolaemia [112] causing accelerated vascular aging and CVDs [113]. For instance, mutation D374Y enhances the protein to self-assemble to form dimers and trimers [44,114], leading to extreme hypercholesterolaemia and severe atherosclerosis [44]. This mutation is therefore widely used as atherosclerotic models in murine studies. Furthermore, S127R and D374Y mutations increase the affinity of PCSK9 towards LDLR as well as increase the level of every apoB100-containing lipoprotein in the plasma [39,115]. Other single nucleotide polymorphisms (SNPs) increase the intima-medial thickness of the arteries [116], cause arterial plaques [117] and stimulate the progression of CVD [107]. Some variants of PCSK9 show both LOF and GOF. For instance, S127R is a mutation that is involved in both apoB synthesis as well as its catabolism [118]. This variant causes increased production of VLDL and IDL apoB-100 and increases the catabolism of LDL independent of its effect on VLDL and IDL [118]. On the other hand, the variants A245T and R272Q undergo more autocatalytic cleavage than the normal PCSK9, but they do not have any functional consequences on the degradation of LDLR [119]. Additionally, variants I474V and E670G cannot be categorised into LOF or GOF mutations, as functional studies for the mutations are not available [120]. Tables 1 and 2 summarise the main LOF and GOF mutations that are currently known and provide an overview of the functional consequences of these variants on the pathology of cardiovascular biology.

Table 1. Loss-of-function mutations.

Name of Mutation + Reference	Cause and Consequences
A443T [121,122]	Presence of a novel PCSK9 O-glycosylation site in the hinge region that promotes furin-cleavage and generates lower circulating levels of LDL-C and lower levels of fasting glucose
A522T [98] T77I [98] V114A [98] P616L [98]	Amino acid substitutions that cause hypocholesterolaemia
Ala68fsLeu82X [98]	Single nucleotide deletion in exon 1 that leads to a frameshift mutation that in turn causes the PCSK9 peptide to be shortened and not functional
C679X (rs28362286) [122–125] Y142X (rs67608943) [124]	SNPs that lead to disruption in the folding of the protein and lower concentrations of Lp(a), LDLC, oxidised phospholipid (OxPL-ApoB), fasting glucose and glycated haemoglobin
G106R [126]	GG/AG genotype in exon 2 that leads to a mutation in the prodomain due to which PCSK9 fails to undergo autocatalytic cleavage and causes an increase in the amount of surface LDLR
G236S [119] Q152D [24]	PCSK9 fails to exit the ER due to abnormal folding of the protein causing hypocholesterolaemia
N157K [126]	Causes hypocholesterolaemia, although studies do not exist on how the mutation causes the condition
N354I [119]	PCSK9 fails to undergo autocatalytic cleavage leading to the production of inactive protein
Arg46Leu [27,127]	Mutation in Exon 1 that leads to amino acid change of R46L and thereby to a lack of circulating PCSK9
Asp301Gly [27]	Mutation in Exon 6 that leads to amino acid change of D301G and thereby to a lack of circulating PCSK9
PCSK9-679X [97]	Elimination of final cysteine in the C-terminal domain that leads to PCSK9 failing to exit the ER after the protein folding is disrupted
PCSK9-FS [24]	C-terminal frameshift by which PCSK9 fails to exit the ER
Q152H (Gln152His) [24,128,129]	Amino acid substitution that prevents the autocatalytic processing of proPCSK9 inducing the reduction in circulating levels of PCSK9 and LDL-C, reduction in risk of developing CVD
R434W [103]	Alteration in the hinge region that impedes PCSK9 retention in the Trans-Golgi network that causes lower secretion levels of PCSK9
rs11206510 [130–132] rs11583680 (A53V) [133] rs2479409 [134] rs151193009 (R93C) [100,134]	SNPs which lead to reduced risk of CAD, peripheral artery disease, abdominal aortic aneurysm, type 2 diabetes, ischemic stroke, dementia, chronic obstructive pulmonary artery disease and cancer
rs11591147 (R46L) [126]	GT/TT genotype in exon 1 that leads to decreased levels of LDL-C and reduced risk of CVDs
S127R [118]	Mutation in pro-domain that causes low binding affinity to LDLR and increased catabolism of LDLC
S386A [126] R237W [126]	Point mutations in the catalytic domain that lead to the failure of PCSK9 to undergo autocatalytic cleavage

Table 2. Gain-of-function mutations.

Name of Mutation + Reference	Cause and Consequences
Arg499His [135] Arg496Trp [136] Asp129Gly [23] Asp374His [23] E32K (Leu108Arg) [23] D374H [137,138]	Variations in C-terminal domain that drive the intracellular degradation of LDLR Causes increased binding affinity to LDLR and hypercholesterolaemia
Asp374Tyr [103,136]	Mutation in the catalytic domain that improves the interaction of PCSK9 with the EGF-A domain of LDLR
Asp35Tyr [23]	Mutation that creates a novel Tyr-sulfation site to enhance the intracellular activity of PCSK9
D129G [103]	Mutation in pro-domain that leads to faster protein mobility from ER to Golgi faster in comparison to normal PCSK9
D374Y (rs137852912) [103,104,126,137,139,140] R496W (rs374603772) [104]	Causes 10–25-fold higher binding capacity to LDLR causing early CAD, atherosclerosis
D377Y [19]	Causes abdominal aortic aneurysm
Phe216Leu [99]	Decreases the circulating LDLR levels due to its destruction with the help of PCSK9 intracellularly
R215H [119] F216L [137,141] R218S [105]	SNPs that abolish furin cleavage
R357H [142]	Mutation in catalytic domain that leads to hypercholesterolaemia
R496Q [126]	Leads to hyperlipoproteinaemia
S386A [141] F216L [104,137,141]	Increases secretion of ApoB100-containing lipoproteins
S127R (rs28942111) [103,104,115,118,137,143,144]	Mutation in pro-domain that leads to increased binding affinity of PCSK9 to VLDLR and high circulating levels of VLDL, IDL and ApoB100-containing lipoproteins
Ser127Arg [136]	Variation in pro-domain that improves the chance of preventing LDLR from entering a closed conformation

6. PCSK9 Activators/Inhibitors

Cholesterol lowering therapies are the gold standard to reduce the risks of cardiovascular mortality and morbidity. One of the most effective ways to lower circulating cholesterol in blood is the usage of statins. Statins increase the activity of LDLRs, thus increasing the catabolism of VLDL, LDL and intermediate-density lipoprotein (IDL). Furthermore, statins also decrease the hepatic and endogenous cholesterol production by inhibiting HMG-CoA reductase [43], but they do not affect Lp(a) [145]. Interestingly, it could be demonstrated that statins also have some detrimental effects, for example by increasing the levels of circulating PCSK9 [146]. Therefore, inhibiting PCSK9 along with statin therapies has become an important and attractive addition in managing hypercholesterolaemia. In this section we will discuss the numerous studies that have identified various mechanisms to inhibit PCSK9.

Studies have identified numerous naturally occurring inhibitors and small molecule inhibitors to rescue PCSK9 induced CVDs. In the context of naturally occurring substances, berberine has the ability to exert inhibitory effects on the transcription and translation of PCSK9 [147]. Berberine is present in the roots and stems of the plant species *Berberis* and other flowering plants such as *Coptis rhizomes* and *Hydratis Canadensis* [148]. Berberine also increases the mRNA and protein levels of LDLR, independent of influencing PCSK9 [148].

Similarly, a protein-rich grain known as Lupin has been shown to decrease the levels of circulating PCSK9 and inhibits the binding of PCSK9 to LDLR [148]. Furthermore, polyphenols which are present in fruits, vegetables, seeds, herbs, tea, red wine, and nuts [148], are also involved in the inhibition of the PCSK9 protein. For example, a polyphenolic compound called resveratrol downregulates the expression of the SREBP-1c pathway and thereby downregulates the expression of PCSK9 [149]. In contrast, another polyphenol called quercetin activates the transcription of SREBP2 to upregulate the expression of the LDLR gene and reduces the expression of PCSK9 [148]. Furthermore, liraglutide is a glucagon-like peptide-1 (GLP-1) receptor agonist that has been used clinically as anti-diabetic and anti-obesity treatment, which has also been identified as a potent suppressor of PCSK9 expression, explaining at least partly its beneficial effect on CVDs [150]. Other than small molecule inhibitors, RNA aptamers also specifically bind to a target protein with high affinity. One study developed a novel RNA aptamer, called PCSK9-binding RNA (PBR), that binds to PCSK9 with a higher affinity than LDLR and thereby reduces the destruction of LDLR [151].

The administration of monoclonal antibodies (mAbs) against PCSK9 is a novel lipid-lowering therapeutic approach, that inhibits the attachment of PCSK9 with LDLR and LDLR-like receptors [152]. They act on the PCSK9 protein itself rather than targeting the gene expression. Additionally, PCSK9 mAb decreases the production rate of hepatic Lp(a) particles [145] on top of increasing the clearance of them [78]. Thus, the use of mAbs have been highly sought out for the treatment of cardiovascular complications. The inhibitors currently in use to hinder the interaction between PCSK9 and LDLR are alirocumab and evolocumab [1]. Both these inhibitors are used alone or in combination with standard lipid-lowering therapies in adults to reduce the risk of several CVDs such as MI, stroke and atherosclerosis [153], and are administered subcutaneously. Long term administration of mAbs against PCSK9 is safe and results in a reduction of inflammation, arterial wall plaques and the risk of cardiovascular events [154]. Even though the efficacy of mAbs is profound, they require a subcutaneous injection once or twice a month and are therefore still rather expensive. The limitation of frequent administration, high production costs and oral unavailability is one of the major points that should be tackled in the upcoming year to enable the large-scale use of mAbs in the clinic.

As an alternative to mAbs, creating an immune response against PCSK9 through vaccination can be used to provide protection against hyperlipidaemia. In fact, several groups have already developed extremely potent vaccines against PCSK9. For instance, PCSK9Qβ-003 is made of Qβ viral particles to decrease total cholesterol and LDL-C levels in $Apoe^{-/-}$ mice that eventually reduces the atherosclerotic lesion sizes and makes the plaques more stable [155]. AT04A is another peptide-based vaccine that generates PCSK9 specific antibodies in mice that have the ability to bind to PCSK9 thereby removing it from the circulation [156]. Peptide based vaccines need to be conjugated with carrier proteins in order to cause sufficient immune responses and they fail to promote recognition of B cells. To rectify these disadvantages, a vaccine that can induce large titres of PCSK9 antibody called the PCSK9 multicopy display nano-vaccine (PMCDN) was developed. This nano-vaccine is constructed by self-assembled carrier proteins, passes through lymph nodes into circulation easily and improves endocytosis of PCSK9 in murine models [157]. Additionally, on the surface of nanoliposomes, a peptide to induce high IgG antibodies called the immunogenic fused PCSK9 tetanus has been developed. This vaccine formulation, termed as Nano-liposomal Immunogenic Fused PCSK9-Tetanus with alum vaccine adjuvant (L-IFPTA+) was tested in mice with dyslipidaemia, where it inhibits the interaction between PCSK9 and LDLR and reduces the progression of atherosclerosis with long term immunity effects and efficacy [158].

Most small molecule inhibitors, mABs and vaccines target extracellular PCSK9, while there are also ways to inhibit PCSK9 intracellularly via modulation of the gene expression [159]. Initially, antisense oligonucleotides (ASOs) were first used in murine and monkey models to inhibit the translation of the PCSK9 mRNA, but studies were terminated due to unknown reasons [160]. Nevertheless recently, a group has developed a highly potent ASO to be orally administered and it was observed that repeated daily dosing in rats, dogs or monkeys reduces dyslipidaemia with great efficacy [161]. Another way to suppress the mRNA is to silence them using small interfering RNA (siRNA) that can be administered via lipoid nanoparticles, for which clinical trials have been successfully carried out. For example, Inclisiran (ALN-PCS) is such synthetic siRNA that specifically targets the synthesis of hepatic PCSK9 to act as a lipid-lowering therapy [162]. ALN-PCS is a synthetic siRNA drug developed by Alnylam Pharmaceuticals in the USA that can be delivered via intravenous administration into the bloodstream [163]. It inhibits the transcription of the PCSK9 gene and thus successfully reduces the plasma PCSK9 and LDL-C levels after just a single dose. Even though the first clinical trials have reported an effective clinical benefit, long term studies are yet to be conducted to assess the safety and efficacy [163]. Furthermore, adenine base editors and CRISPR adenine base editors can be used to insert a splice site mutation in the PCSK9 gene to inhibit and knockdown PCSK9 for therapeutic applications [164,165]. This has already been studied in mouse and macaque models, in order to report that this method can be fitting to treat patients with familial hypercholesterolaemia in the future [164]. PCSK9 expression can also be affected and controlled by influencing various factors involved in its production, like transcription factors. Recently it was also identified that several miRNAs can influence the PCSK9 gene expression. For example, it could be shown that miR-337-3p [166] and miR-483 [167] inhibit the transcription and translation of PCSK9 and thereby promote the uptake of LDL-C, whilst miR-552-3p enhances LDLR protein levels resulting in reduced LDL-C levels [168].

A myriad of studies have discovered synthetic and natural compounds to inhibit the gene expression of PCSK9 aside from hindering the activity of this protein. Despite the extensive research available on monoclonal antibodies to inhibit PCSK9, the end-cost makes it difficult to put this into large-scale clinical use. The development of novel approaches to silence the mRNA of PCSK9 as well as developing immunity against PCSK9 using vaccines looks extremely promising, though large clinical trials are still needed to confirm its efficacy. Further investigation on these novel therapeutic approaches might unveil a big leap in the field of cardiovascular biology.

7. Conclusions

PCSK9 exerts its effect on the cardiovascular system via the degradation of LDLR and by multiple mechanisms that are independent of LDLR. We have already a rather good understanding of the influence of PCSK9 on vascular biology, especially via its LDLR dependent effect, although it is clear that there are still several effects that are yet to be discovered. It has already been shown that PCSK9 inhibitors can be used to reduce cardiovascular complications in patients with well controlled LDL-C plasma levels. Therapeutically, PCSK9 antibodies inhibit the interaction between the EGF-A domain of the LDLR and PCSK9 and reduce LDL-C levels in plasma and have appeared to be used safely in clinical context. However, it remains rather unclear what the effects of PCSK9 manipulation are on LDLR-independent processes, which should therefore be an important focus point of future research. Additionally, based on the effects of the LOF mutations of PCSK9, manipulation of its gene expression might also be an additional approach to treat CVDs.

Author Contributions: Writing—original draft and figure preparation, S.S.S.; review and editing, Y.D. and E.P.C.v.d.V. All authors have read and agreed to the published version of the manuscript.

Funding: This research was funded by a grant from the Interdisciplinary Center for Clinical Research within the faculty of Medicine at the RWTH Aachen University, the DZHK (German Centre for Cardiovascular Research) and the BMBF (German Ministry of Education and Research), and NWO-ZonMw Veni (91619053) to E.P.C.v.d.V.; by the DZHK (German Centre for Cardiovascular Research) and the BMBF (German Ministry of Education and Research) (81X3600216) to Y.D.

Institutional Review Board Statement: Not applicable.

Informed Consent Statement: Not applicable.

Data Availability Statement: Not applicable.

Conflicts of Interest: The authors declare no conflict of interest.

References

1.	Seidah, N.; Awan, Z.; Chrétien, M.; Mbikay, M. PCSK9. *Circ. Res.* **2014**, *114*, 1022–1036. [CrossRef]
2.	Ragusa, R.; Basta, G.; Neglia, D.; De Caterina, R.; Del Turco, S.; Caselli, C. PCSK9 and atherosclerosis: Looking beyond LDL regulation. *Eur. J. Clin. Investig.* **2021**, *51*, e13459. [CrossRef]
3.	Leblond, F.; Seidah, N.G.; Précourt, L.-P.; Delvin, E.; Dominguez, M.; Levy, E. Regulation of the proprotein convertase subtilisin/kexin type 9 in intestinal epithelial cells. *Am. J. Physiol. Liver Physiol.* **2009**, *296*, G805–G815. [CrossRef]
4.	Luquero, A.; Badimon, L.; Borrell-Pages, M. PCSK9 Functions in Atherosclerosis Are Not Limited to Plasmatic LDL-Cholesterol Regulation. *Front. Cardiovasc. Med.* **2021**, *8*, 639727. [CrossRef] [PubMed]
5.	Langhi, C.; Le May, C.; Gmyr, V.; Vandewalle, B.; Kerr-Conte, J.; Krempf, M.; Pattou, F.; Costet, P.; Cariou, B. PCSK9 is expressed in pancreatic δ-cells and does not alter insulin secretion. *Biochem. Biophys. Res. Commun.* **2009**, *390*, 1288–1293. [CrossRef]
6.	Rohrbach, S.; Li, L.; Novoyatleva, T.; Niemann, B.; Knapp, F.; Molenda, N.; Schulz, R. Impact of PCSK9 on CTRP9-Induced Metabolic Effects in Adult Rat Cardiomyocytes. *Front. Physiol.* **2021**, *12*, 593862. [CrossRef] [PubMed]
7.	Artunc, F. Kidney-derived PCSK9—a new driver of hyperlipidemia in nephrotic syndrome? *Kidney Int.* **2020**, *98*, 1393–1395. [CrossRef]
8.	O'Connell, E.M.; Lohoff, F.W. Proprotein Convertase Subtilisin/Kexin Type 9 (PCSK9) in the Brain and Relevance for Neuropsychiatric Disorders. *Front. Neurosci.* **2020**, *14*, 609. [CrossRef] [PubMed]
9.	Persson, L.; Cao, G.; Ståhle, L.; Sjöberg, B.G.; Troutt, J.S.; Konrad, R.J.; Gälman, C.; Wallén, H.; Eriksson, M.; Hafström, I.; et al. Circulating Proprotein Convertase Subtilisin Kexin Type 9 Has a Diurnal Rhythm Synchronous with Cholesterol Synthesis and Is Reduced by Fasting in Humans. *Arterioscler. Thromb. Vasc. Biol.* **2010**, *30*, 2666–2672. [CrossRef] [PubMed]
10.	Cui, Q.; Ju, X.; Yang, T.; Zhang, M.; Tang, W.; Chen, Q.; Hu, Y.; Haas, J.V.; Troutt, J.S.; Pickard, R.T.; et al. Serum PCSK9 is associated with multiple metabolic factors in a large Han Chinese population. *Atherosclerosis* **2010**, *213*, 632–636. [CrossRef]
11.	Zhang, Z.; Wei, T.-F.; Zhao, B.; Yin, Z.; Shi, Q.-X.; Liu, P.-L.; Liu, L.-F.; Zhao, J.-T.; Mao, S.; Rao, M.-M.; et al. Sex Differences Associated With Circulating PCSK9 in Patients Presenting With Acute Myocardial Infarction. *Sci. Rep.* **2019**, *9*, 3113. [CrossRef] [PubMed]
12.	Ferri, N.; Ruscica, M.; Coggi, D.; Bonomi, A.; Amato, M.; Frigerio, B.; Sansaro, D.; Ravani, A.; Veglia, F.; Capra, N.; et al. Sex-specific predictors of PCSK9 levels in a European population: The IMPROVE study. *Atherosclerosis* **2020**, *309*, 39–46. [CrossRef]
13.	Vlachopoulos, C.; Terentes-Printzios, D.; Georgiopoulos, G.; Skoumas, I.; Koutagiar, I.; Ioakeimidis, N.; Stefanadis, C.; Tousoulis, D. Prediction of cardiovascular events with levels of proprotein convertase subtilisin/kexin type 9: A systematic review and meta-analysis. *Atherosclerosis* **2016**, *252*, 50–60. [CrossRef] [PubMed]
14.	Cao, Y.-X.; Jin, J.-L.; Sun, D.; Liu, H.-H.; Guo, Y.-L.; Wu, N.-Q.; Xu, R.-X.; Zhu, C.-G.; Dong, Q.; Sun, J.; et al. Circulating PCSK9 and cardiovascular events in FH patients with standard lipid-lowering therapy. *J. Transl. Med.* **2019**, *17*, 367. [CrossRef] [PubMed]
15.	Navarese, E.P.; Kołodziejczak, M.; Winter, M.-P.; Alimohammadi, A.; Lang, I.M.; Buffon, A.; Lip, G.Y.; Siller-Matula, J.M. Association of PCSK9 with platelet reactivity in patients with acute coronary syndrome treated with prasugrel or ticagrelor: The PCSK9-REACT study. *Int. J. Cardiol.* **2017**, *227*, 644–649. [CrossRef]
16.	Bordicchia, M.; Spannella, F.; Ferretti, G.; Bacchetti, T.; Vignini, A.; Di Pentima, C.; Mazzanti, L.; Sarzani, R. PCSK9 is Expressed in Human Visceral Adipose Tissue and Regulated by Insulin and Cardiac Natriuretic Peptides. *Int. J. Mol. Sci.* **2019**, *20*, 245. [CrossRef] [PubMed]
17.	Panahi, Y.; Ghahrodi, M.S.; Jamshir, M.; Safarpour, M.A.; Bianconi, V.; Pirro, M.; Farahani, M.M.; Sahebkar, A. PCSK9 and atherosclerosis burden in the coronary arteries of patients undergoing coronary angiography. *Clin. Biochem.* **2019**, *74*, 12–18. [CrossRef] [PubMed]
18.	Liu, X.; Suo, R.; Chan, C.Z.Y.; Liu, T.; Tse, G.; Li, G. The immune functions of PCSK9: Local and systemic perspectives. *J. Cell. Physiol.* **2019**, *234*, 19180–19188. [CrossRef] [PubMed]
19.	Lu, H.; Howatt, D.A.; Balakrishnan, A.; Graham, M.J.; Mullick, A.E.; Daugherty, A. Hypercholesterolemia Induced by a PCSK9 Gain-of-Function Mutation Augments Angiotensin II–Induced Abdominal Aortic Aneurysms in C57BL/6 Mice—Brief Report. *Arterioscler. Thromb. Vasc. Biol.* **2016**, *36*, 1753–1757. [CrossRef]
20.	Yuan, F.; Guo, L.; Park, K.; Woollard, J.R.; Taek-Geun, K.; Jiang, K.; Melkamu, T.; Zang, B.; Smith, S.L.; Fahrenkrug, S.C.; et al. Ossabaw Pigs with a PCSK9 Gain-of-Function Mutation Develop Accelerated Coronary Atherosclerotic Lesions: A Novel Model for Preclinical Studies. *J. Am. Heart Assoc.* **2018**, *7*. [CrossRef] [PubMed]
21.	Hedayat, A.F.; Park, K.-H.; Kwon, T.-G.; Woollard, J.R.; Jiang, K.; Carlson, D.; Lerman, A.; Lerman, L.O. Peripheral vascular atherosclerosis in a novel PCSK9 gain-of-function mutant Ossabaw miniature pig model. *Transl. Res.* **2018**, *192*, 30–45. [CrossRef]
22.	Di Taranto, M.D.; Benito-Vicente, A.; Giacobbe, C.; Uribe, K.B.; Rubba, P.; Etxebarria, A.; Guardamagna, O.; Gentile, M.; Martín, C.; Fortunato, G. Identification and in vitro characterization of two new PCSK9 Gain of Function variants found in patients with Familial Hypercholesterolemia. *Sci. Rep.* **2017**, *7*, 1–9. [CrossRef]

23. Abifadel, M.; Guerin, M.; Benjannet, S.; Rabès, J.-P.; Le Goff, W.; Julia, Z.; Hamelin, J.; Carreau, V.; Varret, M.; Bruckert, E.; et al. Identification and characterization of new gain-of-function mutations in the PCSK9 gene responsible for autosomal dominant hypercholesterolemia. *Atherosclerosis* **2012**, *223*, 394–400. [CrossRef]
24. Lebeau, P.; Platko, K.; Al-Hashimi, A.A.; Byun, J.H.; Lhoták, Š.; Holzapfel, N.; Gyulay, G.; Igdoura, S.A.; Cool, D.R.; Trigatti, B.L.; et al. Loss-of-function PCSK9 mutants evade the unfolded protein response sensor GRP78 and fail to induce endoplasmic reticulum stress when retained. *J. Biol. Chem.* **2018**, *293*, 7329–7343. [CrossRef] [PubMed]
25. Ooi, T.C.; A Krysa, J.; Chaker, S.; Abujrad, H.; Mayne, J.; Henry, K.; Cousins, M.; Raymond, A.; Favreau, C.; Taljaard, M.; et al. The Effect of PCSK9 Loss-of-Function Variants on the Postprandial Lipid and ApoB-Lipoprotein Response. *J. Clin. Endocrinol. Metab.* **2017**, *102*, 3452–3460. [CrossRef]
26. Small, A.M.; Huffman, J.E.; Klarin, D.; Lynch, J.A.; Assimes, T.; Duvall, S.; Sun, Y.V.; Shere, L.; Natarajan, P.; Gaziano, M.; et al. PCSK9 loss of function is protective against extra-coronary atherosclerotic cardiovascular disease in a large multi-ethnic cohort. *PLoS ONE* **2020**, *15*, e0239752. [CrossRef]
27. Bayona, A.; Arrieta, F.; Rodríguez-Jiménez, C.; Cerrato, F.; Rodríguez-Nóvoa, S.; Fernández-Lucas, M.; Gómez-Coronado, D.; Mata, P. Loss-of-function mutation of PCSK9 as a protective factor in the clinical expression of familial hypercholesterolemia. *Medicine* **2020**, *99*, e21754. [CrossRef] [PubMed]
28. Iqbal, Z.; Dhage, S.; Mohamad, J.B.; Abdel-Razik, A.; Donn, R.; Malik, R.; Ho, J.H.; Liu, Y.; Adam, S.; Isa, B.; et al. Efficacy and safety of PCSK9 monoclonal antibodies. *Expert Opin. Drug Saf.* **2019**, *18*, 1191–1201. [CrossRef] [PubMed]
29. Schmidt, A.F.; Pearce, L.S.; Wilkins, J.T.; Overington, J.; Hingorani, A.; Casas, J.P. PCSK9 monoclonal antibodies for the primary and secondary prevention of cardiovascular disease. *Cochrane Database Syst. Rev.* **2017**, *4*, CD011748. [CrossRef] [PubMed]
30. Rosenson, R.S.; Hegele, R.A.; Fazio, S.; Cannon, C.P. The Evolving Future of PCSK9 Inhibitors. *J. Am. Coll. Cardiol.* **2018**, *72*, 314–329. [CrossRef]
31. Paton, D. PCSK9 inhibitors: Monoclonal antibodies for the treatment of hypercholesterolemia. *Drugs Today* **2016**, *52*, 183. [CrossRef] [PubMed]
32. Macchi, C.; Ferri, N.; Sirtori, C.R.; Corsini, A.; Banach, M.; Ruscica, M. Proprotein Convertase Subtilisin Kexin Type 9. *Am. J. Pathol.* **2021**. [CrossRef]
33. Tam, J.; Thankam, F.; Agrawal, D.K.; Radwan, M.M. Critical Role of LOX-1-PCSK9 Axis in the Pathogenesis of Atheroma Formation and Its Instability. *Heart Lung Circ.* **2021**. [CrossRef]
34. Barale, C.; Melchionda, E.; Morotti, A.; Russo, I. PCSK9 Biology and Its Role in Atherothrombosis. *Int. J. Mol. Sci.* **2021**, *22*, 5880. [CrossRef] [PubMed]
35. Wiciński, M.; Żak, J.; Malinowski, B.; Popek, G.; Grześk, G. PCSK9 signaling pathways and their potential importance in clinical practice. *EPMA J.* **2017**, *8*, 391–402. [CrossRef]
36. Saavedra, Y.G.L.; Day, R.; Seidah, N.G. The M2 Module of the Cys-His-rich Domain (CHRD) of PCSK9 Protein Is Needed for the Extracellular Low-density Lipoprotein Receptor (LDLR) Degradation Pathway. *J. Biol. Chem.* **2012**, *287*, 43492–43501. [CrossRef] [PubMed]
37. Seidah, N.G.; Prat, A. The biology and therapeutic targeting of the proprotein convertases. *Nat. Rev. Drug Discov.* **2012**, *11*, 367–383. [CrossRef] [PubMed]
38. Shapiro, M.D.; Tavori, H.; Fazio, S. PCSK9. *Circ. Res.* **2018**, *122*, 1420–1438. [CrossRef]
39. Glerup, S.; Schulz, R.; Laufs, U.; Schlüter, K.-D. Physiological and therapeutic regulation of PCSK9 activity in cardiovascular disease. *Basic Res. Cardiol.* **2017**, *112*, 1–23. [CrossRef] [PubMed]
40. Malo, J.; Parajuli, A.; Walker, S.W. PCSK9: From molecular biology to clinical applications. *Ann. Clin. Biochem. Int. J. Lab. Med.* **2019**, *57*, 7–25. [CrossRef]
41. Norata, G.D.; Tavori, H.; Pirillo, A.; Fazio, S.; Catapano, A.L. Biology of proprotein convertase subtilisin kexin 9: Beyond low-density lipoprotein cholesterol lowering. *Cardiovasc. Res.* **2016**, *112*, 429–442. [CrossRef]
42. Oleaga, C.; Hay, J.; Gurcan, E.; David, L.L.; Mueller, P.A.; Tavori, H.; Shapiro, M.D.; Pamir, N.; Fazio, S. Insights into the kinetics and dynamics of the furin-cleaved form of PCSK9. *J. Lipid Res.* **2021**, *62*, 100003. [CrossRef]
43. Nishikido, T.; Ray, K.K. Targeting the peptidase PCSK9 to reduce cardiovascular risk: Implications for basic science and upcoming challenges. *Br. J. Pharmacol.* **2021**, *178*, 2168–2185. [CrossRef]
44. Schulz, R.; Schlüter, K.-D.; Laufs, U. Molecular and cellular function of the proprotein convertase subtilisin/kexin type 9 (PCSK9). *Basic Res. Cardiol.* **2015**, *110*, 1–19. [CrossRef] [PubMed]
45. Tao, R.; Xiong, X.; DePinho, R.; Deng, C.-X.; Dong, X.C. FoxO3 Transcription Factor and Sirt6 Deacetylase Regulate Low Density Lipoprotein (LDL)-cholesterol Homeostasis via Control of the Proprotein Convertase Subtilisin/Kexin Type 9 (Pcsk9) Gene Expression. *J. Biol. Chem.* **2013**, *288*, 29252–29259. [CrossRef]
46. Awan, Z.; Dubuc, G.; Faraj, M.; Dufour, R.; Seidah, N.; Davignon, J.; Rabasa-Lhoret, R.; Baass, A. The effect of insulin on circulating PCSK9 in postmenopausal obese women. *Clin. Biochem.* **2014**, *47*, 1033–1039. [CrossRef]
47. Kappelle, P.J.; Lambert, G.; Dullaart, R.P.F. Plasma proprotein convertase subtilisin–kexin type 9 does not change during 24 h insulin infusion in healthy subjects and type 2 diabetic patients. *Atherosclerosis* **2011**, *214*, 432–435. [CrossRef]
48. Goettsch, C.; Kjolby, M.; Aikawa, E. Sortilin and Its Multiple Roles in Cardiovascular and Metabolic Diseases. *Arterioscler. Thromb. Vasc. Biol.* **2018**, *38*, 19–25. [CrossRef] [PubMed]

49. Gustafsen, C.; Kjolby, M.; Nyegaard, M.; Mattheisen, M.; Lundhede, J.; Buttenschøn, H.; Mors, O.; Bentzon, J.F.; Madsen, P.S.; Nykjaer, A.; et al. The Hypercholesterolemia-Risk Gene SORT1 Facilitates PCSK9 Secretion. *Cell Metab.* **2014**, *19*, 310–318. [CrossRef] [PubMed]
50. Lagace, T.A. PCSK9 and LDLR degradation. *Curr. Opin. Lipidol.* **2014**, *25*, 387–393. [CrossRef]
51. Cunningham, D.; Danley, D.E.; Geoghegan, K.F.; Griffor, M.C.; Hawkins, J.L.; Subashi, T.A.; Varghese, A.H.; Ammirati, M.J.; Culp, J.S.; Hoth, L.R.; et al. Structural and biophysical studies of PCSK9 and its mutants linked to familial hypercholesterolemia. *Nat. Struct. Mol. Biol.* **2007**, *14*, 413–419. [CrossRef]
52. Leren, T.P. Sorting an LDL receptor with bound PCSK9 to intracellular degradation. *Atherosclerosis* **2014**, *237*, 76–81. [CrossRef] [PubMed]
53. Shapiro, M.D.; Fazio, S. PCSK9 and Atherosclerosis—Lipids and Beyond. *J. Atheroscler. Thromb.* **2017**, *24*, 462–472. [CrossRef] [PubMed]
54. Strøm, T.B.; Tveten, K.; Leren, T.P. PCSK9 acts as a chaperone for the LDL receptor in the endoplasmic reticulum. *Biochem. J.* **2013**, *457*, 99–105. [CrossRef]
55. Moens, S.J.B.; Neele, A.E.; Kroon, J.; Van Der Valk, F.M.; Bossche, J.V.D.; Hoeksema, M.; Hoogeveen, R.M.; Schnitzler, J.G.; Baccara-Dinet, M.T.; Manvelian, G.; et al. PCSK9 monoclonal antibodies reverse the pro-inflammatory profile of monocytes in familial hypercholesterolaemia. *Eur. Heart J.* **2017**, *38*, 1584–1593. [CrossRef]
56. Tang, Z.; Jiang, L.; Peng, J.; Ren, Z.; Wei, D.; Wu, C.; Pan, L.; Jiang, Z.; Liu, L. PCSK9 siRNA suppresses the inflammatory response induced by oxLDL through inhibition of NF-κB activation in THP-1-derived macrophages. *Int. J. Mol. Med.* **2012**, *30*, 931–938. [CrossRef] [PubMed]
57. Guo, Y.; Yan, B.; Gui, Y.; Tang, Z.; Tai, S.; Zhou, S.; Zheng, X. Physiology and role of PCSK9 in vascular disease: Potential impact of localized PCSK9 in vascular wall. *J. Cell. Physiol.* **2021**, *236*, 2333–2351. [CrossRef]
58. Falk, E. Pathogenesis of Atherosclerosis. *J. Am. Coll. Cardiol.* **2006**, *47*, C7–C12. [CrossRef] [PubMed]
59. Ding, Z.; Liu, S.; Wang, X.; Deng, X.; Fan, Y.; Sun, C.; Wang, Y.; Mehta, J.L. Hemodynamic Shear Stress via ROS Modulates PCSK9 Expression in Human Vascular Endothelial and Smooth Muscle Cells and Along the Mouse Aorta. *Antioxid. Redox Signal.* **2015**, *22*, 760–771. [CrossRef] [PubMed]
60. Ding, Z.; Pothineni, N.V.K.; Goel, A.; Lüscher, T.F.; Mehta, J.L. PCSK9 and inflammation: Role of shear stress, pro-inflammatory cytokines, and LOX-1. *Cardiovasc. Res.* **2019**, *116*, 908–915. [CrossRef]
61. Liu, S.; Deng, X.; Zhang, P.; Wang, X.; Fan, Y.; Zhou, S.; Mu, S.; Mehta, J.L.; Ding, Z. Blood flow patterns regulate PCSK9 secretion via MyD88-mediated pro-inflammatory cytokines. *Cardiovasc. Res.* **2019**, *116*, 1721–1732. [CrossRef] [PubMed]
62. Tang, Z.-H.; Peng, J.; Ren, Z.; Yang, J.; Li, T.-T.; Li, T.-H.; Wang, Z.; Wei, D.-H.; Liu, L.-S.; Zheng, X.-L.; et al. New role of PCSK9 in atherosclerotic inflammation promotion involving the TLR4/NF-κB pathway. *Atherosclerosis* **2017**, *262*, 113–122. [CrossRef] [PubMed]
63. Kim, Y.U.; Kee, P.; Danila, D.; Teng, B.-B.; Gim, E.; Shim, D.-W.; Hwang, I.; Shin, O.S.; Yu, J.-W. A Critical Role of PCSK9 in Mediating IL-17-Producing T Cell Responses in Hyperlipidemia. *Immune Netw.* **2019**, *19*, e41. [CrossRef] [PubMed]
64. Zeng, J.; Tao, J.; Xi, L.; Wang, Z.; Liu, L. PCSK9 mediates the oxidative low-density lipoprotein-induced pyroptosis of vascular endothelial cells via the UQCRC1/ROS pathway. *Int. J. Mol. Med.* **2021**, *47*, 1. [CrossRef]
65. Ferri, N.; Tibolla, G.; Pirillo, A.; Cipollone, F.; Mezzetti, A.; Pacia, S.; Corsini, A.; Catapano, A.L. Proprotein convertase subtilisin kexin type 9 (PCSK9) secreted by cultured smooth muscle cells reduces macrophages LDLR levels. *Atherosclerosis* **2012**, *220*, 381–386. [CrossRef]
66. Ding, Z.; Liu, S.; Wang, X.; Mathur, P.; Dai, Y.; Theus, S.; Deng, X.; Fan, Y.; Mehta, J.L. Cross-Talk Between PCSK9 and Damaged mtDNA in Vascular Smooth Muscle Cells: Role in Apoptosis. *Antioxid. Redox Signal.* **2016**, *25*, 997–1008. [CrossRef]
67. Ding, Z.; Liu, S.; Wang, X.; Deng, X.; Fan, Y.; Shahanawaz, J.; Reis, R.J.S.; Varughese, K.I.; Sawamura, T.; Mehta, J.L. Cross-talk between LOX-1 and PCSK9 in vascular tissues. *Cardiovasc. Res.* **2015**, *107*, 556–567. [CrossRef]
68. Ferri, N.; Marchianò, S.; Tibolla, G.; Baetta, R.; Dhyani, A.; Ruscica, M.; Uboldi, P.; Catapano, A.L.; Corsini, A. PCSK9 knock-out mice are protected from neointimal formation in response to perivascular carotid collar placement. *Atherosclerosis* **2016**, *253*, 214–224. [CrossRef] [PubMed]
69. Adorni, M.P.; Cipollari, E.; Favari, E.; Zanotti, I.; Zimetti, F.; Corsini, A.; Ricci, C.; Bernini, F.; Ferri, N. Inhibitory effect of PCSK9 on Abca1 protein expression and cholesterol efflux in macrophages. *Atherosclerosis* **2017**, *256*, 1–6. [CrossRef] [PubMed]
70. Tavori, H.; Giunzioni, I.; Predazzi, I.M.; Plubell, D.; Shivinsky, A.; Miles, J.; DeVay, R.M.; Liang, H.; Rashid, S.; Linton, M.F.; et al. Human PCSK9 promotes hepatic lipogenesis and atherosclerosis development via apoE- and LDLR-mediated mechanisms. *Cardiovasc. Res.* **2016**, *110*, 268–278. [CrossRef] [PubMed]
71. Giunzioni, I.; Tavori, H.; Covarrubias, R.; Major, A.S.; Ding, L.; Zhang, Y.; DeVay, R.M.; Hong, L.; Fan, D.; Predazzi, I.M.; et al. Local effects of human PCSK9 on the atherosclerotic lesion. *J. Pathol.* **2016**, *238*, 52–62. [CrossRef]
72. Ricci, C.; Ruscica, M.; Camera, M.; Rossetti, L.; Macchi, C.; Colciago, A.; Zanotti, I.; Lupo, M.G.; Adorni, M.P.; Cicero, A.F.G.; et al. PCSK9 induces a pro-inflammatory response in macrophages. *Sci. Rep.* **2018**, *8*, 1–10. [CrossRef]
73. Ding, Z.; Wang, X.; Liu, S.; Zhou, S.; Kore, R.A.; Mu, S.; Deng, X.; Fan, Y.; Mehta, J.L. NLRP3 inflammasome via IL-1β regulates PCSK9 secretion. *Theranostics* **2020**, *10*, 7100–7110. [CrossRef] [PubMed]
74. Badimon, L.; Luquero, A.; Crespo, J.; Peña, E.; Borrell-Pages, M. PCSK9 and LRP5 in macrophage lipid internalization and inflammation. *Cardiovasc. Res.* **2020**. [CrossRef] [PubMed]

75. Ding, Z.; Liu, S.; Wang, X.; Theus, S.; Deng, X.; Fan, Y.; Zhou, S.; Mehta, J.L. PCSK9 regulates expression of scavenger receptors and ox-LDL uptake in macrophages. *Cardiovasc. Res.* **2018**, *114*, 1145–1153. [CrossRef] [PubMed]

76. Ojha, N.; Dhamoon, A.S. Myocardial Infarction. StatPearls. 2021. Available online: https://www.ncbi.nlm.nih.gov/books/NBK5 37076/ (accessed on 19 May 2021).

77. Zhang, Y.; Liu, J.; Li, S.; Xu, R.-X.; Sun, J.; Tang, Y.; Li, J.-J. Proprotein convertase subtilisin/kexin type 9 expression is transiently up-regulated in the acute period of myocardial infarction in rat. *BMC Cardiovasc. Disord.* **2014**, *14*, 1–7. [CrossRef] [PubMed]

78. Andreadou, I.; Tsoumani, M.; Vilahur, G.; Ikonomidis, I.; Badimon, L.; Varga, Z.V.; Ferdinandy, P.; Schulz, R. PCSK9 in Myocardial Infarction and Cardioprotection: Importance of Lipid Metabolism and Inflammation. *Front. Physiol.* **2020**, *11*, 602497. [CrossRef]

79. Yang, C.L.; Zeng, Y.D.; Hu, Z.X.; Liang, H. PCSK9 promotes the secretion of pro-inflammatory cytokines by macrophages to aggravate H/R-induced cardiomyocyte injury via activating NF-kappaB signalling. *Gen. Physiol. Biophys.* **2020**, *39*, 123–134. [CrossRef]

80. Ding, Z.; Wang, X.; Liu, S.; Shahanawaz, J.; Theus, S.; Fan, Y.; Deng, X.; Zhou, S.; Mehta, J.L. PCSK9 expression in the ischaemic heart and its relationship to infarct size, cardiac function, and development of autophagy. *Cardiovasc. Res.* **2018**, *114*, 1738–1751. [CrossRef] [PubMed]

81. Malakar, A.K.; Choudhury, D.; Halder, B.; Paul, P.; Uddin, A.; Chakraborty, S. A review on coronary artery disease, its risk factors, and therapeutics. *J. Cell. Physiol.* **2019**, *234*, 16812–16823. [CrossRef]

82. Zeller, M.; Lambert, G.; Farnier, M.; Maza, M.; Nativel, B.; Rochette, L.; Vergely, C.; Cottin, Y. PCSK9 levels do not predict severity and recurrence of cardiovascular events in patients with acute myocardial infarction. *Nutr. Metab. Cardiovasc. Dis.* **2021**, *31*, 880–885. [CrossRef] [PubMed]

83. Li, S.; Zhu, C.-G.; Guo, Y.-L.; Xu, R.-X.; Zhang, Y.; Sun, J.; Li, J.-J. The Relationship between the Plasma PCSK9 Levels and Platelet Indices in Patients with Stable Coronary Artery Disease. *J. Atheroscler. Thromb.* **2015**, *22*, 76–84. [CrossRef]

84. Wolf, A.; Kutsche, H.S.; Schreckenberg, R.; Weber, M.; Li, L.; Rohrbach, S.; Schulz, R.; Schlüter, K.-D. Autocrine effects of PCSK9 on cardiomyocytes. *Basic Res. Cardiol.* **2020**, *115*, 1–13. [CrossRef] [PubMed]

85. Nozue, T.; Hattori, H.; Ogawa, K.; Kujiraoka, T.; Iwasaki, T.; Hirano, T.; Michishita, I. Correlation between serum levels of proprotein convertase subtilisin/kexin type 9 (PCSK9) and atherogenic lipoproteins in patients with coronary artery disease. *Lipids Health Dis.* **2016**, *15*, 165. [CrossRef]

86. Yang, S.-H.; Li, S.; Zhang, Y.; Xu, R.-X.; Zhu, C.-G.; Guo, Y.-L.; Wu, N.-Q.; Qing, P.; Gao, Y.; Cui, C.-J.; et al. Analysis of the association between plasma PCSK9 and Lp(a) in Han Chinese. *J. Endocrinol. Investig.* **2016**, *39*, 875–883. [CrossRef]

87. Miñana, G.; Núñez, J.; Bayés-Genís, A.; Revuelta-López, E.; Ríos-Navarro, C.; Núñez, E.; Chorro, F.J.; López-Lereu, M.P.; Monmeneu, J.V.; Lupón, J.; et al. Role of PCSK9 in the course of ejection fraction change after ST-segment elevation myocardial infarction: A pilot study. *ESC Heart Fail.* **2020**, *7*, 118–123. [CrossRef]

88. Battineni, G.; Sagaro, G.; Chintalapudi, N.; Amenta, F.; Tomassoni, D.; Tayebati, S. Impact of Obesity-Induced Inflammation on Cardiovascular Diseases (CVD). *Int. J. Mol. Sci.* **2021**, *22*, 4798. [CrossRef]

89. Roubtsova, A.; Munkonda, M.N.; Awan, Z.; Marcinkiewicz, J.; Chamberland, A.; Lazure, C.; Cianflone, K.; Seidah, N.G.; Prat, A. Circulating Proprotein Convertase Subtilisin/Kexin 9 (PCSK9) Regulates VLDLR Protein and Triglyceride Accumulation in Visceral Adipose Tissue. *Arterioscler. Thromb. Vasc. Biol.* **2011**, *31*, 785–791. [CrossRef] [PubMed]

90. Lebeau, P.F.; Byun, J.H.; Platko, K.; Al-Hashimi, A.A.; Lhoták, Š.; MacDonald, M.E.; Mejia-Benitez, A.; Prat, A.; Igdoura, S.A.; Trigatti, B.; et al. Pcsk9 knockout exacerbates diet-induced non-alcoholic steatohepatitis, fibrosis and liver injury in mice. *JHEP Rep.* **2019**, *1*, 418–429. [CrossRef]

91. Demers, A.; Samami, S.; Lauzier, B.; Rosiers, C.D.; Sock, E.T.N.; Ong, H.; Mayer, G. PCSK9 Induces CD36 Degradation and Affects Long-Chain Fatty Acid Uptake and Triglyceride Metabolism in Adipocytes and in Mouse LiverSignificance. *Arterioscler. Thromb. Vasc. Biol.* **2015**, *35*, 2517–2525. [CrossRef] [PubMed]

92. Liu, M.; Li, W.; Wang, H.; Yin, L.; Ye, B.; Tang, Y.; Huang, C. CTRP9 Ameliorates Atrial Inflammation, Fibrosis, and Vulnerability to Atrial Fibrillation in Post-Myocardial Infarction Rats. *J. Am. Heart Assoc.* **2019**, *8*, e013133. [CrossRef] [PubMed]

93. Poggio, P.; Songia, P.; Cavallotti, L.; Barbieri, S.S.; Zanotti, I.; Arsenault, B.J.; Valerio, V.; Ferri, N.; Capoulade, R.; Camera, M. PCSK9 Involvement in Aortic Valve Calcification. *J. Am. Coll. Cardiol.* **2018**, *72*, 3225–3227. [CrossRef]

94. Perrot, N.; Valerio, V.; Moschetta, D.; Boekholdt, S.M.; Dina, C.; Chen, H.Y.; Abner, E.; Martinsson, A.; Manikpurage, H.D.; Rigade, S.; et al. Genetic and In Vitro Inhibition of PCSK9 and Calcific Aortic Valve Stenosis. *JACC Basic Transl. Sci.* **2020**, *5*, 649–661. [CrossRef]

95. Iida, Y.; Tanaka, H.; Sano, H.; Suzuki, Y.; Shimizu, H.; Urano, T. Ectopic Expression of PCSK9 by Smooth Muscle Cells Contributes to Aortic Dissection. *Ann. Vasc. Surg.* **2018**, *48*, 195–203. [CrossRef]

96. Goettsch, C.; Hutcheson, J.D.; Hagita, S.; Rogers, M.; Creager, M.D.; Pham, T.; Choi, J.; Mlynarchik, A.K.; Pieper, B.; Kjolby, M.; et al. A single injection of gain-of-function mutant PCSK9 adeno-associated virus vector induces cardiovascular calcification in mice with no genetic modification. *Atherosclerosis* **2016**, *251*, 109–118. [CrossRef] [PubMed]

97. Zhao, Z.; Tuakli-Wosornu, Y.; Lagace, T.A.; Kinch, L.; Grishin, N.V.; Horton, J.D.; Cohen, J.C.; Hobbs, H.H. Molecular Characterization of Loss-of-Function Mutations in PCSK9 and Identification of a Compound Heterozygote. *Am. J. Hum. Genet.* **2006**, *79*, 514–523. [CrossRef]

98. Fasano, T.; Cefalu', A.B.; Di Leo, E.; Noto, D.; Pollaccia, D.; Bocchi, L.; Valenti, V.; Bonardi, R.; Guardamagna, O.; Averna, M.; et al. A Novel Loss of Function Mutation of PCSK9 Gene in White Subjects with Low-Plasma Low-Density Lipoprotein Cholesterol. *Arterioscler. Thromb. Vasc. Biol.* **2007**, *27*, 677–681. [CrossRef] [PubMed]

99. El Khoury, P.; Elbitar, S.; Ghaleb, Y.; Khalil, Y.A.; Varret, M.; Boileau, C.; Abifadel, M. PCSK9 Mutations in Familial Hypercholesterolemia: From a Groundbreaking Discovery to Anti-PCSK9 Therapies. *Curr. Atherosclerosis Rep.* **2017**, *19*, 49. [CrossRef] [PubMed]

100. Lee, C.J.; Lee, Y.; Park, S.; Kang, S.-M.; Jang, Y.; Lee, J.H.; Lee, S.-H. Rare and common variants of APOB and PCSK9 in Korean patients with extremely low low-density lipoprotein-cholesterol levels. *PLoS ONE* **2017**, *12*, e0186446. [CrossRef] [PubMed]

101. Hopkins, P.N.; Krempf, M.; Bruckert, E.; Donahue, S.; Yang, F.; Zhang, Y.; DiCioccio, A.T. Pharmacokinetic and pharmacodynamic assessment of alirocumab in patients with familial hypercholesterolemia associated with proprotein convertase subtilisin/kexin type 9 gain-of-function or apolipoprotein B loss-of-function mutations. *J. Clin. Lipidol.* **2019**, *13*, 970–978. [CrossRef]

102. Kaya, E.; Kayıkçıoğlu, M.; Vardarlı, A.T.; Eroğlu, Z.; Payzın, S.; Can, L. PCSK 9 gain-of-function mutations (R496W and D374Y) and clinical cardiovascular characteristics in a cohort of Turkish patients with familial hypercholesterolemia. *Anatol. J. Cardiol.* **2017**, *18*, 266–272. [CrossRef]

103. Poirier, S.; Hamouda, H.A.; Villeneuve, L.; Demers, A.; Mayer, G. Trafficking Dynamics of PCSK9-Induced LDLR Degradation: Focus on Human PCSK9 Mutations and C-Terminal Domain. *PLoS ONE* **2016**, *11*, e0157230. [CrossRef]

104. Eroglu, Z. PCSK9 R496W (rs374603772) and D374Y (rs137852912) mutations in Turkish patients with primary dyslipidemia. *Anatol. J. Cardiol.* **2018**, *19*, 334–340. [CrossRef] [PubMed]

105. Benjannet, S.; Rhainds, D.; Hamelin, J.; Nassoury, N.; Seidah, N. The Proprotein Convertase (PC) PCSK9 Is Inactivated by Furin and/or PC5/6A. *J. Biol. Chem.* **2006**, *281*, 30561–30572. [CrossRef] [PubMed]

106. Cohen, J.C.; Boerwinkle, E.; Mosley, T.H.; Hobbs, H.H. Sequence Variations inPCSK9, Low LDL, and Protection against Coronary Heart Disease. *N. Engl. J. Med.* **2006**, *354*, 1264–1272. [CrossRef]

107. Shyamala, N.; Gundapaneni, K.K.; Galimudi, R.K.; Tupurani, M.A.; Padala, C.; Puranam, K.; Kupsal, K.; Kummari, R.; Gantala, S.R.; Nallamala, K.R.; et al. PCSK9 genetic (rs11591147) and epigenetic (DNA methylation) modifications associated with PCSK9 expression and serum proteins in CAD patients. *J. Gene Med.* **2021**, e3346. [CrossRef]

108. Benn, M.; Tybjærg-Hansen, A.; Nordestgaard, B.G. Low LDL Cholesterol by PCSK9 Variation Reduces Cardiovascular Mortality. *J. Am. Coll. Cardiol.* **2019**, *73*, 3102–3114. [CrossRef] [PubMed]

109. Langsted, A.; Nordestgaard, B.G.; Benn, M.; Tybjærg-Hansen, A.; Kamstrup, P.R. PCSK9 R46L Loss-of-Function Mutation Reduces Lipoprotein(a), LDL Cholesterol, and Risk of Aortic Valve Stenosis. *J. Clin. Endocrinol. Metab.* **2016**, *101*, 3281–3287. [CrossRef]

110. Mostaza, J.M.; Lahoz, C.; Salinero-Fort, M.A.; de Dios, O.; Castillo, E.; González-Alegre, T.; García-Iglesias, F.; Estirado, E.; Laguna, F.; Sabín, C.; et al. R46L polymorphism in the PCSK9 gene: Relationship to lipid levels, subclinical vascular disease, and erectile dysfunction. *J. Clin. Lipidol.* **2018**, *12*, 1039.e3–1046.e3. [CrossRef] [PubMed]

111. Holla, Ø.L.; Laerdahl, J.K.; Strøm, T.B.; Tveten, K.; Cameron, J.; Berge, K.E.; Leren, T.P. Removal of acidic residues of the prodomain of PCSK9 increases its activity towards the LDL receptor. *Biochem. Biophys. Res. Commun.* **2011**, *406*, 234–238. [CrossRef]

112. Abifadel, M.; Varret, M.; Rabès, J.-P.; Allard, D.; Ouguerram, K.; Devillers, M.; Cruaud, C.; Benjannet, S.; Wickham, L.; Erlich, D.; et al. Mutations in PCSK9 cause autosomal dominant hypercholesterolemia. *Nat. Genet.* **2003**, *34*, 154–156. [CrossRef] [PubMed]

113. Myocardial Infarction Genetics Consortium. Genome-wide association of early-onset myocardial infarction with single nucleotide polymorphisms and copy number variants. *Nat. Genet.* **2009**, *41*, 334–341. [CrossRef]

114. Fan, D.; Yancey, P.G.; Qiu, S.; Ding, L.; Weeber, E.J.; Linton, M.F.; Fazio, S. Self-Association of Human PCSK9 Correlates with Its LDLR-Degrading Activity. *Biochemistry* **2008**, *47*, 1631–1639. [CrossRef] [PubMed]

115. Ouguerram, K.; Chetiveaux, M.; Zair, Y.; Costet, P.; Abifadel, M.; Varret, M.; Boileau, C.; Magot, T.; Krempf, M. Apolipoprotein B100 Metabolism in Autosomal-Dominant Hypercholesterolemia Related to Mutations inPCSK9. *Arterioscler. Thromb. Vasc. Biol.* **2004**, *24*, 1448–1453. [CrossRef]

116. Norata, G.D.; Garlaschelli, K.; Grigore, L.; Raselli, S.; Tramontana, S.; Meneghetti, F.; Artali, R.; Noto, D.; Cefalu', A.B.; Buccianti, G.; et al. Effects of PCSK9 variants on common carotid artery intima media thickness and relation to ApoE alleles. *Atherosclerosis* **2010**, *208*, 177–182. [CrossRef] [PubMed]

117. Ferreira, J.P.; Xhaard, C.; Lamiral, Z.; Borges-Canha, M.; Neves, J.S.; Dandine-Roulland, C.; LeFloch, E.; Deleuze, J.; Bacq-Daian, D.; Bozec, E.; et al. PCSK9 Protein and rs562556 Polymorphism Are Associated with Arterial Plaques in Healthy Middle-Aged Population: The STANISLAS Cohort. *J. Am. Heart Assoc.* **2020**, *9*, e014758. [CrossRef] [PubMed]

118. Lalanne, F.; Lambert, G.; Amar, M.J.A.; Chétiveaux, M.; Zaïr, Y.; Jarnoux, A.-L.; Ouguerram, K.; Friburg, J.; Seidah, N., Jr.; Brewer, H.B.; et al. Wild-type PCSK9 inhibits LDL clearance but does not affect apoB-containing lipoprotein production in mouse and cultured cells. *J. Lipid Res.* **2005**, *46*, 1312–1319. [CrossRef] [PubMed]

119. Cameron, J.; Holla, Ø.L.; Laerdahl, J.K.; Kulseth, M.A.; Ranheim, T.; Rognes, T.; Berge, K.E.; Leren, T.P. Characterization of novel mutations in the catalytic domain of the PCSK9 gene. *J. Intern. Med.* **2008**, *263*, 420–431. [CrossRef]

120. Nuglozeh, E.; Fazaludeen, F.; Hasona, N.; Malm, T.; Mayor, L.B.; Al-Hazmi, A.; Ashankyty, I. Genotyping and Frequency of PCSK9 Variations Among Hypercholesterolemic and Diabetic Subjects. *Indian J. Clin. Biochem.* **2019**, *34*, 444–450. [CrossRef]

121. Allard, D.; Amsellem, S.; Abifadel, M.; Trillard, M.; Devillers, M.; Luc, G.; Krempf, M.; Reznik, Y.; Girardet, J.-P.; Fredenrich, A.; et al. Novel mutations of thePCSK9 gene cause variable phenotype of autosomal dominant hypercholesterolemia. *Hum. Mutat.* **2005**, *26*, 497. [CrossRef] [PubMed]

122. Chikowore, T.; Cockeran, M.; Conradie, K.R.; Van Zyl, T. C679X loss-of-function PCSK9 variant lowers fasting glucose levels in a black South African population: A longitudinal study. *Diabetes Res. Clin. Pract.* **2018**, *144*, 279–285. [CrossRef]

123. Sirois, F.; Gbeha, E.; Sanni, A.; Chretien, M.; Labuda, D.; Mbikay, M. Ethnic Differences in the Frequency of the Cardioprotective C679X PCSK9 Mutation in a West African Population. *Genet. Test.* **2008**, *12*, 377–380. [CrossRef]

124. Hooper, A.J.; Marais, A.D.; Tanyanyiwa, D.M.; Burnett, J.R. The C679X mutation in PCSK9 is present and lowers blood cholesterol in a Southern African population. *Atherosclerosis* **2007**, *193*, 445–448. [CrossRef]

125. Chikowore, T.; Sahibdeen, V.; Hendry, L.M.; Norris, S.; Goedecke, J.H.; Micklesfield, L.K.; Lombard, Z.; Chikowore, T. C679X loss-of-function PCSK9 variant is associated with lower fasting glucose in black South African adolescents: Birth to Twenty Plus Cohort. *J. Clin. Transl. Endocrinol.* **2019**, *16*, 100186. [CrossRef] [PubMed]

126. Cameron, J.; Holla, Ø.L.; Ranheim, T.; Kulseth, M.A.; Berge, K.E.; Leren, T.P. Effect of mutations in the PCSK9 gene on the cell surface LDL receptors. *Hum. Mol. Genet.* **2006**, *15*, 1551–1558. [CrossRef] [PubMed]

127. Rimbert, A.; Smati, S.; Dijk, W.; Le May, C.; Cariou, B. Genetic Inhibition of PCSK9 and Liver Function. *JAMA Cardiol.* **2021**, *6*, 353. [CrossRef]

128. Lebeau, P.F.; Wassef, H.; Byun, J.H.; Platko, K.; Ason, B.; Jackson, S.; Dobroff, J.; Shetterly, S.; Richards, W.G.; Al-Hashimi, A.A.; et al. The loss-of-function PCSK9Q152H variant increases ER chaperones GRP78 and GRP94 and protects against liver injury. *J. Clin. Investig.* **2021**, *131*. [CrossRef] [PubMed]

129. Chrétien, M.; Mbikay, M. 60 YEARS OF POMC: From the prohormone theory to pro-opiomelanocortin and to proprotein convertases (PCSK1 to PCSK9). *J. Mol. Endocrinol.* **2016**, *56*, T49–T62. [CrossRef]

130. Yu, Z.; Huang, T.; Zheng, Y.; Wang, T.; Heianza, Y.; Sun, D.; Campos, H.; Qi, L. PCSK9 variant, long-chain n–3 PUFAs, and risk of nonfatal myocardial infarction in Costa Rican Hispanics. *Am. J. Clin. Nutr.* **2017**, *105*, 1198–1203. [CrossRef] [PubMed]

131. Xavier, L.B.; Sóter, M.O.; Sales, M.F.; Oliveira, D.K.; Reis, H.J.; Candido, A.L.; Reis, F.M.; Silva, I.O.; Gomes, K.B.; Ferreira, C.N. Evaluation of PCSK9 levels and its genetic polymorphisms in women with polycystic ovary syndrome. *Gene* **2018**, *644*, 129–136. [CrossRef]

132. Zhang, L.; Yuan, F.; Liu, P.; Fei, L.; Huang, Y.; Xu, L.; Hao, L.; Qiu, X.; Le, Y.; Yang, X.; et al. Association between PCSK9 and LDLR gene polymorphisms with coronary heart disease: Case-control study and meta-analysis. *Clin. Biochem.* **2013**, *46*, 727–732. [CrossRef]

133. Mayne, J.; Ooi, T.C.; Raymond, A.; Cousins, M.; Bernier, L.; Dewpura, T.; Sirois, F.; Mbikay, M.; Davignon, J.; Chrétien, M. Differential effects of PCSK9 loss of function variants on serum lipid and PCSK9 levels in Caucasian and African Canadian populations. *Lipids Health Dis.* **2013**, *12*, 70. [CrossRef]

134. Luo, H.; Zhang, X.; Shuai, P.; Miao, Y.; Ye, Z.; Lin, Y. Genetic variants influencing lipid levels and risk of dyslipidemia in Chinese population. *J. Genet.* **2017**, *96*, 985–992. [CrossRef] [PubMed]

135. Sánchez-Hernández, R.M.; Di Taranto, M.D.; Benito-Vicente, A.; Uribe, K.B.; Lamiquiz-Moneo, I.; Larrea-Sebal, A.; Jebari, S.; Galicia-Garcia, U.; Nóvoa, F.J.; Boronat, M.; et al. The Arg499His gain-of-function mutation in the C-terminal domain of PCSK9. *Atherosclerosis* **2019**, *289*, 162–172. [CrossRef] [PubMed]

136. Martin, W.R.; Lightstone, F.C.; Cheng, F. In Silico Insights into Protein–Protein Interaction Disruptive Mutations in the PCSK9-LDLR Complex. *Int. J. Mol. Sci.* **2020**, *21*, 1550. [CrossRef]

137. Pandit, S.; Wisniewski, D.; Santoro, J.C.; Ha, S.; Ramakrishnan, V.; Cubbon, R.M.; Cummings, R.T.; Wright, S.D.; Sparrow, C.P.; Sitlani, A.; et al. Functional analysis of sites within PCSK9 responsible for hypercholesterolemia. *J. Lipid Res.* **2008**, *49*, 1333–1343. [CrossRef]

138. Bourbon, M.; Alves, A.C.; Medeiros, A.; da Silva, S.P.; Soutar, A. Familial hypercholesterolaemia in Portugal. *Atherosclerosis* **2008**, *196*, 633–642. [CrossRef]

139. Holla, Ø.L.; Cameron, J.; Berge, K.E.; Kulseth, M.A.; Ranheim, T.; Leren, T.P. Low?density lipoprotein receptor activity in Epstein? Barr virus?transformed lymphocytes from heterozygotes for the D374Y mutation in the PCSK9 gene. *Scand. J. Clin. Lab. Investig.* **2006**, *66*, 317–328. [CrossRef] [PubMed]

140. Fasano, T.; Sun, X.-M.; Patel, D.D.; Soutar, A.K. Degradation of LDLR protein mediated by 'gain of function' PCSK9 mutants in normal and ARH cells. *Atherosclerosis* **2009**, *203*, 166–171. [CrossRef]

141. Sun, X.-M.; Eden, E.; Tosi, I.; Neuwirth, C.K.; Wile, D.; Naoumova, R.P.; Soutar, A.K. Evidence for effect of mutant PCSK9 on apolipoprotein B secretion as the cause of unusually severe dominant hypercholesterolaemia. *Hum. Mol. Genet.* **2005**, *14*, 1161–1169. [CrossRef] [PubMed]

142. Alghamdi, R.H.; O'Reilly, P.; Lu, C.; Gomes, J.; Lagace, T.A.; Basak, A. LDL-R promoting activity of peptides derived from human PCSK9 catalytic domain (153–421): Design, synthesis and biochemical evaluation. *Eur. J. Med. Chem.* **2015**, *92*, 890–907. [CrossRef] [PubMed]

143. Homer, V.M.; Marais, A.D.; Charlton, F.; Laurie, A.D.; Hurndell, N.; Scott, R.; Mangili, F.; Sullivan, D.R.; Barter, P.J.; Rye, K.-A.; et al. Identification and characterization of two non-secreted PCSK9 mutants associated with familial hypercholesterolemia in cohorts from New Zealand and South Africa. *Atherosclerosis* **2008**, *196*, 659–666. [CrossRef]

144. Si-Tayeb, K.; Idriss, S.; Champon, B.; Caillaud, A.; Pichelin, M.; Arnaud, L.; Lemarchand, P.; Le May, C.; Zibara, K.; Cariou, B. Urine-sample-derived human induced pluripotent stem cells as a model to study PCSK9-mediated autosomal dominant hypercholesterolemia. *Dis. Model. Mech.* **2015**, *9*, 81–90. [CrossRef] [PubMed]

145. Ying, Q.; Chan, D.C.; Watts, G.F. New Insights into the Regulation of Lipoprotein Metabolism by PCSK9: Lessons From Stable Isotope Tracer Studies in Human Subjects. *Front. Physiol.* **2021**, *12*, 603910. [CrossRef] [PubMed]

146. Katzmann, J.L.; Gouni-Berthold, I.; Laufs, U. PCSK9 Inhibition: Insights from Clinical Trials and Future Prospects. *Front. Physiol.* **2020**, *11*, 595819. [CrossRef]

147. Cameron, J.; Ranheim, T.; Kulseth, M.A.; Leren, T.P.; Berge, K.E. Berberine decreases PCSK9 expression in HepG2 cells. *Atherosclerosis* **2008**, *201*, 266–273. [CrossRef]

148. Adorni, M.P.; Zimetti, F.; Lupo, M.G.; Ruscica, M.; Ferri, N. Naturally Occurring PCSK9 Inhibitors. *Nutrients* **2020**, *12*, 1440. [CrossRef]

149. Jing, Y.; Hu, T.; Lin, C.; Xiong, Q.; Liu, F.; Yuan, J.; Zhao, X.; Wang, R. Resveratrol downregulates PCSK9 expression and attenuates steatosis through estrogen receptor α-mediated pathway in L02 cells. *Eur. J. Pharmacol.* **2019**, *855*, 216–226. [CrossRef]

150. Yang, S.-H.; Xu, R.-X.; Cui, C.-J.; Wang, Y.; Du, Y.; Chen, Z.-G.; Yao, Y.-H.; Ma, C.-Y.; Zhu, C.-G.; Guo, Y.-L.; et al. Liraglutide downregulates hepatic LDL receptor and PCSK9 expression in HepG2 cells and db/db mice through a HNF-1a dependent mechanism. *Cardiovasc. Diabetol.* **2018**, *17*, 1–10. [CrossRef]

151. Ando, T.; Yamamoto, M.; Yokoyama, T.; Horiuchi, D.; Kawakami, T. In vitro selection generates RNA aptamer that antagonizes PCSK9–LDLR interaction and recovers cellular LDL uptake. *J. Biosci. Bioeng.* **2021**, *131*, 326–332. [CrossRef] [PubMed]

152. Sabatine, M.S.; Giugliano, R.P.; Keech, A.C.; Honarpour, N.; Wiviott, S.D.; Murphy, S.A.; Kuder, J.F.; Wang, H.; Liu, T.; Wasserman, S.M.; et al. Evolocumab and Clinical Outcomes in Patients with Cardiovascular Disease. *N. Engl. J. Med.* **2017**, *376*, 1713–1722. [CrossRef] [PubMed]

153. Pokhrel, B.; Yuet, W.C.; Levine, S.N. PCSK9 Inhibitors. StatPearls. 2021. Available online: https://www.ncbi.nlm.nih.gov/books/NBK448100/ (accessed on 24 May 2021).

154. Hoogeveen, R.M.; Opstal, T.S.; Kaiser, Y.; Stiekema, L.C.; Kroon, J.; Knol, R.J.; Bax, W.A.; Verberne, H.J.; Cornel, J.; Stroes, E.S. PCSK9 Antibody Alirocumab Attenuates Arterial Wall Inflammation Without Changes in Circulating Inflammatory Markers. *JACC Cardiovasc. Imaging* **2019**, *12*, 2571–2573. [CrossRef]

155. Wu, D.; Pan, Y.; Yang, S.; Li, C.; Zhou, Y.; Wang, Y.; Chen, X.; Zhou, Z.; Liao, Y.; Qiu, Z. PCSK9Qβ-003 Vaccine Attenuates Atherosclerosis in Apolipoprotein E-Deficient Mice. *Cardiovasc. Drugs Ther.* **2021**, *35*, 141–151. [CrossRef]

156. Landlinger, C.; Pouwer, M.G.; Juno, C.; Van Der Hoorn, J.W.; Pieterman, E.J.; Jukema, J.W.; Staffler, G.; Princen, H.; Galabova, G. The AT04A vaccine against proprotein convertase subtilisin/kexin type 9 reduces total cholesterol, vascular inflammation, and atherosclerosis in APOE*3Leiden.CETP mice. *Eur. Heart J.* **2017**, *38*, 2499–2507. [CrossRef]

157. You, S.; Guo, X.; Xue, X.; Li, Y.; Dong, H.; Ji, H.; Hong, T.; Wei, Y.; Shi, X.; He, B. PCSK9 Hapten Multicopy Displayed onto Carrier Protein Nanoparticle: An Antiatherosclerosis Vaccine. *ACS Biomater. Sci. Eng.* **2019**, *5*, 4263–4271. [CrossRef] [PubMed]

158. Momtazi, A.A.; Jaafari, M.R.; Badiee, A.; Banach, M.; Sahebkar, A. Therapeutic effect of nanoliposomal PCSK9 vaccine in a mouse model of atherosclerosis. *BMC Med.* **2019**, *17*, 1–15. [CrossRef]

159. Stoekenbroek, R.M.; Lambert, G.; Cariou, B.; Hovingh, G.K. Inhibiting PCSK9—Biology beyond LDL control. *Nat. Rev. Endocrinol.* **2018**, *15*, 52–62. [CrossRef] [PubMed]

160. Sobati, S.; Shakouri, A.; Edalati, M.; Mohammadnejad, D.; Parvan, R.; Masoumi, J.; Abdolalizadeh, J. PCSK9: A Key Target for the Treatment of Cardiovascular Disease (CVD). *Adv. Pharm. Bull.* **2020**, *10*, 502–511. [CrossRef]

161. Gennemark, P.; Walter, K.; Clemmensen, N.; Rekić, D.; Nilsson, C.A.; Knöchel, J.; Hölttä, M.; Wernevik, L.; Rosengren, B.; Kakol-Palm, D.; et al. An oral antisense oligonucleotide for PCSK9 inhibition. *Sci. Transl. Med.* **2021**, *13*, eabe9117. [CrossRef]

162. Landmesser, U.; Haghikia, A.; A Leiter, L.; Wright, R.S.; Kallend, D.; Wijngaard, P.; Stoekenbroek, R.; Kastelein, J.J.P.; Ray, K.K. Effect of inclisiran, the small-interfering RNA against proprotein convertase subtilisin/kexin type 9, on platelets, immune cells, and immunological biomarkers: A pre-specified analysis from ORION-1. *Cardiovasc. Res.* **2021**, *117*, 284–291. [CrossRef]

163. Fitzgerald, K.; Frank-Kamenetsky, M.; Shulga-Morskaya, S.; Liebow, A.; Bettencourt, B.R.; E Sutherland, J.; Hutabarat, R.M.; A Clausen, V.; Karsten, V.; Cehelsky, J.; et al. Effect of an RNA interference drug on the synthesis of proprotein convertase subtilisin/kexin type 9 (PCSK9) and the concentration of serum LDL cholesterol in healthy volunteers: A randomised, single-blind, placebo-controlled, phase 1 trial. *Lancet* **2014**, *383*, 60–68. [CrossRef]

164. Rothgangl, T.; Dennis, M.K.; Lin, P.J.C.; Oka, R.; Witzigmann, D.; Villiger, L.; Qi, W.; Hruzova, M.; Kissling, L.; Lenggenhager, D.; et al. In vivo adenine base editing of PCSK9 in macaques reduces LDL cholesterol levels. *Nat. Biotechnol.* **2021**, 1–9. [CrossRef]

165. Musunuru, K.; Chadwick, A.C.; Mizoguchi, T.; Garcia, S.P.; DeNizio, J.E.; Reiss, C.W.; Wang, K.; Iyer, S.; Dutta, C.; Clendaniel, V.; et al. In vivo CRISPR base editing of PCSK9 durably lowers cholesterol in primates. *Nat. Cell Biol.* **2021**, *593*, 429–434. [CrossRef]

166. Xu, X.; Dong, Y.; Ma, N.; Kong, W.; Yu, C.; Gong, L.; Chen, J.; Ren, J. MiR-337-3p lowers serum LDL-C level through targeting PCSK9 in hyperlipidemic mice. *Metabolism* **2021**, *119*, 154768. [CrossRef] [PubMed]

167. Dong, J.; He, M.; Li, J.; Pessentheiner, A.; Wang, C.; Zhang, J.; Sun, Y.; Wang, W.-T.; Zhang, Y.; Liu, J.; et al. microRNA-483 ameliorates hypercholesterolemia by inhibiting PCSK9 production. *JCI Insight* **2020**, *5*. [CrossRef] [PubMed]

168. Ma, N.; Fan, L.; Dong, Y.; Xu, X.; Yu, C.; Chen, J.; Ren, J. New PCSK9 inhibitor miR-552-3p reduces LDL-C via enhancing LDLR in high fat diet-fed mice. *Pharmacol. Res.* **2021**, *167*, 105562. [CrossRef] [PubMed]

 *biomedicines*

Review

Ultrasound and Microbubbles for Targeted Drug Delivery to the Lung Endothelium in ARDS: Cellular Mechanisms and Therapeutic Opportunities

Rajiv Sanwal [1,2,†], Kushal Joshi [1,3,4,†], Mihails Ditmans [1,5,†], Scott S. H. Tsai [1,3,4] and Warren L. Lee [1,2,3,4,5,6,*]

1   Keenan Research Centre for Biomedical Science, St. Michael's Hospital, Unity Health Toronto, Toronto, ON M5B 1T8, Canada; rajiv.sanwal@mail.utoronto.ca (R.S.); kushal.joshi@ryerson.ca (K.J.); misha@ditman.ca (M.D.); scott.tsai@ryerson.ca (S.S.H.T.)
2   Department of Laboratory Medicine and Pathobiology, University of Toronto, Toronto, ON M5S 1A8, Canada
3   Department of Mechanical and Industrial Engineering, Ryerson University, Toronto, ON M5B 2K3, Canada
4   Institute of Biomedical Engineering, Science and Technology (iBEST), Toronto, ON M5B 1T8, Canada
5   Biomedical Engineering Graduate Program, Ryerson University, Toronto, ON M5B 2K3, Canada
6   Interdepartmental Division of Critical Care Medicine, University of Toronto, Toronto, ON M5S 1A1, Canada
*   Correspondence: warren.lee@unityhealth.to; Tel.: +416-864-6060 (ext. 77655)
†   Contributed equally.

Citation: Sanwal, R.; Joshi, K.; Ditmans, M.; Tsai, S.S.H.; Lee, W.L. Ultrasound and Microbubbles for Targeted Drug Delivery to the Lung Endothelium in ARDS: Cellular Mechanisms and Therapeutic Opportunities. *Biomedicines* **2021**, 9, 803. https://doi.org/10.3390/biomedicines9070803

Academic Editors: Harry Karmouty-Quintana and Alberto Ricci

Received: 19 May 2021
Accepted: 7 July 2021
Published: 12 July 2021

Abstract: Acute respiratory distress syndrome (ARDS) is characterized by increased permeability of the alveolar–capillary membrane, a thin barrier composed of adjacent monolayers of alveolar epithelial and lung microvascular endothelial cells. This results in pulmonary edema and severe hypoxemia and is a common cause of death after both viral (e.g., SARS-CoV-2) and bacterial pneumonia. The involvement of the lung in ARDS is notoriously heterogeneous, with consolidated and edematous lung abutting aerated, less injured regions. This makes treatment difficult, as most therapeutic approaches preferentially affect the normal lung regions or are distributed indiscriminately to other organs. In this review, we describe the use of thoracic ultrasound and microbubbles (USMB) to deliver therapeutic cargo (drugs, genes) preferentially to severely injured areas of the lung and in particular to the lung endothelium. While USMB has been explored in other organs, it has been under-appreciated in the treatment of lung injury since ultrasound energy is scattered by air. However, this limitation can be harnessed to direct therapy specifically to severely injured lungs. We explore the cellular mechanisms governing USMB and describe various permutations of cargo administration. Lastly, we discuss both the challenges and potential opportunities presented by USMB in the lung as a tool for both therapy and research.

Keywords: acute respiratory distress syndrome; ultrasound; microbubbles; endothelial cells; vascular leak; drug and gene delivery

1. Introduction

Endothelial cells line the entire vascular system and are the interface between blood and tissue [1]. A primary role of the microvascular endothelium is to regulate the flux of molecules between the vascular lumen and the surrounding tissue parenchyma. A characteristic feature of the endothelium is heterogeneity between different-sized vessels (e.g., macrovascular, such as the aorta, vs. microvascular in capillaries) and between different tissues. For instance in the lung, microvascular endothelial cell permeability is relatively low and tightly regulated by cell junctional complexes, of which major protein components include vascular endothelial (VE)-cadherin, claudin-5 and occludins. By comparison, the endothelial cells in the liver sinusoids have intercellular gaps that permit the free movement of fluid and circulating components in and out of the vascular lumen.

The role of endothelial dysfunction and increased permeability in the pathogenesis of inflammatory disease has been appreciated for decades [2], but interest in its potential modulation as a therapy has grown with the COVID-19 pandemic; there has also been a broader realization of the importance of endothelial leakage in sepsis and lung injury [3,4]. Therapies that decrease endothelial permeability would be of particular interest in the lung, where acute respiratory distress syndrome (ARDS, also called acute lung injury) is characterized by disrupted alveolar endothelial and epithelial barriers, leading to pulmonary edema and arterial hypoxemia [5]. It occurs in nearly 10% of all intensive care unit (ICU) admissions and results in a mortality rate of up to 40% despite the best supportive care [6]. ARDS is usually caused by an excessive host inflammatory response to an infectious (e.g., pneumonia, sepsis) or non-infectious insult (e.g., trauma), resulting in endothelial and/or epithelial damage and fluid extravasation [7].

Treatment of ARDS is difficult due to an incomplete understanding of its pathophysiology but also because of an inability to target severely injured regions of the lung. In this article, we first briefly review some fundamental determinants of endothelial activation and permeability; subsequently, we focus our discussion on how the combination of thoracic ultrasound and intravenously administered microbubbles (USMB) may permit preferential delivery of therapeutic cargoes to the lung endothelium in the most damaged regions of the lung [3,8].

1.1. Endothelial Activation and Leakage during Inflammation

1.1.1. Pulmonary Endothelial Inflammation and Vascular Permeability

Endothelial cells undergo phenotypic changes after the binding of pro-inflammatory cytokines such as interleukin-1β (IL-1β) and tumor necrosis factor-α (TNF-α) to their receptors [9]. The most notable changes are an upregulation of cellular adhesion molecules such as intercellular adhesion molecules (ICAM), vascular cell adhesion molecules (VCAM), and endothelial cell-leukocyte adhesion molecules (ELAM), which facilitate the binding of leukocytes [10,11]. ICAM-1 has been well documented to be increased in the serum and lung tissue of ARDS patients [12,13]. Its knockout also showed an attenuation of edema in a septic mouse model attributed to impaired leukocyte recruitment to the site of injury [14]. The levels of E-selectin, the endothelial selectin, are also elevated in ARDS patients [15]. Despite their involvement in the disease, however, therapies targeting the selectins have yielded mixed results. Baboons with sepsis-induced ARDS did not respond to anti-E-selectin and L-selectin antibody therapy, despite an antibody targeting both selectins being protective in pigs [16,17].

Endothelial cells also contribute to inflammation by secreting pro-inflammatory cytokines themselves, directing immune cells to the site of injury and leading to an increase in vascular leakage [18,19]. Increased cytokines have been found in the alveolar lavage fluid of ARDS patients [20,21] and, while necessary for the immune response, may contribute to tissue injury. Cytokines such as IL-6 can induce endothelial permeability [22]; others, such as IL-1β and TNF-α, are potent chemoattractants for leukocytes. The sensing of endogenous or exogenous inflammatory signals (e.g., bacterial lipopolysaccharide, viral nucleic acid) triggers the formation and activation of the inflammasome in both immune cells and lung parenchymal tissue. The inflammasome is a multiprotein signaling complex that catalyzes cleavage and release of IL-1β and IL-18 and has been shown to contribute to ARDS (as reviewed in [23]); inflammasomes consist of sensor, adaptor and effector components such as the NOD-like receptor, pyrin domain containing 3 (NLRP3) protein which detects intracellular stress (as reviewed in [24]). Activation of the NLRP3-inflammasome has been postulated to contribute to lung injury in COVID-19, and blockade of pro-inflammatory cytokines with antibodies is a promising therapeutic strategy [25–29].

Excessive inflammation, as seen in ARDS, causes barrier disruption and fluid leakage into the interstitial space. This is mediated through various mechanisms including the previously mentioned cytokines, an increase in reactive oxygen species production by activated leukocytes and the production of neutrophil extracellular traps (NETs) (e.g., in patients suffering from critical illness due to COVID-19) [30–32].

Much of the barrier integrity in the continuous microvascular endothelium is conferred by the adherens junctions, in which homophilic interactions between VE-cadherin hold adjacent cells together tightly [33,34] (Figure 1). A decrease in cell-surface VE-cadherin is an important step for the extravasation of leukocytes through selective tyrosine phosphorylation and internalization [11,35,36]. In the context of lung injury, a decrease in VE-cadherin is mediated by multiple signals, including TNF-α, thrombin, and vascular endothelial growth factor (VEGF) [34]. Other non-inflammatory signaling molecules can decrease permeability. Angiopoietin-2 is a well-characterized negative regulatory ligand for receptor tyrosine kinase Tie2, inhibiting its phosphorylation and increasing endothelial permeability [37]. Levels of angiopoietin-2 act as a marker of clinical outcome and injury severity in trauma and sepsis patients [38,39]. The angiogenic cytokine VEGF is also higher in critically ill patients; paradoxically, while higher VEGF levels are associated with a poor prognosis, it is not predictive of lung edema [39,40].

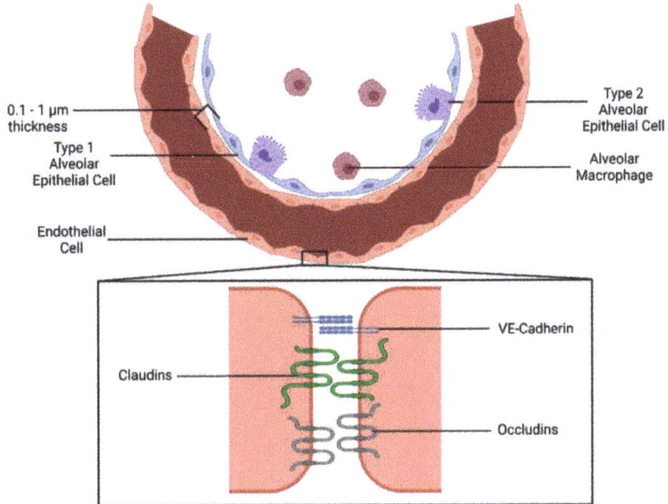

Figure 1. Anatomy of the alveolus. The alveolus is the basic functional unit of the lung. An extremely thin alveolar wall allows for gas exchange between air and blood. The outer layer is composed of pulmonary epithelial cells (also known as pneumocytes), of which there are two types. Type I epithelial cells permit gas exchange, while type II epithelial cells produce pulmonary surfactant. Pulmonary endothelial cells form the inner layer of the wall and line the pulmonary microvasculature. Endothelial cells form a continuous layer formed through tight and adherens cellular junctions between neighboring cells, thus preventing fluid leakage out of the lung. Vascular endothelial cadherin (VE-cadherin), claudins, and occludins form these junctions through homophilic and/or heterophilic interactions, restricting junction width to 2–5 nm. Created with BioRender.com.

More recently, additional players have been identified as being involved in the disrupted lung endothelium in ARDS [41]. The forkhead box protein M1, an important transcription factor involved in cell proliferation, is key in repairing the pulmonary endothelium in multiple models of murine lung injury [42,43]. Various microRNAs (miRs) are also of interest due to their crucial roles in the regulation of the expression of multiple genes: miR-150 levels are inversely correlated with the severity of disease in ARDS patients, and it has been demonstrated to decrease the severity of lung injury by repairing cellular junctions [44,45]. miR-26a-5p, through the reduction in connective tissue growth factor, protected mice with lung injury from lipopolysaccharide (LPS, i.e., endotoxin) by decreasing the severity of inflammation [46]. Finally, fibroblast growth factor 2 and phospholipase D2 have been reported to stabilize VE-cadherin—the former by inhibiting pro-inflammatory pathways, and the latter by preventing VE-cadherin phosphorylation—thereby improving pulmonary endothelial barrier permeability [47,48].

Despite this accruing knowledge as to the regulation of lung microvascular endothelial permeability, there has been relatively little progress in its therapeutic manipulation.

1.1.2. Pulmonary Endothelial Dysfunction—Coagulation

While endothelial cells normally inhibit coagulation, when activated they initiate the coagulation cascade by the expression of tissue factor. Ultimately, the cascade produces thrombin and fibrin. Tissue factor-induced coagulation is a significant contributor to mortality in critical illness and play a large role in ARDS [49,50]. In addition to being a direct pro-coagulant, thrombin interferes with the endothelial barrier and exacerbates vascular leakage through binding to its receptor proteinase-activated receptor 1 (PAR1) [51]. Activated endothelial cells also exhibit an inhibition of their normally pro-fibrinolytic state due to an upregulation of the plasminogen activator inhibitor-1 (PAI-1) [52]. Indeed, endothelial cells collected from the lungs of ARDS patients demonstrate an increase in both thrombin and PAI-1 [53]. Of note, microvascular thrombosis has been recognized as playing a role in the pathology of patients with severe COVID-19 [54]. A study comparing ARDS lung autopsies between COVID-19 patients and influenza patients found a nine-fold higher incidence of microvascular thrombosis in the lungs of COVID-19 patients [55].

1.2. Current Therapies Targeting the Endothelium

Given the central role of the lung microvascular endothelium in the pathogenesis of ARDS and sepsis, numerous approaches to modulate endothelial cell activation and decrease vascular leakage are being investigated [56]. For example, the sphingolipid sphingosine-1-phosphate (S1P) is a potent enhancer of the endothelial barrier while also acting to regulate systemic cytokine levels [22,57]. Very recently, Akhter et al. used an endothelial-specific S1P receptor-1 (S1PR1) knockout animal model to investigate the role of S1P in pulmonary endothelial regeneration in an endotoxemia injury model [58]. They found that S1P1R deletion resulted in increased endothelial permeability. Furthermore, they determined that S1PR1-expressing endothelial cells are required for barrier repair, and that these cells were reprogrammed to increase the production of S1P, suggesting therapeutic potential for the S1P pathway in endothelial injury.

There has also been extensive interest in targeting the angiopoietin receptor Tie2 and its signaling axis as a means to restore the endothelial barrier. For example, the Tie2 agonist vasculotide was shown to decrease vascular leakage, tissue edema and organ dysfunction in multiple animal models of sepsis or lung injury [59–61]. The simultaneous sequestration of the inhibitory ligand angiopoietin-2 and the activation of Tie2 was shown to restore endothelial barrier integrity and reduce mortality in septic animals [62]. Targeting the Tie2 pathway combined with decreasing circulating VEGF rescued septic mice by reducing ICAM-1 levels, decreasing inflammation, and improving endothelial barrier strength [63]. Increasing the activity of Tie2 by means of the inhibition of VE-protein tyrosine phosphatase, a Tie2-inactivating protein, also enhanced the endothelial barrier in the lungs and retinas of mice [64,65].

More generally, it is important to point out that a number of groups have now independently shown that decreasing vascular leakage can be accomplished without impairing innate immunity, highlighting the fact that tissue edema and leukocyte recruitment can be controlled separately [2].

Despite these conceptual and therapeutic advances, targeting the endothelium for therapy remains challenging, especially in the injured lung. Nanoparticle-enhanced delivery systems are the subject of intense research [66]. Inhalation of nanoparticles did show transit into the vascular space, suggesting that the endothelial barrier is reached [67]. The specific composition of nanoparticles can regulate their delivery, as demonstrated by one group showing the specificity of their nanoparticles for the alveolar endothelium, with or without cargo [68–70]. To increase endothelial specificity even further, antibodies and ligands for endothelial-specific receptors have been added to the nanoparticles. For example, peptides against sialic acid, ephrin type-A receptor 2, and ICAM have all been described [71–74].

Nonetheless, a major issue is how to preferentially deliver therapeutic cargo to the most injured regions of the lung in ARDS. This is particularly problematic given the heterogeneity of the lung in ARDS, in which severely consolidated areas in the lungs of a given patient may be abutting relatively aerated regions [75,76]. Intravenous administration of drugs (including nanoparticles) may result in undesirable off-target effects in other organs such as the kidney or liver. Hypoxic vasoconstriction (the normal narrowing of vessels in the lung in response to hypoxia) may complicate this further by decreasing the amount of drug able to reach the tissue. Inhaled agents are preferentially distributed to the best ventilated areas of the lung, potentially neglecting the most injured regions; furthermore, they make first contact with the lung epithelium, potentially decreasing access to the lung endothelium. Thus, there is a need for a delivery method which can achieve a high rate of lung endothelial delivery and preferentially targets the most injured lung areas.

2. Introduction to Ultrasound-Microbubble Mediated Therapy

Over the past two decades, ultrasound-microbubble (USMB)-mediated intracellular drug and gene delivery has emerged as a promising therapeutic approach for the treatment of the endothelium. One of the first reports on USMB-mediated intracellular delivery appeared in the late 1990s when ultrasound and commercial ultrasound imaging contrast microbubbles (Albunex®) were used for the transfection of cultured Chinese hamster ovary cells with a pGL2 luciferase reporter plasmid [77]. Since then, USMB-mediated delivery has been used for a variety of in vitro and in vivo applications such as delivering drugs across the blood–brain barrier, transfecting liver cancer cells with genes for therapy as well as delivering drugs to endothelial cells [78–81]. The principle of the method is as follows:

When circulating microbubbles are exposed to ultrasound, they grow and shrink in response to the alternating low- and high-pressure portions of the acoustic wave (i.e., cavitation). These oscillations of microbubbles exert mechanical forces on the endothelium, which increases local vascular permeability and facilitates delivery of external molecules into targeted tissues [82] (Figure 2). Depending on the intrinsic properties of the microbubbles and applied ultrasound parameters, microbubbles may undergo stable oscillations (i.e., stable cavitation) or violent oscillations and collapse (i.e., inertial cavitation) [83]. Microbubble oscillations induce acoustic microstreaming flows in the surrounding medium while violent collapse leads to the formation of high-velocity microjets and shockwaves which induce shear stresses on the endothelium, leading to intracellular and trans-endothelial delivery of drugs or plasmids [84,85]. Since microbubbles undergo cavitation only in the presence of an ultrasound field, the delivery is highly targeted towards the specific area of insonation, thus reducing off-target effects.

Microbubbles used for this application often contain a high molecular weight perfluorocarbon gas core surrounded by a protective shell made from lipids, proteins or polymers which improve their stability and lifetime [86]. The microbubbles can be co-administered intravenously along with drugs or other cargo, or can be designed to carry the therapeutic cargo inside the bubble or bound to the microbubble surface [87].

Due to its ease of application, non-invasive and highly localized nature, USMB-mediated drug/gene delivery is an attractive option for the treatment of the endothelium. The endothelium is known to play a crucial role in several vascular pathologies including arterial restenosis, arteriosclerosis, thrombosis, and hypertension [88]. To date, several research groups have demonstrated the utility of USMB-mediated delivery for endothelial treatment. For example, USMB has been utilized for successful intra-endothelial delivery of plasmid DNA encoding a phosphomimetic variant of the endothelial nitric oxide synthase (eNOS S1177D) gene, enhancing the production of nitric oxide (NO), which is a crucial mediator of anti-atherosclerotic effects [89]. Another study utilized USMB-mediated delivery of rapamycin (an antiproliferative drug) for successful inhibition of neointima formation, a condition resulting from endothelial denudation, in rat carotid arteries [90].

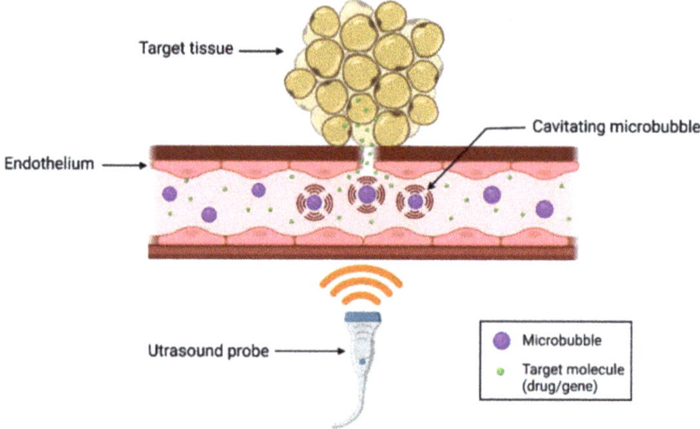

Figure 2. Basic principle of ultrasound-microbubble (USMB)-mediated drug and gene delivery. When microbubbles flowing through the blood are exposed to the ultrasound field, they undergo rapid changes in size and shape (i.e., cavitation), inducing high shear stresses on the surrounding endothelium. This increases vascular permeability and cellular internalization (e.g., endocytosis), leading to targeted delivery of drugs/genes in the area of insonation. The result is enhanced delivery of therapeutic cargo to the local endothelium itself and to the organ being perfused. Created with BioRender.com.

The cellular mechanisms of how USMB enhances drug delivery include sonoporation and endocytosis; these will be briefly reviewed here.

2.1. Mechanisms of USMB-Mediated Intracellular Delivery

2.1.1. Sonoporation

The term 'sonoporation' refers to the formation of transient pores in the cell membrane upon exposure to ultrasound-activated microbubbles. The formation of pores enhances membrane permeability, thus facilitating intracellular delivery of external molecules. Several investigations have been carried out to characterize these membrane pores. Using indirect molecular probing and direct visualization using scanning-electron microscopy (SEM), one study revealed the formation of membrane pores in the size range of 75–100 nm following USMB treatment of MAT B III cells [91]. Another study utilized transmembrane current measurements of *Xenopus laevis* oocytes for real-time monitoring of sonoporation, following treatment with ultrasound and commercial Definity® microbubbles. The results indicate the formation of membrane pores of radius 110 ± 40 nm [92]. Several other investigations have been performed to characterize the membrane pore size following USMB treatment using advanced atomic force microscopy and electron microscopy techniques [93–97].

The membrane pores formed by sonoporation are transient and reseal within a few seconds or minutes. Few researchers have investigated the temporal dynamics of membrane resealing following sonoporation by cavitating microbubbles. Using live-cell fluorescence microscopy and the fluorescent dyes Fura-2 and propidium iodide, one study observed membrane resealing in endothelial cells within 5 s of exposure to single-shot pulsed ultrasound with microbubbles [98]. Another recent investigation utilized real-time confocal fluorescence microscopy to visualize membrane resealing and observed that the resealing process could be completed with 1 min, depending on the initial pore size [99]. All these studies have provided strong visual evidence of membrane perforation and resealing following USMB treatment, suggesting sonoporation as a major mechanism of USMB-mediated cargo delivery.

Why USMB induces sonoporation has been explained by physical phenomena such as acoustic microstreaming, shock waves and microjets generated by the cavitating microbubbles in the surrounding medium; these will be briefly discussed in the next section.

Acoustic Microstreaming

Ultrasound-driven, oscillating microbubbles undergoing stable cavitation are known to generate steady vortical flows in the surrounding liquid (i.e., swirling motion of surrounding liquid), frequently referred as acoustic microstreaming [100]. It is suggested that these microstreaming flows induce a shear stress on the nearby cells resulting in tension and stretching of the membrane, thereby inducing transient membrane perforation [101,102]. One of the earliest experimental studies on microstreaming-induced sonoporation was performed with an ultrasonic horn transducer (also known as Mason horn, sonotrode, ultrasonic homogenizer or disintegrator) vibrating at 21.4 kHz inside a Jurkat lymphocyte suspension [103,104]. These experiments suggested that microstreaming flows generated by the vibrating Mason horn induce a shear stress of 12 Pa, which is sufficient to induce sonoporation in Jurkat lymphocytes upon exposure for 7 min. Theoretical calculations performed by the same research group indicated that acoustic microstreaming flows generated by microbubbles could induce shear stresses of similar or higher magnitude, thus facilitating sonoporation [101]. Using a combination of particle image velocimetry (PIV) and theoretical analysis, other researchers have shown that the shear stresses induced by acoustic microstreaming flows around microbubbles are significantly higher (~19 Pa) than the shear stresses induced by normal blood flow (~0.5–2 Pa), resulting in sonoporation [102]. These and several other theoretical and experimental studies showed that acoustic microstreaming may be a major mechanism of sonoporation [105–107]. Furthermore, microstreaming flow pattern is strongly dependent on the driving ultrasound frequency, microbubble size, pressure amplitude, properties of surrounding media as well as oscillation mode of microbubbles, which might explain why sonoporation and intracellular delivery of molecules are strongly affected by these parameters, as noted in several other studies [102,108–110].

Shock Waves and Liquid Microjets

Besides acoustic microstreaming, shockwaves and liquid microjets erupting from a collapsing microbubble undergoing inertial cavitation are suggested as possible mechanisms of sonoporation. Cavitating microbubbles undergoing symmetrical bubble collapse generate a strong shock wave in the surrounding medium which may exert a significant shear stress to perforate the membrane. Some experimental studies show that the amplitude of a shock wave generated by a single collapsing bubble can be as high as 1 GPa; however, this shock dies down rapidly and can only make an impact at distances comparable to the initial microbubble radius [100]. However, in the presence of multiple microbubbles, shockwaves emitted by several collapsing bubbles may combine and make an impact over larger distances [100]. Some molecular dynamics simulation studies have shown that a lipid bilayer membrane exposed to shockwaves may undergo compression and rebound, resulting in the formation of membrane pores during the reorganization of the lipid bilayer [111–113]. Though this study did not consider shockwaves erupting from collapsing bubbles, the mechanism may be similar for sonoporation induced by collapsing bubble shockwaves.

Fundamental fluid mechanics studies have revealed that a cavitating microbubble near a wall or a solid boundary collapses asymmetrically, ejecting fast liquid microjets towards the wall [114,115]. It has been hypothesized that when a microbubble undergoes inertial cavitation near the walls of blood vessels, the microjets ejected from asymmetrical collapse exert high shear stresses on the endothelial cells, causing membrane poration [116].

Some experimental studies conducted in the late 2000s have supported this hypothesis [94,98,117].

2.1.2. Endocytosis

For several years, sonoporation was thought to be the major mechanism of USMB-mediated intracellular delivery. However, mounting evidence now suggests that endocytosis may also play a major role. Experiments with primary endothelial cells exposed to USMB in the presence of fluorescent dextrans (4.4 to 500 kDa) have shown localization of 155 and 500 kDa dextrans in distinct vesicles and uniform distribution of 4.4 and 70 kDa dextrans throughout the cytosol [118]. Furthermore, this study showed a significant decrease in the intracellular delivery of all dextran molecules after independent inhibition of caveolin-mediated endocytosis, clathrin-mediated endocytosis and micropinocytosis pathways. Additionally, adenosine triphosphate (ATP) depletion showed a reduced uptake of 4.4 kDa dextran with no uptake of 500 kDa after USMB treatment, consistent with the inhibition of endocytosis (which requires ATP) rather than an effect on sonoporation. Using 3D fluorescence microscopy, this study also demonstrated colocalization of 500 kDa dextran vesicles with caveolin-1 and clathrin [118]. Together, this evidence suggests that endocytosis is a major mechanism of USMB-mediated intracellular delivery, especially for larger molecules.

In another study, C6 rat glioma cells were treated with either chlorpromazine (a non-specific inhibitor of clathrin-mediated endocytosis) or genistein (used to inhibit caveolae-mediated endocytosis) before exposure to USMB in the presence of the fluorescent dye SYTOX green [119]. The results showed a 2.5-fold increase in the SYTOX uptake time constant for the chlorpromazine-treated group, and a 1.1-fold increase in the uptake time constant for the genistein-treated group, indicating the dominance of the clathrin-mediated pathway in USMB-mediated endocytosis. Unfortunately, the lack of specificity of the inhibitors limits the interpretation of this study [119]. Similar results were obtained in another investigation which showed increased clathrin content per clathrin-coated pit and enhanced clathrin-mediated endocytosis within 5 min of USMB exposure [120]. Some researchers have also studied the influence of ultrasound parameters on the uptake mechanisms (i.e., sonoporation vs. endocytosis). When human melanoma cells were subjected to USMB in the presence of fluorescein isothiocyanate (FITC)-dextran at different acoustic pressures (100–500 kPa), endocytosis, as evidenced by the presence of dextrans in vesicles, appeared to be a dominant mechanism of cellular uptake at low acoustic pressures. In contrast, sonoporation appeared to be more dominant at higher acoustic pressures [121]. The dependency of uptake mechanisms on applied acoustic pressures may be linked to the acoustic behavior of microbubbles, which change at different acoustic pressures.

Biological Mechanisms of USMB-Induced Endocytosis

Several biological mechanisms of USMB-induced endocytosis have been suggested in the literature [122]. Some studies have shown that when cells are exposed to USMB or just ultrasound, a strong influx of Ca^{2+} ions is induced [118,123,124]. It is believed that the high intracellular Ca^{2+} concentration stimulates endocytosis; however, the exact mechanism is unclear [122]. It has also been suggested that the increase in reactive oxygen species (ROS) production upon exposure to ultrasound stimulates endocytosis by a ROS-dependent mechanism such as caveolae internalization due to ROS-induced caveolin-1 phosphorylation or by increasing the calcium influx [122,123,125]. However, the detailed mechanism of ROS-activated endocytosis remains elusive. According to another hypothesis, the exocytosis of lysosomes plays a crucial role in the regulation of clathrin-dependent endocytosis induced by USMB treatment [122]. The evidence for this comes from the observation that the lysosomal marker protein (Lamp-1) accumulates in the plasma membrane at the same time as USMB-induced endocytosis occurs, while the inhibition of lysosome exocytosis significantly reduces the activity of the transferrin receptor, which plays a crucial role in clathrin-dependent endocytosis [120,122]. At present, however, the precise biological mechanisms of USMB-induced endocytosis are not fully understood and more study is necessary.

Physical Mechanisms of USMB-Induced Endocytosis

Endocytosis is thought to be initiated by direct physical interactions between microbubbles and the cells or through certain biological pathways triggered by these physical interactions [122]. Though the exact physical mechanisms responsible for triggering endocytosis remain elusive, certain hypotheses have been proposed. As discussed earlier, oscillating or collapsing microbubbles produce microstreaming flows, shockwaves or liquid microjets in the surrounding liquid. The shear stresses produced by these physical phenomena may induce membrane deformations and trigger endocytosis, as noted in some earlier investigations [126,127]. A second hypothesis states that the forces exerted on the plasma membrane due to microbubble–cell interactions can be transmitted downstream to the cytoskeleton, leading to cytoskeleton remodeling [122]. This process may in turn activate mechano-sensors such as integrins or stretch-activated ion channels, triggering certain endocytotic pathways [122]. Another hypothesis suggests that endocytosis is not an independent phenomenon but is triggered by sonoporation caused by exposure to USMB [122]. According to this hypothesis, formation of lysosomal patches and exocytosis is essential for sealing large membrane pores caused by USMB treatment. The lysosomal acid sphingomyelinase released during the exocytosis process converts sphingomyelin in the plasma membrane to ceramide, leading to inward budding and formation of vesicles, which is considered to be a prelude to the endocytosis process [122]. Though multiple hypotheses exist, there is no consensus on the physical mechanisms triggering USMB-induced endocytosis and more investigations need to be performed in this area.

In summary, sonoporation and endocytosis both lead to intracellular delivery of molecules; however, it is difficult to say which one of the two dominates. The size of the molecules and the acoustic parameters are likely critical determinants [118]. Endocytosis may also be an unavoidable consequence of sonoporation and may be required for resealing of membrane pores [122]. Relative to sonoporation, endocytosis is an active and regulated pathway for the delivery of molecules and is therefore considered by some to be potentially safer for cargo delivery [122].

2.1.3. Cargo Delivery Using USMB—On, Inside, or Around?

Different approaches for delivering cargo using microbubbles have been described. Here, we will discuss the three most common approaches used for various USMB applications, namely—binding the cargo to the microbubble shell using electrostatic or covalent interactions, embedding cargo inside the microbubble shell and co-administration of cargo and microbubbles. Though the approaches discussed here are not specific to the treatment of endothelium, the same strategies could be useful for treating lung endothelial dysfunction using USMB.

Electrostatic or Covalent Binding of Cargo to the Microbubble Shell

Several attempts have been made to bind cargo (i.e., drugs or plasmids) to the microbubble shell using electrostatic or covalent interactions [128]. Due to the proximity of the cargo to the microbubble shell, it is speculated that this approach leads to enhanced delivery efficacy due to cavitation effects [129]. In addition, the release of cargo could be limited to the area of insonation, thus reducing any adverse side-effects related to free circulating drug [129]. Plasmids, which have an inherent negative charge, can be bound onto the shells of cationic lipid microbubbles using electrostatic interactions. This approach ensures that the plasmid DNA is protected from degradation by endonucleases, thereby increasing local DNA concentration for USMB-mediated transfection [130]. Cationic microbubbles bound with plasmid DNA have been used to transfect click beetle luciferase (CBLuc) plasmids into endothelial cells using USMB treatment in vitro [130]. In another investigation, researchers used cationic microbubbles loaded with short hairpin RNA interference therapy targeting prolyl hydroxylase-2 (shPHD2) plasmid for USMB-assisted myocardial transfection to protect the heart from myocardial infarction [131]. Cationic microbubbles are usually made by incorporating cationic lipids such as 1,2-dioctadecanoyl-3-trimethylammonium-propane (DSTAP) into the shells of lipid microbubbles or by coating the shell with cationic polymers such as poly-(allylamine hydrochloride) when using albumin microbubbles [130,132]. The plasmid loading capacity is limited by the surface area of microbubbles. The maximum plasmid loading capacity usually ranges from 0.001 to 0.005 pg/μm^2 [133]. In order to increase the plasmid loading capacity, a layer-by-layer construction of microbubbles with alternate plasmid (anionic) and poly-L-lysine (cationic) layers has been proposed [133]. Using this technique, the plasmid loading capacity has been reported to increase over 10-fold using five paired layers [133].

Apart from plasmids, various other cargos such as drug-loaded liposomes or nanoparticles have been bound to the surface of microbubbles. For example, doxorubicin (DOX)-loaded liposomes were covalently attached to lipid-shelled microbubbles via thiol-maleimide linkages and then used for USMB-mediated delivery to human glioblastoma cells [134]. The results indicated a four-fold decrease in cell viability with DOX-liposome loaded microbubbles compared to free DOX-liposomes or DOX alone [134].

Embedding Cargo Inside the Microbubble Shell

Similar to electrostatic or covalent binding, embedding the cargo inside the microbubble shell could have a protective effect on the cargo, preventing its early degradation and prolonging its half-life inside the body [129]. Different types of cargo have been embedded inside the microbubble shell. In one study, rapamycin, an anti-proliferative drug, was embedded inside lipidshells of microbubbles for USMB-mediated attenuation of smooth muscle cell proliferation [90]. In vitro assays indicated that the delivery efficacy of rapamycin-loaded microbubbles was significantly higher than microbubbles co-administered with free rapamycin [90]. In another investigation, plasmid DNA encoding for the LacZ gene was incorporated into albumin microbubble shell and then used for USMB-mediated vascular gene transfection [89]. It has been demonstrated that the in vitro efficiency of transfection using plasmid-DNA embedded microbubbles is significantly higher than the transfection efficiency of the co-administration of microbubbles and plasmid [135].

Though both the approaches (namely binding the cargo to the microbubble shell (Section 2.1.3) and embedding the cargo inside the microbubble shell) prevent early degradation of cargo inside the body, there is no study which directly compares the cargo delivery efficacy of these two approaches. Therefore, it is difficult to say which of these two approaches is better.

Co-Administration of Cargo and Microbubbles

Co-administration of cargo with microbubbles is one of the simplest approaches for cargo delivery using USMB. In the method, a mixture of free drugs/plasmids and microbubbles is injected into the circulation or directly at the site of insonation. For example, in a recent investigation, a mixture of lipid-coated microbubbles and propidium iodide (PI) was utilized for USMB-mediated delivery of PI into endothelial cells cultured in microfluidic channels [136]. In another study, mixtures of microbubbles and evans blue dye/fluorescent-dextran were used to study USMB-mediated drug delivery to the brain in mice [137]. A mixture of commercial SonoVue microbubbles and enhanced green fluorescent protein plasmid (pEGFP)were used to study USMB-assisted gene transfection in prostate cancer cells in vitro and in vivo in a recent investigation [138].

2.1.4. Modification of the Microbubble to Enhance Delivery

Further methods have been developed to specifically target the area of endothelial inflammation or endothelial dysfunction. By covalently attaching receptors for markers expressed by endothelial cells during inflammation to the microbubbles, the bubbles attach and remain adjacent to the cells that require therapeutic cargo delivery. Activated endothelial cells have increased expression of ICAM-1, allowing it to be used as a binding target for microbubbles [139]. Conjugation of an ICAM-1 antibody to the microbubble surface resulted in significant binding to activated endothelial cells in culture under shear flow compared to no binding to healthy endothelial cells [139]. Conjugating VCAM-1 and E-selectin antibodies to microbubbles through a biotin-streptavidin reaction has also been evaluated for binding to activated endothelial cells [140]. Microbubbles conjugated to any of the three antibodies had significantly increased binding to human endothelial cells activated with 6 h or TNF-α treatment compared to untreated cells [140]. Microscopic imaging observed improved binding of antibody-conjugated microbubbles to activated human and mouse endothelial cells following exposure to the bubbles under flow conditions compared to non-conjugated microbubbles [140]. A potential method to improve the targeting of activated endothelial cells would be dual targeting with several of the previously mentioned antibodies. While Barriero et al. demonstrated dual targeting with endothelial markers (CD9 and ICAM-1), it has not yet been tested with both markers only present on activated endothelial cells and not healthy cells [141].

Targeted microbubbles have been especially useful in delivering therapeutic cargo to injured heart tissue. Using cationic microbubbles conjugated to P-selectin antibodies for targeting ischemic areas in a murine ischemia model and bound to a plasmid coding for hVEGF165, Shentu et al. delivered genetic cargo to the cardiac endothelium [142]. The study observed enhanced deposition of the genetic cargo resulting in improvement to cardiac function compared to without the targeting ligand [142]. A similar study successfully delivered the angiopoietin-1 gene using cationic microbubbles targeted with an ICAM-1 antibody to improve endothelial function and stimulate angiogenesis in an ischemic murine model [143].

2.2. A New Frontier? Ultrasound and Microbubble Treatment of the Injured Lung

Despite the abundant literature describing ultrasound and microbubbles for cargo delivery across the blood–brain barrier or to tumors, there are very few reports of the technique in the lung. Early use of ultrasound in the lung focused on the study of pleural disease rather than lung tissue itself [144], since ultrasound is scattered by the air in healthy air-filled lung tissue. This has likely also delayed recognition of the potential of USMB to enhance drug delivery to the injured lung. In fact, lung injury or ARDS may be ideally suited to USMB-mediated therapy, since fluid leaking into the alveoli and the loss of air allows ultrasound waves to selectively penetrate the most injured lung regions [144]. Normal (aerated) or less-diseased regions of the lung would scatter the ultrasound energy, preventing or reducing off-target effects [145]. USMB is also likely to be effective for targeting densely consolidated lung (as in lobar pneumonia), even in the absence of diffuse disease (Figure 3).

The amount of penetration of ultrasound in the most injured, fluid-filled regions of the lung and normal, air-filled regions of the lung can be quantitatively estimated using acoustic impedance values of lung tissue, air and blood (Table 1) [146].

Table 1. Acoustic impedance values of lung, air, and blood.

Medium	Acoustic Impedance
Lung	0.18×10^6 Rayls
Air	0.0004×10^6 Rayls
Blood	1.65×10^6 Rayls

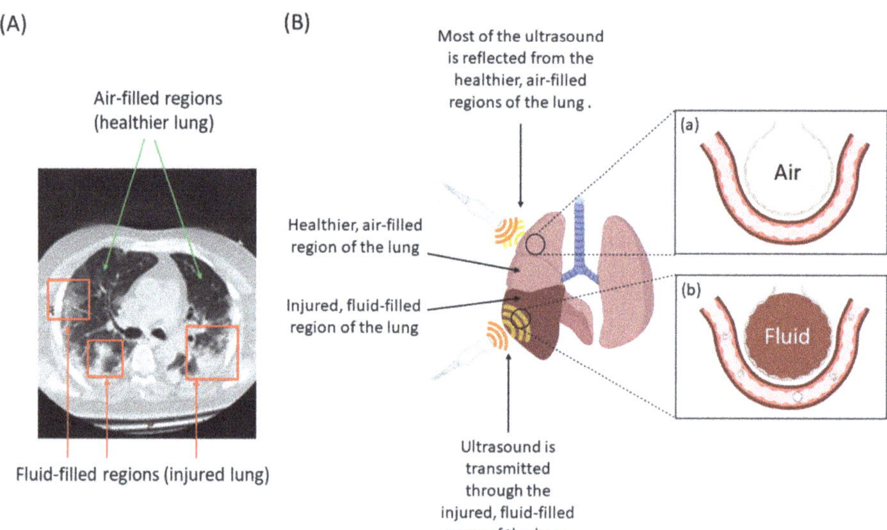

Figure 3. Courtesy of Laurent J. Brochard, St. Michael's Hospital, Toronto (**A**) Lung CT scan (cross-sectional view) from a patient with ARDS showing heterogenous distribution of injured, fluid-filled regions of the lung (**B**) USMB treatment of the lung targets the most injured and edematous regions. (**a**) Ultrasound energy is unable to penetrate the relatively healthy aerated sections of the lung, reflecting instead. Microbubbles in this region flow through the capillaries unaffected by the ultrasound waves. (**b**) Ultrasound energy is able to penetrate when the alveoli are filled with fluid in edematous or de-aerated regions, sonicating the microbubbles. Cavitation of the microbubbles is induced, resulting in cellular uptake of therapeutic cargo preferentially in the most injured area of the lungs. Created with BioRender.com.

The amount of reflection and transmission of incident ultrasound wave at the lung–air interface and lung–fluid (i.e., blood) interface can be calculated using ultrasound reflection and transmission coefficients.

$$R = \left(\frac{Z_2 - Z_1}{Z_2 + Z_1} \right)^2 \tag{1}$$

$$T = \frac{4Z_2 Z_1}{(Z_2 + Z_1)^2} \tag{2}$$

where R is the ultrasound reflection coefficient and T is the ultrasound transmission coefficient. Here, Z_1 and Z_2 are acoustic impedance values of medium 1 and medium 2, respectively, at the interface that is exposed to the ultrasound wave. For the lung–air interface, Z_1 can be assumed as the acoustic impedance of the lung while Z_2 can be assumed as the acoustic impedance of the air. Similarly, for the lung–blood interface, Z_1 can be considered as the acoustic impedance of the lung and Z_2 can be considered as the acoustic impedance of the blood.

Using the above equations, it can be observed that almost the entire incident ultrasound wave (99.1%) is reflected back at the lung–air interface, while only 0.9% penetrates in the air-filled regions of the lung. On the contrary, 35.5% of the incident ultrasound penetrates the lung–blood interface, which is significantly higher than the penetration at the lung–air interface. These numbers quantitatively show that ultrasound energy penetrates the most injured areas of lungs, showing its potential for targeted drug and gene delivery. This remarkable selectivity of USMB for the most injured regions of the lung sets it apart from other potential endothelial-targeted therapeutic strategies such as nanoparticles or microparticles.

The first instance of USMB treatment for the delivery of therapeutic cargo to injured lung tissue was recently published by our group (2018), demonstrating delivery of the aminoglycoside antibiotic gentamycin in an E.coli-induced murine pneumonia model [147]. The study demonstrated an almost ten-fold reduction in bacterial colony-forming units following USMB treatment with gentamycin compared to gentamycin alone, an intriguing finding given that aminoglycoside antibiotics do not normally distribute well to the lung [147]. In fact, the dose of gentamicin that was administered was too low to inhibit bacterial growth in the absence of concomitant microbubbles and thoracic ultrasound [147]. USMB treatment significantly increased gentamicin concentrations in both bronchoalveolar lavage fluid and lung lysates [147]. In this study, USMB was also shown to have no detrimental effect on oxygenation or on the degree of lung injury scored in a blinded fashion by a lung pathologist [147]. These results of enhanced delivery to injured lung tissue were later validated in a rabbit model by a French group, who reported enhanced delivery of another aminoglycoside antibiotic, amikacin, to fluid-filled lung tissue [148]. Flooding the rabbits' lungs with saline allowed the group to apply USMB treatment while administering amikacin at two doses, comparing delivery to sonicated and non-sonicated lung tissue [148]. USMB treatment significantly increased amikacin concentration in the sonicated lung tissue compared to the non-sonicated lung tissue, with a greater degree of enhancement observed at the lower amikacin dose [148]. Both of these studies highlight the enhancement of therapeutic cargo delivery by USMB at doses below the physiologically effective dosage.

The therapeutic benefits of delivering pulmonary surfactant (sinapultide) using USMB has also been investigated; delivery of sinapultide by insonation of sinapultide-loaded microbubbles reduced the severity of injury and levels of the inflammatory cytokines IL-6 and TNF-α in an LPS-induced model of lung injury [149]. Finally, USMB treatment has also been used to explore the therapeutic effects of delivering a VEGF antagonist (soluble fms-like tyrosine kinase-1) encapsulated in the microbubbles to a murine model of LPS-induced lung injury [150]. The lung injury group receiving USMB with soluble fms-like tyrosine kinase-1 exhibited improved P_aO_2, reduced lung injury score and wet-to-dry ratio, and a lower 7-day mortality rate compared to injured counterparts receiving USMB treatment with empty microbubbles [150]. This treatment was found to enhance endothelial barrier function by inhibiting the endothelial permeability caused by VEGF while avoiding off-target effects [150].

Despite these intriguing examples, however, the use of USMB in the injured lung still remains largely unexplored. Specifically, it remains to be seen whether other therapeutic cargoes such as plasmids and non-coding RNA can be harnessed with USMB to treat lung injury.

3. Future Directions—Challenges and Opportunities

3.1. Trade off of Increased Leakage vs. Cargo Delivery

One concern with USMB in the lung is the risk of increasing endothelial leakage or inflammation through sonoporation. Although sonoporation is transient, in theory it could aggravate tissue damage in the injured lung; on the other hand, sonoporation will also further enhance the delivery of therapeutic cargoes. Inflammation of the blood–brain barrier has been observed when treating with a microbubble dose 10 times higher than the clinical imaging dose; no damage was observed with the clinical dose [151]. The high microbubble dose also corresponded to increased contrast agent uptake in the surrounding tissue, indicating that increased inflammation and leakage allows for more cargo delivery [151]. Steroids such as dexamethasone have been investigated as a countermeasure to USMB-inflicted inflammation in the blood–brain barrier, with beneficial results [152]. Real-time microscopy of USMB-induced cultured endothelial barrier disruption observed the effect of shear stress on barrier permeability and how it can be manipulated through ultrasound frequency and microbubble oscillation dynamics [153]. Large microbubble-induced shear stress was determined to induce larger pores which could cause gap formation between cells [153]. Ultrasound settings, microbubble dosage and dynamics, and how often treatment is applied can all affect the degree of penetration and inflammation due to USMB treatment, as well as the amount of cargo delivery. While increasing the penetration has been linked to increased cargo delivery to underlying tissue, the safety of the procedure must be weighed against the effectiveness. Optimization of all of these parameters is recommended for the application of safe therapy while inducing cargo delivery with USMB.

3.2. Tissue (Depth) Penetration and Specificity for the Lung

Another feasibility issue is the degree to which thoracic ultrasound can penetrate the edematous or consolidated lung. Thoracic ultrasound has a maximum tissue penetration of about 10 cm, which would theoretically permit access to most of the average-sized human thorax (e.g., a chest circumference of 38 inches is a radius of $\approx$ 15 cm) [154]. In the event that surface ultrasound does not permit sufficient access, endobronchial ultrasound exists and would theoretically permit treatment of injured areas of the lung that are deep under the surface. For USMB to be used in clinical practice in humans, one could envisage development of a specific clam-shell-shaped ultrasound transducer that would permit simultaneous treatment of both the ventral and dorsal thorax. Finally, USMB is likely to be most effective in only the most severely injured lung regions; residual air in less severely injured areas of the lung is likely to block the ultrasound energy. Specifically, whether the technique will work in areas where interstitial (rather than alveolar) edema predominates is uncertain.

3.3. Optimizing Bubble Size and Charge for Delivery

Efficient delivery of cargo is affected by the properties of the microbubbles [155]. A particularly important variable is bubble surface charge. In lipid-shelled bubbles, the lipids chosen can impart a charge property on the bubble (e.g., 1,2-stearoyl-3-trimethylammonium-propane will lead to the generation of cationic microbubbles), and these properties have direct and measurable effects on the success of delivery. For example, genetic material in the systemic circulation is rapidly degraded, necessitating large amounts of genetic material to be injected, which may be unfeasible [156]. However, when nucleic acids were delivered with cationic microbubbles, their lifespan in the circulation was extended, leading to a lower required nucleic acid dose and a higher level of transfection [157–162]. This is a result of negatively charged nucleic acids coupling to the bubbles [163].

Bubble size is another important property that contributes to delivery capacity. In comparison to larger bubbles, smaller bubbles require higher ultrasound frequencies to undergo inertial cavitation [164]. Larger bubbles were found to achieve greater penetration depth of Evan's blue dye and ascorbyl tetraisopalmitate in skin samples, suggesting that bigger bubbles are better for delivery [165]. However, large bubbles are more likely to cause detrimental blockage of the circulation; larger bubbles also result in more local tissue damage [166]. These consequences must be considered when choosing the right bubble size for a given application.

On the other end of the spectrum are nanobubbles, which are bubbles that are less than 1 micron in diameter [167]. Due to their very small size, they are able to traverse intercellular gaps and deposit deep into tissues, perhaps making them less desirable to endothelial transfection. They are desirable for applications such as tumor targeting, where they are able to infiltrate deep into the tumor for treatment [168]. They were also shown to be safer to cells than microbubbles, causing transfection in an in vitro model without impacting cell viability [169]. However, a major limitation of nanobubbles is the possibility of extravasation at other vascular sites, thus decreasing the amount deposited at the tissue of interest [170]. Nonetheless, their increased stability makes them an interesting avenue for endothelial delivery.

3.4. Emerging Techniques to Control Bubble Sizes and Charge

Most of the current USMB-mediated drug/gene delivery studies use commercial microbubble ultrasound contrast agents (such as SonoVue®, DEFINITY®, OPTISON®) [171].

Some studies also use custom-made bubbles which are prepared by either sonication or a mechanical agitation method [171]. This method of preparation often leads to polydisperse microbubbles with varying mean sizes [171]. The resonance frequency of microbubbles (i.e., ultrasound frequency at which the maximum amplitude of oscillation occurs) depends on the size of the microbubbles [86]. For example, smaller bubbles have a higher resonance frequency compared to larger bubbles, and the maximum amplitude of oscillation of smaller bubbles is less than larger bubbles because of increased damping [86]. Therefore, when polydisperse microbubbles are insonated, only a small fraction of the total population microbubbles resonate, which may negatively affect the drug/gene delivery efficacy [86]. One way to solve this problem is to use emerging microfluidic techniques which are capable of generating monodisperse microbubble populations. Using carefully designed microchannels and appropriate flow rates, monodisperse microbubbles of varying mean sizes can be generated at a high throughput [172–174]. Moreover, the shell properties (such as composition, charge) of the microbubbles can be varied by using different lipid mixtures in the continuous phase. Apart from microbubbles, microfluidics can also be used to generate monodisperse nanobubbles, as reported in a recent study [175]. Microfluidics thus provides a promising alternative to conventional agitation/sonication techniques for generating monodisperse bubbles, which could maximize the efficiency of USMB-mediated drug/gene delivery.

3.5. Clinical Trials of USMB Treatment for ARDS

The use of USMB treatment to enhance drug delivery for tissues in the digestive tract has been supported in recent clinical trials. Improvement in delivery of chemotherapeutic drug to target tissue has been demonstrated in pancreatic cancer, as well as various malignant tumors of hepatic and pancreatic organs [176,177]. These studies concluded that USMB-enhanced treatment improved patient outcome without additional toxicity or side effects compared to chemotherapeutic treatment alone. However, given the novelty of USMB treatment for injured lung tissue (first published in 2018), USMB has not yet been attempted in clinical trials for ARDS or lobar pneumonia [147]. Once sufficient in vitro and in vivo evidence has accumulated on the effectiveness and safety of USMB treatment for injured lung tissue, clinical trials are likely to follow [178].

4. Conclusions—A New Technique Provides New Opportunities

ARDS is a major cause of death after respiratory infection, whether from existing or emerging viral and bacterial pathogens. The pulmonary endothelium plays a significant role in the development and progression of ARDS, where loss of barrier integrity and excessive inflammation drive pulmonary edema. Current treatments are unable to specifically target the most injured pulmonary endothelium due to the heterogeneity of lung damage in any given patient. Because air scatters ultrasound energy, USMB-mediated cargo delivery is an attractive method that will preferentially target the most injured areas of the lung, sparing relatively aerated regions. Even in the absence of diffuse disease, USMB could also be used in the setting of a dense, lobar consolidation. In principle, this technique could be used to deliver small molecule or genetic agents that enhance endothelial barrier integrity, stimulate endothelial repair, and prevent excessive endothelial cell activation.

Finally, the ability to deliver drugs and potentially genetic material preferentially to non-aerated lung in vivo is likely to be useful in determining the pathogenesis of lung diseases even beyond ARDS. For instance, this technique might facilitate research into the pathogenesis and treatment of lung fibrosis or lung malignancy. Although the field is young, ultrasound-microbubble-mediated drug and gene delivery has the potential to be a valuable tool to understand and to treat the severely injured lung.

5. Patents

W.L.L. is listed as a co-inventor on a patent related to this work and is the Chief Scientific Officer of a spin-off company related to this field.

Author Contributions: Conceptualization, writing, original draft preparation, review and editing, R.S., K.J., M.D., S.S.H.T. and W.L.L. All authors have read and agreed to the published version of the manuscript.

Funding: Work in W.L.L.'s lab is funded by a Collaborative Health Research Grant (CHRP) from the Canadian Institutes of Health Research (CIHR) and the Natural Sciences and Engineering Research Council (NSERC) of Canada. W.L.L. holds a Canada Research Chair in Mechanisms of Endothelial Permeability. Work in S.S.H.T.'s lab is funded by the Natural Sciences and Engineering Research Council (NSERC) Discovery grant (RGPIN-2019-04618).

Institutional Review Board Statement: Not applicable.

Informed Consent Statement: Not applicable.

Data Availability Statement: No new data were created or analyzed in this study. Data sharing is not applicable to this article.

Conflicts of Interest: W.L.L. is a co-inventor on a patent for the use of Vasculotide for influenza and previously (2016–2018) served on the Scientific Advisory Board for Vasomune.

References

1. Sturtzel, C. Endothelial Cells. *Adv. Exp. Med. Biol.* **2017**, *1003*, 71–91. [CrossRef]
2. Filewod, N.C.; Lee, W.L. Inflammation without Vascular Leakage. Science Fiction No Longer? *Am. J. Respir. Crit. Care Med.* **2019**, *200*, 1472–1476. [CrossRef]
3. Goldenberg, N.M.; Steinberg, B.E.; Slutsky, A.S.; Lee, W.L. Broken Barriers: A New Take on Sepsis Pathogenesis. *Sci. Transl. Med.* **2011**, *3*, 88ps25. [CrossRef]
4. Lee, W.L.; Slutsky, A.S. Sepsis and Endothelial Permeability. *N. Engl. J. Med.* **2010**, *363*, 689–691. [CrossRef] [PubMed]
5. ARDS Definition Task Force; Ranieri, V.M.; Rubenfeld, G.D.; Thompson, B.T.; Ferguson, N.D.; Caldwell, E.; Fan, E.; Camporota, L.; Slutsky, A.S. Acute Respiratory Distress Syndrome: The Berlin Definition. *JAMA* **2012**, *307*, 2526–2533. [CrossRef]
6. Bellani, G.; Laffey, J.G.; Pham, T.; Fan, E.; Brochard, L.; Esteban, A.; Gattinoni, L.; van Haren, F.; Larsson, A.; McAuley, D.F.; et al. Epidemiology, Patterns of Care, and Mortality for Patients with Acute Respiratory Distress Syndrome in Intensive Care Units in 50 Countries. *JAMA* **2016**, *315*, 788–800. [CrossRef] [PubMed]
7. Vassiliou, A.G.; Kotanidou, A.; Dimopoulou, I.; Orfanos, S.E. Endothelial Damage in Acute Respiratory Distress Syndrome. *Int. J. Mol. Sci.* **2020**, *21*, 8793. [CrossRef]
8. Latreille, E.; Lee, W.L. Interactions of Influenza and SARS-CoV-2 with the Lung Endothelium: Similarities, Differences, and Implications for Therapy. *Viruses* **2021**, *13*, 161. [CrossRef] [PubMed]
9. Hunt, B.J.; Jurd, K.M. Endothelial Cell Activation. A Central Pathophysiological Process. *BMJ* **1998**, *316*, 1328–1329. [CrossRef]
10. Gavard, J.; Gutkind, J.S. VEGF Controls Endothelial-Cell Permeability by Promoting the Beta-Arrestin-Dependent Endocytosis of VE-Cadherin. *Nat. Cell Biol.* **2006**, *8*, 1223–1234. [CrossRef]
11. Wessel, F.; Winderlich, M.; Holm, M.; Frye, M.; Rivera-Galdos, R.; Vockel, M.; Linnepe, R.; Ipe, U.; Stadtmann, A.; Zarbock, A.; et al. Leukocyte Extravasation and Vascular Permeability Are Each Controlled in Vivo by Different Tyrosine Residues of VE-Cadherin. *Nat. Immunol.* **2014**, *15*, 223–230. [CrossRef]
12. Moss, M.; Gillespie, M.K.; Ackerson, L.; Moore, F.A.; Moore, E.E.; Parsons, P.E. Endothelial Cell Activity Varies in Patients at Risk for the Adult Respiratory Distress Syndrome. *Crit. Care Med.* **1996**, *24*, 1782–1786. [CrossRef] [PubMed]
13. Müller, A.M.; Cronen, C.; Müller, K.-M.; Kirkpatrick, C.J. Heterogeneous Expression of Cell Adhesion Molecules by Endothelial Cells in ARDS. *J. Pathol.* **2002**, *198*, 270–275. [CrossRef] [PubMed]
14. van Griensven, M.; Probst, C.; Müller, K.; Hoevel, P.; Pape, H.-C. Leukocyte-Endothelial Interactions via ICAM-1 Are Detrimental in Polymicrobial Sepsis. *Shock* **2006**, *25*, 254–259. [CrossRef] [PubMed]
15. Donnelly, S.C.; Haslett, C.; Dransfield, I.; Robertson, C.E.; Carter, D.C.; Ross, J.A.; Grant, I.S.; Tedder, T.F. Role of Selectins in Development of Adult Respiratory Distress Syndrome. *Lancet* **1994**, *344*, 215–219. [CrossRef]
16. Carraway, M.S.; Welty-Wolf, K.E.; Kantrow, S.P.; Huang, Y.C.; Simonson, S.G.; Que, L.G.; Kishimoto, T.K.; Piantadosi, C.A. Antibody to E- and L-Selectin Does Not Prevent Lung Injury or Mortality in Septic Baboons. *Am. J. Respir. Crit. Care Med.* **1998**, *157*, 938–949. [CrossRef]
17. Ridings, P.C.; Windsor, A.C.; Jutila, M.A.; Blocher, C.R.; Fisher, B.J.; Sholley, M.M.; Sugerman, H.J.; Fowler, A.A. A Dual-Binding Antibody to E- and L-Selectin Attenuates Sepsis-Induced Lung Injury. *Am. J. Respir. Crit. Care Med.* **1995**, *152*, 247–253. [CrossRef]
18. Zhang, C. The Role of Inflammatory Cytokines in Endothelial Dysfunction. *Basic Res. Cardiol.* **2008**, *103*, 398–406. [CrossRef]
19. Sprague, A.H.; Khalil, R.A. Inflammatory Cytokines in Vascular Dysfunction and Vascular Disease. *Biochem. Pharmacol.* **2009**, *78*, 539–552. [CrossRef]
20. Meduri, G.U.; Kohler, G.; Headley, S.; Tolley, E.; Stentz, F.; Postlethwaite, A. Inflammatory Cytokines in the BAL of Patients with ARDS. Persistent Elevation over Time Predicts Poor Outcome. *Chest* **1995**, *108*, 1303–1314. [CrossRef]

21. Park, W.Y.; Goodman, R.B.; Steinberg, K.P.; Ruzinski, J.T.; Radella, F.; Park, D.R.; Pugin, J.; Skerrett, S.J.; Hudson, L.D.; Martin, T.R. Cytokine Balance in the Lungs of Patients with Acute Respiratory Distress Syndrome. *Am. J. Respir. Crit. Care Med.* **2001**, *164*, 1896–1903. [CrossRef]

22. Teijaro, J.R.; Walsh, K.B.; Cahalan, S.; Fremgen, D.M.; Roberts, E.; Scott, F.; Martinborough, E.; Peach, R.; Oldstone, M.B.A.; Rosen, H. Endothelial Cells Are Central Orchestrators of Cytokine Amplification during Influenza Virus Infection. *Cell* **2011**, *146*, 980–991. [CrossRef]

23. McVey, M.J.; Steinberg, B.E.; Goldenberg, N.M. Inflammasome Activation in Acute Lung Injury. *Am. J. Physiol. Lung Cell. Mol. Physiol.* **2021**, *320*, L165–L178. [CrossRef]

24. Yang, Y.; Wang, H.; Kouadir, M.; Song, H.; Shi, F. Recent Advances in the Mechanisms of NLRP3 Inflammasome Activation and Its Inhibitors. *Cell Death Dis.* **2019**, *10*, 128. [CrossRef]

25. The REMAP-CAP Investigators. Interleukin-6 Receptor Antagonists in Critically Ill Patients with Covid. *N. Engl. J. Med.* **2021**, *384*, 1491–1502. [CrossRef] [PubMed]

26. Conti, P.; Caraffa, A.; Gallenga, C.E.; Ross, R.; Kritas, S.K.; Frydas, I.; Younes, A.; Ronconi, G. Coronavirus-19 (SARS-CoV-2) Induces Acute Severe Lung Inflammation via IL-1 Causing Cytokine Storm in COVID-19: A Promising Inhibitory Strategy. *J. Biol. Regul. Homeost. Agents* **2020**, *34*, 1971–1975. [CrossRef] [PubMed]

27. Conti, P.; Pregliasco, F.E.; Calvisi, V.; Calvisi, V.; Caraffa, A.; Gallenga, C.E.; Kritas, S.K.; Ronconi, G. Monoclonal Antibody Therapy in COVID. *J. Biol. Regul. Homeost. Agents* **2021**, *35*, 423–427. [CrossRef]

28. Kelley, N.; Jeltema, D.; Duan, Y.; He, Y. The NLRP3 Inflammasome: An Overview of Mechanisms of Activation and Regulation. *Int. J. Mol. Sci.* **2019**, *20*, 3328. [CrossRef]

29. Freeman, T.L.; Swartz, T.H. Targeting the NLRP3 Inflammasome in Severe COVID. *Front. Immunol.* **2020**, *11*, 1518. [CrossRef]

30. Grunwell, J.R.; Stephenson, S.T.; Mohammad, A.F.; Jones, K.; Mason, C.; Opolka, C.; Fitzpatrick, A.M. Differential Type I Interferon Response and Primary Airway Neutrophil Extracellular Trap Release in Children with Acute Respiratory Distress Syndrome. *Sci. Rep.* **2020**, *10*, 19049. [CrossRef] [PubMed]

31. Middleton, E.A.; He, X.-Y.; Denorme, F.; Campbell, R.A.; Ng, D.; Salvatore, S.P.; Mostyka, M.; Baxter-Stoltzfus, A.; Borczuk, A.C.; Loda, M.; et al. Neutrophil Extracellular Traps Contribute to Immunothrombosis in COVID-19 Acute Respiratory Distress Syndrome. *Blood* **2020**, *136*, 1169–1179. [CrossRef] [PubMed]

32. Yildiz, C.; Palaniyar, N.; Otulakowski, G.; Khan, M.A.; Post, M.; Kuebler, W.M.; Tanswell, K.; Belcastro, R.; Masood, A.; Engelberts, D.; et al. Mechanical Ventilation Induces Neutrophil Extracellular Trap Formation. *Anesthesiology* **2015**, *122*, 864–875. [CrossRef] [PubMed]

33. Vestweber, D.; Winderlich, M.; Cagna, G.; Nottebaum, A.F. Cell Adhesion Dynamics at Endothelial Junctions: VE-Cadherin as a Major Player. *Trends Cell Biol.* **2009**, *19*, 8–15. [CrossRef] [PubMed]

34. Corada, M.; Mariotti, M.; Thurston, G.; Smith, K.; Kunkel, R.; Brockhaus, M.; Lampugnani, M.G.; Martin-Padura, I.; Stoppacciaro, A.; Ruco, L.; et al. Vascular Endothelial-Cadherin Is an Important Determinant of Microvascular Integrity in Vivo. *Proc. Natl. Acad. Sci. USA* **1999**, *96*, 9815–9820. [CrossRef]

35. Broermann, A.; Winderlich, M.; Block, H.; Frye, M.; Rossaint, J.; Zarbock, A.; Cagna, G.; Linnepe, R.; Schulte, D.; Nottebaum, A.F.; et al. Dissociation of VE-PTP from VE-Cadherin Is Required for Leukocyte Extravasation and for VEGF-Induced Vascular Permeability in Vivo. *J. Exp. Med.* **2011**, *208*, 2393–2401. [CrossRef] [PubMed]

36. Schulte, D.; Küppers, V.; Dartsch, N.; Broermann, A.; Li, H.; Zarbock, A.; Kamenyeva, O.; Kiefer, F.; Khandoga, A.; Massberg, S.; et al. Stabilizing the VE-Cadherin-Catenin Complex Blocks Leukocyte Extravasation and Vascular Permeability. *EMBO J.* **2011**, *30*, 4157–4170. [CrossRef]

37. Brindle, N.P.J.; Saharinen, P.; Alitalo, K. Signaling and Functions of Angiopoietin-1 in Vascular Protection. *Circ. Res.* **2006**, *98*, 1014–1023. [CrossRef]

38. Dekker, N.A.M.; van Leeuwen, A.L.I.; van Strien, W.W.J.; Majolée, J.; Szulcek, R.; Vonk, A.B.A.; Hordijk, P.L.; Boer, C.; van den Brom, C.E. Microcirculatory Perfusion Disturbances Following Cardiac Surgery with Cardiopulmonary Bypass Are Associated with in Vitro Endothelial Hyperpermeability and Increased Angiopoietin-2 Levels. *Crit. Care* **2019**, *23*, 117. [CrossRef]

39. van der Heijden, M.; van NieuwAmerongen, G.P.; Koolwijk, P.; van Hinsbergh, V.W.M.; Groeneveld, A.B.J. Angiopoietin-2, Permeability Oedema, Occurrence and Severity of ALI/ARDS in Septic and Non-Septic Critically Ill Patients. *Thorax* **2008**, *63*, 903–909. [CrossRef] [PubMed]

40. Yano, K.; Liaw, P.C.; Mullington, J.M.; Shih, S.-C.; Okada, H.; Bodyak, N.; Kang, P.M.; Toltl, L.; Belikoff, B.; Buras, J.; et al. Vascular Endothelial Growth Factor Is an Important Determinant of Sepsis Morbidity and Mortality. *J. Exp. Med.* **2006**, *203*, 1447–1458. [CrossRef] [PubMed]

41. Evans, C.E.; Iruela-Arispe, M.L.; Zhao, Y.-Y. Mechanisms of Endothelial Regeneration and Vascular Repair and Their Application to Regenerative Medicine. *Am. J. Pathol.* **2021**, *191*, 52–65. [CrossRef]

42. Rajput, C.; Tauseef, M.; Farazuddin, M.; Yazbeck, P.; Amin, M.-R.; Avin, V.B.; Sharma, T.; Mehta, D. MicroRNA-150 Suppression of Angiopoetin-2 Generation and Signaling Is Crucial for Resolving Vascular Injury. *Arterioscler. Thromb. Vasc. Biol.* **2016**, *36*, 380–388. [CrossRef]

43. Li, P.; Yao, Y.; Ma, Y.; Chen, Y. MiR-150 Attenuates LPS-Induced Acute Lung Injury via Targeting AKT. *Int. Immunopharmacol.* **2019**, *75*, 105794. [CrossRef]

44. Zhao, Y.-Y.; Gao, X.-P.; Zhao, Y.D.; Mirza, M.K.; Frey, R.S.; Kalinichenko, V.V.; Wang, I.-C.; Costa, R.H.; Malik, A.B. Endothelial Cell-Restricted Disruption of FoxM1 Impairs Endothelial Repair Following LPS-Induced Vascular Injury. *J. Clin. Investig.* **2006**, *116*, 2333–2343. [CrossRef]
45. Huang, X.; Zhang, X.; Zhao, D.X.; Yin, J.; Hu, G.; Evans, C.E.; Zhao, Y.-Y. Endothelial Hypoxia-Inducible Factor-1α Is Required for Vascular Repair and Resolution of Inflammatory Lung Injury through Forkhead Box Protein M. *Am. J. Pathol.* **2019**, *189*, 1664–1679. [CrossRef] [PubMed]
46. Li, H.; Yang, T.; Fei, Z. MiR-26a-5p Alleviates Lipopolysaccharide-Induced Acute Lung Injury by Targeting the Connective Tissue Growth Factor. *Mol. Med. Rep.* **2021**, *23*. [CrossRef]
47. Pan, X.; Xu, S.; Zhou, Z.; Wang, F.; Mao, L.; Li, H.; Wu, C.; Wang, J.; Huang, Y.; Li, D.; et al. Fibroblast Growth Factor-2 Alleviates the Capillary Leakage and Inflammation in Sepsis. *Mol. Med.* **2020**, *26*, 108. [CrossRef]
48. Fu, P.; Ramchandran, R.; Shaaya, M.; Huang, L.; Ebenezer, D.L.; Jiang, Y.; Komarova, Y.; Vogel, S.M.; Malik, A.B.; Minshall, R.D.; et al. Phospholipase D2 Restores Endothelial Barrier Function by Promoting PTPN14-Mediated VE-Cadherin Dephosphorylation. *J. Biol. Chem.* **2020**, *295*, 7669–7685. [CrossRef]
49. Pawlinski, R.; Pedersen, B.; Schabbauer, G.; Tencati, M.; Holscher, T.; Boisvert, W.; Andrade-Gordon, P.; Frank, R.D.; Mackman, N. Role of Tissue Factor and Protease-Activated Receptors in a Mouse Model of Endotoxemia. *Blood* **2004**, *103*, 1342–1347. [CrossRef]
50. Frantzeskaki, F.; Armaganidis, A.; Orfanos, S.E. Immunothrombosis in Acute Respiratory Distress Syndrome: Cross Talks between Inflammation and Coagulation. *Respiration* **2017**, *93*, 212–225. [CrossRef] [PubMed]
51. Vogel, S.M.; Gao, X.; Mehta, D.; Ye, R.D.; John, T.A.; Andrade-Gordon, P.; Tiruppathi, C.; Malik, A.B. Abrogation of Thrombin-Induced Increase in Pulmonary Microvascular Permeability in PAR-1 Knockout Mice. *Physiol. Genom.* **2000**, *4*, 137–145. [CrossRef] [PubMed]
52. Ware, L.B.; Bastarache, J.A.; Wang, L. Coagulation and Fibrinolysis in Human Acute Lung Injury—New Therapeutic Targets? *Keio J. Med.* **2005**, *54*, 142–149. [CrossRef]
53. Grau, G.E.; de Moerloose, P.; Bulla, O.; Lou, J.; Lei, Z.; Reber, G.; Mili, N.; Ricou, B.; Morel, D.R.; Suter, P.M. Haemostatic Properties of Human Pulmonary and Cerebral Microvascular Endothelial Cells. *Thromb. Haemost.* **1997**, *77*, 585–590. [CrossRef] [PubMed]
54. Katneni, U.K.; Alexaki, A.; Hunt, R.C.; Schiller, T.; DiCuccio, M.; Buehler, P.W.; Ibla, J.C.; Kimchi-Sarfaty, C. Coagulopathy and Thrombosis as a Result of Severe COVID-19 Infection: A Microvascular Focus. *Thromb. Haemost.* **2020**, *120*, 1668–1679. [CrossRef] [PubMed]
55. Ackermann, M.; Verleden, S.E.; Kuehnel, M.; Haverich, A.; Welte, T.; Laenger, F.; Vanstapel, A.; Werlein, C.; Stark, H.; Tzankov, A.; et al. Pulmonary Vascular Endothelialitis, Thrombosis, and Angiogenesis in Covid. *N. Engl. J. Med.* **2020**, *383*, 120–128. [CrossRef]
56. Juffermans, N.P.; van den Brom, C.E.; Kleinveld, D.J.B. Targeting Endothelial Dysfunction in Acute Critical Illness to Reduce Organ Failure. *Anesth. Analg.* **2020**, *131*, 1708–1720. [CrossRef]
57. Garcia, J.G.; Liu, F.; Verin, A.D.; Birukova, A.; Dechert, M.A.; Gerthoffer, W.T.; Bamberg, J.R.; English, D. Sphingosine 1-Phosphate Promotes Endothelial Cell Barrier Integrity by Edg-Dependent Cytoskeletal Rearrangement. *J. Clin. Investig.* **2001**, *108*, 689–701. [CrossRef]
58. Akhter, M.Z.; Joshi, J.C.; Balaji, R.V.A.; Maienschein-Cline, M.; Richard, L.P.; Asrar, B.M. Mehta Dolly Programming to S1PR1+ Endothelial Cells Promote Restoration of Vascular Integrity. *Circ. Res.* **2021**, *129*, 221–236. [CrossRef]
59. Trieu, M.; van Meurs, M.; van Leeuwen, A.L.I.; Van Slyke, P.; Hoang, V.; Geeraedts, L.M.G.; Boer, C.; van den Brom, C.E. Vasculotide, an Angiopoietin-1 Mimetic, Restores Microcirculatory Perfusion and Microvascular Leakage and Decreases Fluid Resuscitation Requirements in Hemorrhagic Shock. *Anesthesiology* **2018**, *128*, 361–374. [CrossRef]
60. Kumpers, P.; Gueler, F.; David, S.; Slyke, P.V.; Dumont, D.J.; Park, J.-K.; Bockmeyer, C.L.; Parikh, S.M.; Pavenstadt, H.; Haller, H.; et al. The Synthetic Tie2 Agonist Peptide Vasculotide Protects against Vascular Leakage and Reduces Mortality in Murine Abdominal Sepsis. *Crit. Care* **2011**, *15*, R261. [CrossRef]
61. Sugiyama, M.G.; Armstrong, S.M.; Wang, C.; Hwang, D.; Leong-Poi, H.; Advani, A.; Advani, S.; Zhang, H.; Szaszi, K.; Tabuchi, A.; et al. The Tie2-Agonist Vasculotide Rescues Mice from Influenza Virus Infection. *Sci. Rep.* **2015**, *5*, 11030. [CrossRef]
62. Han, S.; Lee, S.-J.; Kim, K.E.; Lee, H.S.; Oh, N.; Park, I.; Ko, E.; Oh, S.J.; Lee, Y.-S.; Kim, D.; et al. Amelioration of Sepsis by TIE2 Activation–Induced Vascular Protection. *Sci. Transl. Med.* **2016**, *8*, 335ra55. [CrossRef] [PubMed]
63. Hauschildt, J.; Schrimpf, C.; Thamm, K.; Retzlaff, J.; Idowu, T.O.; von Kaisenberg, C.; Haller, H.; David, S. Dual Pharmacological Inhibition of Angiopoietin-2 and VEGF-A in Murine Experimental Sepsis. *J. Vasc. Res.* **2020**, *57*, 34–45. [CrossRef] [PubMed]
64. Frye, M.; Dierkes, M.; Küppers, V.; Vockel, M.; Tomm, J.; Zeuschner, D.; Rossaint, J.; Zarbock, A.; Koh, G.Y.; Peters, K.; et al. Interfering with VE-PTP Stabilizes Endothelial Junctions in Vivo via Tie-2 in the Absence of VE-Cadherin. *J. Exp. Med.* **2015**, *212*, 2267–2287. [CrossRef]
65. Shen, J.; Frye, M.; Lee, B.L.; Reinardy, J.L.; McClung, J.M.; Ding, K.; Kojima, M.; Xia, H.; Seidel, C.; Lima e Silva, R.; et al. Targeting VE-PTP Activates TIE2 and Stabilizes the Ocular Vasculature. *J. Clin. Investig.* **2014**, *124*, 4564–4576. [CrossRef]
66. Deng, Z.; Kalin, G.T.; Shi, D.; Kalinichenko, V.V. Nanoparticle Delivery Systems with Cell-Specific Targeting for Pulmonary Diseases. *Am. J. Respir. Cell. Mol. Biol.* **2021**, *64*, 292–307. [CrossRef]
67. Miller, M.R.; Raftis, J.B.; Langrish, J.P.; McLean, S.G.; Samutrtai, P.; Connell, S.P.; Wilson, S.; Vesey, A.T.; Fokkens, P.H.B.; Boere, A.J.F.; et al. Inhaled Nanoparticles Accumulate at Sites of Vascular Disease. *ACS Nano* **2017**, *11*, 4542–4552. [CrossRef]

68. Dunn, A.W.; Kalinichenko, V.V.; Shi, D. Highly Efficient In Vivo Targeting of the Pulmonary Endothelium Using Novel Modifications of Polyethylenimine: An Importance of Charge. *Adv. Health Mater.* **2018**, *7*, e1800876. [CrossRef]

69. Bolte, C.; Ustiyan, V.; Ren, X.; Dunn, A.W.; Pradhan, A.; Wang, G.; Kolesnichenko, O.A.; Deng, Z.; Zhang, Y.; Shi, D.; et al. Nanoparticle Delivery of Proangiogenic Transcription Factors into the Neonatal Circulation Inhibits Alveolar Simplification Caused by Hyperoxia. *Am. J. Respir. Crit. Care Med.* **2020**, *202*, 100–111. [CrossRef] [PubMed]

70. Pradhan, A.; Dunn, A.; Ustiyan, V.; Bolte, C.; Wang, G.; Whitsett, J.A.; Zhang, Y.; Porollo, A.; Hu, Y.-C.; Xiao, R.; et al. The S52F FOXF1 Mutation Inhibits STAT3 Signaling and Causes Alveolar Capillary Dysplasia. *Am. J. Respir. Crit. Care Med.* **2019**, *200*, 1045–1056. [CrossRef]

71. Elwakil, M.M.A.; Khalil, I.A.; Elewa, Y.H.A.; Kusumoto, K.; Sato, Y.; Shobaki, N.; Kon, Y.; Harashima, H. Lung-Endothelium-Targeted Nanoparticles Based on a PH-Sensitive Lipid and the GALA Peptide Enable Robust Gene Silencing and the Regression of Metastatic Lung Cancer. *Adv. Funct. Mater.* **2019**, *29*, 1807677. [CrossRef]

72. Jiang, S.; Li, S.; Hu, J.; Xu, X.; Wang, X.; Kang, X.; Qi, J.; Lu, X.; Wu, J.; Du, Y.; et al. Combined Delivery of Angiopoietin-1 Gene and Simvastatin Mediated by Anti-Intercellular Adhesion Molecule-1 Antibody-Conjugated Ternary Nanoparticles for Acute Lung Injury Therapy. *Nanomedicine* **2019**, *15*, 25–36. [CrossRef] [PubMed]

73. Roki, N.; Tsinas, Z.; Solomon, M.; Bowers, J.; Getts, R.C.; Muro, S. Unprecedently High Targeting Specificity toward Lung ICAM-1 Using 3DNA Nanocarriers. *J.Control. Release* **2019**, *305*, 41–49. [CrossRef]

74. Muro, S.; Mateescu, M.; Gajewski, C.; Robinson, M.; Muzykantov, V.R.; Koval, M. Control of Intracellular Trafficking of ICAM-1-Targeted Nanocarriers by Endothelial Na+/H+ Exchanger Proteins. *Am. J. Physiol. Lung. Cell. Mol. Physiol.* **2006**, *290*, L809–L817. [CrossRef]

75. Ashbaugh, D.G.; Bigelow, D.B.; Petty, T.L.; Levine, B.E. Acute Respiratory Distress in Adults. *Lancet* **1967**, *2*, 319–323. [CrossRef]

76. Katzenstein, A.L.; Bloor, C.M.; Leibow, A.A. Diffuse Alveolar Damage—The Role of Oxygen, Shock, and Related Factors. A Review. *Am. J. Pathol.* **1976**, *85*, 209–228.

77. Bao, S.; Thrall, B.D.; Miller, D.L. Transfection of a Reporter Plasmid into Cultured Cells by Sonoporation in Vitro. *Ultrasound Med. Biol.* **1997**, *23*, 953–959. [CrossRef]

78. Song, K.-H.; Harvey, B.K.; Borden, M.A. State-of-the-Art of Microbubble-Assisted Blood-Brain Barrier Disruption. *Theranostics* **2018**, *8*, 4393–4408. [CrossRef]

79. Rinaldi, L.; Folliero, V.; Palomba, L.; Zannella, C.; Isticato, R.; Di Francia, R.; Berretta, M.; de Sio, I.; Adinolfi, L.E.; Morelli, G.; et al. Sonoporation by Microbubbles as Gene Therapy Approach against Liver Cancer. *Oncotarget* **2018**, *9*, 32182–32190. [CrossRef]

80. van Wamel, A.; Kooiman, K.; Harteveld, M.; Emmer, M.; ten Cate, F.J.; Versluis, M.; de Jong, N. Vibrating Microbubbles Poking Individual Cells: Drug Transfer into Cells via Sonoporation. *J. Control Release* **2006**, *112*, 149–155. [CrossRef]

81. Kooiman, K.; Foppen-Harteveld, M.; van der Steen, A.F.W.; de Jong, N. Sonoporation of Endothelial Cells by Vibrating Targeted Microbubbles. *J. Control Release* **2011**, *154*, 35–41. [CrossRef]

82. Mullick Chowdhury, S.; Lee, T.; Willmann, J.K. Ultrasound-Guided Drug Delivery in Cancer. *Ultrasonography* **2017**, *36*, 171–184. [CrossRef]

83. Kooiman, K.; Vos, H.J.; Versluis, M.; de Jong, N. Acoustic Behavior of Microbubbles and Implications for Drug Delivery. *Adv. Drug Deliv. Rev.* **2014**, *72*, 28–48. [CrossRef]

84. Doinikov, A.A.; Bouakaz, A. Acoustic Microstreaming around a Gas Bubble. *J.Acoust. Soc. Am.* **2010**, *127*, 703–709. [CrossRef]

85. Brujan, E.A.; Ikeda, T.; Matsumoto, Y. Jet Formation and Shock Wave Emission during Collapse of Ultrasound-Induced Cavitation Bubbles and Their Role in the Therapeutic Applications of High-Intensity Focused Ultrasound. *Phys. Med. Biol.* **2005**, *50*, 4797–4809. [CrossRef]

86. Roovers, S.; Segers, T.; Lajoinie, G.; Deprez, J.; Versluis, M.; De Smedt, S.C.; Lentacker, I. The Role of Ultrasound-Driven Microbubble Dynamics in Drug Delivery: From Microbubble Fundamentals to Clinical Translation. *Langmuir* **2019**, *35*, 10173–10191. [CrossRef]

87. Presset, A.; Bonneau, C.; Kazuyoshi, S.; Nadal-Desbarats, L.; Mitsuyoshi, T.; Bouakaz, A.; Kudo, N.; Escoffre, J.-M.; Sasaki, N. Endothelial Cells, First Target of Drug Delivery Using Microbubble-Assisted Ultrasound. *Ultrasound Med. Biol.* **2020**, *46*, 1565–1583. [CrossRef]

88. Meijering, B.D.M.; Henning, R.H.; Van Gilst, W.H.; Gavrilovic, I.; Van Wamel, A.; Deelman, L.E. Optimization of Ultrasound and Microbubbles Targeted Gene Delivery to Cultured Primary Endothelial Cells. *J. Drug. Target.* **2007**, *15*, 664–671. [CrossRef] [PubMed]

89. Teupe, C.; Richter, S.; Fisslthaler, B.; Randriamboavonjy, V.; Ihling, C.; Fleming, I.; Busse, R.; Zeiher, A.M.; Dimmeler, S. Vascular Gene Transfer of Phosphomimetic Endothelial Nitric Oxide Synthase (S1177D) Using Ultrasound-Enhanced Destruction of Plasmid-Loaded Microbubbles Improves Vasoreactivity. *Circulation* **2002**, *105*, 1104–1109. [CrossRef] [PubMed]

90. Phillips, L.C.; Dhanaliwala, A.H.; Klibanov, A.L.; Hossack, J.A.; Wamhoff, B.R. Focused Ultrasound-Mediated Drug Delivery from Microbubbles Reduces Drug Dose Necessary for Therapeutic Effect on Neointima Formation–Brief Report. *Arterioscler. Thromb. Vasc. Biol.* **2011**, *31*, 2853–2855. [CrossRef] [PubMed]

91. Mehier-Humbert, S.; Bettinger, T.; Yan, F.; Guy, R.H. Plasma Membrane Poration Induced by Ultrasound Exposure: Implication for Drug Delivery. *J. Control Release* **2005**, *104*, 213–222. [CrossRef] [PubMed]

92. Zhou, Y.; Kumon, R.E.; Cui, J.; Deng, C.X. The Size of Sonoporation Pores on The Cell Membrane. *Ultrasound Med. Biol.* **2009**, *35*, 1756–1760. [CrossRef] [PubMed]

93. Zhao, Y.-Z.; Luo, Y.-K.; Lu, C.-T.; Xu, J.-F.; Tang, J.; Zhang, M.; Zhang, Y.; Liang, H.-D. Phospholipids-Based Microbubbles Sonoporation Pore Size and Reseal of Cell Membrane Cultured in Vitro. *J. Drug. Target.* **2008**, *16*, 18–25. [CrossRef] [PubMed]

94. Prentice, P.; Cuschieri, A.; Dholakia, K.; Prausnitz, M.; Campbell, P. Membrane Disruption by Optically Controlled Microbubble Cavitation. *Nat. Phys.* **2005**, *1*, 107–110. [CrossRef]

95. Zhou, Y.; Yang, K.; Cui, J.; Ye, J.Y.; Deng, C.X. Controlled Permeation of Cell Membrane by Single Bubble Acoustic Cavitation. *J. Control Release* **2012**, *157*, 103–111. [CrossRef]

96. Qiu, Y.; Luo, Y.; Zhang, Y.; Cui, W.; Zhang, D.; Wu, J.; Zhang, J.; Tu, J. The Correlation between Acoustic Cavitation and Sonoporation Involved in Ultrasound-Mediated DNA Transfection with Polyethylenimine (PEI) in Vitro. *J. Control Release* **2010**, *145*, 40–48. [CrossRef]

97. Schlicher, R.K.; Radhakrishna, H.; Tolentino, T.P.; Apkarian, R.P.; Zarnitsyn, V.; Prausnitz, M.R. Mechanism of Intracellular Delivery by Acoustic Cavitation. *Ultrasound Med. Biol.* **2006**, *32*, 915–924. [CrossRef] [PubMed]

98. Kudo, N.; Okada, K.; Yamamoto, K. Sonoporation by Single-Shot Pulsed Ultrasound with Microbubbles Adjacent to Cells. *Biophys. J.* **2009**, *96*, 4866–4876. [CrossRef]

99. Hu, Y.; Wan, J.M.F.; Yu, A.C.H. Membrane Perforation and Recovery Dynamics in Microbubble-Mediated Sonoporation. *Ultrasound Med. Biol.* **2013**, *39*, 2393–2405. [CrossRef] [PubMed]

100. Bouakaz, A.; Zeghimi, A.; Doinikov, A.A. Sonoporation: Concept and Mechanisms. *Adv. Exp. Med. Biol.* **2016**, *880*, 175–189. [CrossRef]

101. Wu, J. Theoretical Study on Shear Stress Generated by Microstreaming Surrounding Contrast Agents Attached to Living Cells. *Ultrasound Med. Biol.* **2002**, *28*, 125–129. [CrossRef]

102. Novell, A.; Collis, J.; Doinikov, A.A.; Ooi, A.; Manasseh, R.; Bouakazaz, A. Theoretical and Experimental Evaluation of Microstreaming Created by a Single Microbubble: Application to Sonoporation. In Proceedings of the 2011 IEEE International Ultrasonics Symposium, Orlando, FL, USA, 18–21 October 2011; pp. 1482–1485.

103. Wu, J.; Ross, J.P.; Chiu, J.-F. Reparable Sonoporation Generated by Microstreaming. *J.Acoust. Soc. Am.* **2002**, *111*, 1460–1464. [CrossRef]

104. Žnidarčič, A.; Mettin, R.; Cairós, C.; Dular, M. Attached Cavitation at a Small Diameter Ultrasonic Horn Tip. *Phys. Fluids* **2014**, *26*, 023304. [CrossRef]

105. Liu, X.; Wu, J. Acoustic Microstreaming around an Isolated Encapsulated Microbubble. *J. Acoust. Soc. Am.* **2009**, *125*, 1319–1330. [CrossRef] [PubMed]

106. Doinikov, A.A.; Palanchon, P.; Kaddur, K.; Bouakaz, A. Theoretical Exploration of Shear Stress Generated by Oscillating Microbubbles on the Cell Membrane in the Context of Sonoporation. In Proceedings of the 2009 IEEE International Ultrasonics Symposium, Roma, Italy, 20–23 September 2009; pp. 1215–1218.

107. Mobadersany, N.; Sarkar, K. Acoustic Microstreaming near a Plane Wall Due to a Pulsating Free or Coated Bubble: Velocity, Vorticity and Closed Streamlines. *J. Fluid Mech.* **2019**, *875*, 781–806. [CrossRef]

108. Karshafian, R.; Bevan, P.D.; Williams, R.; Samac, S.; Burns, P.N. Sonoporation by Ultrasound-Activated Microbubble Contrast Agents: Effect of Acoustic Exposure Parameters on Cell Membrane Permeability and Cell Viability. *Ultrasound Med. Biol.* **2009**, *35*, 847–860. [CrossRef]

109. Choi, J.J.; Feshitan, J.A.; Baseri, B.; Wang, S.; Tung, Y.-S.; Borden, M.A.; Konofagou, E.E. Microbubble-Size Dependence of Focused Ultrasound-Induced Blood-Brain Barrier Opening in Mice in Vivo. *IEEE Trans Biomed. Eng.* **2010**, *57*, 145–154. [CrossRef] [PubMed]

110. Qin, P.; Jin, L.; Li, F.; Han, T.; Du, L.; Yu, A.C.H. The Relationship between Microbubble Size and Heterogeneous Sonoporation at the Single-Cell Level. In Proceedings of the 2016 IEEE International Ultrasonics Symposium (IUS), Tours, France, 18–21 September 2016; pp. 1–4.

111. Koshiyama, K.; Kodama, T.; Yano, T.; Fujikawa, S. Structural Change in Lipid Bilayers and Water Penetration Induced by Shock Waves: Molecular Dynamics Simulations. *Biophys. J.* **2006**, *91*, 2198–2205. [CrossRef]

112. Koshiyama, K.; Kodama, T.; Yano, T.; Fujikawa, S. Molecular Dynamics Simulation of Structural Changes of Lipid Bilayers Induced by Shock Waves: Effects of Incident Angles. *Biochim. Biophys. Acta (BBA) Biomembr.* **2008**, *1778*, 1423–1428. [CrossRef]

113. Koshiyama, K.; Yano, T.; Kodama, T. Self-Organization of a Stable Pore Structure in a Phospholipid Bilayer. *Phys. Rev. Lett.* **2010**, *105*, 018105. [CrossRef]

114. Vogel, A.; Lauterborn, W.; Timm, R. Optical and Acoustic Investigations of the Dynamics of Laser-Produced Cavitation Bubbles near a Solid Boundary. *J. Fluid Mech.* **1989**, *206*, 299–338. [CrossRef]

115. Ohl, C.; Kurz, T.; Geisler, R.; Lindau, O.; Lauterborn, W. Bubble Dynamics, Shock Waves and Sonoluminescence. *Philos. Trans. R. Soc. London. Ser. A Math. Phys. Eng. Sci.* **1999**, *357*, 269–294. [CrossRef]

116. Lentacker, I.; De Cock, I.; Deckers, R.; De Smedt, S.C.; Moonen, C.T.W. Understanding Ultrasound Induced Sonoporation: Definitions and Underlying Mechanisms. *Adv. Drug. Deliv. Rev.* **2014**, *72*, 49–64. [CrossRef] [PubMed]

117. Ohl, C.-D.; Arora, M.; Ikink, R.; de Jong, N.; Versluis, M.; Delius, M.; Lohse, D. Sonoporation from Jetting Cavitation Bubbles. *Biophys. J.* **2006**, *91*, 4285–4295. [CrossRef] [PubMed]

118. Meijering, B.D.M.; Juffermans, L.J.M.; van Wamel, A.; Henning, R.; Zuhorn, I.; Emmer, M.; Versteilen, A.M.G.; Paulus, W.J.; van Gilst, W.; Kooiman, K.; et al. Ultrasound and Microbubble-Targeted Delivery of Macromolecules Is Regulated by Induction of Endocytosis and Pore Formation. *Circ. Res.* **2009**, *104*, 679–687. [CrossRef] [PubMed]

119. Derieppe, M.; Rojek, K.; Escoffre, J.-M.; de Senneville, B.D.; Moonen, C.; Bos, C. Recruitment of Endocytosis in Sonopermeabilization-Mediated Drug Delivery: A Real-Time Study. *Phys. Biol.* **2015**, *12*, 046010. [CrossRef]

120. Fekri, F.; Delos Santos, R.C.; Karshafian, R.; Antonescu, C.N. Ultrasound Microbubble Treatment Enhances Clathrin-Mediated Endocytosis and Fluid-Phase Uptake through Distinct Mechanisms. *PLoS ONE* **2016**, *11*, e0156754. [CrossRef] [PubMed]

121. De Cock, I.; Zagato, E.; Braeckmans, K.; Luan, Y.; de Jong, N.; De Smedt, S.C.; Lentacker, I. Ultrasound and Microbubble Mediated Drug Delivery: Acoustic Pressure as Determinant for Uptake via Membrane Pores or Endocytosis. *J. Control Release* **2015**, *197*, 20–28. [CrossRef]

122. Qin, P.; Han, T.; Yu, A.C.H.; Xu, L. Mechanistic Understanding the Bioeffects of Ultrasound-Driven Microbubbles to Enhance Macromolecule Delivery. *J. Control Release* **2018**, *272*, 169–181. [CrossRef]

123. Paula, D.M.B.; Valero-Lapchik, V.B.; Paredes-Gamero, E.J.; Han, S.W. Therapeutic Ultrasound Promotes Plasmid DNA Uptake by Clathrin-Mediated Endocytosis. *J. Gene Med.* **2011**, *13*, 392–401. [CrossRef]

124. Hauser, J.; Ellisman, M.; Steinau, H.-U.; Stefan, E.; Dudda, M.; Hauser, M. Ultrasound Enhanced Endocytotic Activity of Human Fibroblasts. *Ultrasound Med. Biol.* **2009**, *35*, 2084–2092. [CrossRef] [PubMed]

125. Basta, G.; Venneri, L.; Lazzerini, G.; Pasanisi, E.; Pianelli, M.; Vesentini, N.; Del Turco, S.; Kusmic, C.; Picano, E. In Vitro Modulation of Intracellular Oxidative Stress of Endothelial Cells by Diagnostic Cardiac Ultrasound. *Cardiovasc. Res.* **2003**, *58*, 156–161. [CrossRef]

126. Apodaca, G. Modulation of Membrane Traffic by Mechanical Stimuli. *Am. J. Physiol. Renal. Physiol.* **2002**, *282*, F179–F190. [CrossRef] [PubMed]

127. Vlahakis, N.E.; Schroeder, M.A.; Pagano, R.E.; Hubmayr, R.D. Deformation-Induced Lipid Trafficking in Alveolar Epithelial Cells. *Am. J. Physiol. Lung Cell Mol. Physiol.* **2001**, *280*, L938–L946. [CrossRef] [PubMed]

128. Kim, D.; Lee, S.S.; Moon, H.; Park, S.Y.; Lee, H.J. PD-L1 Targeting Immune-Microbubble Complex Enhances Therapeutic Index in Murine Colon Cancer Models. *Pharmaceuticals* **2021**, *14*, 6. [CrossRef]

129. Bettinger, T.; Tranquart, F. Design of Microbubbles for Gene/Drug Delivery. *Adv. Exp. Med. Biol.* **2016**, *880*, 191–204. [CrossRef]

130. Wang, D.S.; Panje, C.; Pysz, M.A.; Paulmurugan, R.; Rosenberg, J.; Gambhir, S.S.; Schneider, M.; Willmann, J.K. Cationic versus Neutral Microbubbles for Ultrasound-Mediated Gene Delivery in Cancer. *Radiology* **2012**, *264*, 721–732. [CrossRef]

131. Zhang, L.; Sun, Z.; Ren, P.; You, M.; Zhang, J.; Fang, L.; Wang, J.; Chen, Y.; Yan, F.; Zheng, H.; et al. Localized Delivery of ShRNA against PHD2 Protects the Heart from Acute Myocardial Infarction through Ultrasound-Targeted Cationic Microbubble Destruction. *Theranostics* **2017**, *7*, 51–66. [CrossRef]

132. Lentacker, I.; De Geest, B.G.; Vandenbroucke, R.E.; Peeters, L.; Demeester, J.; De Smedt, S.C.; Sanders, N.N. Ultrasound-Responsive Polymer-Coated Microbubbles That Bind and Protect DNA. *Langmuir* **2006**, *22*, 7273–7278. [CrossRef]

133. Borden, M.A.; Caskey, C.F.; Little, E.; Gillies, R.J.; Ferrara, K.W. DNA and Polylysine Adsorption and Multilayer Construction onto Cationic Lipid-Coated Microbubbles. *Langmuir* **2007**, *23*, 9401–9408. [CrossRef]

134. Escoffre, J.-M.; Mannaris, C.; Geers, B.; Novell, A.; Lentacker, I.; Averkiou, M.; Bouakaz, A. Doxorubicin Liposome-Loaded Microbubbles for Contrast Imaging and Ultrasound-Triggered Drug Delivery. *IEEE Trans Ultrason. Ferroelectr. Freq. Control* **2013**, *60*, 78–87. [CrossRef] [PubMed]

135. Frenkel, P.A.; Chen, S.; Thai, T.; Shohet, R.V.; Grayburn, P.A. DNA-Loaded Albumin Microbubbles Enhance Ultrasound-Mediated Transfection in Vitro. *Ultrasound Med. Biol.* **2002**, *28*, 817–822. [CrossRef]

136. Juang, E.K.; De Cock, I.; Keravnou, C.; Gallagher, M.K.; Keller, S.B.; Zheng, Y.; Averkiou, M. Engineered 3D Microvascular Networks for the Study of Ultrasound-Microbubble-Mediated Drug Delivery. *Langmuir* **2019**, *35*, 10128–10138. [CrossRef]

137. Omata, D.; Hagiwara, F.; Munakata, L.; Shima, T.; Kageyama, S.; Suzuki, Y.; Azuma, T.; Takagi, S.; Seki, K.; Maruyama, K.; et al. Characterization of Brain-Targeted Drug Delivery Enhanced by a Combination of Lipid-Based Microbubbles and Non-Focused Ultrasound. *J. Pharm. Sci.* **2020**, *109*, 2827–2835. [CrossRef] [PubMed]

138. Zhang, W.; Nan, S.-L.; Bai, W.-K.; Hu, B. Low-Frequency Ultrasound Combined with Microbubbles Improves Gene Transfection in Prostate Cancer Cells in Vitro and in Vivo. *Asia Pac. J. Clin. Oncol.* **2021**. [CrossRef] [PubMed]

139. Villanueva, F.S.; Jankowski, R.J.; Klibanov, S.; Pina, M.L.; Alber, S.M.; Watkins, S.C.; Brandenburger, G.H.; Wagner, W.R. Microbubbles Targeted to Intercellular Adhesion Molecule-1 Bind to Activated Coronary Artery Endothelial Cells. *Circulation* **1998**, *98*, 1–5. [CrossRef]

140. Ahmed, M.; Gustafsson, B.; Aldi, S.; Dusart, P.; Egri, G.; Butler, L.M.; Bone, D.; Dähne, L.; Hedin, U.; Caidahl, K. Molecular Imaging of a New Multimodal Microbubble for Adhesion Molecule Targeting. *Cel. Mol. Bioeng.* **2019**, *12*, 15–32. [CrossRef]

141. Barreiro, O.; Aguilar, R.J.; Tejera, E.; Megías, D.; de Torres-Alba, F.; Evangelista, A.; Sánchez-Madrid, F. Specific Targeting of Human Inflamed Endothelium and in Situ Vascular Tissue Transfection by the Use of Ultrasound Contrast Agents. *JACC Cardiovasc. Imaging* **2009**, *2*, 997–1005. [CrossRef]

142. Shentu, W.-H.; Yan, C.-X.; Liu, C.-M.; Qi, R.-X.; Wang, Y.; Huang, Z.-X.; Zhou, L.-M.; You, X.-D. Use of Cationic Microbubbles Targeted to P-Selectin to Improve Ultrasound-Mediated Gene Transfection of HVEGF165 to the Ischemic Myocardium. *J. Zhejiang Univ. Sci. B* **2018**, *19*, 699–707. [CrossRef] [PubMed]

143. Zhou, Q.; Deng, Q.; Hu, B.; Wang, Y.-J.; Chen, J.-L.; Cui, J.-J.; Cao, S.; Song, H.-N. Ultrasound Combined with Targeted Cationic Microbubble-mediated Angiogenesis Gene Transfection Improves Ischemic Heart Function. *Exp. Ther. Med.* **2017**, *13*, 2293–2303. [CrossRef]

144. Baston, C.; West, T.E. Lung Ultrasound in Acute Respiratory Distress Syndrome and Beyond. *J.Thorac. Dis.* **2016**, *8*, E1763–E1766. [CrossRef]
145. Sperandeo, M.; Varriale, A.; Sperandeo, G.; Filabozzi, P.; Piattelli, M.L.; Carnevale, V.; Decuzzi, M.; Vendemiale, G. Transthoracic Ultrasound in the Evaluation of Pulmonary Fibrosis: Our Experience. *Ultrasound Med. Biol.* **2009**, *35*, 723–729. [CrossRef]
146. Chan, V.; Perlas, A. Basics of Ultrasound Imaging. In *Atlas of Ultrasound-Guided Procedures in Interventional Pain Management*; Springer: Cham, Switzerland, 2011; pp. 13–19.
147. Sugiyama, M.G.; Mintsopoulos, V.; Raheel, H.; Goldenberg, N.M.; Batt, J.E.; Brochard, L.; Kuebler, W.M.; Leong-Poi, H.; Karshafian, R.; Lee, W.L. Lung Ultrasound and Microbubbles Enhance Aminoglycoside Efficacy and Delivery to the Lung in Escherichia Coli-Induced Pneumonia and Acute Respiratory Distress Syndrome. *Am. J. Respir. Crit. Care Med.* **2018**, *198*, 404–408. [CrossRef] [PubMed]
148. Espitalier, F.; Darrouzain, F.; Escoffre, J.-M.; Ternant, D.; Piver, E.; Bouakaz, A.; Remerand, F. Enhanced Amikacin Diffusion with Ultrasound and Microbubbles in a Mechanically Ventilated Condensed Lung Rabbit Model. *Front. Pharmacol.* **2020**, *10*, 1562. [CrossRef]
149. Liu, D.; Chen, Y.; Li, F.; Chen, C.; Wei, P.; Xiao, D.; Han, B. Sinapultide-Loaded Microbubbles Combined with Ultrasound to Attenuate Lipopolysaccharide-Induced Acute Lung Injury in Mice. *Drug Des. Devel. Ther.* **2020**, *14*, 5611–5622. [CrossRef]
150. Zhang, Z.; Lu, D.-S.; Zhang, D.-Q.; Wang, X.; Ming, Y.; Wu, Z.-Y. Targeted Antagonism of Vascular Endothelial Growth Factor Reduces Mortality of Mice with Acute Respiratory Distress Syndrome. *Curr. Med. Sci.* **2020**, *40*, 671–676. [CrossRef]
151. McMahon, D.; Hynynen, K. Acute Inflammatory Response Following Increased Blood-Brain Barrier Permeability Induced by Focused Ultrasound Is Dependent on Microbubble Dose. *Theranostics* **2017**, *7*, 3989–4000. [CrossRef] [PubMed]
152. McMahon, D.; Oakden, W.; Hynynen, K. Investigating the Effects of Dexamethasone on Blood-Brain Barrier Permeability and Inflammatory Response Following Focused Ultrasound and Microbubble Exposure. *Theranostics* **2020**, *10*, 1604–1618. [CrossRef] [PubMed]
153. Helfield, B.; Chen, X.; Watkins, S.C.; Villanueva, F.S. Biophysical Insight into Mechanisms of Sonoporation. *Proc. Natl. Acad. Sci. USA* **2016**, *113*, 9983–9988. [CrossRef] [PubMed]
154. Papadakos, P.J.; Gestring, M.L. (Eds.) Lung Ultrasound. In *Encyclopedia of Trauma Care*; Springer: Berlin/Heidelberg, Germany, 2015; p. 896.
155. Upadhyay, A.; Dalvi, S.V. Microbubble Formulations: Synthesis, Stability, Modeling and Biomedical Applications. *Ultrasound Med. Biol.* **2019**, *45*, 301–343. [CrossRef] [PubMed]
156. Endo-Takahashi, Y.; Negishi, Y. Microbubbles and Nanobubbles with Ultrasound for Systemic Gene Delivery. *Pharmaceutics* **2020**, *12*, 964. [CrossRef] [PubMed]
157. Chen, Z.-Y.; Liang, K.; Qiu, R.-X. Targeted Gene Delivery in Tumor Xenografts by the Combination of Ultrasound-Targeted Microbubble Destruction and Polyethylenimine to Inhibit Survivin Gene Expression and Induce Apoptosis. *J. Exp. Clin. Cancer Res.* **2010**, *29*, 152. [CrossRef] [PubMed]
158. Tan, J.-K.Y.; Pham, B.; Zong, Y.; Perez, C.; Maris, D.O.; Hemphill, A.; Miao, C.H.; Matula, T.J.; Mourad, P.D.; Wei, H.; et al. Microbubbles and Ultrasound Increase Intraventricular Polyplex Gene Transfer to the Brain. *J. Control Release* **2016**, *231*, 86–93. [CrossRef]
159. Devulapally, R.; Lee, T.; Barghava-Shah, A.; Sekar, T.V.; Foygel, K.; Bachawal, S.V.; Willmann, J.K.; Paulmurugan, R. Ultrasound-Guided Delivery of Thymidine Kinase-Nitroreductase Dual Therapeutic Genes by PEGylated-PLGA/PIE Nanoparticles for Enhanced Triple Negative Breast Cancer Therapy. *Nanomedicine* **2018**, *13*, 1051–1066. [CrossRef]
160. Liufu, C.; Li, Y.; Tu, J.; Zhang, H.; Yu, J.; Wang, Y.; Huang, P.; Chen, Z. Echogenic PEGylated PEI-Loaded Microbubble As Efficient Gene Delivery System. *IJN* **2019**, *14*, 8923–8941. [CrossRef]
161. Panje, C.M.; Wang, D.S.; Pysz, M.A.; Paulmurugan, R.; Ren, Y.; Tranquart, F.; Tian, L.; Willmann, J.K. Ultrasound-Mediated Gene Delivery with Cationic versus Neutral Microbubbles: Effect of DNA and Microbubble Dose on in Vivo Transfection Efficiency. *Theranostics* **2012**, *2*, 1078–1091. [CrossRef]
162. Sirsi, S.R.; Hernandez, S.L.; Zielinski, L.; Blomback, H.; Koubaa, A.; Synder, M.; Homma, S.; Kandel, J.J.; Yamashiro, D.J.; Borden, M.A. Polyplex-Microbubble Hybrids for Ultrasound-Guided Plasmid DNA Delivery to Solid Tumors. *J. Control Release* **2012**, *157*, 224–234. [CrossRef]
163. Xie, A.; Wu, M.D.; Cigarroa, G.; Belcik, J.T.; Ammi, A.; Moccetti, F.; Lindner, J.R. Influence of DNA-Microbubble Coupling on Contrast Ultrasound-Mediated Gene Transfection inMuscleand Liver. *J. Am. Soc. Echocardiogr.* **2016**, *29*, 812–818. [CrossRef]
164. Ferrara, K.W.; Borden, M.A.; Zhang, H. Lipid-Shelled Vehicles: Engineering for Ultrasound Molecular Imaging and Drug Delivery. *Acc. Chem. Res.* **2009**, *42*, 881–892. [CrossRef] [PubMed]
165. Liao, A.-H.; Ho, H.-C.; Lin, Y.-C.; Chen, H.-K.; Wang, C.-H. Effects of Microbubble Size on Ultrasound-Induced Transdermal Delivery of High-Molecular-Weight Drugs. *PLoS ONE* **2015**, *10*, e0138500. [CrossRef] [PubMed]
166. Miller, D.L.; Lu, X.; Fabiilli, M.; Dou, C. Influence of Microbubble Size and Pulse Amplitude on Hepatocyte Injury Induced by Contrast-Enhanced Diagnostic Ultrasound. *Ultrasound Med. Biol.* **2019**, *45*, 170–176. [CrossRef]
167. Cavalli, R.; Bisazza, A.; Lembo, D. Micro- and Nanobubbles: A Versatile Non-Viral Platform for Gene Delivery. *Int. J. Pharm.* **2013**, *456*, 437–445. [CrossRef] [PubMed]
168. Tayier, B.; Deng, Z.; Wang, Y.; Wang, W.; Mu, Y.; Yan, F. Biosynthetic Nanobubbles for Targeted Gene Delivery by Focused Ultrasound. *Nanoscale* **2019**, *11*, 14757–14768. [CrossRef]

169. Cavalli, R.; Bisazza, A.; Trotta, M.; Argenziano, M.; Civra, A.; Donalisio, M.; Lembo, D. New Chitosan Nanobubbles for Ultrasound-Mediated Gene Delivery: Preparation and in Vitro Characterization. *Int. J. Nanomedicine* **2012**, *7*, 3309–3318. [CrossRef] [PubMed]
170. du Toit, L.C.; Govender, T.; Pillay, V.; Choonara, Y.E.; Kodama, T. Investigating the Effect of Polymeric Approaches on Circulation Time and Physical Properties of Nanobubbles. *Pharm. Res.* **2011**, *28*, 494–504. [CrossRef]
171. Al-Jawadi, S.; Thakur, S.S. Ultrasound-Responsive Lipid Microbubbles for Drug Delivery: A Review of Preparation Techniques to Optimise Formulation Size, Stability and Drug Loading. *Int. J. Pharm.* **2020**, *585*, 119559. [CrossRef] [PubMed]
172. Gnyawali, V.; Moon, B.-U.; Kieda, J.; Karshafian, R.; Kolios, M.C.; Tsai, S.S.H. Honey, I Shrunk the Bubbles: Microfluidic Vacuum Shrinkage of Lipid-Stabilized Microbubbles. *Soft. Matter.* **2017**, *13*, 4011–4016. [CrossRef]
173. Matsumi, C.T.; Da Silva, W.J.; Schneider, F.K.; Maia, J.M.; Morales, R.E.M.; Filho, W.D.A. Micropipette-Based Microfluidic Device for Monodisperse Microbubbles Generation. *Micromachines* **2018**, *9*, 387. [CrossRef]
174. Pulsipher, K.W.; Hammer, D.A.; Lee, D.; Sehgal, C.M. Engineering Theranostic Microbubbles Using Microfluidics for Ultrasound Imaging and Therapy: A Review. *Ultrasound Med. Biol.* **2018**, *44*, 2441–2460. [CrossRef]
175. Xu, J.; Salari, A.; Wang, Y.; He, X.; Kerr, L.; Darbandi, A.; de Leon, A.C.; Exner, A.A.; Kolios, M.C.; Yuen, D.; et al. Microfluidic Generation of Monodisperse Nanobubbles by Selective Gas Dissolution. *Small* **2021**, e2100345. [CrossRef]
176. Dimcevski, G.; Kotopoulis, S.; Bjånes, T.; Hoem, D.; Schjøtt, J.; Gjertsen, B.T.; Biermann, M.; Molven, A.; Sorbye, H.; McCormack, E.; et al. A Human Clinical Trial Using Ultrasound and Microbubbles to Enhance Gemcitabine Treatment of Inoperable Pancreatic Cancer. *J. Control Release* **2016**, *243*, 172–181. [CrossRef] [PubMed]
177. Wang, Y.; Li, Y.; Yan, K.; Shen, L.; Yang, W.; Gong, J.; Ding, K. Clinical Study of Ultrasound and Microbubbles for Enhancing Chemotherapeutic Sensitivity of Malignant Tumors in Digestive System. *Chin. J. Cancer Res.* **2018**, *30*, 553–563. [CrossRef] [PubMed]
178. Rubenfeld, G.D. Confronting the Frustrations of Negative Clinical Trials in Acute Respiratory Distress Syndrome. *Ann. ATS* **2015**, *12*, S58–S63. [CrossRef] [PubMed]

Review

High Na$^+$ Salt Diet and Remodeling of Vascular Smooth Muscle and Endothelial Cells

Ghassan Bkaily *, Yanick Simon, Ashley Jazzar, Houssein Najibeddine, Alexandre Normand and Danielle Jacques

Department of Immunology and Cell Biology, Faculty of Medicine and Health Sciences, Université de Sherbrooke, Sherbrooke, QC J1H 5N4, Canada; yanick.simon@usherbrooke.ca (Y.S.); ashley.jazzar@usherbrooke.ca (A.J.); houssein.najibeddine@usherbrooke.ca (H.N.); alexandre.normand@usherbrooke.ca (A.N.); danielle.jacques@usherbrooke.ca (D.J.)
* Correspondence: ghassan.bkaily@usherbrooke.ca

Abstract: Our knowledge on essential hypertension is vast, and its treatment is well known. Not all hypertensives are salt-sensitive. The available evidence suggests that even normotensive individuals are at high cardiovascular risk and lower survival rate, as blood pressure eventually rises later in life with a high salt diet. In addition, little is known about high sodium (Na$^+$) salt diet-sensitive hypertension. There is no doubt that direct and indirect Na$^+$ transporters, such as the Na/Ca exchanger and the Na/H exchanger, and the Na/K pump could be implicated in the development of high salt-induced hypertension in humans. These mechanisms could be involved following the destruction of the cell membrane glycocalyx and changes in vascular endothelial and smooth muscle cells membranes' permeability and osmolarity. Thus, it is vital to determine the membrane and intracellular mechanisms implicated in this type of hypertension and its treatment.

Keywords: Na$^+$ salt; hypertension; Na$^+$ salt sensitive hypertension; vascular endothelial cells; vascular smooth muscle cells; glycocalyx; Na/Ca exchanger; Na/H exchanger

Citation: Bkaily, G.; Simon, Y.; Jazzar, A.; Najibeddine, H.; Normand, A.; Jacques, D. High Na$^+$ Salt Diet and Remodeling of Vascular Smooth Muscle and Endothelial Cells. *Biomedicines* **2021**, *9*, 883. https://doi.org/10.3390/biomedicines9080883

Academic Editor: Byeong Hwa Jeon

Received: 31 May 2021
Accepted: 21 July 2021
Published: 24 July 2021

Publisher's Note: MDPI stays neutral with regard to jurisdictional claims in published maps and institutional affiliations.

1. The Vascular System

The vascular system is a closed transport network with a rigid structure that contracts and relaxes. Under the impulse of the cardiac pump, this system ensures the transport of blood to supply the cells with oxygen and necessary nutrients and to eliminate their waste products for their proper functioning and the maintenance of their homeostasis [1].

From a macroscopic perspective, this system is divided into three major categories: veins (afferent vessels operating in a low-pressure system), arteries (efferent blood vessels operating in a high-pressure system), and capillaries (connecting veins and arteries). Microscopically, except the capillaries, artery and vein histological organization and their tissue composition are similar and divided into three major tunics or layers (Figure 1) [2].

On the surface, there is the tunica externa or adventitia, which is composed of loosely intertwined collagen fibers and fibroblasts (Figure 1) [2]. Between the circulating blood and the vascular wall is the tunica interna or intima, consisting primarily of a monolayer of vascular endothelial cells (VECs) (Figure 1) [2]. Finally, between the intima and the adventitia lies the tunica media or media composed of abundant elastic fibers and particularly contractile vascular smooth muscle cells (VSMCs), which control the vessel tone (Figure 1). VECs and VSMCs are heterogeneous. VSMCs are a fusiform cell population of approximately 200 µm in length and 5 µm in diameter. However, VECs are roughly 30–50 µm in length, 10–30 µm wide, and a thickness of 0.1–10 µm. Both cell types are of multiple embryonic origins (neural crests, mesoderm) [2,3].

Due to the complexity of their origin during early embryogenesis, understanding the differentiation of these cells remains a significant challenge [2,4]. Under physiological conditions, contractile VSMCs in mature blood vessels exhibited a low rate of synthesis of extracellular matrix components and was characterized by a series of highly regulated smooth muscle markers, such as cytoskeletal and contractile proteins, which include smooth muscle actin, myosin heavy chain, calponin, and smooth muscle 22 alpha (SM22α), as well as signaling molecules [2,4–6]. All are required for the primary function of VSMCs [4,7]. However, because contractile proteins and several transcription factors are Ca^{2+} dependents, the contractility of VSMCs is determined primarily by intracellular Ca^{2+}. Therefore, to regulate the various Ca^{2+}-dependent functions under normal conditions and during excitation-contraction coupling, VSMCs use various ion transporters and must keep intracellular Na^+ at a low concentration [8,9]. VECs markers are numerous, but the most used is the von Willebrand factor. They do not possess L-type Ca^{2+} channels, and the Ca^{2+} influx takes place via the R-type Ca^{2+} channels [10] and the Na/Ca exchanger. In both cell types, Na^+ influx takes place mainly via the Na/Ca and Na/H exchangers (Figures 2 and 3). These two membrane exchangers must effectively control intracellular Ca^{2+} and Na^+ homeostasis [5,6].

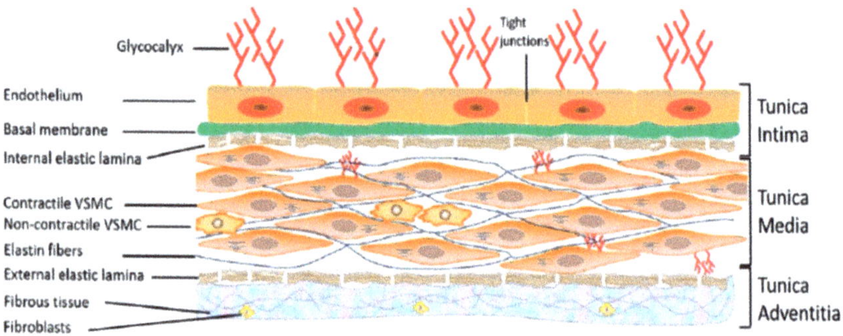

Figure 1. Structure of the vascular wall. Schematic representation showing the three layers of the vascular wall: tunica intima, tunica media, and tunica adventitia, as well as the components of each layer. VSMC: vascular smooth muscle cell. From Bkaily et al., 2021 [2].

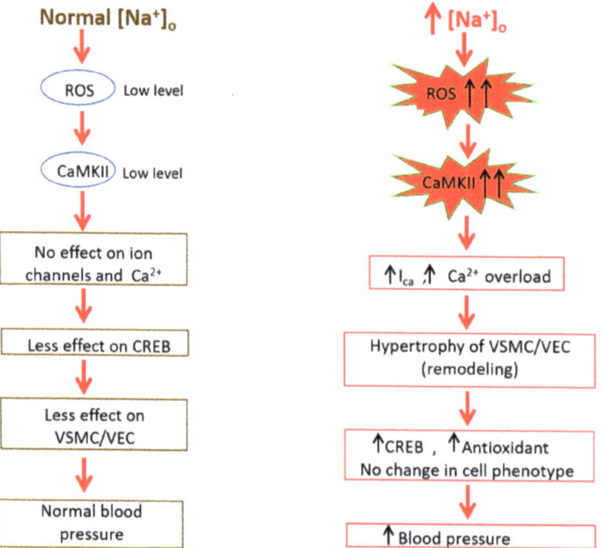

Figure 2. Schematic representation summarizing the literature in the field showing that chronic high salt induced an increase in the intracellular levels of ROS (reactive oxygen species) and activation of CaMKII (calmodulin kinase II) and cyclic AMP response element binding protein (CREB). I_{Ca}: L-type Ca^{2+} channels; [Na^+]$_o$: extracellular Na^+ concentration.

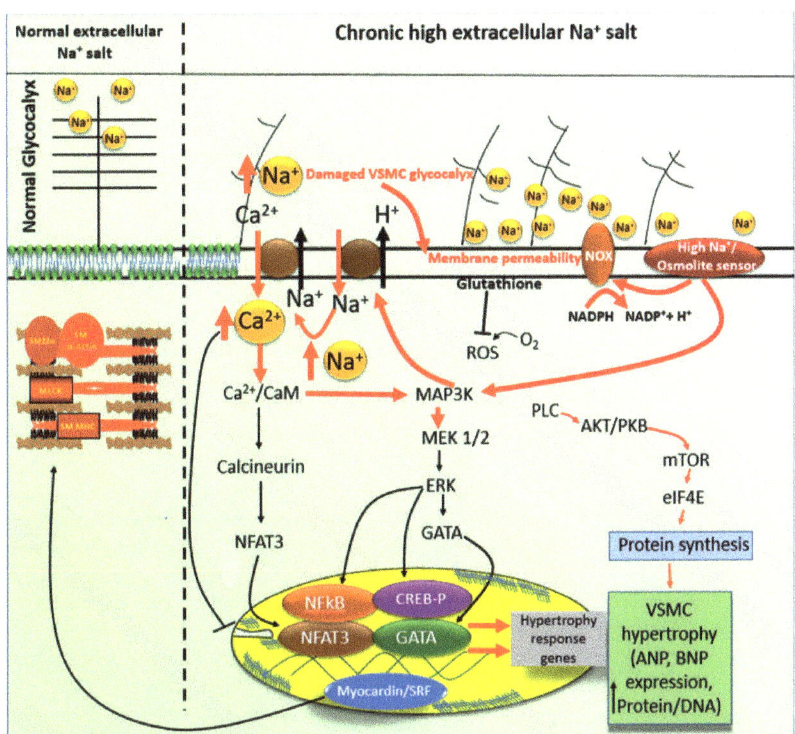

Figure 3. Schematic representation summarizing the literature on the glycocalyx, plasma membrane Na⁺ transport, and Ca²⁺-dependent signaling and transcription factors that could be implicated in chronic high salt induced vascular smooth muscle and endothelial cells. ROS: reactive oxygen species; Ca²⁺/CaM: calcium calmodulin; VSMC: vascular smooth muscle cells. Modified from Bkaily et al., 2021 [2].

2. Vascular Remodeling

The concept of vascular remodeling was first described by Baumbach, based on observations in arterioles of hypertensive rats [11]. It is an active process of structural adaptation of VSMCs and VECs to hemodynamic changes or to long-term vascular damage [12]. Depending on the type of hemodynamic changes and VECs and/or VSMCs injury, vascular remodeling is characterized structurally by hypertrophy (wall thickening), eutrophy (constant wall thickness), or hypotrophy (wall thinning) [12–15]. These structural changes may be eccentric (increased remodeled arterial lumen) to accommodate reduced luminal space due to atherogenic lesions or post-intraluminal restenosis, to maintain adequate blood flow [14]. However, these changes may be concentric (reduction in remodeled arterial lumen) due to prolonged wall tension and vasoconstriction as in hypertension [14,16,17].

Among the different cellular and molecular mechanisms observed, the media is the most active layer [18]. As a result, VSMCs contribute largely to the phenomenon of vascular remodeling by undergoing morphological changes characterized mainly by hypertrophy and hyperplasia [19] without changes in the contractile phenotype of the cells [2]. Little is known about the structural and morphological remodeling of VECs and, more particularly, in humans.

As all muscle cells, VSMCs also undergo hypertrophy in response to local stimuli [20–22]. This hypertrophic process is a cell growth response characterized by increased cell size with or without an increase in protein synthesis or changes in the cell phenotype [2,20,23]. In general, this cell remodeling may be due, on the one hand, to physiological conditions as in pregnancy [24] or secondary to physical exercise [25]. On the other hand, it can be due to pathological conditions, such as hypertension [26] and high sodium salt diet [20]. During these pathologies, an elevation of intracellular Ca^{2+} level of VSMCs and an increase in sensitivity to vasoactive stimuli have been reported [20,27–29]. These pieces of information indicate that hypertrophied VSMCs, similar to cells with a contractile phenotype, retain the ability to interact with stimuli and increase intracellular Ca^{2+} levels following a signaling pathway different from normal calcium dynamics [30,31]. Indeed, the mechanism underlying VSMCs hypertrophy is poorly studied and may occur due to the degradation of protein inhibitors or protein synthesis stimulation [19,32]. Atef and Anand-Srivastava, including other authors, have demonstrated that the Gqα protein-related signaling pathway is a classical pathway of the hypertrophic response of VSMCs [33,34]. Generally, this pathway can be triggered by growth factors, or by one or more varieties of vasoactive substances, including angiotensin II, endothelin-1and neuropeptide Y [19,21,32,35].

3. Sodium and Sodium Transport in Vascular Smooth Muscle and Endothelial Cells

Intracellular free Ca^{2+} and Na^+ are not homogeneously distributed in excitable and non-excitable cells [36–38]. When VSMCs or VECs are at rest, the concentration of free Ca^{2+} in the cytoplasm, perinucleoplasm, and nucleoplasm are 50, 600, and 300 mmol/L, respectively [36,37]. Furthermore, the concentration of Na^+ in these three compartments is 10, 40, and 20 mmol/L, respectively [10,36]. As in resting cells, in response to increased Na^+ or Ca^{2+} influxes across the sarcolemma membrane, the concentration of nucleoplasmic free Ca^{2+} and Na^+ are higher than that of the cytoplasm and lower than that of the perinucleoplasm [10,20,36,38].

As mentioned previously, in VSMCs and VECs, there are several types of Na^+ transporters: the Na^+/H^+ exchanger (NHE1), the Na^+/Ca^{2+} exchanger (NCX) and the Na^+/K^+ pump (Figures 2 and 3). Functionally, NHE-1 is involved in cytoskeletal organization, cell volume regulation [39], differentiation, proliferation [40,41], cell migration, and even apoptosis [42]. It plays a primary role in the pH regulation at both the cytosolic and nuclear levels [38,43,44]. In the nucleus, it is known to be involved in the activation of chromatin [45] and nuclear pore functioning [38,46]. Undoubtedly, this exchanger's activity can directly affect gene expression and perinucleoplasmic, nucleoplasmic, and cytoplasmic homeostasis of Na^+ and Ca^{2+} under normal and pathological conditions [37,38,43]. During intracellular acidosis, the H^+ outflux will induce Na^+ influx through this exchanger and contribute to an intracellular increase of Na^+ (Figures 2 and 3) [2,20,47].

Another important Na^+ transporter that continually cross-talk with the Na^+/H^+ exchanger is the sodium/calcium exchanger isoform 1 (NCX1) (Figures 2 and 3). The NCX1 is a transmembrane protein that was cloned in 1990 and is expressed at the cytoplasmic membrane, nuclear membrane, and in the mitochondria of a variety of cells, such as hepatocytes, cardiomyocytes, endothelial cells, VECs, and VSMCs [37,43,48–51]. It has an amino-terminal portion composed of 5 transmembrane domains and a carboxy-terminal portion consisting of 4 transmembrane domains [49,50]. These portions are separated by a large cytosolic loop containing an endogenous XIP (exchanger inhibitor peptide) region, a binding site for Ca^{2+} regulation, and a region where alternative splicing occurs [50–52].

Functionally, this exchanger involves at least $1Ca^{2+}$ for $3Na^+$ [48,53]. Depending on the Na^+ concentration gradient, this system can also be reversed [48,53]. In the presence of intracellular Ca^+ overload and the absence of an increase of intracellular Na^+, this exchanger excludes Ca^{2+} from the cell [54]. Thus, it regulates Na^+ homeostasis and indirectly Ca^{2+} homeostasis at the cytosolic, perinucleoplasmic, and nucleoplasmic levels [43].

In sum, these two ion transporters are essential for Na^+ and Ca^{2+} homeostasis [36,43]. At the cytosolic level, this homeostasis is directly regulated by the nucleus [36,55]. Furthermore, independent of the cytosol and like a cell within a cell, the nucleus is regulated by an auto nuclear mechanism that protects it from trauma or damage [36,43,55]. However, an alteration in the cytosolic and/or nuclear compartment of Na^+ and Ca^{2+} homeostasis could affect excitation-contraction (VSMCs) and excitation-secretion (VECs) coupling, cell function, and survival resulting in vascular remodeling [37,38].

There is several other Na^+ transporters that are less known in VECs and VSMCs, such as the Na^+-bicarbonate (NBCn), the Na^+-Cl^--K^+ (NKCC), and the taurine-Na^+ symporters.

Recently, several reports suggest the presence of an epithelial Na^+ channel in renal vessels from male C57BL/6J mice [56] and rat mesenteric VSMCs [57,58], as well as in rat mesenteric artery endothelial cells [58,59]. However, this type of channel in VSMCs and VECs, and, more particularly, from the healthy human origin, is still a matter of debate because of the absence of a specific blocker of this channel.

4. High Sodium Salt-Induced Salt-Sensitive Memory

Extracellular Na^+ is a primary determinant of plasma osmolarity and VSMCs tone [60–62]. It is finely maintained under normal conditions and in humans at a concentration between 135 and 145 mmol/L [60]. However, the latter depends in particular on the daily salt intake estimated at less than five gr per day (Na^+: 2400 and Cl^-: 3000) [63–65]. However, in humans, the physiological salt requirement is less than 1 gr per day (Na^+: 400 and Cl^-: 600) [66–68].

With salt as a preservative and to improve the organoleptic character of foods, humans have dramatically increased their consumption to over 9.6 gr per day (Na^+: 3840 and Cl^-: 5760) [69,70]. Studies by Chauveau et al. on the proportion of average daily salt consumption in industrialized countries showed that 75% of the salt consumed is found in preserved foods, 15% is related to kitchen preparation, and only 10% is naturally present in foods [69]. This excess consumed salt cannot be eliminated by the kidneys. It can lead to an accumulation of extracellular Na^+ ($[Na^+]_o$) of 2 to 4 mmol/L [63,65,68,71–74].

It is well known that Na^+ can have a direct effect on blood pressure [59,62,75,76]. Studies in hypertensive patients have shown that $[Na^+]_o$ increases by 1 to 3 mmol/L [63,77]. According to Suckling and colleagues, ingestion of 6 gr per day of salt in a healthy subject can increase extracellular Na^+ of 2 mmol/L and osmolarity of 4 mosm/L [78]. Other studies in healthy subjects have shown that an increase in extracellular Na^+ of 3 mmol/L can lead to alteration of ion transporters and movement of fluid from intracellular to extracellular space associated with secretory excitation of certain hormones, such as aldosterone, renin, and vasopressin [63,73,77]. Changes of only 1% in plasma osmolarity are sufficient to cause a significant increase in plasma vasopressin [79]. The latter can increase blood pressure following tonus contraction of VSMCs [80]. By causing an alteration in intracellular ion homeostasis, prolonged accumulation of extracellular Na^+ may also promote cellular hyperosmotic stress and contribute to salt-sensitive hypertension [63,80,81].

Kawasaki et al. and Weinberger et al. were among the first to recognize the heterogeneity of the blood pressure response to a sodium-rich diet and to develop the concept of salt sensitivity in humans [82,83]. According to clinical findings, it is defined as a factor contributing to an increase in blood pressure of at least 10% [72]. According to De la Sierra and colleagues, salt sensitivity contributes to the rise in mean pressure of more than four mmHg (24-h ambulatory blood pressure monitoring) [84]. Depending on the definitions and measurement methods used, salt sensitivity is observed in 25% to 50% of normotensive subjects (BP < 120/80) and 40% to 75% of hypertensive patients [85]. However, this prevalence appears to be more pronounced in elderly, obese, renal failure, and African American subjects [71].

To date, the mechanism responsible for salt sensitivity remains controversial [63,86]. For a long time, and even today, some authors attribute it to renal malfunction [63,87]. According to them, after a regular hypersodium diet, these salt-sensitive patients show decreased renal blood flow, an increase in renal vascular resistance, and intraglomerular pressure [88]. According to Kawasaki, the inability of the kidneys to excrete excess sodium may be either secondary to primary hyperaldosteronism or renal pathology or genetic [63,82,83,85]. Recently, attention is no longer directed solely to the kidneys but primarily to the vascular system [59,62,65,81,89,90]. Thus, particular interest has been focused on the first barrier located on the surface of the endothelium and vascular smooth muscle cells, the glycocalyx [2,74,91,92].

The VSMCs and VECs glycocalyx (Figure 1) is a negatively charged anionic biopolymer layer at hundreds of nanometer thicknesses [68,93]. This thin layer in VECs acts as a barrier and prevents nonspecific adhesion of circulating blood cells to the endothelium, slows blood flow in the capillary system [94] and selectively controls VECs and VSMCs cell membrane Na^+ permeability and vascular permeability [68,95]. When this layer is exposed to a chronic concentration of 5% NaCl above the standard physiological value, a significant reduction of negatively charged heparan sulfate residues occurs [96]. The loss of these surface charges renders VECs and VSMCs (second protective barrier) vulnerable to unwanted intruders, including excessive sodium, leading to VECs shrinkage to more than 25% [74,91,92]. This may promote increased vascular permeability, the release of VECs vasoactive substances (angiotensin II, endothelin-1) [81,89], which lead to direct exposure of VSMCs to these substances and sodium overload. Such changes in VSMCs extracellular space induce remodeling associated with an increase in muscle tension which leads to hypertension.

According to various epidemiological studies and animal models developed and used, it is well established that dietary salt intake is the most common and important risk factor for developing essential hypertension [56,59,81,82,89]. It is also well known that a chronic concentration of 2 to 4 mmol/L extracellular Na^+ can directly affect blood pressure in both normal and hypertensive subjects [63]. However, the cellular and molecular mechanisms underlying salt sensitivity and associated vascular disorders are not fully understood [81]. Some suggest renal dysfunction, while others hypothesize vascular dysfunction by demonstrating impairment of both vascular barriers (endothelial glycocalyx and endothelium) following chronic exposure above 145 mM NaCl [74,91,92].

Knowing that VSMCs are the central contracting cells of the vascular wall, this slight increase in plasma sodium may lead to altered intracellular Na^+ and Ca^{2+} homeostasis promoting hypertrophy and/or hyperplasia of these cells (Figure 3) [2,97–99]. Eventually, this would promote increased peripheral resistance and blood pressure (Figure 3) [62,91]. However, the mechanisms responsible for the increase in blood pressure still obscure.

5. High Sodium Salt-Induced VSMCs and VECs Stress

Hyperosmotic stress is an often-overlooked process that potentially contributes to the pathogenesis and progression of various human pathologies (hypertension, diabetes, atherosclerosis, and other cardiovascular diseases) [80]. It is the increase in extracellular osmolarity above the average physiological value (280–300 mOsm/kg H_2O) [100] that can be observed in various cell types, such as T and B cells, macrophages, neurons, epithelial cells, renal cells, myoblasts, fibroblasts, and, especially, VSMCs and VECs (Figures 2 and 3) [101–103].

Depending on the cell type, responses to hyperosmotic stress can be variable, and the signaling pathways involved differ from cell to cell [80].

However, in all cell types, hyperosmotic stress is characterized by shrinkage of cell volume, increased oxidative stress (Figure 3) [100], protein carbonylation, mitochondrial depolarization, DNA double-strand breaks caused mainly by activation of p53 and/or p38, and cell cycle arrest [80,104,105]. Depending on the duration of exposure and NaCl concentration, cell cycle arrest is short-lived to prevent cell death by apoptosis and give the cell time to adapt to the increased osmolarity [105]. To date, there are very few studies that specify the different signaling pathways involved [80]. Studies have shown activation of the mitogen-activated protein kinase (MAP kinase) and c-Jun N-terminal kinases (JNK pathway) (Figures 2 and 3) [104,105], including expression of the aquaporin 1 (AQP1) and aquaporin 5 (AQP5) genes, facilitate water movement [106,107]. Besides, increased calcium influx leading to nuclear factor of activated T-cells 5 (NFAT5) activation and nuclear translocation has been observed (Figures 2 and 3) [101,104,105]. The latter leads to the subsequent regulation of target genes, including those associated with osmolyte transport and synthesis, antioxidant defense, and numerous molecular chaperones [80] (Figures 2 and 3).

6. High Na$^+$ Salt-Induced Glycocalyx Remodeling

Several reviews are available concerning glycocalyx and main, particularly in VEC [108]. In 1940, the cell biologist James Danielli (who discovered that the membrane is a lipid bilayer) hypothesized that a layer of proteins covered the vascular system's inner walls [108,109]. Later on, this plasma membrane layer was given the name glycocalyx [108,110]. Several groups have highlighted the physiological role of the endothelial glycocalyx, given the importance of this structure as a shear stress sensor of blood flow, thus contributing to blood pressure regulation [58,68,95].

The glycocalyx is formed by two essential components: proteoglycans, syndecans, glypicans, and glycoproteins. [108]. The transmembrane proteoglycans are the critical element in all cell types and, more particularly, in the endothelial glycocalyx. There are four known syndecans in vertebrates: syndecans 1, 2, 3, and 4; however, the glycocalyx endothelial contains primarily syndecan-1 [108].

The glycocalyx also plays a role in mechanotransduction. Since glycocalyx is negatively charged, it acts to buffer positively charged substances and, more particularly, Na$^+$, the most abundant positively charged ion. Thus, it contributes to the modulation of extracellular surface charge and acts as a buffer of extracellular Na$^+$, and controls its cell membrane permeability. However, chronic loading of the glycocalyx with Na$^+$ at the heparan sulfate residues leads to collapse [110] and a decrease of glycocalyx [20]. Such destruction of the glycocalyx affects plasma membrane charges, which will affect the level of membrane potential, ligand-receptor binding, and ionic transporters. Therefore, damaging the glycocalyx affects the excitation-secretion coupling of VECs [101,108] and excitation-contraction coupling of VSMCs, as well as intracellular homeostasis of Ca^{2+}, Na$^+$, and ROS (Figure 3) [20]. A deterioration of the glycocalyx seems to occur during aging [111–113]. This particular aspect needs to be verified in healthy humans to determine whether this age-dependent deterioration of the glycocalyx is independent of renal dysfunction. Besides, it is not clear in the literature whether the vascular remodeling in a chronic high salt diet is due totally to deterioration and/or decrease in glycocalyx is the main contributor to the physiopathology of chronic high salt diet.

7. Adaptive Responses to High Sodium Salt Induced VSMCs Hypertrophy

In response to hyperosmotic stress, cells develop several compensatory and adaptive mechanisms [80,114]. When a small perturbation in extracellular osmolarity occurs, an accumulation of inorganic osmolytes (K$^+$, Cl$^-$, Na$^+$) increases cell volume [115–117]. During this process, an increase in NHE-1 activation and an increase in sodium influx have been observed in some studies [105,115]. The increase of ion transporters, in particular for Na$^+$, constitutes a double-edged sword by preventing cell volume shrinkage, on the one hand, and severely disrupting intracellular ion homeostasis, on the other hand (Figures 2 and 3) [80]. Secondly, this will lead to an overexpression of genes (SLC2A4, SLC5A3, SLC6A8, SLC9A1) involved in the synthesis (Figure 3) and transport of compatible organic osmolytes (betaine, sorbitol, taurine, choline, creatine, myoinositol, glucose) [80,118,119]. The latter are small molecules concentrated inside the cell that usually have cytoprotective properties, such as antioxidation, and structural stabilization of proteins by acting as chemical chaperones [80,119]. In general, these compatible osmolytes utilize the Na$^+$ gradient across the plasma membrane as an electromotive force (Figure 3) [117]. In addition, high extracellular activate an osmolyte sensor, which stimulates NOXs and induces ROS generation (Figures 2 and 3). Taurine is known to stimulate protein synthesis, promote gene expression, and promote increased sodium and calcium influx [2,54,118,119]. Osmolyte accumulation promotes cytoskeletal rearrangement, an essential adaptation in response to increased extracellular osmolarity by allowing the cell to maintain its volume and enhance its structural integrity [80,119].

Despite the various adaptive mechanisms, cells adapted to hyperosmotic stress differ from normal cells [105]. They can remodel by hypertrophying [2,20,115] and or proliferating. According to various studies, they enter a state where several changes are associated with multiple persistent lesions, such as DNA double-strand breaks, oxidation of DNA bases and proteins, and cytoskeleton remodeling [80,105]. The mechanism implicated in high salt-induced VSMCs hypertrophy remains to be explored in VECs.

8. Na$^+$ Salt-Sensitive Hypertension

The role of Na$^+$ in the physiological evolution of animals is nicely reviewed by Natochin in 2007 [120]. The relation between high Na$^+$ salt consummation and hypertension was first reported in 1904 by Ambard and Beaujard [121,122] and then confirmed by the groups of Dahl [3,120] and Freis [123].

Hypertension is among the most critical public health problems worldwide despite advances in prevention, detection, treatment, and blood pressure control [124]. Its prevalence depends on the study population's racial composition and its criteria. It also depends on sex, gender, and age. In general, it is defined as systolic blood pressure (SBP) > 140 mmHg and/or diastolic blood pressure (DBP) > 90 mmHg at rest 9 stage 2 and on multiple occasions [124,125]. It is a multifactorial disease clinically classified as stage 1 to 3, depending on the severity of blood pressure [124,126]. From the etiological point of view, it is classified into secondary arterial hypertension and essential or primary arterial hypertension. However, in response to salt intake, it has been observed that hypertensive subjects can be salt-resistant (increase in blood pressure less than 10% after salt intake) or salt-sensitive (an increase of BP more than 10% after salt intake) [82,83].

Intracellular Na$^+$ homeostasis in VEC and VSMC depends partly on the level of plasma circulating Na$^+$ [64,127] and on the type and density of plasma membrane Na$^+$ transporters. In humans, the physiological need for sodium salts should not exceed 2.5 g per day [68]. Today, salt overconsumption is a significant health issue [127]. Occasional consumption of high Na$^+$ salt by healthy humans has no significant effect on blood pressure since the kidneys eliminate it. However, due to the limited capacity of the kidneys to eliminate chronic high Na$^+$, the excess of this ion accumulates in the circulation leading to the development of Na$^+$ salt-sensitive hypertension [58,110,127,128]. The accumulation of circulating Na$^+$ is aggravated in the presence of renal dysfunction [127], with chronic levels reportedly elevated by 2–4 mM beyond normal resting values [63,127]. This chronic increase in circulating Na$^+$ was reported to affect only VECs [80] and damages their glycocalyx, leading to remodeling of VECs [20,68,110,128–131]. However, an increase in vascular permeability will allow VSMC interstitial Na$^+$ overload, affecting VSMC glycocalyx (Figures 2 and 3). Our group recently reported the latter aspect to occur together with morphological remodeling and an increase in basal intracellular Na$^+$ and Ca^{2+} (Figures 2 and 3). Such a remodeling of VSMCs may lead to hypertension [63,78,83,132,133]. The chronic increase in Na$^+$ salts promotes epigenetic 'salt memory' programming [110,134], which predisposes the patient to Na$^+$ salt-sensitive hypertension [110,127]. This salt memory programming is transmitted to by the parents to their children. Thus, chronic high salt my induce permanent remodeling at the gene level. What is gene implicated? This still to be clarified.

9. Implication of ROS/RNS in Na$^+$ Sensitive Hypertension

Several review papers highlighted the role of reactive oxygen species (ROS) and reactive nitrogen species (RNS) in the development of several pathologies [135–138], including hypertension [135]. Cellular radicals (hydroxyl, nitric oxide, nitrogen dioxide and superoxide anion) and nonradicals (hydrogen peroxide, hypochlorous acid, and peroxynitrite) are generated by mechanisms present in the cell, such as: the plasma membrane, the endoplasmic reticulum, the mitochondria, the peroxisomes, and the cytosol [135,136], as well as the nuclear envelop membranes' and the nucleoplasm [37,38,139]. Na$^+$ sensitive hypertension also seems to implicate oxidative stress in many animal models [138,140,141]. This increase in oxidative stress was attributed to inflammation, as well as to renal epithelial cell damage, by activation of NADPH oxidase [136,138,140,142] NOX2/NOX4-derived ROS [138,142]. Such an increase in oxidative stress was reported to activate nuclear factor-kappa B [141], as well as several mechanisms implicated in cell membrane ionic transporters, in addition to aquaporin 1 sensitive transmembrane transporter of hydrogen peroxide [136,143,144] and osmolyte sensor (Figure 3). However, recent literature in the field showed that high -salt-sensitive hypertension is due, at least in part, to damage of the glycocalyx of both vascular endothelial [136] and vascular smooth muscle [20] cells. It is logical to mention that sustained high salt would first induce damage to endothelial cells, and then the inflammation would follow. It is worth noting that the effect of high salt could be relatively less important in one particular vascular cell type compared to another due to the relative density and presence of different types of NOXs, as well as to the different basal levels of oxidant and anti-oxidant factors [20]. Although the mitochondria play an essential role in ROS generation, we should not forget that the nucleus may also contribute to the ROS generation in high-salt sensitive hypertension via probably activation of the calcium-dependent NOX5 [139]. Thus, it is imperative to revisit the nature of essential ROS/RNS generation in high-salt sensitive hypertension and, more particularly, in human vascular endothelial and smooth muscle cells.

10. Conclusions

Although Na$^+$ sensitive hypertension was reported for the first time in 1904, our knowledge of this disease is still limited and, there is no yet cure for such vascular illness yet. In addition, the mechanisms implicated in developing a memory of high Na$^+$ salt-sensitive hypertension is still obscure. There still a lot to be done in this field that is to be rediscovered. Several questions need to be answered: is the damage to the glycocalyx the main most important factor contributing to the development of hypertension and Na$^+$ salt sensitivity? Is the chronic increase of extracellular osmolarity induced permanent remodeling of VECs and/or VSMCs? What is the mechanism implicated in the development of salt-sensitive hypertension? To answer these questions and other questions, we need to develop more representee in vivo and/or in vitro models that help us better explore this type of hypertension and develop specific treatment.

Author Contributions: G.B. and D.J. were responsible for the conception, design, and written of the review; Y.S., A.J., H.N., and A.N. conducted a literature search and reading of the manuscript. All authors have read and agreed to the published version of the manuscript.

Funding: The work was supported by the National Sciences and Engineering Research Council of Canada (NSERC).

Institutional Review Board Statement: Not applicable.

Informed Consent Statement: Not applicable.

Conflicts of Interest: The authors declare that the research was conducted in the absence of any commercial of financial relationships that could be construed as a potential conflict of interest.

References

1. Pugsley, M.K.; Tabrizchi, R. The vascular system: An overview of structure and function. *J. Pharmacol. Toxicol.* **2000**, *44*, 333–340. [CrossRef]
2. Bkaily, G.; Abou Abdallah, N.; Simon, Y.; Jazzar, A.; Jacques, D. Vascular smooth muscle remodeling in health and disease. *Can. J. Physiol. Pharmacol.* **2021**, *99*, 171–178. [CrossRef] [PubMed]
3. Bennett, H.S. Morphological aspects of extracellular polysaccharides. *J. Histochem. Cytochem.* **1963**, *11*, 14–23. [CrossRef]
4. Wang, G.; Jacquet, L.; Karamariti, E.; Xu, Q. Origin and differentiation of vascular smooth muscle cells. *J. Physiol.* **2015**, *593*, 3013–3030. [CrossRef] [PubMed]
5. Brozovich, F.V.; Nicholson, C.J.; Degen, C.V.; Gao, Y.Z.; Aggarwal, M.; Morgan, K.G. Mechanisms of Vascular Smooth Muscle Contraction and the Basis for Pharmacologic Treatment of Smooth Muscle Disorders. *Pharmacol. Rev.* **2016**, *68*, 476–532. [CrossRef] [PubMed]
6. Chistiakov, D.A.; Orekhov, A.N.; Bobryshev, Y.V. Vascular smooth muscle cell in atherosclerosis. *Acta Physiol.* **2015**, *214*, 33–50. [CrossRef] [PubMed]

7. Rzucidlo, E.M.; Martin, K.A.; Powell, R.J. Regulation of vascular smooth muscle cell differentiation. *J. Vasc. Surg.* **2007**, *45*, A25–A32. [CrossRef]
8. Friedman, S.M. Sodium in blood vessels. *J. Vasc. Res.* **1979**, *16*, 2–16. [CrossRef]
9. Owens, G.K. Control of hypertrophic versus hyperplastic growth of vascular smooth muscle cells. *Am. J. Physiol. Heart Circ. Physiol.* **1989**, *257*, H1755–H1765. [CrossRef] [PubMed]
10. Bkaily, G.; Jaalouk, D.; Haddad, G.; Gros-Louis, N.; Simaan, M.; Naik, R.; Pothier, P. Modulation of cytosolic and nuclear Ca^{2+} and Na^{+} transport by taurine in heart cells. *Mol. Cell. Biochem.* **1997**, *170*, 1–8. [CrossRef]
11. Baumbach, G.L.; Heistad, D.D. Remodeling of cerebral arterioles in chronic hypertension. *Hypertension* **1989**, *13*, 968–972. [CrossRef] [PubMed]
12. Gibbons, G.H.; Dzau, V.J. The emerging concept of vascular remodeling. *N. Engl. J. Med.* **1994**, *330*, 1431–1438. [PubMed]
13. Feihl, F.; Liaudet, L.; Waeber, B.; Levy, B.I. Hypertension: A disease of the microcirculation? *Hypertension* **2006**, *48*, 1012–1017. [CrossRef]
14. Renna, N.F.; de las Heras, N.; Miatello, R.M. Pathophysiology of vascular remodeling in hypertension. *Int. J. Hypertens.* **2013**, *2013*, 808353. [CrossRef] [PubMed]
15. Sharma, R.; Dever, D.P.; Lee, C.M.; Azizi, A.; Pan, Y.; Camarena, J.; Köhnke, T.; Bao, G.; Porteus, M.H.; Majeti, R. The trace-seq method tracks recombination alleles and identifies clonal reconstitution dynamics of gene targeted human hematopoietic stem cells. *Nat. Commun.* **2021**, *12*, 472. [CrossRef]
16. Lai, E.Y.; Onozato, M.L.; Solis, G.; Aslam, S.; Welch, W.J.; Wilcox, C.S. Myogenic responses of mouse isolated perfused renal afferent arterioles: Effects of salt intake and reduced renal mass. *Hypertension* **2010**, *55*, 983–989. [CrossRef] [PubMed]
17. Raffetto, J.D.; Ligi, D.; Maniscalco, R.; Khalil, R.A.; Mannello, F. Why Venous Leg Ulcers Have Difficulty Healing: Overview on Pathophysiology, Clinical Consequences, and Treatment. *J. Clin. Med.* **2020**, *24*, 29. [CrossRef]
18. Boutouyrie, P.; Laurent, S. Remodelage des grosses et petites artères dans l'hypertension artérielle. *Sang Thromb. Vaiss.* **2004**, *16*, 81–89.
19. Berk, B.C. Vascular smooth muscle growth: Autocrine growth mechanisms. *Physiol. Rev.* **2001**, *81*, 999–1030. [CrossRef]
20. Bkaily, G.; Simon, Y.; Menkovic, I.; Bkaily, C.; Jacques, D. High salt-induced hypertrophy of human vascular smooth muscle cells associated with a decrease in glycocalyx. *J. Mol. Cell. Cardiol.* **2018**, *125*, 1–5. [CrossRef]
21. Jacques, D.; Bkaily, G. Endocardial endothelial cell hypertrophy takes place during the development of hereditary cardiomyopathy. *Mol. Cell. Biochem.* **2019**, *453*, 157–161. [CrossRef]
22. Shiraishi, Y.; Ishigami, N.; Kujiraoka, T.; Sato, A.; Fujita, M.; Ido, Y.; Adachi, T. Deletion of Superoxide Dismutase 1 Blunted Inflammatory Aortic Remodeling in Hypertensive Mice under Angiotensin II Infusion. *Antioxidants* **2021**, *10*, 471. [CrossRef] [PubMed]
23. Walsh, K.; Shiojima, I.; Gualberto, A. DNA replication and smooth muscle cell hypertrophy. *J. Clin. Investig.* **1999**, *104*, 673–674. [CrossRef] [PubMed]
24. Osol, G.; Moore, L.G. Maternal uterine vascular remodeling during pregnancy. *Microcirculation* **2014**, *21*, 38–47. [CrossRef] [PubMed]
25. Hellsten, Y.; Nyberg, M. Cardiovascular adaptations to exercise training. *Compr. Physiol.* **2015**, *15*, 1–32.
26. Daou, G.B.; Srivastava, A.K. Reactive oxygen species mediate Endothelin-1-induced activation of ERK1/2, PKB, and Pyk2 signaling, as well as protein synthesis, in vascular smooth muscle cells. *Free Radic. Biol. Med.* **2004**, *37*, 208–215. [CrossRef]
27. Bohr, D.F.; Webb, R.C. Vascular smooth muscle membrane in hypertension. *Annu. Rev. Pharmacol. Toxicol.* **1988**, *28*, 389–409. [CrossRef] [PubMed]
28. Noon, J.P.; Rice, P.J.; Baldessarini, R.J. Calcium leakage as a cause of the high resting tension in vascular smooth muscle from the spontaneously hypertensive rat. *Proc. Natl. Acad. Sci. USA* **1978**, *75*, 1605–1607. [CrossRef] [PubMed]
29. Orlov, S.; Resink, T.J.; Bernhardt, J.; Ferracin, F.; Buhler, F.R. Vascular smooth muscle cell calcium fluxes. Regulation by angiotensin II and lipoproteins. *Hypertension* **1993**, *21*, 195–203. [CrossRef]
30. Wamhoff, B.R.; Bowles, D.K.; Owens, G.K. Excitation–transcription coupling in arterial smooth muscle. *Circ. Res.* **2006**, *98*, 868–878. [CrossRef]
31. Wamhoff, B.R.; Bowles, D.K.; Dietz, N.J.; Hu, Q.; Sturek, M. Exercise training attenuates coronary smooth muscle phenotypic modulation and nuclear Ca^{2+} signaling. *Am. J. Physiol. Heart Circ. Physiol.* **2002**, *283*, H2397–H2410. [CrossRef]
32. Geisterfer, A.A.; Peach, M.J.; Owens, G.K. Angiotensin II induces hypertrophy, not hyperplasia, of cultured rat aortic smooth muscle cells. *Circ. Res.* **1988**, *62*, 749–756. [CrossRef]
33. Atef, M.E.; Anand-Srivastava, M.B. Enhanced expression of Gqα and PLC-β1 proteins contributes to vascular smooth muscle cell hypertrophy in SHR: Role of endogenous angiotensin II and endothelin-1. *Am. J. Physiol. Cell Physiol.* **2014**, *307*, C97–C106. [CrossRef] [PubMed]
34. Ohtsu, H.; Higuchi, S.; Shirai, H.; Eguchi, K.; Suzuki, H.; Hinoki, A.; Brailoiu, E.; Eckhart, A.D.; Frank, G.D.; Eguchi, S. Central role of Gq in the hypertrophic signal transduction of angiotensin II in vascular smooth muscle cells. *Endocrinology* **2008**, *149*, 3569–3575. [CrossRef]
35. Jacques, D.; Abdel-Karim Abdel-Malak, N.; Abou Abdallah, N.; Al-Khoury, J.; Bkaily, G. Difference in the response to angiotensin II between left and right ventricular endocardial endothelial cells. *Can. J. Physiol. Pharmacol.* **2017**, *95*, 1271–1282. [CrossRef] [PubMed]

36. Bkaily, G. The possible role of Ca^{2+} and K^+ channels in VSM pathophysiology. In *Ion Channels in Vascular Smooth Muscle*; Bkaily, G., Ed.; R.G. Landers Company: Austin, TX, USA, 1994; pp. 103–113.

37. Bkaily, G.; Nader, M.; Avedanian, L.; Choufani, S.; Jacques, D.; D'Orléans-Juste, P.; Gobeil, F.; Chemtob, S.; Al-Khoury, J. G-protein-coupled receptors, channels, and Na^+–H^+ exchanger in nuclear membranes of heart, hepatic, vascular endothelial, and smooth muscle cells. *Can. J. Physiol. Pharmacol.* **2006**, *84*, 431–441. [CrossRef]

38. Bkaily, G.; Avedanian, L.; Al-Khoury, J.; Ahmarani, L.; Perreault, C.; Jacques, D. Receptors and ionic transporters in nuclear membranes: New targets for therapeutical pharmacological interventions. *Can. J. Physiol. Pharmacol.* **2012**, *90*, 953–965. [CrossRef]

39. Alexander, R.T.; Grinstein, S. Na^+/H^+ exchangers and the regulation of volume. *Acta Physiol.* **2006**, *187*, 159–167. [CrossRef]

40. Stock, C.; Schwab, A. Role of the Na^+/H^+ exchanger NHE1 in cell migration. *Acta Physiol.* **2006**, *187*, 149–157. [CrossRef] [PubMed]

41. Pouysségur, J.; Sardet, C.; Franchi, A.; L'Allemain, G.; Paris, S. A specific mutation abolishing Na/H antiport activity in hamster fibroblasts precludes growth at neutral and acidic ph. *Proc. Nat. Acad. Sci. USA* **1984**, *81*, 4833–4837. [CrossRef] [PubMed]

42. Pedersen, S.F. The Na^+/H^+ exchanger NHE1 in stress-induced signal transduction: Implications for cell proliferation and cell death. *Pflüg. Arch. Eur. J. Physiol.* **2006**, *45*, 249–259. [CrossRef]

43. Bkaily, G.; Avedanian, L.; Jacques, D. Nuclear membranes' receptors and channels as targets for drug development in cardiovascular diseases. *Can. J. Physiol. Pharmacol.* **2009**, *87*, 108–119. [CrossRef] [PubMed]

44. Cingolani, H.E.; Irene, L. Sodium-Hydrogen exchanger, cardiac overload, and myocardial hypertrophy. *Circulation* **2007**, *115*, 1090–1100. [CrossRef] [PubMed]

45. Oberleithner, H.; Schuricht, B.; Wünsch, S.; Schneider, S.; Püschel, B. Role of H^+ ions in volume and voltage of epithelial cell nuclei. *Pflug. Arch. Eur. J. Phsysiol.* **1993**, *423*, 88–96. [CrossRef] [PubMed]

46. Oberleithner, H.; Schiller, H.; Wilhelmi, M.; Butzke, D.; Danker, T. Nuclear pores collapse in response to CO_2 imaged with atomic force microscopy. *Pflug. Arch. Eur. J. Physiol.* **2000**, *439*, 251–255. [CrossRef] [PubMed]

47. Karmazyn, M.; Kilić, A.; Javadov, S. The role of NHE-1 in myocardial hypertrophy and remodeling. *J. Mol. Cell. Cardiol.* **2008**, *44*, 647–653. [CrossRef] [PubMed]

48. Nicoll, D.A.; Longoni, S.; Philipson, K.D. Molecular cloning and functional expression of the cardiac sarcolemmal Na^+-Ca^{2+} exchanger. *Science* **1990**, *250*, 562–565. [CrossRef] [PubMed]

49. Blaustein, M.P.; Lederer, W.J. Sodium/calcium exchange: Its physiological implications. *Physiol. Rev.* **1999**, *79*, 763–854. [CrossRef] [PubMed]

50. Brini, M.; Carafoli, E. The plasma membrane Ca^{2+} ATPase and the plasma membrane sodium calcium exchanger cooperate in the regulation of cell calcium. *Cold Spring Harb. Perspect. Biol.* **2011**, *3*, 1–15. [CrossRef] [PubMed]

51. O'Donnell, M.E.; Owen, N.E. Regulation of ion pumps and carriers in vascular smooth muscle. *Physiol. Rev.* **1994**, *74*, 683–722. [CrossRef]

52. Philipson, K.D.; Nicoll, D.A. Sodium-Calcium Exchange: À Molecular Perspective. *Annu. Rev. Physiol.* **2000**, *62*, 111–133. [CrossRef]

53. Gill, D.L.; Grollman, E.F.; Kohn, L.D. Calcium transport mechanisms in membrane vesicles from guinea pig brain synaptosomes. *J. Biol. Chem.* **1981**, *256*, 184–192. [CrossRef]

54. Bkaily, G.; Jaalouk, D.; Sader, S.; Shbaklo, H.; Pothier, P.; Jacques, D.; D'Orléans-Juste, P.; Cragoe, E.J., Jr.; Bose, R. Taurine indirectly increases [Ca]$_i$ by inducing Ca^{2+} influx through the Na^+-Ca^{2+} Exchanger. In *Molecular and Cellular Effects of Nutrition on Disease Processes*; Pierce, G.N., Izumi, T., Rupp, H., Grynberg, A., Eds.; Springer: Boston, MA, USA, 1998; pp. 187–197.

55. Bkaily, G.; Gros-Louis, N.; Naik, R.; Jaalouk, D.; Pothier, P. Implication of the nucleus in excitation contraction coupling of heart cells. *Mol. Cell. Biochem.* **1996**, *154*, 113–121. [CrossRef]

56. Jernigan, N.L.; Drummond, H.A. Vascular ENaC proteins are required for renal myogenic constriction. *Am. J. Renal Physiol.* **2005**, *289*, F891–F901. [CrossRef] [PubMed]

57. Jernigan, N.L.; Lamarca, B.; Speed, J.; Galmiche, L.; Granger, J.P.; Drummond, H.A. Dietary salt enhances benzamil-sensitive component of myogenic constriction in mesenteric arteries. *Am. J. Physiol. Heart Circ. Physiol.* **2008**, *294*, H409–H420. [CrossRef]

58. Kushe-Vihrog, K.; Jeggle, P.; Oberleithner, H. The role of ENaC in vascular endothelium. *Pflug. Arch. Eur. J. Physiol.* **2014**, *466*, 851–859. [CrossRef]

59. Yang, X.; Niu, N.; Liang, C.; Wu, M.M.; Tang, L.L.; Wang, Q.S. Stimulation of epithelial sodium channels in endothelial cells by bone morphogenetic protein-4 contributes to salt-sensitive hypertension in rats. *Oxid. Med. Cell. Longev.* **2020**, *2020*, 3921897. [CrossRef] [PubMed]

60. Reynolds, R.M.; Padfield, P.L.; Seckl, J.R. Disorders of sodium balance. *BMJ* **2006**, *332*, 702–705. [CrossRef]

61. Rose, B.D. New approach to disturbances in the plasma sodium concentration. *Am. J. Med.* **1986**, *81*, 1033–1040. [CrossRef]

62. Shenouda, N.; Ramick, G.M.; Lennon, L.S.; Farquhar, B.W.; Edwards, G.D. High dietary sodium augments vascular tone and attenuates low-flow mediated constriction in salt-resistant adults. *Eur. J. Appl. Physiol.* **2020**, *120*, 1383–1389. [CrossRef] [PubMed]

63. He, F.J.; MacGregor, G.A. Plasma sodium and hypertension. *Kidney Int.* **2004**, *66*, 2454–2466. [CrossRef] [PubMed]

64. Uzan, A.; Delaveau, P. The salt content of food: A public health problem. *Ann. Pharm. Fr.* **2009**, *67*, 291–294. [CrossRef] [PubMed]

65. Chen, Y.; Wang, Z.; Yang, Z.; Yang, Y.; Yang, J.; Han, H.; Yang, H. The effect of different dietary levels of sodium and chloride on performance, blood parameters and excreta quality in goslings at 29 to 70 days of age. *J. Anim. Physiol. Anim. Nutr.* **2021**. [CrossRef]

66. Eaton, S.B.; Konner, M. Paleolithic nutrition. A consideration of its nature and current implications. *N. Engl. J. Med.* **1985**, *312*, 283–289. [CrossRef] [PubMed]

67. Meneton, P.; Jeunemaitre, X.; de Wardener, H.E.; Macgregor, G.A. Links between dietary salt intake, renal salt handling, blood pressure, and cardiovascular diseases. *Physiol. Rev.* **2005**, *85*, 679–715. [CrossRef] [PubMed]

68. Oberleithner, H. Vascular endothelium: A vulnerable transit zone for merciless sodium. *Nephrol. Dial. Transplant.* **2013**, *29*, 240–246. [CrossRef]

69. Chauveau, P.; Fouque, D.; Combe, C.; Aparicio, M. Évolution de l'alimentation du paléolithique à nos jours: Progression ou régression? *Néphrologie Thérapeutique* **2013**, *9*, 202–208. [CrossRef]

70. Powles, J.; Fahimi, S.; Micha, R.; Khatibzadeh, S.; Shi, P.; Ezzati, M.; Engell, R.E.; Lim, S.S.; Danaei, G.; Mozaffarian, D. Global, regional and national sodium intakes in 1990 and 2010: A systematic analysis of 24 h urinary sodium excretion and dietary surveys worldwide. *BMJ Open* **2013**, *3*, e003733. [CrossRef] [PubMed]

71. Luft, F.C.; Rankin, L.I.; Bloch, R.; Weyman, A.E.; Willis, L.R.; Murray, R.H.; Grim, C.E.; Weinberger, M.H. Cardiovascular and humoral responses to extremes of sodium intake in normal black and white men. *Circulation* **1979**, *60*, 697–706. [CrossRef]

72. Sullivan, J.M.; Ratts, T.E.; Taylor, J.C.; Kraus, D.H.; Barton, B.R.; Patrick, D.R.; Reed, S.W. Hemodynamic effects of dietary sodium in man: A preliminary report. *Hypertension* **1980**, *2*, 506–514. [CrossRef]

73. Sagnella, G.A.; Markandu, N.D.; Buckley, M.G.; Miller, M.A.; Singer, D.R.; MacGregor, G.A. Hormonal responses to gradual changes in dietary sodium intake in humans. *Am. J. Physiol.* **1989**, *256*, R1171–R1175. [CrossRef]

74. Kliche, K.; Jeggle, P.; Pavenstädt, H.; Oberleithner, H. Role of cellular mechanics in the function and life span of vascular endothelium. *Pflüg. Arch. Eur. J. Physiol.* **2011**, *462*, 209–217. [CrossRef] [PubMed]

75. Elliott, P.; Walker, L.L.; Little, M.P.; Blair-West, J.R.; Shade, R.E.; Lee, D.R.; Rouquet, P.; Leroy, E.; Jeunemaitre, X.; Ardaillou, R.; et al. Change in salt intake affects blood pressure of chimpanzees: Implications for human populations. *Circulation* **2007**, *116*, 1563–1568. [CrossRef]

76. He, F.J.; MacGregor, G.A. Reducing population salt intake worldwide: From evidence to implementation. *Prog. Cardiovasc. Dis.* **2010**, *52*, 363–382. [CrossRef] [PubMed]

77. He, F.J.; Markandu, N.D.; Sagnella, G.A.; de Wardener, H.E.; MacGregor, G.A. Plasma Sodium: Ignored and Underestimated. *Hypertension* **2004**, *4*, 98–102. [CrossRef]

78. Suckling, R.J.; He, F.J.; Markandu, N.D.; MacGregor, G.A. Dietary salt influences postprandial plasma sodium concentration and systolic blood pressure. *Kidney Int.* **2012**, *81*, 407–411. [CrossRef] [PubMed]

79. Verbalis, J.G. Disorders of body water homeostasis. *Best Pract. Res. Clin. Endocrinol. Metab.* **2003**, *17*, 471–503. [CrossRef]

80. Brocker, C.; Thompson, D.C.; Vasiliou, V. The role of hyperosmotic stress in inflammation and disease. *Biomol. Concepts* **2012**, *3*, 345–364. [CrossRef]

81. Choi, H.Y.; Park, H.C.; Ha, S.K. Salt sensitivity and hypertension: A paradigm shift from kidney malfunction to vascular endothelial dysfunction. *Electrolyte Blood Press.* **2015**, *13*, 7–16. [CrossRef] [PubMed]

82. Kawasaki, T.; Delea, C.S.; Bartter, F.C.; Smith, H. The effect of high-sodium and low-sodium intakes on blood pressure and other related variables in human subjects with idiopathic hypertension. *Am. J. Med.* **1978**, *64*, 193–198. [CrossRef]

83. Weinberger, M.H. Salt Sensitivity of Blood Pressure in Humans. *Hypertension* **1996**, *27*, 481–490. [CrossRef]

84. De la Sierra, A.; Giner, V.; Bragulat, E.; Coca, A. Lack of correlation between two methods for the assessment of salt sensitivity in essential hypertension. *J. Hum. Hypertens.* **2002**, *16*, 255–260. [CrossRef]

85. The GenSalt Collaborative Research Group. Genetic epidemiology network of salt sensitivity (GenSalt): Rationale, design, methods, and baseline characteristics of study participants. *J. Hum. Hypertens.* **2007**, *21*, 639–646. [CrossRef]

86. Ando, K.; Fujita, T. Pathophysiology of salt sensitivity hypertension. *Ann. Med. Hels.* **2012**, *44*, S119–S126. [CrossRef]

87. De Wardener, H.E. The primary role of the kidney and salt intake in the etiology of essential hypertension: Part I. *Clin. Sci.* **1990**, *79*, 193–200. [CrossRef]

88. Campese, V.M.; Parise, M.; Karubian, F.; Bigazzi, R. Abnormal renal hemodynamics in black salt-sensitive patients with hypertension. *Hypertension* **1991**, *18*, 805–812. [CrossRef] [PubMed]

89. Denton, D.; Weisinger, R.; Mundy, N.I.; Wickings, E.J.; Dixson, A.; Moisson, P.; Pingard, A.M.; Shade, R.; Carey, D.; Ardaillou, R. The effect of increased salt intake on blood pressure of chimpanzees. *Nat. Med.* **1995**, *1*, 1009–1016. [CrossRef] [PubMed]

90. Paudel, P.; Mcdonald, J.F.; Fronius, M. The δ subunit of epithelial sodium channel in humans—A potential player in vascular physiology. *Am. J. Physiol. Heart Circ. Physiol.* **2021**, *320*, H487–H493. [CrossRef]

91. Kusche-Vihrog, K.; Oberleithner, H. An emerging concept of vascular salt sensitivity. *F1000 Biol. Rep.* **2012**, *4*, 20. [CrossRef] [PubMed]

92. Vestweber, D. Novel insights into leukocyte extravasation. *Curr. Opin. Hematol.* **2012**, *19*, 212–217. [CrossRef]

93. Bevan, J.A. Flow regulation of vascular tone. Its sensitivity to changes in sodium and calcium. *Hypertension* **1993**, *22*, 273–281. [CrossRef] [PubMed]

94. Pries, A.R.; Secomb, T.W.; Gaehtgens, P. The endothelial surface layer. *Pflüg. Arch. Eur. J. Physiol.* **2000**, *440*, 653–666. [CrossRef]

95. Oberleithner, H. A physiological concept unmasking vascular salt sensitivity in man. *Pflug. Arch. Eur. J. Physiol.* **2012**, *464*, 287–293. [CrossRef]

96. Oberleithner, H.; Peters, W.; Kusche-Vihrog, K.S.; Schillers, H.; Kliche, K.; Oberleithner, K. Salt overload damages the glycocalyx sodium barrier of vascular endothelium. *Pflüg. Arch. Eur. J. Physiol.* **2011**, *462*, 519–528. [CrossRef]

97. Lee, R.; Garfield, R.E.; Forrest, J.B.; Daniel, E.E. Morphometric study of structural changes in the mesenteric blood vessels of spontaneously hypertensive rats. *J. Vasc. Res.* **1983**, *20*, 57–71. [CrossRef] [PubMed]

98. Mulvany, M.J.; Baandrup, U.; Gundersen, H.J.G. Evidence for hyperplasia in mesenteric resistance vessels of spontaneously hypertensive rats using a three-dimensional disector. *Circ. Res.* **1985**, *57*, 794–800. [CrossRef] [PubMed]

99. Schiffrin, E.L. Vascular Remodeling in Hypertension. *Hypertension* **2012**, *59*, 367–374. [CrossRef]

100. Kuper, C.; Beck, F.-X.; Neuhofer, W. Osmoadaptation of Mammalian cells-an orchestrated network of protective genes. *Curr. Genom.* **2007**, *8*, 209–218. [CrossRef] [PubMed]

101. Halterman, J.A.; Kwon, H.M.; Zargham, R.; Bortz, P.D.S.; Wamhoff, B.R. Nuclear factor of activated T cells 5 regulates vascular smooth muscle cell phenotypic modulation. *Arterioscler. Thromb. Vasc. Biol.* **2011**, *31*, 2287–2296. [CrossRef] [PubMed]

102. Neuhofer, W. Role of NFAT5 in inflammatory disorders associated with osmotic stress. *Curr. Genom.* **2010**, *11*, 584–590. [CrossRef]

103. Trama, J.; Go, W.Y.; Ho, S.N. The osmoprotective function of the NFAT5 transcription factor in T cell development and activation. *J. Immunol.* **2002**, *169*, 5477–5488. [CrossRef]

104. Reinehr, R.; Häussinger, D. Hyperosmotic activation of the CD95 death receptor system. *Acta Physiol.* **2006**, *187*, 199–203. [CrossRef] [PubMed]

105. Burg, M.B.; Ferraris, J.D.; Dmitrieva, N.I. Cellular response to hyperosmotic stresses. *Physiol. Rev.* **2007**, *87*, 1441–1474. [CrossRef] [PubMed]

106. Maruyama, T.; Kadowaki, H.; Okamoto, N.; Nagai, A.; Naguro, I.; Matsuzawa, A.; Shibuya, H.; Tanaka, K.; Murata, S.; Takeda, K. CHIP-dependent termination of MEKK2 regulates temporal ERK activation required for proper hyperosmotic response. *EMBO J.* **2010**, *29*, 2501–2514. [CrossRef]

107. Umenishi, F.; Yoshihara, S.; Narikiyo, T.; Schrier, R.W. Modulation of hypertonicity-induced aquaporin-1 by sodium chloride, urea, betaine, and heat shock in murine renal medullary cells. *J. Am. Soc. Nephrol.* **2005**, *16*, 600–607. [CrossRef] [PubMed]

108. Gaudette, S.; Hughes, D.; Boller, M. The endothelial glycocalyx: Structure and function in health and critical illness. *J. Vet. Emerg. Crit. Care* **2020**, *30*, 117–134. [CrossRef]

109. Danielli, J.F. Capillary permeability and oedema in the perfused frog. *J. Physiol.* **1940**, *98*, 109–129. [CrossRef]

110. Kusche-Vihrog, K.; Schmitz, B.; Brand, E. Salt controls endothelial and vascular phenotype. *Pflug. Arch. Eur. J. Physiol.* **2015**, *467*, 499–512. [CrossRef] [PubMed]

111. Machin, D.R.; Bloom, S.I.; Campbell, R.A.; Phuong, T.T.; Gates, P.E.; Lesniewski, L.A.; Rondina, M.T.; Donato, A.J. Advanced age results in a diminished endothelial glycocalyx. *Am. J. Physiol. Heart Circ. Physiol.* **2018**, *315*, H531–H539. [CrossRef]

112. Machin, D.R.; Phuong, T.T.; Donato, A.J. The role of the endothelial glycocalyx in advanced age and cardiovascular disease. *Curr. Opin. Pharmacol.* **2019**, *45*, 66–71. [CrossRef]

113. Osuka, A.; Kusuki, H.; Yoneda, K.; Matsuura, H.; Matsumoto, H.; Ogura, H.; Ueyama, M. Glycocalyx shedding is enhanced by age and correlates with increased fluid requirement in patients with major burns. *Shock* **2018**, *50*, 60–65. [CrossRef]

114. Dmitrieva, N.I.; Michea, L.F.; Rocha, G.M.; Burg, M.B. Cell cycle delay and apoptosis in response to osmotic stress. *Comp. Biochem. Physiol. Part A Mol. Integr. Physiol.* **2001**, *130*, 411–420. [CrossRef]

115. Schaffer, S.; Takahashi, K.; Azuma, J. Role of osmoregulation in the actions of taurine. *Amino Acids* **2000**, *19*, 527–546. [CrossRef]

116. Wehner, F.; Tinel, H. Osmolyte and Na$^+$ transport balances of rat hepatocytes as a function of hypertonic stress. *Pflüg. Arch. Eur. J. Physiol.* **2000**, *441*, 12–24. [CrossRef] [PubMed]

117. Wehner, F.; Olsen, H.; Tinel, H.; Kinne-Saffran, E.; Kinne, R.K. Cell volume regulation: Osmolytes, osmolyte transport, and signal transduction. *Rev. Physiol. Biochem. Pharmacol.* **2003**, *148*, 1–80.

118. Alfieri, R.R.; Bonelli, M.A.; Cavazzoni, A.; Brigotti, M.; Fumarola, C.; Sestili, P.; Mozzoni, P.; De Palma, G.; Mutti, A.; Carnicelli, D. Creatine as a compatible osmolyte in muscle cells exposed to hypertonic stress. *J. Physiol.* **2006**, *576*, 391–401. [CrossRef] [PubMed]

119. Yancey, P.H. Organic osmolytes as compatible, metabolic and counteracting cytoprotectants in high osmolarity and other stresses. *J. Exp. Biol.* **2005**, *208*, 2819–2830. [CrossRef] [PubMed]

120. Natochin, Y.V. The physiological evolution of animals: Sodium is the clue to resolving contradictions. *Her. Russ. Acad. Sci.* **2007**, *77*, 581–591. [CrossRef]

121. Ambard, L.; Beaujard, E. Causes of arterial hypertension. *Arch. Gen. Med.* **1904**, *1*, 520–533.

122. Amiri, M.; Kelishadi, R. Can salt hypothesis explain the trends of mortality from stroke and stomach cancer in western Europe? *Int. J. Prev. Med.* **2012**, *3*, 377–378.

123. Rapp, J.P. Dahl Salt-Susceptible and Salt-Resistant Rats. *Hypertension* **1982**, *4*, 753–763. [CrossRef]

124. Freis, E.D. The role of salt in hypertension. *Blood Press.* **1992**, *1*, 196–200. [CrossRef]

125. Rivasi, G.; Fedorowski, A. Hypertension, hypotension and syncope. *Minerva Med.* **2021**. [CrossRef]

126. Lim, S.S.; Vos, T.; Flaxman, A.D.; Danaei, G.; Shibuya, K.; Adair-Rohani, H.; Amann, M.; Anderson, H.R.; Andrews, K.G.; Aryee, M.; et al. A comparative risk assessment of burden of disease and injury attributable to 67 risk factors and risk factor clusters in 21 regions, 1990–2010: A systematic analysis for the Global Burden of Disease Study 2010. *Lancet* **2012**, *380*, 2224–2260. [CrossRef]

127. Chobanian, A.V.; Bakris, G.L.; Black, H.R.; Cushman, W.C.; Green, L.A.; Izzo, J.L.; Jones, D.W.; Materson, B.J.; Oparil, S.; Wright, J.T. Seventh report of the joint national committee on prevention, detection, evaluation, and treatment of high blood pressure. *Hypertension* **2003**, *42*, 1206–1252. [CrossRef]

128. Marketou, M.E.; Maragkoudakis, S.; Anastasiou, I.; Nakou, H.; Plataki, M.; Vardas, P.E.; Parthenakis, F.I. Salt-induced effects on microvascular function: A critical factor in hypertension mediated organ damage. *J. Clin. Hypertens.* **2019**, *21*, 749–757. [CrossRef]

129. Nijst, P.; Verbrugge, F.H.; Grieten, L.; Dupont, M.; Steels, P.; Tang, W.H.W.; Mullens, W. The pathophysiological role of interstitial sodium in heart failure. *J. Am. Coll. Cardiol.* **2015**, *65*, 378–388. [CrossRef] [PubMed]

130. Dogné, S.L.; Flamion, B.; Caron, N. Endothelial Glycocalyx as a Shield Against Diabetic Vascular Complications: Involvement of Hyaluronan and Hyaluronidases. *Arterioscler. Thromb. Vasc. Biol.* **2018**, *8*, 1427–1439. [CrossRef] [PubMed]

131. Gu, J.W.; Anand, V.; Shek, E.W.; Moore, M.C.; Brady, A.L.; Kelly, W.C.; Adair, T.H. Sodium induces hypertrophy of cultured myocardial myoblasts and vascular smooth muscle cells. *Hypertension* **1998**, *31*, 1083–1087. [CrossRef] [PubMed]

132. Zhou, X.; Naguro, I.; Ichijo, H.; Watanabe, K. Mitogen-activated protein kinases as key players in osmotic stress signaling. *Biochim. Biophys. Acta* **2016**, *1860*, 2037–2052. [CrossRef] [PubMed]

133. Li, J.; White, J.; Guo, L.; Zhao, X.; Wang, J.; Smart, E.J.; Li, X.A. Salt inactivates endothelial nitric oxide synthase in endothelial cells. *J. Nutr.* **2009**, *139*, 447–451. [CrossRef]

134. Touyz, R.M.; Schiffrin, E.L. Signal transduction mechanisms mediating the physiological and pathophysiological actions of angiotensin II in vascular smooth muscle cells. *Pharmacol. Rev.* **2000**, *52*, 639–672. [PubMed]

135. Oguchi, H.; Sasamura, H.; Shinoda, K.; Morita, S.; Kono, H.; Nakagawa, K.; Ishiguro, K.; Hayashi, K.; Nakamura, M.; Azegami, T.; et al. Renal arteriolar injury by salt intake contributes to salt memory for the development of hypertension. *Hypertension* **2014**, *64*, 784–791. [CrossRef]

136. Di Meo, S.; Reed, T.T.; Venditti, P.; Victor, V.M. Role of ROS and RNS sources in physiological and pathological conditions. *Oxid. Med. Cell. Longev.* **2016**, *2016*, 1245049. [CrossRef] [PubMed]

137. Ahmarani, L.; Avedanian, L.; Al-Khoury, J.; Perreault, C.; Jacques, D.; Bkaily, G. Whole-cell and nuclear NADPH oxidases levels and distribution in human endocardial endothelial, vascular smooth muscle, and vascular endothelial cells. *Can. J. Physiol. Pharmacol.* **2013**, *91*, 71–79. [CrossRef]

138. Bayorh, M.A.; Ganafa, A.A.; Socci, R.R.; Silvestrov, N.; Abukhalaf, I.K. The role of oxidative stress in salt-induced hypertension. *Am. J. Hypertens.* **2004**, *17*, 31–36.

139. Cai, H. Hydrogen peroxide regulation of endothelial function: Origins, mechanisms, and consequences. *Cardiovasc. Res.* **2005**, *68*, 26–36. [CrossRef] [PubMed]

140. Li, L.; Lai, E.Y.; Luo, Z.; Solis, G.; Mendonca, M.; Griendling, K.K.; Wellstein, A.; Welch, W.J.; Wilcox, C.S. High Salt Enhances Reactive Oxygen Species and Angiotensin II Contractions of Glomerular Afferent Arterioles from Mice with Reduced Renal Mass. *Hypertension* **2018**, *72*, 1208–1216. [CrossRef] [PubMed]

141. Koga, Y.; Hirooka, Y.; Araki, S.; Nozoe, M.; Kishi, T.; Sunagawa, K. High salt intake enhances blood pressure increase during development of hypertension via oxidative stress in rostral ventrolateral medulla of spontaneously hypertensive rats. *Hypertens. Res.* **2008**, *31*, 2075–2083. [CrossRef]

142. Dornas, W.C.; Cardoso, L.M.; Silva, M.; Machado, N.L.S.; Chianca, D.A., Jr.; Alzamora, A.C.; Lima, W.G.; Lagente, V.; Silva, M.E. Oxidative stress causes hypertension and activation of nuclear factor-κB after high-fructose and salt treatments. *Sci. Rep.* **2017**, *11*, 46051. [CrossRef]

143. Abais-Battad, J.M.; Lund, H.; Dasinger, J.H.; Fehrenbach, D.J.; Cowley, A.W., Jr.; Mattson, D.L. NOX2-derived reactive oxygen species in immune cells exacerbates salt-sensitive hypertension. *Free Radic. Biol. Med.* **2020**, *146*, 333–339. [CrossRef] [PubMed]

144. Montiel, V.; Bella, R.; Michel, L.Y.M.; Esfahani, H.; Mulder, D.D.; Robinson, E.L.; Deglasse, J.P.; Tiburcy, M.; Chow, P.H.; Jonas, J.C.; et al. Inhibition of aquaporin-1 prevents myocardial remodeling by blocking the transmembrane transport of hydrogen peroxide. *Sci. Transl. Med.* **2020**, *12*, eaay2176. [CrossRef] [PubMed]

Review

COVID-19 Pathogenesis: From Molecular Pathway to Vaccine Administration

Francesco Nappi [1,*], **Adelaide Iervolino** [2] and **Sanjeet Singh Avtaar Singh** [3]

[1] Department of Cardiac Surgery, Centre Cardiologique du Nord, 93200 Saint-Denis, France
[2] Department of Cardiovascular Sciences, Fondazione Policlinico Universitario Agostino Gemelli IRCSS, 00168 Rome, Italy; adelaide.iervolino@libero.it
[3] Department of Cardiothoracic Surgery, Golden Jubilee National Hospital, Glasgow G81 4DY, UK; sanjeetsinghtoor@gmail.com
* Correspondence: francesconappi2@gmail.com; Tel.: +33-149-334-104; Fax: +33-149-334-119

Abstract: The Coronavirus 2 (SARS-CoV-2) infection is a global pandemic that has affected millions of people worldwide. The advent of vaccines has permitted some restitution. Aside from the respiratory complications of the infection, there is also a thrombotic risk attributed to both the disease and the vaccine. There are no reliable data for the risk of thromboembolism in SARS-CoV-2 infection in patients managed out of the hospital setting. A literature review was performed to identify the pathophysiological mechanism of thrombosis from the SARS-CoV-2 infection including the role of Angiotensin-Converting Enzyme receptors. The impact of the vaccine and likely mechanisms of thrombosis following vaccination were also clarified. Finally, the utility of the vaccines available against the multiple variants is also highlighted. The systemic response to SARS-CoV-2 infection is still relatively poorly understood, but several risk factors have been identified. The roll-out of the vaccines worldwide has also allowed the lifting of lockdown measures and a reduction in the spread of the disease. The experience of the SARS-CoV-2 infection, however, has highlighted the crucial role of epidemiological research and the need for ongoing studies within this field.

Keywords: SARS-CoV-2; COVID-19; thromboembolism; ACE inhibition; pathophysiology

Citation: Nappi, F.; Iervolino, A.; Avtaar Singh, S.S. COVID-19 Pathogenesis: From Molecular Pathway to Vaccine Administration. *Biomedicines* **2021**, *9*, 903. https://doi.org/10.3390/biomedicines9080903

Academic Editor: Byeong Hwa Jeon

Received: 2 June 2021
Accepted: 26 July 2021
Published: 27 July 2021

Publisher's Note: MDPI stays neutral with regard to jurisdictional claims in published maps and institutional affiliations.

1. Introduction

Patients with severe acute respiratory syndrome Coronavirus 2 (SARS-CoV-2) infection may develop associated arterial and venous thrombotic complications. Data reported in the 2019 U.S. Coronavirus Disease Patient Registry (COVID-19) recorded a 2.6% rate of thrombotic complications in the 299 patients who required non-critical hospitalization compared to the rate of 35.3% of the 170 patients hospitalized in critical care units [1,2]. Klok et al. confirmed a remarkably high 31% incidence of thrombotic complications in ICU patients with COVID-19 infections[2]. These results supported the recommendation to use drug prophylaxis for thrombosis in all COVID-19 patients admitted to the ICU. The data strongly support increased drug prophylaxis dosage, even in the absence of randomized trials.

To date, there are no reliable data to establish the risk of thromboembolism in SARS-CoV-2 infection in patients whose clinical conditions do not require hospitalization. Several studies reported that patients admitted to the hospital with COVID-19 disease experienced thrombotic complications involving the heart, brain, and peripheral vascular system, which mainly led to myocardial infarction (MI), ischemic stroke, and venous thromboembolism (VTE) [3–5].

During the initial months of the pandemic acceleration, several autopsy studies [6,7] highlighted the presence of systemic microthrombi in many organs, including lungs, heart, and kidneys, thereby suggesting how thrombosis could contribute to the frequent and often fatal multisystem organ failure in patients with severe COVID-19 disease [8,9].

1.1. Angiotensin-Converting Enzyme (ACE) 2 Receptor: The Evolutionary Stage of Infection from Himalayan Palm Civet and Bat Coronavirus to SARS-CoV2 Infection

The gateway for SARS-CoV-2 to target cells is the angiotensin-converting enzyme (ACE) 2 receptor, which is mostly expressed by epithelial cells of the lung, heart, blood vessels, kidneys, and intestines. The ACE family of receptors includes both ACE and ACE2 which, although they both are dipeptidyl mono-carboxydipeptidases have distinct physiological functions.

1.2. Structure of the ACE as Ligand-Binding Receptors

SARS-CoV-2 uses common cellular transmission which is based on the binding of ligands to specific cell surface receptors. ACE2 is a G protein-coupled receptor (GPCR) and belongs to a category of receptors that play a central role in the initiation and regulation of cellular processes [10]. The GPCR constitutes the most prominent class of receptors implicated in pathological disorders of the cardiovascular, respiratory, endocrine, immune, and neural systems. Activation of GPCRs is also common in neoplastic pathologies. The function that GPCRs exert is mediated by responses to specific interactions with hormones, neurotransmitters, pathogens, metabolites, ions, fatty acids, and drugs [11,12].

GPCRs are crucial modulators of transmission between the internal and external environment of cells. GPCRs are integral membrane proteins with an extracellular N-terminal and seven transmembrane (TM) helical domains, from TM1 to TM7, connected via link regions. Evidence suggested that GPCRs have a more complex role than originally considered. The binding of GPCRs to very different types of extracellular stimuli leads to conformational changes of the TM domain with the consequent structural remodeling of the protein [13–17]. Inter alia, these conformation changes induce coupling with cytoplasmic proteins and subsequently the activation of enzymes that lead to the generation of a second messenger. Once the second messenger is formed it can activate a sequence of signals inside the cell [14]. This specific role of GPCRs results in increasing levels of intracellular cyclic Adenosine Monophosphate (cAMP) and represents the pivotal pathway in response to ligands, such as signaling of the renin–angiotensin system (RAS). It is important to underline that the levels of cAMP production in the cellular domain are modulated by several factors. Multidrug Resistance Proteins (MRPs) allow the efflux of cAMP from the inside of the cell to the extracellular fluid, thus maintaining homeostatic intracellular concentrations. The role of transporters activated by MRPs serves to regulate the balance of cAMP within the cell.

Lu et al. [15] reported concern about the structural conformation of the ACE/GPCR complex and its interaction with SARS-CoV by focusing on lipid rafts. The structure, activation, and signaling of the ACE/GPCR complex are strongly influenced by the bilayer domain with specific membrane-GPCR interactions [16]. It has been shown that some subsets of GPCR are preferentially isolated in distinct regions of the membrane defined as lipid rafts [17–19]. Cholesterol partitions preferentially into lipid rafts which contain 3 to 5-fold the amount of cholesterol found in the surrounding bilayer. Evidence has shown that lipid rafts serve as an entry site for SARS-CoV. For example, lipid rafts in Vero E6 cells were involved in the "entry" of the coronavirus of the severe acute respiratory syndrome (SARS-CoV). As has been clarified by the tests after SARS-CoV infection, the integrity of the lipid rafts was a necessary requirement to produce the pseudotyped SARS-CoV infection. If plasma membrane cholesterol depletion was induced using the relocalized MbetaCD marker on the caveolin raft the SARS-CoV, ACE2 receptor was not significantly modified. Although the surface expression of ACE2 still allowed binding to the virus, treatment with MbetaCD inhibited the infectivity of the pseudotyped SARS-CoV by 90%. The observed data concern the ectodomain of the SARS-CoV protein S (S1188HA) which can be associated with lipid rafts. The spike protein, after binding to its receptor, colocalized with the ganglioside marker GM1 residing on the raft. The study found that S1188HA binding was not affected by plasma membrane cholesterol depletion supporting the conclusion that lipid rafts serve as a gateway for SARS-CoV [20–24].

1.3. Function of ACE Receptor

The function of ACE is to split angiotensin I into angiotensin II, which binds and activates the type 1 angiotensin II receptor. This activation triggers a series of pathophysiological mechanisms that ultimately have vasoconstrictor, proinflammatory, and pro-oxidative effects. It is important to underline that among the functions of ACE2 is the hydrolytic degradation of angiotensin II to angiotensin 1–7 and angiotensin I to angiotensin 1–9. Once angiotensin 1–9 is generated, it binds to the Mas receptor, producing anti-inflammatory, antioxidant, and vasodilatory reactions. From a pathophysiological point of view, it is important to distinguish the two forms of ACE2 receptors. The first is a type 1 integral transmembrane protein with structural features representing the extracellular domain that acts as a receptor for the SARS-CoV-2 spike protein. The second is a soluble form representing circulating ACE2. To date, our knowledge is limited on the relationship that is established between SARS-CoV-2 and the two forms of the receptor. A better understanding of this relationship may more precisely define the operational adaptive or maladaptive processes that sustained COVID-19 infection [25,26].

1.4. ACE Receptor and Binding to Human Coronary Viruses

The knowledge we have on the interaction between the human ACE2 receptor (hACE2) and the Himalayan palm civet receptor (cACE2) with SARS-CoV derives from the usage of the receptor by the Human (hSARS-CoV) and Himalayan palm civet coronavirus (cSARS-CoV) [27]. The hSARS- CoV can bind both hACE2 and cACE2 receptors while the palm civet coronavirus has no interaction with the ACE2 receptor expressed in humans. It is known that the adaptation of c SARS-CoV to humans was determined by two-point mutations, recognized as K479N and S487T, in the binding domain of the SARS-CoV spike protein (SARS-CoV-S) [26].

The mutations that have recently characterized SARS-CoV-2 led to more aggressive variants of the virus and the concept of adaptive mutations (as noted by Wu et al. [28]) with strengthened receptor binding and tropism (RBT). The authors demonstrated that adaptive mutations of RBT led to the identification of genetic mutations of the virus that enhanced interaction with human or palm civet ACE2. The genetic adaptation processes that took place between hSARS-CoV and cSARS-CoV could also be recorded in SARS-CoV-like viruses that have been isolated in bats [28]. A previous study found that the pathways in which bat coronaviruses infected host cells did not occur through the interaction of the ACE2 receptor with expressed SARS-CoVS and remained a mystery. However, the important finding remains that the substitution of the amino acid sequence found between residues 323 and 505 of the corresponding sequence of the SARS-CoV-S/RBT is sufficient to allow the activation of the human ACE2 receptor [28].

Coronaviruses can enter target cells effortlessly due to their ability to exploit many cell surface molecules such as proteins and carbohydrates. Lectins play a fundamental role in this process. For example, host calcium-dependent (type C) lectins have been recognized to play a central role in SARS-CoV-2 infection. Evidence suggests a specific intercellular role exerted by non-integrin 3-grabbing adhesion molecule (DC-SIGN) of dendritic cells. This is a type C lectin expressed on macrophages and dendritic cells that functions to recognize the high-mannose glycosylation patterns commonly found on viral and bacterial pathogens. Coronavirus protein S is highly glycosylated, thus, providing the virus with the opportunity to interact with host lectins such as Dendritic Cell (DC)/Liver/lymph node-specific intercellular adhesion molecule-3-grabbing integrin (L-SIGN). L-SIGN, which is expressed on liver and lung endothelial cells and has been reported as an alternative receptor for SARS-CoV and bat coronavirus type HCoV-229E [29,30].

The first demonstration of the possibility that SARS-CoV-2 interacts with the human ACE2 receptor is reported in the landmark study from the University of North Carolina at Chapel Hill [31,32]. The authors reported the substantially high risk of SARS-like bat coronavirus disease named SHC014-CoV circulating in Chinese horseshoe bat populations. This type of coronavirus has a high binding affinity with the ACE2 receptor [33,34]

The new SARS-CoV-2 virus expressed the bat coronavirus SHC014 spike in mouse-adapted SARS-CoV backbones.

Menachery et al. created a chimeric virus starting with the RsSHC014-CoV sequence that was isolated from Chinese horseshoe bats [34]. The chimeric virus encoded a different, zoonotic CoV spike protein in the context of the SARS-CoV mouse-adapted backbone. This new SARS-CoV-2 virus expressed the bat coronavirus SHC014 spike. Through the hybrid virus, the authors were able to evaluate the ability of the new spike protein to cause disease independently of other necessary adaptive mutations in its natural backbone [32].

The evidence showed at least two very interesting findings. The first was that group 2b viruses encoding the SHC014 peak in a wild-type backbone could efficiently use more orthologs of the human angiotensin II converting enzyme (ACE2) receptor than the unmodified SARS virus. The second was that group 2b viruses could replicate efficiently in primary human airway cells and that it was also possible to obtain in vitro viral titers equivalent to the epidemic strains of SARS-CoV. Once these results were translated in vivo, replication of the chimeric virus in the mouse lung demonstrated considerable pathogenesis. This led to the trials of immunotherapeutic and prophylactic modalities to cope with the SARS-CoV infection which had poor outcomes. In fact, both monoclonal antibodies and the vaccine approach failed to neutralize and protect against CoV infection using the new SARS-CoVS. Based on these results, the authors synthetically re-derived an infectious full-length recombinant SHC014 virus and demonstrated robust viral replication both in vitro and in vivo settings. This landmark report suggested 6 years ago that there was a potential risk of SARS-CoV re-emergence from viruses circulating in bat populations.

Recently the same group coordinated by Ralph Baric [35] studied the critical determinants of the ACE2 receptor that support SARS-CoV-2-ACE2 interactions during infection and replication of the preemergent 2B coronavirus (WIV). The Authors identified the key changes that lead to infection by creating a humanized murine ACE2 receptor (hmACE2) and provided evidence for the potential pan-virus capabilities of this chimeric receptor.

SARS-CoV-2 cannot infect mice due to incompatibility between its receptor-binding domain (RBD) and the murine ACE2 (mACE2) receptor. Since the mouse models of human ACE2 (hACE2) and viruses adapted to mice have shown limitations, the researchers developed another model that would allow evaluation of the pathogenetic phenomena that occur in human SARS-CoV-2 infection. For example, hACE2 transgenic mice are susceptible to unadapted SARS-CoV-2 viruses, but the pathogenesis observed in these mice showed that virus-induced encephalitis and multi-organ infection were not comparable to that observed in humans. Thus, Adams et al., to map the SARS-CoV-2 RBD and mACE2 interaction network, created a panel of mACE2 receptors, which have increasing levels of humanizing mutations. The study used predictive structural models that allowed identification of the minimal changes needed to restore replication [35].

The ACE2 receptor has structurally critical sites whose integrity determines its activity. The investigators worked at the level of three hot spots that determine ACE2 interaction: position K353 interconnects with SARS-CoV-2 binding residues G496, N501, and Y505, position K31 which forms a salt bridge with ACE2 residue K353 and links with SARS-CoV-2 Q493 and Y489, and position M82 which interconnects with RBD residues F486, N487, and Y489. These aforementioned interface hotspots are the critical molecular sites for the interaction between SARS-CoV2 and the receptor leading to virus entry. The authors demonstrated that divergent residue modifications in these hot spots significantly reduce the binding between humanized murine ACE2 (hmACE2) [33] and SARS-CoV-2 RBD. They recorded that five amino acid changes (N30D, N31K, S82M, F83Y, and H353K) in the SARS-CoV-2 RBD-ACE2 interaction hot spots lead to the modulation of infection and can re-establish infection in the hmACE2 models [35].

This study is crucial for the following reasons. The first is related to the fact that mouse models are essential for understanding the pathogenesis of coronaviruses and are a key resource for the preclinical development of vaccines and antiviral therapies. The second is that a detailed analysis of this study will allow the development of model systems to screen for emerging coronaviruses and to develop new treatments to combat infections [35].

1.5. The Role of ACE2 in COVID-19 Pathogenesis

The ACE2 receptor has been implicated in the pathogenesis of COVID-19, especially with regards to its potential effects on the most vulnerable patients presenting with cardiovascular co-morbidities. COVID-19 does not have the same impact on all members of the population. An exponential increase in the severity of the disease as well as mortality, due to devastating thromboembolic complications, occurs in patients over the sixth decade of life with comorbidities such as cardiovascular disease and diabetes.

The angiotensin-converting-enzyme 2 receptors (ACE2) serve as the attachment site of the SARS-CoV-2 spike protein to enter the lung epithelial cells [36]. Upregulation with increased ACE2 expression has been demonstrated in patients with cardiovascular disease and diabetes treated with angiotensin-converting enzyme (ACEI) inhibitors and angiotensin receptor blockers (ARBs). However, whether treatment with these agents can lead to greater COVID-19 severity has not been fully clarified.

Discussions related to the use of ACEI/ARBs have surfaced regarding the need to continue therapy in patients taking these drugs. The current recommendations are to discontinue the administration of these drugs, despite diverging opinions, which were not universally endorsed by experts due to the lack of strong evidence [37]. ACE2 not only plays a role in the pathogenesis of COVID-19 but also as a component of renin–angiotensin system signaling (RAS) localized throughout the body. Although the evidence has conclusively revealed that ACE2 receptors allow SARS-CoV-2 to enter cells, ACE2 plays a central anti-inflammatory role in RAS signaling by converting angiotensin II, responsible for the inflammatory process, into angiotensin 1–7, which leads to its anti-inflammatory effects [38]. A study performed on rodent lungs [39] showed that the reduced expression of ACE2 leads to a sequence of major proinflammatory processes, that are exacerbated by age, and result in dysregulation of RAS signaling throughout the body [40]. It is important to note that this typical inflammatory profile, even in accentuated forms, supports pathophysiological processes that represent the main feature of hypertension and diabetes, as well as being very widespread in old age [36]. The upregulation of the ACE2 receptor in subjects with diabetes and hypertension treated with ACI/ARBs must be seen as a restorative substrate that has a physiological function. The process that unfolds during SARS-CoV2 infection sees ACE2 receptors as a gateway for the virus to enter cells, while the reduction of ACE2 protective features in older people and those with CVD can potentially predispose them to more severe forms of the disease. The ACE2 receptor facilitates SARS-CoV2 infection while the fundamental anti-inflammatory function, linked to RAS signaling, is reduced because it is compromised in patients who develop COVID-19. In fact, data provided by the first SARS epidemic in 2003 demonstrated the double role of the ACE2 receptor, thus delineating the factors predisposing to the occurrence of the disease and its severity [41,42].

In SARS-CoV2 infection it is plausible that the higher expression of ACE2 leads to a greater predisposition to experience the disease. Epidemiological data from the South Korean population, where genetic testing has been widely used in individuals, reported higher numbers of infected among young adults [41] and those with increased ACE2 levels. In this regard, an Italian study, examining the severity of COVID-19 disease in the elderly population with CVD, hypothesized that a reduction in ACE2 levels due to aging and CVD coupled to the upregulation of the proinflammatory angiotensin II pathway are factors that likely predispose older individuals to severe forms of COVID-19. Therefore, younger people are more susceptible to viral infection, but older people are more likely to have severe disease manifestations [42].

SARS-CoV2 uses the ACE2 receptor in carrying out its infectious manifestation, thereby leading to a reduced expression of ACE2 on the cell surface and an upregulation of angiotensin II signaling in the lungs which results in the development of acute damage [38]. The consequence of these morphofunctional and biochemical changes can predispose elderly individuals with CVD, who have reduced levels of ACE2 compared to young people, to exaggerated inflammation and further reduction of ACE2 expression in the context of COVID-19. In these cases, the disease manifests itself with greater severity [43]. Observations suggest that older individuals, especially those with hypertension and diabetes, have reduced ACE2 expression and upregulation of proinflammatory angiotensin II signaling. Therefore, the morphofunctional and biochemical changes can be corrected by the increase in ACE2 levels induced by ACEI/ARB treatment [43].

First, it is possible to hypothesize that in COVID-19 disease, the binding of SARS-CoV-2 to ACE2 receptors acutely exacerbates this proinflammatory background, predisposing these subpopulations to greater severity and mortality of COVID-19 disease. Second, considering this hypothesis credible, a protective role of the antagonistic action of angiotensin II against acute lung injury associated with sepsis could be effective. This supports the use of continuous therapy with ACEI/ARB [1,44,45] (Figure 1).

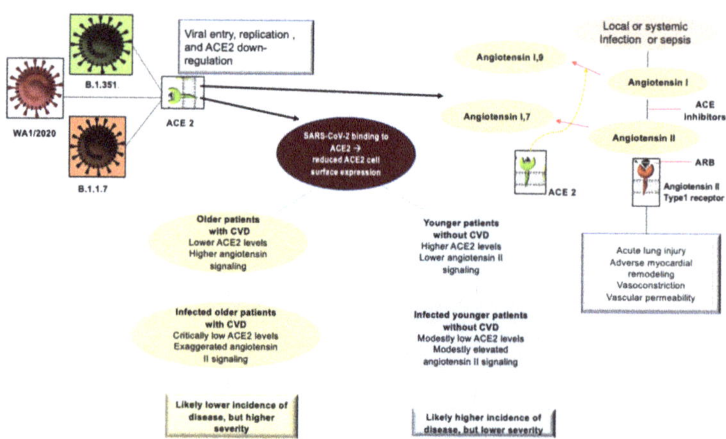

Figure 1. Depicts the interaction of SARS-CoV2 with the ACE2 receptor and the inflammatory profile pattern before and after Coronavirus 2019 (COVID-19) infection in patients with or without CVD. The initial entry of severe acute respiratory syndrome coronavirus 2 (SARS-CoV-2) into cells is shown with involvement mainly of type II pneumocytes. SARS-CoV-2 binds to its functional receptor, the angiotensin-converting enzyme 2 (ACE2). After endocytosis of the viral complex, surface ACE2 is further down-regulated, resulting in unobstructed accumulation of angiotensin II. Local activation of the renin–angiotensin–aldosterone system may mediate lung injury responses to viral insults. The elderly and the young may present with different pathophysiological profiles. The simplified scheme of the pre-infection inflammatory profile among predisposed older individuals compared to their younger counterparts is illustrated. Abbreviations: ACE2, angiotensin-converting enzyme 2; ARB, angiotensin-receptor blocker; CVD, cardiovascular disease; SARS-CoV-2, severe acute respiratory syndrome coronavirus 2.

Third, the aforementioned biomechanical modifications of the receptor, plausible with aging, should be investigated. Therefore, experiments on the functioning, the regulatory mechanisms of RAS, and the biomechanics of the receptors involved in these functions should be implemented. Specifically, the biophysical mechanisms underlying the associated remodeling of the lipid membrane remain to be clarified. They may be useful in the prevention of fatal lung complications caused by genetic variants of the Wuhan virus [26–28,32,34,35,46].

1.6. ACE Inhibitors and Angiotensin Type II Blockers Role in COVID-19 Severity

Tetlow et al. [47] did not identify any associations between ACE-I/ARB use and AKI, macrovascular thrombi, or mortality. Other studies [48,49] also supported the continuation of these drugs during hospitalization from COVID-19. Among those hospitalized, a large percentage are likely being administered either ACE inhibitors or Angiotensin II blockers, since epidemiological data reveal that cardiomyopathies, diabetes mellitus, and hypertension are the most frequent comorbidities found among those patients [50].

Although the upregulation of ACE2 expression, which can be altered by drug administration, has not been defined, it has nevertheless been associated with disease severity. Several preceding studies have demonstrated that the risk of developing COVID-19 after the administration of ACEi and ARBs increased significantly. This could be an indirect effect of overproduction of the circulating ACE2 transcripts in the cells [51,52]. As an example, Enalapril, which is a frequently used ACEi, was reported to increase ACE2 expression in the kidney [53].

Concerning possible therapeutical targets, ACE2 blockers have been developed, such as the small synthetic inhibitor N-(2-aminoethyl)-1aziridine-ethanamine (NAAE) [54]. It is able to bind ACE2 in its closed conformation so that molecular interaction between the viral particle and the receptor cannot be possible and fusion does not happen. Thus, NAAE could exert dual inhibitory effects: one on ACE2 catalytic activity and another on SARS binding [55]. Despite this, current research drives opinions towards a cautionary use of this agent.

2. Pathophysiology of Arterial and Venous Thrombosis

To date, the complete pathophysiology profile of arterial and venous thrombosis during COVID-19 disease has not yet been fully clarified. The literature reports prothrombotic abnormalities in patients with COVID-19. In a Chinese study [56] performed in the first phase of the SARS-CoV2 epidemic, 19 patients with COVID-19, who presented with critical clinical conditions, had elevated levels of markers of hypercoagulability such as D-dimer found in 100%, fibrinogen in 74%, and factor VIII in 100%. The dysregulation of the coagulation process included the presence of antiphospholipid antibodies in 53% of the population studied. Reduced levels of protein C, protein S, and antithrombin were noted in all patients. Complications such as stroke, arterial ischaemia, and VTE accompanied the coagulation disorder.

Zaid et al. studied 115 patients with COVID-19 disease reporting that SARS-CoV2 directly interferes with platelets. Viral RNA and high platelet-associated cytokine levels were found in the platelets of all study participants. These abnormalities were not related to the severity of the disease because in 71 infected individuals the disease manifested in a non-serious manner while for 44 patients' hospitalization was required for critical clinical conditions. Specific tests performed on platelets showed aggregation at lower than expected thrombin concentrations [57].

Nicolai et al. examined the autopsy findings of 38 individuals who died with COVID-19 which showed that histopathological changes in coagulation were marked in the vessel microcirculation. The abnormalities recorded were microvascular thrombotic formations, and neutrophil extracellular traps characterized by networks of extracellular neutrophil-derived DNA and polymorphonuclear neutrophil (PMN)-platelet aggregates [58]. The authors compared the peripheral blood of patients with COVID-19 with that of healthy patients. In vitro responses on peripheral blood samples from the three infected patients exhibited excessive platelet and neutrophil activation, as assessed by degranulation and integrin IIb-IIIa activation and immunofluorescence, compared to healthy control patients' samples.

2.1. The Inflammatory Response during SARS-CoV2 Infection and Thrombotic Complication

Histopathology of SARS infection Cov2 is distinguished from that caused by other viruses with tropism for the respiratory tract. SARS-CoV2 leads to direct damage of endothelial cells characterized by dense perivascular infiltration of T lymphocytes combined with aberrant activation of macrophages. The excessive and uncontrolled inflammatory response, endothelial cell apoptosis, thrombotic microangiopathy, and angiogenesis are other distinctive histopathological features that denote the aggressiveness of SARS-CoV2, which may be responsible for clinically severe forms of COVID-19 thus conferring disease characteristics not comparable to any other viral respiratory disorder [59]. One significant finding that emerges in the evaluation of the pathophysiology of thromboembolism in COVID-19 versus non-COVID-19 disorders is the possibility that the coagulation alterations are mediated more by platelet-dependent activation and intrinsically related to viral-mediated endothelial inflammation. As noted a distinguishing feature of thrombosis during SARS-CoV2 infection is the exacerbated hypercoagulability associated with increased concentrations of coagulation factors, acquired antiphospholipid antibodies, and reduced concentrations of endogenous anticoagulant proteins [56].

Patients with COVID-19 who develop more severe systemic inflammation and more critical respiratory dysfunction have a higher prevalence of thrombotic complications. Lodigiani et al. reported 388 patients hospitalized with COVID-19 including 16% with serious clinical conditions. Despite the use of low molecular weight heparin (LMWH) for thromboprophylaxis in all patients in the ICU and 75% of those not in the ICU, symptomatic VTE occurred in 4.4% of patients, ischemic stroke in 2.5%, and MI in 1.1% [60].

Given the knowledge we have, there is still no clarity on the extent to which SARS-CoV-2 increases the risk of thromboembolism. A study performed in the United Kingdom compared 1877 patients discharged from hospital after COVID-19 disease and 18,159 hospitalized for a non-COVID-19 disease reported no difference in hospital-associated VTE rates (4.8/1000 vs. 3.1/1000; odds ratio, 1.6 [95% CI, 0.77–3.1]; $p = 0.20$) [61]. One point to clarify is whether the high rate of VTE is specific to patients who develop COVID-19 or if VTE is mainly occurring in patients as a complication associated with severe critical disease [61]. These results are in line with a recent meta-analysis that included 41,768 patients in whom VTE was assessed in COVID-19 versus non-COVID-19 cohorts. The authors did not record a significant statistical difference for overall risk of VTE (RR 1.18; 95%CI 0.79–1.77; $p = 0.42$; $I^2 = 54$%), pulmonary embolism (RR 1.25; 95%CI 0.77–2.03; $p = 0.36$; $I^2 = 52$%) and deep venous thrombosis (RR 0.92; 95%CI 0.52–1.65; $p = 0.78$; $I^2 = 0$%). A difference was reported after analyzing the subgroups of patients who were admitted to the intensive care unit (ICU). Critically ill patients had an increased risk of VTE in the COVID-19 cohort compared to non-COVID-19 patients admitted to the ICU (RR 3.10; 95% CI 1.54–6, 23), which was not observed in cohorts of non-ICU patients (RR 0.95; 95% CI 0.81–1.11) (P interaction = 0.001) [62].

2.2. Management of Thrombosis in COVID-19 Patients

There are no international guidelines that direct the prevention and treatment of thrombotic complications in COVID-19 patients. Both published and ongoing studies testing interventions to prevent thrombosis complications in COVID-19 are based on the evidence reported in current clinical guidelines about VTE prophylaxis in acute COVID-19 infections. Therefore, pending the results to be provided by the completion of ongoing trials, guidelines for the treatment of thrombotic complications in patients with COVID-19 disease are derived from medical recommendations in the coagulation disorder populations (Figure 2). However, the crucial point that remains to be clarified is whether these guidelines are also optimal for the treatment of thrombosis due to COVID-19 [63–65].

Guidelines from the American College of Chest Physicians (ACCP) suggest (in the absence of contraindications) prophylaxis with LMWH or fondaparinux rather than unfractionated heparin or direct oral anticoagulants (DOACs) for all hospitalized COVID-19 patients [63] Clearly the optimum choice of the drug to be taken is constrained by the incomplete knowledge of the possible interference of CoV 2 SARS with the medicament. So, the 40 mg dose of LMWH for injection once a day and the 2.5 mg dose of fondaparinux are preferred over the administration of unfractionated heparin injected subcutaneously 2–3 times a day thus limiting the caregiver's contact with infected patients. In addition, these drugs are preferred over DOACs because of drug–drug interactions with antiviral agents. Both are substrates of the P-glycoprotein and/or cytochrome P450-based metabolic pathways.

Thus, concomitant administration of DOACs and antiviral drugs has the potential to sharply increase DOAC anticoagulant plasma levels, thus increasing hemorrhagic risk.

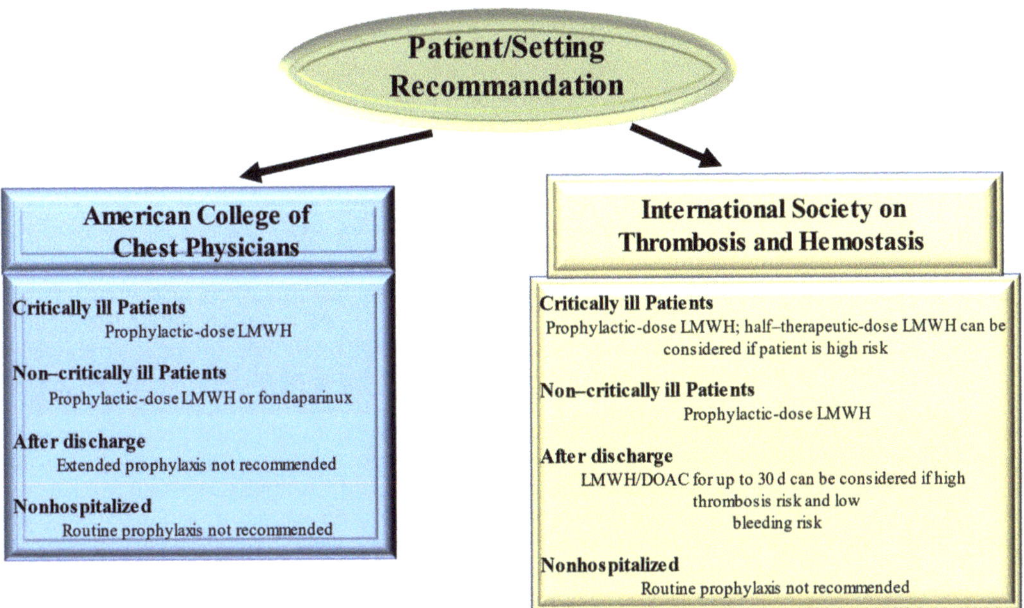

Figure 2. Current Guideline Recommendations for Venous Thromboembolism Prevention in patients With Coronavirus Disease 2019. Abbreviations: DOAC, direct oral anticoagulant; LMWH, low-molecular-weight heparin.

Given the high incidence of VTE, the proposed therapeutic dose to be used for standard thromboprophylaxis in critically ill patients with COVID-19 was either double or single-dose administration of LMWH. The ACCP guidelines suggest the standard dose LWMH in the absence of new clinical trial data [64]. Guideline-Directed Medical Therapy (GDMT), which was established by the International Society on Thrombosis and Hemostasis (ISTH), suggested that half-therapeutic-dose LMWH (1 mg/kg daily) can be considered for prophylaxis in high-risk patients with COVID-19. A 50% higher dose can be considered in patients with severe obesity (BMI $\geq$ 40 kg/m^2). However, it remains to be clarified which is the best dosage for optimal prophylactic therapy [65]. The results of ongoing randomized controlled trials (REMAP-CAP, ACTIV-4, and ATTACC), comparing therapeutic-intensity anticoagulation with prophylactic-intensity anticoagulation for patients with COVID-19-related critical illness, are awaited to establish optimal antithrombotic prophylactic therapy [66]. Considering the pathophysiology of thromboembolism in COVID-19 is characterized by platelet hyperreactivity, another point under discussion with RCTs initiated is the evaluation of administering an antiplatelet agent for therapeutic prophylaxis.

High-risk patients hospitalized for COVID-19 have a high possibility of developing VTE that persists after discharge [65]. However, for the latter, no specific recommendations for post-discharge thromboprophylaxis have been established by the ACCP [64]. In contrast, the ISTH recommendations for post-discharge thromboprophylaxis suggest the use of LMWH or a DOAC for all high-risk hospitalized patients with COVID-19 who have a low risk of bleeding. Patients with COVID-19 considered to be at high risk include those with age $\geq$ 65 years, presence of critical illness, cancer, previous VTE, thrombophilia, severe immobility, and elevated D-dimer (>2 times the upper limit of normal). ISTH recommendations suggest a duration of 14 to 30 days for post-discharge thromboprophylaxis, although the ideal administration period remains to be clarified [65].

For patients with COVID-19 disease, no diagnostic protocols have been established for thromboembolic complications, such as pulmonary embolism and MI, so the methods to be used should be those validated for patients without COVID-19. Given the lack of evidence to support the benefit, routine ultrasound checks for VTE surveillance are not recommended. For patients with COVID-19 diagnosed with arterial or venous thrombosis, we recommend treatment according to current established guidelines. These recognize the benefits of LMWH administration in hospitalized patients. In the outpatient setting, DOAC administration is recommended [58]. There are currently no recommendations issued by ISTH and ACCP to support measuring the D-dimer to screen for VTE or to establish the intensity of prophylaxis or treatment [64,65].

3. COVID-19 Vaccines Administration vs. Thrombosis and Variant. The New Challenge

3.1. Nucleoside-Modified RNA Encoding the SARS-CoV-2 Spike

Several studies have demonstrated the substantial role of the SARS-CoV-2 spike protein that binds to ACE2 receptors on target cells during viral entry. Studies performed on convalescent patients have highlighted the central role that the spike protein plays as an immunodominant antigen triggering the host response, mediated by both antibody and T lymphocytes [67].

Concerns about the rapid spread of the COVID-19 pandemic have favored the registration of numerous randomized clinical trials (RCTs) using different vaccination platforms in order to evaluate their efficacy and safety. Evidence has shown that the use of rapid response genetic platforms mRNA [67,68], adenoviral vector vaccines [69,70], inactivated viruses [71,72], and adjuvanted spike glycoprotein [73] resulted in neutralizing antibody responses after immunization.

The particular biological characteristics of mRNA synthesized in vitro may explain the superior efficacy of an ideal non-viral gene replacement tool leading to many intrinsic benefits [67,68]. First, quick protein assembly and well-regulated primary cell transduction. Second, mRNA-based therapy avoids harmful side effects, such as its incorporation into the cellular genetic substrate, which can ultimately limit the clinical application of most virus- and DNA-based vectors [67,68].

Although the use of in vivo gene transfer therapy was first applied almost twenty years ago [74,75], its usage as a vector for introducing genetic material into animals or even into cultured cells has been very limited. Indeed, the reports of Gilboa [76] and Pascolo [77] focused on the use of mRNA for vaccination purposes mainly directed at the development of cellular and humoral immune responses through antigen-encoding transcripts that were administered in vivo or delivered to dendritic cells (DC) ex vivo.

Several studies conducted at the beginning of the year 2000 [78–82] have shown that RNA interferes with the cell-mediated adaptive immune response (pathogen and antigen-specific response) by activating the cells of the innate immune system (non-specific response). In particular, the action of the mRNA is directed towards the Toll-like receptors (TLR) and specifically for the cellular subgroups TLR3, TLR7, and TLR8. It should be noted that compared to gene replacement, RNA showed greater immunogenicity and was associated with greater efficacy, highlighting a key role in the immune response.

Only 1 published study compared the in vitro immune response between nucleosides and modified nucleosides. The use of incorporated pseudouridine (Ψ), 5-methylcytidine (m5C), N6-methyladenosine (m6A), 5-methyluridine (m5U) or 2-thiouridine (s2U) in the transcript affected the immune response of most TLRs with a substantial loss of their activation [83]. Progressing to testing nucleoside-modified mRNAs to evaluate their translation potentials and immune characteristics in vivo, Hornung et al. [83] demonstrated that the $5'$-triphosphate end of RNA produced by viral polymerases is accountable for retinoic acid-inducible protein I (RIG-I)-mediated detection of RNA molecules. Identification of $5'$-triphosphate RNA is repealed by capping the $5'$-triphosphate end or by nucleoside modification of RNA, including the use of s2U and Ψ.

The major implication of these findings led to in vitro transcripts containing nucleoside modifications not only translatable but also capable of activating an immune response in vivo. Therefore, it was possible to subsequently develop mRNA with the function of a dual therapeutic tool for both gene replacement and vaccination. Evidence reported by Kariko et al. [84,85] on the in vitro incorporation of pseudouridine, a modified nucleoside present in mRNA, has suggested that it improves RNA translation capacity but also suppresses RNA-mediated immune activation in vitro and in vivo.

3.2. RNA Vaccine Platform

In January 2020 the RNA sequence of the new coronavirus SARS-CoV2 was introduced in the RNA vaccine platform to allow rapid development of the vaccine in response to the worsening spread of the pandemic. The advantages of vaccines that use RNA are manifold and related to the incorporation of pseudouridine [84,85]. They offer both greater flexibility during vaccine antigen design and expression as well as the ability to imitate viral antigen structure and expression during native infection. One of the characteristics that make them innovative is the lack of integration of viral RNA, which is necessary for protein synthesis, in the cell's genome. In fact, the viral genome is transiently expressed, then metabolized and eliminated by the natural mechanisms of the organism, giving these vaccines greater safety [68–72]. As a rule, vaccination with RNA can stimulate a vigorous innate immune response eliciting B and T cell-dependent activity. RNA leads to the expression of the vaccine antigen in host cells and, as demonstrated in specific mRNA vaccines, could address considerable medical demand in the area of influenza prophylaxis [86].

The immunogenic benefits associated with the in vivo administration of 1-methyl pseudouridine-containing mRNA including superior translational capacity and biological stability were established in a landmark paper from Kariko et al. [84,85]. The same research paper demonstrated lower serum levels of interferon-α (IFN-α) elicited by modified mRNA in mice models with respect to unmodified ones, thus potentially reducing exaggerated systemic inflammation and increasing safety. The improved efficacy of nucleoside-modified RNA (modRNA) encoding the SARS-CoV-2 full-length spike modified by two proline mutations is almost certainly due to its superior immune response [85]. Studies conducted in the United States and Germany have reported substantially higher elicited SARS-CoV-2 neutralizing antibody titers and robust antigen-specific CD8+ and Th1-type CD4+ T-cell responses against nucleoside modified mRNA delivered in lipid nanoparticles [86–88].

There are two widely administered mRNA vaccines, BNT162b2 (Pfizer–BioNTech) and mRNA-1273 (Moderna). Administration of a two-dose regimen of BNT162b2 conferred 95% protection against COVID-19 in phase 3 trial participants (n = 21,720), aged 16 years and older (95%; confidence interval, 90.3 to 97.6). The group of participants assigned to receive BNT162b2 recorded 8 cases of COVID-19 disease with onset at least 7 days after the second dose while in the group of individuals assigned to placebo there were 162 cases of COVID-19 disease. Evidence observed from the analysis of subgroups defined by age, sex, race, ethnicity, baseline body mass index, and the presence of coexisting conditions reported similar efficacy of the vaccine with percentages between 90 and 100%. It is important to note that among 10 cases of severe COVID-19 with onset after administration of the first dose, 9 were reported in recipients of the placebo dose and 1 recipient with the BNT162b2 dose. The safety profile of BNT162b2 was very high as evidenced by the short-term appearance of mild-to-moderate pain at the injection site, fatigue, and headache. The side effects were low and comparable to those recorded in the placebo dose recipients. Furthermore, they were equivalent for a median of 2 months when comparing BNT162b2 with that of other viral vaccines [89].

Similar results were reported in phase 3 of the randomized, observer-blind, placebo-controlled trial after administration of mRNA-1273 (100 μg in 15,210 participants) [90]. The efficacy of the mRNA-1273 vaccine was recorded at 94.1% (95% CI, 89.3 to 96.8%; $p < 0.001$) in the prevention of COVID-19 disease, including severe disease. A double dose of mRNA-1273 was administered to more than 96% of vaccine recipients and only

2.2% had serological, virological, or both evidence of SARS-CoV-2 infection. Symptomatic COVID-19 disease was confirmed in 185 recipients of the placebo dose versus 11 recipients of mRNA-1273 dose.

Recipients of mRNA-1273 reported only transient local and systemic side effects and no safety concerns were recorded. A critical illness from COVID-19 occurred in the 30 recipients of the placebo dose (with one death) and in no participant who was administered the mRNA-1273. It is important to clarify that current data on the messenger, derived from laboratory studies, have demonstrated the efficacy of RNA (mRNA) vaccines against SARS-CoV-2 variants. Researchers exposed serum samples from immunized individuals to genetically modified versions of related variants and then measured neutralizing antibody titers [90].

3.3. SARS-CoV-2 and Vaccine

We searched PubMed for research articles published by the launch of the database until April 30, 2021, indicating no language restrictions and using the terms "SARS-CoV-2", "vaccine", and "clinical trial". We identified published clinical trial data on seven SARS-CoV-2 studies and vaccines.

Four recombinant viral vectored vaccines have been tested in phase I/II clinical trials [91–98]. Phase I and II trials were represented in the same study by two parts with different patients subsets. The vaccine ChAdOx1 nCoV-19 (AZD1222), known as AstraZeneca vaccine, was developed by the Oxford University and is constituted with an adenoviral vector inactivated (unable to replicate) chimpanzee ChAdOx1 replication, containing the antigen glycoprotein gene structural surface SARS-CoV-2 (protein spike; nCoV-19). This vaccine is one of the more extensively studied following the first UK Phase 1 clinical trial published on 23 April 2020 [92]. To date three more randomized controlled trials of the candidate vaccine have been initiated in the UK (COV002), Brazil (COV003), and South Africa (COV005). Recently a further phase 1/2 study was carried out in Kenya.

A pooled interim analysis of four trials (COV004) showed the safety and efficacy of the ChAdOx1 nCoV-19 vaccine (Oxford-AstraZeneca COVID-19 vaccine), (Covishield or Vaxzevria). In recipients of two standard doses of Vaxzevria, the vaccine efficacy was 62.1% and in recipients given a low dose followed by a standard dose, the efficacy was 90.0%. The overall efficacy of the vaccine after administration of both doses in the population studied was 70.4%. There were ten cases of COVID-19 that required hospitalization 21 days after the first dose, all in the control population. Two patients were in serious condition and one died. The authors recorded 175 serious adverse events that occurred in 168 participants, of which 84 events occurred in recipients of the Oxford-AstraZeneca COVID-19 vaccine and 91 in the control group. Concern relating to clot formation or the occurrence of bleeding episodes were not suggested across the analysis of these 4 RCTs [91].

The immune response after vaccine administration in participants who received two doses of the vaccine was very effective. In particular, the specific objectives of phase 3 RCT were the evaluation of humoral and cellular safety and immunogenicity concerning both a single dose and two-dose regimen in adults over 55 years of age. Median peak anti- SARS-CoV-2 IgG responses 28 days after the boost dose were similar across the three cohorts (including two groups of patients aged 18–55 and one group enrolling >55 years old patients). Furthermore, neutralizing antibody titers after a boost dose were similar across age groups. Within 14 days of boost dose administration, a total of 208 of 209 (>99%) recipients of the booster dose of ChAdOx1 nCoV-19 had neutralizing antibody responses. T cell responses peaked on day 14 following a standard single dose of ChAdOx1 nCoV-19 [90,91].

The Ad26.COV2. S vaccine, known as Johnson&Johnson/Janssen COVID-19 vaccine, is composed of a recombinant, replication-incompetent human adenovirus type 26 (Ad26) vector that encodes a full-length, membrane-bound SARS-CoV-2 spike protein in a prefusion-stabilized conformation [95–97]. At least 14 days after single-dose administration, Ad26.COV2. S conferred protection against both symptomatic COVID-19 infection and asymptomatic SARS-CoV-2 infection. The level of efficacy remained stable at 28 days after administration with an efficacy of 66.1% (adjusted 95% CI, 55.0 to 74.8). COVID-19 disease occurred in 66 recipients of the administration dose of Ad26.COV2. S compared to 193 for the placebo dose [98]. The results of the administration of Ad26.COV2. S vaccine has demonstrated efficacy against COVID-19 clinical disease with severe-critical manifestation, including hospitalization and death. Evidence suggested a high level of efficacy after administration of Ad26.COV2, which was greater against severe-critical COVID-19 disease and reached a rate of 76.7% for onset at $\geq$14 days [adjusted 95% CI, from 54.6 at 89.1]. At 28 days in participants receiving the single dose the reported efficacy was 85.4% [adjusted 95% CI, 54.2 to 96.9] for onset at $\geq$28 days).

However, the unexpected data were related to the immunogenic response to Ad26.COV2 vaccine against the South African variant 20H/501Y.V2. Out of 91 cases of patients in which the virus variant was sequenced and confirmed, 86 (94.5%), showed vaccine efficacy against moderate to severe-critical COVID-19 that reached 52.0% and 64.0% with onset of at least 14 days and at least 28 days after dosing, respectively. The efficacy against severe-critical COVID-19 disease reached 73.1% and 81.7% at 14 days and 28 days respectively after the single dose of Ad26.COV2. Evidence supported a level of safety comparable to that of other COVID-19 vaccines that progressed to phase 3 studies. The reactogenicity was greater with the administration of Ad26.COV2. S compared to the placebo dosage; however, it was generally mild to moderate and transient. Note that the incidence of severe adverse events was similar between the two populations of participants (vaccine and placebo group) with three deaths occurring in the vaccine group (but none of them was related to COVID-19 infection). No episodes attributable to thrombotic or haemorrhagic phenomena were reported [98].

Vector-based adenovirus (Ad) 5 (CanSino Biological/Beijing Institute of Biotechnology, China) [69,93] was administered in a single dose and resulted in the production of neutralizing antibodies which increased significantly on day 14 and peaked 28 days after vaccination. A specific T-cell response in a dose-dependent manner peaked at day 14 after vaccination. However, of note, the vaccine demonstrated lower immunogenicity in participants over the age of 55. Administration of adenovirus type-5 vectored COVID-19 recorded no serious adverse event within 28 days post-vaccination. An equal rate of side effects was reported in the three groups studied. Reactogenicity was evident in the first 7 days after administration in 30 (83%) recipients of a low dose vaccine, in 30 (83%) recipients of a medium dose, and in 27 (75%) recipients of a high dose, respectively [93].

A heterologous recombinant adenovirus (rAd26 and rAd5)-based vaccine, Gam-COVID-Vac (Sputnik V) [70,94] showed efficacy and safety from the interim analysis of a phase 3 RCT. Gam-COVID-Vac is a combined vector vaccine because it consists of rAd type 26 (rAd26) and rAd type 5 (rAd5). Both adenoviruses carry the gene for the SARS-CoV-2 full-length glycoprotein S (rAd26-S and rAd5-S). The administration of rAd26-S and rAd5-S is carried out (intramuscularly) separately with an interval of 21 days. The results of the Phase 1/2 clinical trials showed that the Gam-COVID-Vac vaccine was well tolerated and highly immunogenic in healthy participants. Vaccine efficacy of Gam-COVID-Vac reached 91.6% (95% CI 85.6–95.2) with few tolerable side effects (7485 [94.0%] of 7966 total events). Although 45 (0.3%) of recipients of this vaccine (n = 16,427) and 23 (0.4%) of recipients of the placebo dose recorded serious adverse events; however, none were deemed associated with vaccination. There were four deaths during the study. Three (<0.1% of a total of 16,427) were participants from the vaccinated population and one (<0.1% of the total of 5435) received a placebo. None of these deaths were considered related to the vaccine.

Chinese researchers worked on two inactivated viral vaccines [71,72]. Two SinoPharm vaccines demonstrated the neutralization of antibody responses in participants, aged 18–59, who received the first SinoPharm vaccine (Wuhan Institute Biological Products Co Ltd /SinoPharm, Wuhan, China). The immune response was dose-dependent. The second SinoPharm product (Beijing Institute of Biological Products-Sinopharm-China National Biotec Group Co, Beijing, China) elicited neutralizing antibody response in adults aged

Table 1. Efficacies of COVID-19 vaccines are compared according to disease and infection prevention.

Anti-SARS-CoV-2 Vaccine Type	Efficacy at Preventing Disease (D614G and B.1.1.7.)	Efficacy at Preventing Infection (D614G and B.1.1.7.)
BNT162b2	91%	86%
mRNA-1273 (Moderna)	94%	85%
ChAdOx-1 nCov-2	75%	52%
Ad26.COV2-S (Janssen)	72%	72%
CoronaVac (Sinovac)	50%	43%
Sputnik V	92%	80%
Novavax	89%	77%
Sinopharm	73%	63%
Tianjin CanSino	66%	57%

Highlighted are efficacies exceeding the 75% of potency which we set as a point of comparison. Some variables are estimations available from literature studies (UK SIREN study). The general source is IHME—Institute for Health Metrics and Evaluation documents on data summaries. Data are updated until 26 April 2021 [105].

Indeed, the opposite has happened. Individuals who had been vaccinated were hospitalized because they had COVID-19 infection caused by the mutated virus [106]. A frustrating situation is suggested by the discouraging results from the Phase 2 trials of the Oxford-AstraZeneca vaccine administered in South Africa. In the study participants, it was recorded that the vaccine did not prevent mild to moderate COVID-19 disease caused by variant B.1.351 [107]. Following the findings, the vaccination schedule based on the administration of AstraZeneca was modified [108]. Analyzing this report it emerges that it was not designed to determine if the vaccine leads to protection against severe forms of COVID-19 or not. A potential bias was the number of participants (n = 2000) who were young, mean age 31, healthy and at low risk of developing severe COVID-19 disease regardless of inclusion in the vaccine or non-vaccine group.

Studies by Novavax and Janssen have provided more evidence from the administration of their vaccines in South Africa than the Oxford vaccine/AstraZeneca. The results demonstrate lower efficacy rates for Novavax and Janssen in participants enrolled in randomized controlled trials in South Africa compared to studies in other countries. Nevertheless, the phase 3 results in recipients of the Janssen vaccine showed that the likelihood of being hospitalized for a severe form of COVID-19 was lower than those who received the placebo dose. The phase 3 results of the Novavax vaccine RCT, not yet published, could point in the same direction [73,109].

The reference point for an analysis of the real situation on the progression of COVID-19, both in reference to the spread of the infection/number of cases and the need for hospitalization or intensive care, is Israel [110] but the Kingdom of Bhutan a land locked country in the Eastern Himalayas has been a case study for low viral diffusion. As of 11 May 2021, the World Health Organization WHO reports 20 new cases per day, 1241 total cases (PCR-positive COVID-19 patients, both symptomatic and asymptomatic), and 1 death.

In Israel, the country which is the world leader in percentage of the population vaccinated, the number of COVID-19 patients began to decline in mid-January 2021. The effects of the mass vaccination program are evident in the drastic reduction in the need for hospitalization and a reduction in the absolute number of infections in older individuals who were the highest priority for vaccination [111]. An analysis of the hospitalization trend shows that in one week the percentage of patients requiring hospitalization for a severe form of COVID-19 decreased from 36% to 29%, compared to the previous 3 weeks. As variant B.1.1.7, first isolated in the UK, is now the dominant variant of SARS-CoV-2 in Israel as well as in the United Kingdom it is evident that this variant does not seem to influence the production of neutralizing antibodies after the administration of the Pfitzer Biointech vaccine to the same extent as it did for B.1.351 [111].

The same outcomes were recorded from a report performed in the United Kingdom in which researchers compared the Pfizer-BioNT and Oxford-AstraZeneca vaccines. They demonstrated that the latter achieved efficacy in preventing COVID-19 disease in 94% of recipients compared to 85% of Pfizer-BioNT vaccine recipients. Hospital admissions were therefore reduced in the 28–34 days after a single dose. The results of this study suggest a postponement of the administration of the second dose until 12 weeks after the first dose [112].

Generally, human vaccines have the characteristic of reducing the manifestations of the disease and in general also the transmission. This ability has not yet been fully demonstrated for vaccines against COVID-19. Therefore, another critical issue could involve asymptomatic vaccinated individuals infected by variants which represent a reservoir of contagion and diffusion of the variants. The increased risk of these individuals lies in the fact that the absence of symptoms does not prevent the spread of the infection because they are capable of infecting unvaccinated individuals.

Recent evidence suggests a reduction in transmission. An Israeli study conducted at the Hadassah Hebrew University Medical Center (HHUMC) evaluated viral transmission in healthcare professionals. Individuals were immunized with two doses, 21 days apart, of Pfizer-BioNTech vaccine and were subjected to regular testing with a PCR test and two rapid tests to identify the percentage of infected individuals in the populations of asymptomatic and symptomatic patients. The results revealed a 70% reduction in infection in the two populations after 21 days from the administration of the first dose of vaccine and an 85% reduction after administration of the second dose [113]. Vaccines efficacy was also demonstrated to be active during B.1.1.7 variant surge, constituting 80% of PCR-positive cases, as reported by the Ministry of Health of Israel in COVID-19 research reports [113]. These findings could be supported by a Pfizer-BioNTech-sponsored study not yet peer-reviewed in which the vaccine was 94% effective in reducing transmission of asymptomatic SARS-CoV-2 infection [114,115].

Uncomfortable data emerge after the identification of SARS-CoV-2 variant B.1.617.2 (Delta). This viral variant was identified in India in late 2020 and was subsequently detected in 60 other countries. The main feature of variant B.1.617.2 is the potentially higher transmission speed than other variants.

On 12–18 May 2021, the Oklahoma State Department of Health (OSDH) Acute Disease Service (ADS) registered the presence of the delta variant. A total of 21 SARS-CoV-2 delta specimens, temporally and geographically clustered in central Oklahoma, were sequenced by the OSDH Public Health Laboratory (PHL). The data that emerged from public health surveillance indicated that people infected with delta viral variants were associated with a local gymnastics facility [116].

The checks put into place by the OSDH ADS and by members of the staff of the local health department led to the identification of people who tested positive for the delta variant with the search for contacts. Forty-seven cases of COVID-19 emerged that developed in an age group between 5 and 58 years. Of these 21 delta variant cases belonged to the primary cluster and 26 other cases were epidemiologically linked and were associated with the primary outbreak. Distribution was essentially restricted to 10 of 16 gymnast cohorts* and three staff members. However, the spread of the delta variant was recorded in seven (33%) of the 26 families interviewed. It is important to note that forty (85%) cases of COVID-19 associated with the outbreak had never received any dose of COVID-19 vaccine; three persons (6%) were partially vaccinated having administered 1 dose of Moderna or Pfizer-BioNTech $\geq$14 days prior to a positive PCR test result but they had not received the second dose; four were (9%) were fully vaccinated because they have had 2 doses of Moderna or Pfizer-BioNTech or a single dose of Janssen (Johnson & Johnson, New Brunswick, NJ, USA) vaccine $\geq$14 days prior to a positive PCR test result.

These results suggest that the delta variant is highly transmissible in indoor sports settings and within families. This suggests multi-component prevention strategies, including vaccination, that remain important to reduce the spread of SARS-CoV-2, with a focus on people who play indoor sports † and their contacts [117].

The most heated debate concerns whether for COVID-19 there will be the same periodicity for new vaccinations as in the case of influenza with the difference that for the latter, despite being an infectious disease, vaccination is not mandatory. Undoubtedly, mRNA-based technology opens up new possibilities such as creating a vaccine that protects against most variants of SARS-CoV2. The biggest challenge appears to be to make enough changes to the mRNA platform vaccines to address the emerging variants. Pfizer and BioNTech have raised the possibility of administering a third dose of the vaccine BNT162b2 to increase immunological activity, to confer greater safety and efficacy against SARS-CoV-2 variants. One study was designed to make specific changes to BNT162b2 directed specifically against variant B.1.351 [117].

Modifications of mRNA-1273 with a booster dose for variant B.1.351 are under study [118]. In regard to the Novavax vaccine, the first generation of which has not yet been authorized in the United States, scientists are working on a booster dose or a bivalent combination vaccine, to increase the degree of protection from variants [119]. Recently, the 768 NVX-CoV2373 vaccine was efficacious in preventing severe COVID-19 disease due to the B.1.351 variant, resulting in mild to moderate manifestations of the disease due to the B.1.351 variant [119].

The main challenge today lies in conducting studies comparing the degree of production of neutralizing antibodies against SARS-CoV2 elicited by prototype vaccines and engineered on the Wuhan-hu-1 variant from elicited antibody responses against the new variants of SARS-CoV2. In another study, scientists could use serum samples from people previously vaccinated with a prototype vaccine who are given an experimental booster dose against the more contagious variants (Figures 3 and 4).

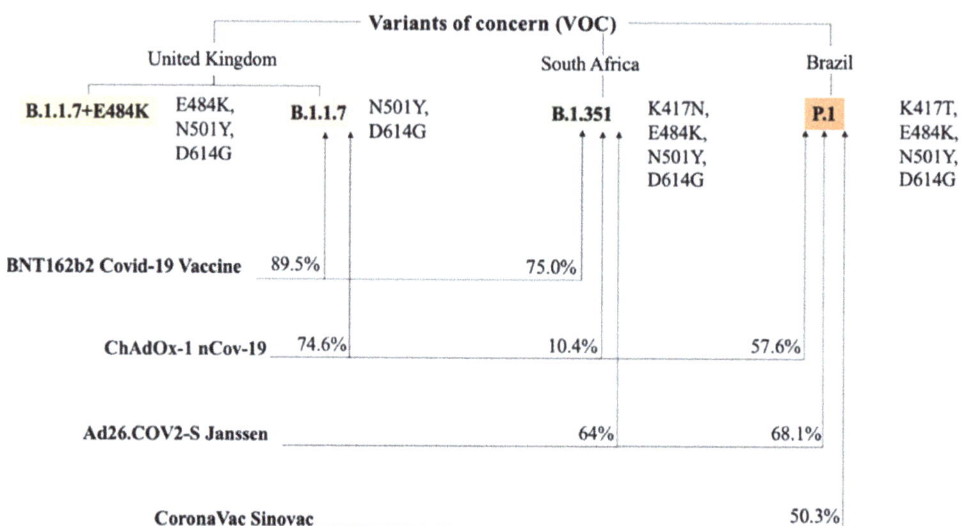

Figure 3. SARS-CoV-2 variants and the respective efficacy of most administered vaccines are shown [120]. B.1.1.7. i.e., the UK variant has recently been studied by investigators: 89.5% and 74.6% efficacies have been demonstrated against the variants by the B1.1.7 + E484K and ChadOx-1 nCoV-19 vaccines respectively.

Percentages of efficacy from the other vaccines versus the main variants of concern (VOC) are also presented. Numbers mainly reflect efficacies against the symptomatic non-severe infection by SARS-CoV-2. Spike protein mutations are pictured on the side of each variant to which they belong [105].

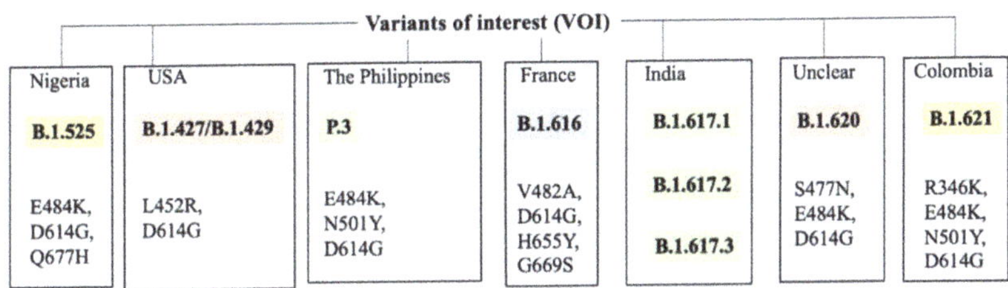

Figure 4. Variants of Interest (VOI) are illustrated. Every box reflects a country the first variant was discovered in. As for variants of concern, the Spike protein mutations associated with the variant are displayed. The efficacy of vaccines has not been reported due to the lack of data in terms of both literature studies and official national reports. Data are updated until 6 May 2021 [121].

3.7. SARS-CoV-2 Adenoviral Vector Vaccines and the Risk to Develop Thrombosis

SARS-CoV-2 vaccines were reported as safe and effective before their marketing and worldwide distribution by first, second, third phase clinical trials and pooled analyses.

In February 2021, large-scale epidemiological data started to raise suspicion of coagulopathies, after adenoviral vector-based vaccines reached millions of administered doses both in Europe and United States. In particular, the ChAdOx1 nCov-19, marketed by AstraZeneca and subsequently renamed Vaxzevria vaccine, has been considered responsible for the development of arterial and venous thromboembolism in a selected population, mostly of the female sex, under the age of 60 years [122].

A single dose (0.5 mL) of the vaccine has been formulated to contain about 2.5×10^8 infectious units (Inf.U) of Chimpanzee Adenovirus encoding the SARS-CoV-2 Spike glycoprotein (ChAdOx1-S). The generation of the final product derives from the genetically modified human embryonic kidney (HEK) 293 cells. Recombinant DNA technology was also used for this intent [122].

Despite being rare events, around 1/100 000 recipients, further considerations are warranted and justified the European Medicines Agency (EMA) examination of those cases [123].

In April 2021, Greinacher et al. [124] reported 11 cases of thrombosis, including 10 multiple ones, involving cerebral and splanchnic veins with one death. Pulmonary emboli (PE) were also frequent in this subset of patients. Other conditions that required medical treatment included disseminated intravascular coagulation DIC and severe thrombocytopenia. The etiology has not been fully understood but it has become clear that anti-platelet factor 4 (PF4) antibodies in those patients' serum were implicated and considered to have a fundamental role in what was initially described as similar to heparin-induced thrombocytopenia (HIT), with the difference being that heparin was not administered in all cases and antibody binding was successful even in the absence of heparin.

Under the assumption that immune mechanisms drove the syndrome, it was named VITT (vaccine-induced thrombotic thrombocytopenia). Of note is the time of development of the first symptoms, with an average of 1–2 weeks after the administration of the first dose. This is the reason why European governments started to reassure the population who already underwent the first dose of the ChAdOx1 nCov-19 vaccine to also receive the second one if no side effects were reported.

Schultz et al. [125] also published clinical cases of patients, mostly women, presenting with venous thrombosis, which included 2 sigmoid cerebral sinuses thromboses, a portal vein branch thrombosis, cerebral vein thrombosis, and a right cerebellar haemorrhagic infarction. In all of those cases, anti-PF4 antibodies were identified in patient plasma.

3.8. SARS-CoV-2 Vaccine-Induced Thrombotic Thrombocytopenia

Platelet factor 4 (PF4) is a platelet-derived cytokine of the CXC (chemokine) family. It is released by activated platelets to promote coagulation via neutralization of heparin-like molecules on endothelial cells. In concert with polyanionic proteoglycans (PGs) derived from endothelial cells, they create complex autoantibodies direct to these components. They have been identified by enzyme-linked immunosorbent assay (ELISA) and assays based on platelet activation, which, when tested, were enhanced by the addition of PF4. Thus, IgGs were found responsible for directly stimulating coagulation via FcγRIIA receptor-dependent mechanisms.

Goldman et al. [120] recently proposed a model according to which the first activation of platelets may lead to PF4 release in the circulation. In turn, PF4 complexing with PGs stimulates extrafollicular B cells to produce anti-PF4 IgGs. From this point, the consequent molecular effects would then resemble heparin-induced thrombocytopenia (Figure 5).

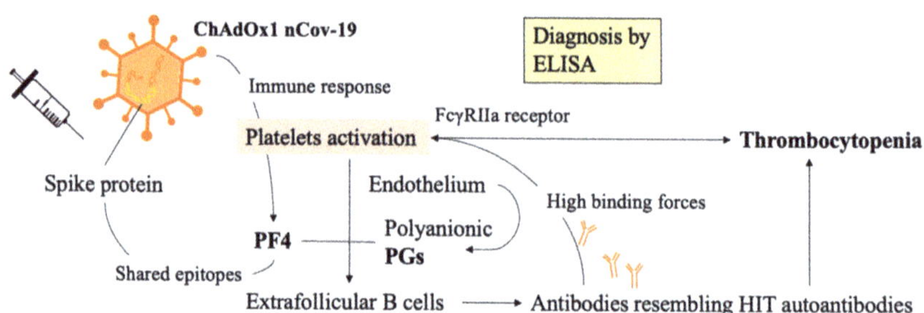

Figure 5. The proposed mechanism of autoantibodies generation is described. Following administration of adenoviral vector encoding the spike protein, a subsequent inflammatory cascade, stimulated by the individual immune response, activates platelets to generate platelet-factor 4 (PF4). In complexes with polyanionic PGs, derived from endothelial cells, they stimulate extrafollicular B cells to antibody production which, in turn, would exert positive feedback on platelet activation. The antibodies generated would resemble HIT autoantibodies from that point on with respect to thrombogenesis and hemostasis disorders. PF4 shares some epitopes with the Spike protein but, despite this, they are not sufficient to induce cross-reactivity. PF4: platelet factor 4, HIT: heparin-induced thrombocytopenia, PGs: proteoglycans.

Interestingly, Greinacher et al. have discarded the hypothesis of molecular mimicry mechanisms, due to the absence of cross-reaction with SARS-CoV-2 derived Spike protein [124]. The authors analyzed the Spike protein sequence and found 3 similar immunogenic epitopes, the part of an antigen molecule to which an antibody attaches itself with PF4. Prediction tools and 3D modeling software including IMED and SIM were used to compare them. Subsequently, they collected blood sera from 222 patients which tested positive for PCR analysis of SARS-CoV-2 infection and tested them for the presence of PF4/heparin ELISA, as well as heparin-dependent and PF4-dependent platelet activation assays [124]. Their results reinforce the hypothesis that the Spike protein is not inducing VITT.

Only 19 of 222 patients tested positive for PF4/heparin ELISA but those patients did not show any platelet hyperactivation sign. Furthermore, anti-PF4 and anti-PF4/heparin antibodies from two VITT patients were tested. They did not show any cross-reactivity to the recombinant SARS-CoV-2 spike protein [124].

The same research group 4 [121] had been studying the mechanisms underlying autoimmunity of HIT. By examining the binding forces expressed as pN (picoNewton) which are applied by antibodies in their binding to the target, they found substantial differences in terms of platelet activation. In particular, they divided groups of antibodies according to classes of force: group 1 showing $pN < 60$ antibodies (most of which do not activate platelets), group 2 with binding forces of $60 \leq pN \leq 100$, which were found to activate platelets in the presence of polyanions, and group 3 with binding forces $pN > 100$, which bind to PF4 even without polyanions present. A statistical difference was recorded in the second ($60 \leq pN \leq 100$; $p < 0.001$) and third groups ($pN \geq 100$; $p = 0.006$) [125]. Therefore, those higher forces were able to cluster PF4-molecules forming antigenic complexes which allow binding of polyanion-dependent anti-PF4/polyanion antibodies (anti-PF4/P-ABS). That induced massive platelet activation in the absence of heparin [121].

3.9. Updates from the European Medicines Agency

Since 7 May 2021, the PRAC committee has released more suspected reports of hemostatic and non-hemostatic conditions in other anti-COVID-19 vaccines under evaluation [126].

Myocarditis and pericarditis cases have been associated, even though a causative link has never been proven, with Comirnaty (the new brand name for Pfizer and BioNTech's COVID-19 vaccine, BNT162b2). EMA has requested analyses of data from marketing authorization holders for both Comirnaty and Moderna [127]. Recent investigations include interesting cases of Guillain–Barre syndrome after vaccination with the Vaxzevria vaccine. Further detailed monitoring is necessary [127].

Some cases of thrombocytopenia have been reported for mRNA vaccines too, in particular Comirnaty and Moderna vaccines. Epidemiological studies are confirming the rate of those cases is lower than the general population incidence rate [128]. The BNT162b2 vaccine has also been the subject of EMA evaluations for cases of facial swelling in people with a history of injections with dermal fillers [127].

The Pharmacovigilance Risk Assessment Committee (PRAC) has decided to validate this adverse effect by inserting it in Cominarty's product information. This was possible thanks to literature revision and the European database for suspected side effects EudraVigilance [127].

3.10. Public Health Challenges: Do Benefits Outweigh Risks?

Since the first cases of thrombosis for the Vaxzevria vaccine were reported [129], several studies tried to compare the incidence of those events with that of the general population. Pottegard and colleagues [128] assessed rates of hemostatic events in a cohort of 148,792 vaccinated Danes and 132,472 vaccinated Norwegians and compared them with general population cohorts from Denmark and Norway. Results showed 59 venous thromboembolic events in the vaccinated cohort compared with 30 expected based on general incidence rates. In particular, higher rates of cerebral venous thrombosis were observed: the standardized morbidity ratio, calculated as the ratio between observed and expected events, was demonstrated to be 20.25 (8.14 to 41.73). Interestingly, the rate of death was higher for the general population: 44 cases versus 15 for the vaccinated group [128]. However, it should be noted that the absolute risk of developing venous thromboembolic events was small (11 excess events/100,000 vaccines) and the results should be interpreted considering the proven beneficial effects of the vaccine. These data emerge clearly when the analysis was undertaken to investigate subgroup effects stratified according to gender and to the young versus middle-aged adults (age categories 18–44 years and 45–65 years). For example, when the analysis was restricted to women, no excess rate of thrombocytopenia/coagulation disorders was observed [128].

Increased surveillance could reinforce these findings and prove the beneficial effects of Vaxzevria vaccine outweigh the risks. Notably, the European Medicines Agency (EMA) started to investigate thrombotic events through the PRAC to review all conditions related to haemostatic alterations, (including thrombocytopenia and bleeding) after vaccines administration [127]. According to the pharmacovigilance legislation, additional monitoring is mentioned with a label on the package whenever it is needed. Furthermore, the medical literature is continuously monitored by EMA to guarantee suspect adverse reactions are correctly addressed. Thus, the status under which vaccines have been assigned by EMA is "conditional marketing, authorization granted". This is true for Vaxzevria, Moderna, Janssen, as well as Pfizer and BioNTech's Comirnaty vaccines [127].

3.11. Janssen Adenoviral Vector Vaccine and Cases of Thrombosis

The Ad26.COV2 vaccine has been associated with cases of cerebral venous sinus thrombosis (CVST), severe thrombocytopenia, and disseminated intravascular coagulation [129]. Muir et al. [130] described a patient who showed signs of autoimmune heparin-induced thrombocytopenia. The patient, a 48-year-old female, presented with general malaise and abdominal pain. From peripheral blood tests, thrombocytopenia with schistocytes on the blood smear was revealed. Coagulation tests also confirmed a low fibrinogen level, prolonged activated partial thromboplastin time, and an elevation in the D-dimer level. Factor V Leiden and Prothrombin G20210A gene mutation assessments were negative. Additionally, hepatitis, HIV, and lupus were tested for, but later ruled out. Investigators concluded the diagnostic workup with a computed tomographic (CT) scan of the abdomen and pelvis which demonstrated massive splanchnic venous thrombosis. After the patient developed a new-onset headache, a cerebral CT scan revealed cerebral venous sinus thrombosis involving the right transverse and straight sinuses.

The patient was further tested for anti-PF4-heparin antibodies which returned negative even though she was previously treated with unfractionated heparin. When evaluated for anti-PF4-polyanions antibodies, the patient tested positive (2.550 optical-density units [upper limit of the normal range], ≤ 0.399). She was therefore switched to therapy with argatroban. Thus, the case of thrombosis described by Muir et al. [130–132], demonstrates an update in literature reports associated with the Oxford-AstraZeneca vaccine [127]. Nevertheless, several authors, including Sadoff et al. [129], concluded that the Janssen vaccine and Vaxzevria, despite sharing some common features have different intrinsic structures.

Notably, the Ad26.COV2. S vaccine is composed of a human Ad26-based vector and Ad26 is from Ad species D. Its cellular receptor, is different from the Oxford-AstraZeneca vaccine which uses the Coxsackie and adenovirus receptor (CAR), CD46. The ChAdOx1 nCoV-19 vaccine is a chimpanzee adenovirus-based vector and Ad26 is from Ad species E. Thus, their biological characteristics are strongly different, even though the mechanism of immunologic response in the host is similar.

4. Conclusions

The SARS-CoV-2 infection has highlighted the importance of preventative medicine. The systemic effects of this are still relatively poorly understood, but several risk factors have been identified. The worldwide roll-out of vaccines has allowed the lifting of lockdown measures and reduced the spread of the disease. Ongoing epidemiological research is needed to monitor the variant strains and ascertain the ongoing efficacy of the vaccines against the virus.

Author Contributions: Conceptualization, F.N. and A.I.; methodology, F.N.; software, A.I.; validation, F.N., A.I. and S.S.A.S.; formal analysis, F.N.; investigation, F.N.; data curation, F.N. and A.I.; writing—original draft preparation, F.N.; writing—review and editing, F.N., A.I., S.S.A.S.; visualization, F.N. and A.I.; supervision, F.N., A.I. and S.S.A.S. All authors have read and agreed to the published version of the manuscript.

Funding: This research received no external funding.

Institutional Review Board Statement: Not applicable.

Informed Consent Statement: Not applicable.

Data Availability Statement: Not applicable.

Conflicts of Interest: The authors declare no conflict of interest.

References

1. Imai, Y.; Kuba, K.; Rao, S.; Huan, Y.; Guo, F.; Guan, B.; Yang, P.; Sarao, R.; Wada, T.; Leong-Poi, H.; et al. Angiotensin-converting enzyme 2 protects from severe acute lung failure. *Nature* **2005**, *436*, 112–116. [CrossRef]
2. Wang, D.; Chai, X.-Q.; Magnussen, C.G.; Zosky, G.; Shu, S.-H.; Wei, X.; Hu, S.-S. Renin-angiotensin-system, a potential pharmacological candidate, in acute respiratory distress syndrome during mechanical ventilation. *Pulm. Pharmacol. Ther.* **2019**, *58*, 101833. [CrossRef]
3. Van Tendeloo, V.F.; Ponsaerts, P.; Berneman, Z.N. mRNA-based gene transfer as a tool for gene and cell therapy. *Curr. Opin. Mol. Ther.* **2007**, *9*, 423–431. [PubMed]
4. Piazza, G.; Campia, U.; Hurwitz, S.; Snyder, J.E.; Rizzo, S.M.; Pfeferman, M.B.; Morrison, R.B.; Leiva, O.; Fanikos, J.; Nauffal, V.; et al. Registry of Arterial and Venous Thromboembolic Complications in Patients With COVID-19. *J. Am. Coll. Cardiol.* **2020**, *76*, 2060–2072. [CrossRef]
5. Klok, F.A.; Kruip, M.J.H.A.; Van der Meer, N.J.M.; Arbous, M.S.; Gommers, D.A.M.P.J.; Kant, K.M.; Kaptein, F.H.J.; van Paassen, J.; Stals, M.A.M.; Huisman, M.V.; et al. Incidence of thrombotic complications in critically ill ICU patients with COVID-19. *Thromb. Res.* **2020**, *191*, 145–147. [CrossRef]
6. Bikdeli, B.; Madhavan, M.V.; Jimenez, D.; Chuich, T.; Dreyfus, I.; Driggin, E.; Der Nigoghossian, C.; Ageno, W.; Madjid, M.; Guo, Y.; et al. COVID-19 and Thrombotic or Thromboembolic Disease: Implications for Prevention, Antithrombotic Therapy, and Follow-Up. *J. Am. Coll. Cardiol.* **2020**, *75*, 2950–2973. [CrossRef] [PubMed]
7. Driggin, E.; Madhavan, M.V.; Bikdeli, B.; Chuich, T.; Laracy, J.; Biondi-Zoccai, G.; Brown, T.S.; Der Nigoghossian, C.; Zidar, D.A.; Haythe, J.; et al. Cardiovascular considerations for patients, health care workers, and health sys-tems during the coronavirus disease 2019 (COVID-19) pandemic. *J. Am. Coll. Cardiol.* **2020**, *75*, 2352–2371. [CrossRef]
8. Bonow, R.O.; Fonarow, G.C.; O'Gara, P.T.; Yancy, C.W. Association of coronavirus disease 2019 (COVID-19) with myocardial in-jury and mortality. *JAMA Cardiol.* **2020**, *5*, 751–753. [CrossRef]
9. Fox, S.E.; Akmatbekov, A.; Harbert, J.L.; Li, G.; Brown, J.Q.; Heide, R.S.V. Pulmonary and cardiac pathology in African American patients with COVID-19: An autopsy series from New Orleans. *Lancet Respir. Med.* **2020**, *8*, 681–686. [CrossRef]
10. Wichmann, D.; Sperhake, J.-P.; Lütgehetmann, M.; Steurer, S.; Edler, C.; Heinemann, A.; Heinrich, F.; Mushumba, H.; Kniep, I.; Schröder, A.S.; et al. Autopsy Findings and Venous Thromboembolism in Patients With COVID-19. *Ann. Intern. Med.* **2020**, *173*, 268–277. [CrossRef] [PubMed]
11. Tang, N.; Li, D.; Wang, X.; Sun, Z. Abnormal coagulation parameters are associated with poor prognosis in patients with novel coronavirus pneumonia. *J. Thromb. Haemost.* **2020**, *18*, 844–847. [CrossRef] [PubMed]
12. Cui, S.; Chen, S.; Li, X.; Liu, S.; Wang, F. Prevalence of venous thromboembolism in patients with severe novel coronavirus pneumonia. *J. Thromb. Haemost.* **2020**, *18*, 1421–1424. [CrossRef]
13. Allard, J.F.; Dushek, O.; Coombs, D.; van der Merwe, P. Mechanical Modulation of Receptor-Ligand Interactions at Cell-Cell Interfaces. *Biophys. J.* **2012**, *102*, 1265–1273. [CrossRef] [PubMed]
14. Mary, S.; Fehrentz, J.; Damian, M.; Vedié, P.; Martinez, J.; Marie, K.; Baneres, J. How ligands and signaling pro-teins affect G-protein-coupled receptors' conformational landscape. *Biochem. Soc. Trans.* **2013**, *41*, 144–147. [CrossRef]
15. Unal, H.; Karnik, S.S. Domain coupling in GPCRs: The engine for induced conformational changes. *Trends Pharmacol. Sci.* **2012**, *33*, 79–88. [CrossRef] [PubMed]
16. Kobilka, B.K.; Deupi, X. Conformational complexity of G-protein-coupled receptors. *Trends Pharmacol. Sci.* **2007**, *28*, 397–406. [CrossRef] [PubMed]
17. Lefkowitz, R.J. Seven transmembrane receptors: Something old, something new. *Acta Physiol.* **2007**, *190*, 9–19. [CrossRef]
18. Lu, Y.; Liu, D.X.; Tam, J.P. Lipid rafts are involved in SARS-CoV entry into vero E6 cells. *Biochem. Biophys. Res. Commun.* **2008**, *369*, 344–349. [CrossRef]
19. Oates, J.; Watts, A. Uncovering the intimate relationship between lipids, cholesterol and GPCR activation. *Curr. Opin. Struct. Biol.* **2011**, *21*, 802–807. [CrossRef]
20. Shukla, A. (Ed.) *G Protein-Coupled Receptors. Signaling, Trafficking and Regulation*, 1st ed.; Methods in Cell Biology; Elsevier: Amsterdam, The Netherlands, 2016; Volume 132.
21. Villar, V.; Cuevas, S.; Zheng, X.; Jose, P. Chapter 1: Localization and signaling of GPCRs in lipid rafts. *Methods Cell Biol.* **2016**, *136*, 3–23.
22. Watkins, E.; Miller, C.; Majewski, J.; Kuhl, T. Membrane texture induced by specific protein binding and receptor clustering: Active roles for lipids in cellular function. *Proc. Natl. Acad. Sci. USA* **2011**, *108*, 6975–6980. [CrossRef]
23. Lu, Y.; Liu, D.X.; Tam, J.P. Lipid rafts play an important role in the early stage of severe acute respiratory syn-drome-coronavirus life cycle. *Microbes Infect.* **2007**, *9*, 96–102.
24. Lingwood, D.; Simons, K. Lipid Rafts As a Membrane-Organizing Principle. *Science* **2009**, *327*, 46–50. [CrossRef]
25. Li, W.; Moore, M.J.; Vasilieva, N.; Sui, J.; Wong, S.K.; Berne, M.A.; Somasundaran, M.; Sullivan, J.L.; Luzuriaga, K.; Greenough, T.C.; et al. Angiotensin-converting enzyme 2 is a functional receptor for the SARS coronavirus. *Nature* **2003**, *426*, 450–454. [CrossRef] [PubMed]
26. Li, W.; Zhang, C.; Sui, J.; Kuhn, J.H.; Moore, M.J.; Luo, S.; Wong, S.-K.; Huang, I.-C.; Xu, K.; Vasilieva, N.; et al. Receptor and viral determinants of SARS-coronavirus adaptation to human ACE2. *EMBO J.* **2005**, *24*, 1634–1643. [CrossRef] [PubMed]

27. Song, H.-D.; Tu, C.-C.; Zhang, G.-W.; Wang, S.-Y.; Zheng, K.; Lei, L.-C.; Chen, Q.-X.; Gao, Y.-W.; Zhou, H.-Q.; Xiang, H.; et al. Cross-host evolution of severe acute respiratory syndrome coronavirus in palm civet and human. *Proc. Natl. Acad. Sci. USA* **2005**, *102*, 2430–2435. [CrossRef]

28. Wu, K.; Peng, G.; Wilken, M.; Geraghty, R.J.; Li, F. Mechanisms of host receptor adaptation by severe acute respiratory syndrome coronavirus. *J. Biol. Chem.* **2012**, *287*, 8904–8911. [CrossRef] [PubMed]

29. Menachery, V.D.; Yount, B.L., Jr.; Debbink, K.; Agnihothram, S.; Gralinski, L.E.; Plante, J.A.; Graham, R.L.; Scobey, T.; Ge, X.Y.; Donaldson, E.F.; et al. A SARS-like cluster of circulating bat coronaviruses shows potential for human emergence. *Nat. Med.* **2015**, *21*, 1508–1513. [CrossRef]

30. Menachery, V.D.; Yount, B.L.; Debbink, K.; Agnihothram, S.; Gralinski, L.E.; Plante, J.A.; Graham, R.L.; Scobey, T.; Ge, X.-Y.; Donaldson, E.F.; et al. Author Correction: A SARS-like cluster of circulating bat coronaviruses shows potential for human emergence. *Nat. Med.* **2020**, *26*, 1146. [CrossRef]

31. Ren, W.; Qu, X.; Li, W.; Han, Z.; Yu, M.; Zhou, P.; Zhang, S.Y.; Wang, L.F.; Deng, H.; Shi, Z. Difference in receptor usage be-tween severe acute respiratory syndrome (SARS) coronavirus and SARS-like coronavirus of bat origin. *J. Virol.* **2008**, *82*, 1899–1907. [CrossRef]

32. Ge, X.-Y.; Li, J.; Yang, X.-L.; Chmura, A.; Zhu, G.; Epstein, J.H.; Mazet, J.K.; Hu, B.; Zhang, W.; Peng, C.; et al. Isolation and characterization of a bat SARS-like coronavirus that uses the ACE2 receptor. *Nature* **2013**, *503*, 535–538. [CrossRef] [PubMed]

33. Adams, L.E.; Dinnon, K.H., 3rd.; Hou, Y.J.; Sheahan, T.P.; Heise, M.T.; Baric, R.S. Critical ACE2 Determinants of SARS-CoV-2 and Group 2B Coronavirus Infection and Replication. *mBio* **2021**, *12*, e03149-20. [CrossRef]

34. Jeffers, S.A.; Tusell, S.M.; Gillim-Ross, L.; Hemmila, E.M.; Achenbach, J.E.; Babcock, G.J.; Thomas, W.D.; Thackray, L.B.; Young, M.D.; Mason, R.J.; et al. CD209L (L-SIGN) is a receptor for severe acute respiratory syndrome coronavirus. *Proc. Natl. Acad. Sci. USA* **2004**, *101*, 15748–15753. [CrossRef]

35. Jeffers, S.A.; Hemmila, E.M.; Holmes, K.V. Human Coronavirus 229E can Use CD209L (L-Sign) to Enter Cells. *Adv. Exp. Med. Biol.* **2006**, *581*, 265–269. [CrossRef]

36. Kuba, K.; Imai, Y.; Rao, S.; Gao, H.; Guo, F.; Guan, B.; Huan, Y.; Yang, P.; Zhang, Y.; Deng, W.; et al. A crucial role of angiotensin converting enzyme 2 (ACE2) in SARS coronavirus–induced lung injury. *Nat. Med.* **2005**, *11*, 875–879. [CrossRef] [PubMed]

37. Bozkurt, B.; Kovacs, R.; Harrington, B. HFSA/ACC/AHA Statement Addresses Concerns Re: Using RAAS Antagonists in COVID-19. Available online: https://professional.heart.org/professional/ScienceNews/UCM_505836_HFSAAC-CA-HA-statement-addresses-concerns-reusing-RAAS-antagonists-in-COVID-19.jsp?utm_campaign=sciencenews19-20&utm_source=science-news&utm_medium=email&utm_content=phd03-17-20 (accessed on 22 March 2020).

38. Imai, Y.; Kuba, K.; Penninger, J.M. The discovery of angiotensin-converting enzyme 2 and its role in acute lung injury in mice. *Exp. Physiol.* **2008**, *93*, 543–548. [CrossRef] [PubMed]

39. Xudong, X.; Junzhu, C.; Xingxiang, W.; Furong, Z.; Yanrong, L. Age- and gender-related difference of ACE2 expression in rat lung. *Life Sci.* **2006**, *78*, 2166–2171. [CrossRef]

40. Lakatta, E.G. The reality of getting old. *Nat. Rev. Cardiol.* **2018**, *15*, 499–500. [CrossRef] [PubMed]

41. Liang, W.; Zhu, Z.; Guo, J.; Liu, Z.; He, X.; Zhou, W.; Chin, D.P.; Schuchat, A.; Beijing Joint SARS Expert Group. Severe Acute Respiratory Syndrome, Beijing, 2003. *Emerg. Infect. Dis.* **2004**, *10*, 25–31. [CrossRef]

42. Backhaus, A. Coronavirus: Why It's So Deadly in Italy. Available online: https://www.researchgate.net/publication/342014166_Coronavirus_Why_it\T1\textquoterights_so_deadly_in_Italy (accessed on 22 March 2020).

43. Rodrigues Prestes, T.R.; Rocha, N.P.; Miranda, A.S.; Teixeira, A.L.; Simoes-E-Silva, A.C. The anti-inflammatory potential of ACE2/angiotensin-(1-7)/Mas receptor axis: Evidence from basic and clinical research. *Curr. Drug Targets* **2017**, *18*, 1301–1313. [CrossRef] [PubMed]

44. Imai, Y.; Kuba, K.; Penninger, J.M. Angiotensin-converting enzyme 2 in acute respiratory distress syndrome. *Cell. Mol. Life Sci.* **2007**, *64*, 2006–2012. [CrossRef] [PubMed]

45. Zhang, X.; Li, S.; Niu, S. ACE2 and COVID-19 and the resulting ARDS. *Postgrad. Med. J.* **2020**, *96*, 403–407. [CrossRef] [PubMed]

46. Fallahi-Sichani, M.; Linderman, J. Lipid raft-mediated regulation of G-protein coupled receptor signaling by ligands which influence receptor dimerization: Acomputational study. *PLoS ONE* **2009**, *4*, e6604. [CrossRef]

47. Tetlow, S.; Segiet-Swiecicka, A.; O'Sullivan, R.; O'Halloran, S.; Kalb, K.; Brathwaite-Shirley, C.; Alger, L.; Ankuli, A.; Baig, M.S.; Catmur, F.; et al. ACE inhibitors, angiotensin receptor blockers and endothelial injury in COVID-19. *J. Intern. Med.* **2020**, *289*, 688–699. [CrossRef] [PubMed]

48. Cannata, F.; Chiarito, M.; Reimers, B.; Azzolini, E.; Ferrante, G.; My, I.; Viggiani, G.; Panico, C.; Regazzoli, D.; Ciccarelli, M.; et al. Continuation versus discontinuation of ACE inhibitors or angiotensin II receptor blockers in COVID-19: Effects on blood pressure control and mortality. *Eur. Heart J. Cardiovasc. Pharmacother.* **2020**, *6*, 412–414. [CrossRef]

49. Derington, C.G.; Cohen, J.B.; Mohanty, A.F.; Greene, T.H.; Cook, J.; Ying, J.; Wei, G.; Herrick, J.S.; Stevens, V.W.; Jones, B.E.; et al. Angiotensin II receptor blocker or angiotensin-converting enzyme inhibitor use and COVID-19-related outcomes among US Veterans. *PLoS ONE* **2021**, *16*, e0248080. [CrossRef]

50. Parit, R.; Jayavel, S. Association of ACE inhibitors and angiotensin type II blockers with ACE2 overexpression in COVID-19 comorbidities: A pathway-based analytical study. *Eur. J. Pharmacol.* **2021**, *896*, 173899. [CrossRef]

51. Fang, L.; Karakiulakis, G.; Roth, M. Are patients with hypertension and diabetes mellitus at increased risk for COVID-19 infection? *Lancet Respir. Med.* **2020**, *8*, e21. [CrossRef]

52. Watkins, J. Preventing a covid-19 pandemic. *BMJ* **2020**, *368*, m810. [CrossRef]
53. Abuohashish, H.M.; Ahmed, M.M.; Sabry, D.; Khattab, M.M.; Al-Rejaie, S.S. ACE-2/Ang1-7/Mas cascade mediates ACE inhibitor, captopril, protective effects in estrogen-deficient osteoporotic rats. *Biomed. Pharmacother.* **2017**, *92*, 58–68. [CrossRef]
54. Saponaro, F.; Rutigliano, G.; Sestito, S.; Bandini, L.; Storti, B.; Bizzarri, R.; Zucchi, R. ACE2 in the Era of SARS-CoV-2: Controversies and Novel Perspectives. *Front. Mol. Biosci.* **2020**, *7*, 588618. [CrossRef]
55. Adedeji, A.; Sarafianos, S.G. Antiviral drugs specific for coronaviruses in preclinical development. *Curr. Opin. Virol.* **2014**, *8*, 45–53. [CrossRef]
56. Zhang, Y.; Cao, W.; Jiang, W.; Xiao, M.; Li, Y.; Tang, N.; Liu, Z.; Yan, X.; Zhao, Y.; Li, T.; et al. Profile of natural anticoagulant, coagulant factor and anti-phospholipid antibody in critically ill COVID-19 patients. *J. Thromb. Thrombolysis* **2020**, *50*, 580–586. [CrossRef] [PubMed]
57. Zaid, Y.; Puhm, F.; Allaeys, I.; Naya, A.; Oudghiri, M.; Khalki, L.; Limami, Y.; Zaid, N.; Sadki, K.; Ben El Haj, R.; et al. Platelets can associate with SARS-Cov-2 RNA and are hyperactivated in COVID-19. *Circ Res.* **2020**, *127*, 1404–1418. [CrossRef]
58. Nicolai, L.; Leunig, A.; Brambs, S.; Kaiser, R.; Weinberger, T.; Weigand, M.; Muenchhoff, M.; Hellmuth, J.C.; Ledderose, S.; Schulz, H.; et al. Immunothrombotic Dysregulation in COVID-19 Pneumonia Is Associated With Respiratory Failure and Coagulopathy. *Circulation* **2020**, *142*, 1176–1189. [CrossRef] [PubMed]
59. Ackermann, M.; Verleden, S.; Kuehnel, M.; Haverich, A.; Welte, T.; Laenger, F.; Vanstapel, A.; Werlein, C.; Stark, H.; Tzankov, A.; et al. Pulmonary Vascular Endothelialitis, Thrombosis, and Angiogenesis in COVID-19. *N. Engl. J. Med.* **2020**, *383*, 120–128. [CrossRef] [PubMed]
60. Girardi, F.M.; Barra, M.B.; Zettler, C.G. Papillary thyroid carcinoma: Does the association with Hashimoto's thyroiditis affect the clinicopathological characteristics of the disease? *Braz. J. Otorhinolaryngol.* **2015**, *81*, 283–287. [CrossRef] [PubMed]
61. Roberts, L.N.; Whyte, M.B.; Georgiou, L.; Giron, G.; Czuprynska, J.; Rea, C.; Vadher, B.; Patel, R.K.; Gee, E.; Arya, R. Postdischarge venous thromboembolism following hospital admission with COVID-19. *Blood* **2020**, *136*, 1347–1350. [CrossRef] [PubMed]
62. Mai, V.; Tan, B.K.; Mainbourg, S.; Potus, F.; Cucherat, M.; Lega, J.-C.; Provencher, S. Venous thromboembolism in COVID-19 compared to non-COVID-19 cohorts: A systematic review with meta-analysis. *Vasc. Pharmacol.* **2021**, *139*, 106882. [CrossRef] [PubMed]
63. Patell, R.; Bogue, T.; Koshy, A.; Bindal, P.; Merrill, M.; Aird, W.C.; Bauer, K.A.; Zwicker, J.I. Postdischarge thrombosis and hemorrhage in patients with COVID-19. *Blood* **2020**, *136*, 1342–1346. [CrossRef]
64. Moores, L.K.; Tritschler, T.; Brosnahan, S.; Carrier, M.; Collen, J.F.; Doerschug, K.; Holley, A.B.; Jimenez, D.; LeGal, G.; Rali, P.; et al. Prevention, diagnosis and treatment of venous thromboembolism in patients with COVID-19: CHEST Guideline and Expert Panel Report. *Chest* **2020**, *158*, 1143–1163. [CrossRef] [PubMed]
65. Spyropoulos, A.C.; Levy, J.H.; Ageno, W.; Connors, J.M.; Hunt, B.J.; Iba, T.; Levi, M.; Samama, C.M.; Thachil, J.; Giannis, D.; et al. Scientific and Standardization Committee communication: Clinical guidance on the diagnosis, prevention, and treatment of venous thromboembolism in hospitalized patients with COVID-19. *J. Thromb. Haemost.* **2020**, *18*, 1859–1865. [CrossRef] [PubMed]
66. Cuker, A.; Tseng, E.K.; Nieuwlaat, R.; Angchaisuksiri, P.; Blair, C.; Dane, K.; Davila, J.; DeSancho, M.T.; Diuguid, D.; Griffin, D.O.; et al. American Society of Hematology 2021 guidelines on the use of anticoagulation for thromboprophylaxis in patients with COVID-19. *Blood Adv.* **2021**, *5*, 872–888. [CrossRef]
67. Anderson, E.J.; Rouphael, N.G.; Widge, A.T.; Jackson, L.A.; Roberts, P.C.; Makhene, M.; Chappell, J.D.; Denison, M.R.; Stevens, L.J.; Pruijssers, A.J.; et al. Safety and immunogenicity of SARS-CoV-2 mRNA-1273 vaccine in older adults. *N. Engl. J. Med.* **2020**, *383*, 2427–2438. [CrossRef] [PubMed]
68. Walsh, E.E.; Frenck, J.R.W.; Falsey, A.R.; Kitchin, N.; Absalon, J.; Gurtman, A.; Lockhart, S.; Neuzil, K.; Mulligan, M.J.; Bailey, R.; et al. Safety and Immunogenicity of Two RNA-Based COVID-19 Vaccine Candidates. *N. Engl. J. Med.* **2020**, *383*, 2439–2450. [CrossRef]
69. Zhu, F.C.; Guan, X.H.; Li, Y.H.; Huang, J.Y.; Jiang, T.; Hou, L.H.; Li, J.X.; Yang, B.F.; Wang, L.; Wang, W.J.; et al. Immunogenicity and safety of are combinant adenovirus type-5-vectored COVID-19 vaccine in healthy adults aged 18 years or older: A randomised, double-blind, placebo-controlled, phase 2 trial. *Lancet* **2020**, *396*, 479–488. [CrossRef]
70. Logunov, D.Y.; Dolzhikova, I.V.; Zubkova, O.V.; Tukhvatullin, A.I.; Shcheblyakov, D.V.; Dzharullaeva, A.S.; Grousova, D.M.; Erokhova, A.S.; Kovyrshina, A.V.; Botikov, A.G.; et al. Safety and immunogenicity of an rAd26 and rAd5 vector-based heterologous prime-boost COVID-19 vaccine in two formulations: Two open, non-randomized phase 1/2 studies from Russia. *Lancet* **2020**, *396*, 887–897. [CrossRef]
71. Xia, S.; Zhang, Y.; Wang, Y.; Wang, H.; Yang, Y.; Gao, G.F.; Tan, W.; Wu, G.; Xu, M.; Lou, Z.; et al. Safety and immunogenicity of an inactivated SARS-CoV-2 vaccine, BBIBP-CorV: A randomised, double-blind, placebo-controlled, phase 1/2 trial. *Lancet Infect Dis.* **2021**, *21*, 39–51. [CrossRef]
72. Xia, S.; Duan, K.; Zhang, Y.; Zhao, D.; Zhang, H.; Xie, Z.; Li, X.; Peng, C.; Zhang, Y.; Zhang, W.; et al. Effect of an inactivated vaccine against SARS-CoV-2 on safety and immunogenicity outcomes: Interim analysis of 2 randomized clinical trials. *JAMA* **2020**, *324*, 951–960. [CrossRef]
73. Keech, C.; Albert, G.; Cho, I.; Robertson, A.; Reed, P.; Neal, S.; Plested, J.S.; Zhu, M.; Cloney-Clark, S.; Zhou, H.; et al. Phase 1-2 Trial of a SARS-CoV-2 recombinant spike protein nanoparticle vaccine. *N. Engl. J. Med.* **2020**, *383*, 2320–2332. [CrossRef]

74. Hacein-Bey-Abina, S.; Von Kalle, C.; Schmidt, M.; McCormack, M.; Wulffraat, N.; Leboulch, P.; Lim, A.; Osborne, C.; Pawliuk, R.; Morillon, E.; et al. LMO2-Associated Clonal T Cell Proliferation in Two Patients after Gene Therapy for SCID-X1. *Science* **2003**, *302*, 415–419. [CrossRef] [PubMed]

75. Wolff, J.A.; Malone, R.W.; Williams, P.; Chong, W.; Acsadi, G.; Jani, A.; Felgner, P.L. Direct Gene Transfer into Mouse Muscle In Vivo. *Science* **1990**, *247*, 1465–1468. [CrossRef]

76. Gilboa, E.; Vieweg, J. Cancer immunotherapy with mRNA-transfected dendritic cells. *Immunol Rev.* **2004**, *199*, 251–263. [CrossRef]

77. Pascolo, S. Vaccination with Messenger RNA (mRNA). *Immunol. Rev.* **2008**, *199*, 221–235. [CrossRef]

78. Alexopoulou, L.; Holt, A.C.; Medzhitov, R.; Flavell, R.A. Recognition of double-stranded RNA and activation of NF-κB by Toll-like receptor 3. *Nature* **2001**, *413*, 732–738. [CrossRef] [PubMed]

79. Diebold, S.S.; Kaisho, T.; Hemmi, H.; Akira, S.; e Sousa, C.R. Innate antiviral responses by means of TLR7-mediated recognition of single-stranded RNA. *Science* **2004**, *303*, 1529–1531. [CrossRef]

80. Heil, F.; Hemmi, H.; Hochrein, H.; Ampenberger, F.; Kirschning, C.; Akira, S.; Lipford, G.; Wagner, H.; Bauer, S. Species-Specific Recognition of Single-Stranded RNA via Toll-like Receptor 7 and 8. *Science* **2004**, *303*, 1526–1529. [CrossRef] [PubMed]

81. Karikó, K.; Buckstein, M.; Ni, H.; Weissman, D. Suppression of RNA recognition by Toll-like receptors: The impact of nu-cleoside modification and the evolutionary origin of RNA. *Immunity* **2005**, *23*, 165–175. [CrossRef]

82. Karikó, K.; Weissman, D. Naturally occurring nucleoside modifications suppress the immunostimulatory activity of RNA: Implication for therapeutic RNA development. *Curr. Opin. Drug Discov. Dev.* **2007**, *10*, 523.

83. Hornung, V.; Ellegast, J.; Kim-Hellmuth, S.; Brzózka, K.; Jung, A.; Kato, H.; Poeck, H.; Akira, S.; Conzelmann, K.-K.; Schlee, M.; et al. 5′-Triphosphate RNA Is the Ligand for RIG-I. *Science* **2006**, *314*, 994–997. [CrossRef]

84. Sioud, M.; Furset, G.; Cekaite, L. Suppression of immunostimulatory siRNA-driven innate immune activation by 2′-modified RNAs. *Biochem. Biophys. Res. Commun.* **2007**, *361*, 122–126. [CrossRef]

85. Karikó, K.; Muramatsu, H.; Welsh, F.A.; Ludwig, J.; Kato, H.; Akira, S.; Weissman, D. Incorporation of Pseudouridine Into mRNA Yields Superior Nonimmunogenic Vector With Increased Translational Capacity and Biological Stability. *Mol. Ther.* **2008**, *16*, 1833–1840. [CrossRef]

86. Wrapp, D.; Wang, N.; Corbett, K.S.; Goldsmith, J.A.; Hsieh, C.-L.; Abiona, O.; Graham, B.S.; McLellan, J.S. Cryo-EM structure of the 2019-nCoV spike in the prefusion conformation. *Science* **2020**, *367*, 1260–1263. [CrossRef]

87. Sahin, U.; Muik, A.; Vogler, I.; Derhovanessian, E.; Kranz, L.M.; Vormehr, M.; Quandt, J.; Bidmon, N.; Ulges, A.; Baum, A.; et al. BNT162b2 vaccine induces neutralizing antibodies and poly-specific T cells in humans. *Nature* **2021**, *595*, 572–577. [CrossRef] [PubMed]

88. Pardi, N.; Tuyishime, S.; Muramatsu, H.; Kariko, K.; Mui, B.L.; Tam, Y.K.; Madden, T.D.; Hope, M.J.; Weissman, D. Expression kinetics of nucleoside modified mRNA delivered in lipid nanopar-ticles to mice by various routes. *J. Control. Release* **2015**, *217*, 345–351. [CrossRef]

89. Polack, F.P.; Thomas, S.J.; Kitchin, N.; Absalon, J.; Gurtman, A.; Lockhart, S.; Perez, J.L.; Marc, G.P.; Moreira, E.D.; Zerbini, C.; et al. Safety and Efficacy of the BNT162b2 mRNA COVID-19 Vaccine. *N. Engl. J. Med.* **2020**, *383*, 2603–2615. [CrossRef] [PubMed]

90. Baden, L.R.; El Sahly, H.M.; Essink, B.; Kotloff, K.; Frey, S.; Novak, R.; Diemert, D.; Spector, S.A.; Rouphael, N.; Creech, C.B.; et al. Efficacy and Safety of the mRNA-1273 SARS-CoV-2 Vaccine. *N. Engl. J. Med.* **2021**, *384*, 403–416. [CrossRef] [PubMed]

91. Voysey, M.; Clemens, S.A.C.; Madhi, S.A.; Weckx, L.Y.; Folegatti, P.M.; Aley, P.K.; Angus, B.; Baillie, V.L.; Barnabas, S.L.; Bhorat, Q.E.; et al. Safety and efficacy of the ChAdOx1 nCoV-19 vaccine (AZD1222) against SARS-CoV-2: An interim analysis of four randomised controlled trials in Brazil, South Africa, and the UK. *Lancet* **2021**, *397*, 99–111. [CrossRef]

92. Ramasamy, M.N.; Minassian, A.M.; Ewer, K.J.; Flaxman, A.L.; Folegatti, P.M.; Owens, D.R.; Voysey, M.; Aley, P.K.; Angus, B.; Babbage, G.; et al. Safety and immunogenicity of ChA-dOx1 nCoV-19 vaccine administered in a prime-boost regimen in young and old adults (COV002): A single-blind, randomised, controlled, phase 2/3 trial. *Lancet* **2021**, *396*, 1979–1993. [CrossRef]

93. Zhu, F.C.; Li, Y.H.; Guan, X.H.; Hou, L.H.; Wang, W.J.; Li, J.X.; Wu, S.P.; Wang, B.S.; Wang, Z.; Wang, L.; et al. Safety, tolerability, and immunogenicity of a recombinant ade-novirus type-5 vectored COVID-19 vaccine: A dose-escalation, open-label, non-randomised, first-in-human trial. *Lancet* **2020**, *395*, 1845–1854. [CrossRef]

94. Logunov, D.Y.; Dolzhikova, I.V.; Shcheblyakov, D.V.; Tukhvatulin, A.I.; Zubkova, O.V.; Dzharullaeva, A.S.; Kovyrshina, A.V.; Lubenets, N.L.; Grousova, D.M.; Erokhova, A.S.; et al. Safety and efficacy of an rAd26 and rAd5 vector-based heterologous prime-boost COVID-19 vaccine: An interim analysis of a randomised controlled phase 3 trial in Russia. *Lancet* **2021**, *397*, 671–681. [CrossRef]

95. Abbink, P.; Lemckert, A.A.; Ewald, B.A.; Lynch, D.M.; Denholtz, M.; Smits, S.; Holterman, L.; Damen, I.; Vogels, R.; Thorner, A.R.; et al. Comparative seroprevalence and immunogenicity of six rare serotype recom-binant adenovirus vaccine vectors from subgroups B and D. *J. Virol.* **2007**, *81*, 4654–4663. [CrossRef] [PubMed]

96. Bos, R.; Rutten, L.; van der Lubbe, J.E.; Bakkers, M.J.; Hardenberg, G.; Wegmann, F.; Zuijdgeest, D.; de Wilde, A.H.; Koornneef, A.; Verwilligen, A.; et al. Ad26 vector-based COVID-19 vaccine encoding a prefusion-stabilized SARS-CoV-2 spike immunogen induces potent humoral and cellular immune responses. *NPJ Vaccines* **2020**, *5*, 91. [CrossRef] [PubMed]

97. Sadoff, J.; Le Gars, M.; Shukarev, G.; Heerwegh, D.; Truyers, C.; de Groot, A.M.; Stoop, J.; Tete, S.; Van Damme, W.; Leroux-Roels, I.; et al. Interim results of a phase 1–2a trial of Ad26.COV2.S COVID-19 vaccine. *N. Engl. J. Med.* **2021**, *384*, 1824–1835. [CrossRef]

98.   Sadoff, J.; Gray, G.; Vandebosch, A.; Cárdenas, V.; Shukarev, G.; Grinsztejn, B.; Goepfert, P.A.; Truyers, C.; Fennema, H.; Spiessens, B.; et al. Safety and Efficacy of Single-Dose Ad26.COV2.S Vaccine against COVID-19. *N. Engl. J. Med.* **2021**, *384*, 2187–2201. [CrossRef] [PubMed]

99.   ECDC-European Center for Disease Control. SARS-CoV-2 Variants of Concern as of 6 May 2021. Available online: https://www.ecdc.europa.eu/en/covid-19/variants-concern (accessed on 5 June 2021).

100. Wu, F.; Zhao, S.; Yu, B.; Chen, Y.M.; Wang, W.; Song, Z.G.; Hu, Y.; Tao, Z.W.; Tian, J.H.; Pei, Y.Y.; et al. A new coronavirus associated with human respiratory disease in China. *Nature* **2020**, *579*, 265–269. [CrossRef]

101. Shinde, V.; Bhikha, S.; Hossain, Z.; Archary, M.; Bhorat, Q.; Fairlie, L.; Lalloo, U.; Masilela, M.L.S.; Moodley, D.; Hanley, S.; et al. Preliminary Efficacy of the NVX-CoV2373 COVID-19 Vaccine Against the B.1.351 Variant. *MedRxiv* **2021**. [CrossRef]

102. Liu, Y.; Liu, J.; Xia, H.; Zhang, X.; Fontes-Garfias, C.R.; Swanson, K.A.; Cai, H.; Sarkar, R.; Chen, W.; Cutler, M.; et al. Neutralizing Activity of BNT162b2-Elicited Serum. *N. Engl. J. Med.* **2021**, *384*, 1466–1468. [CrossRef]

103. Wu, K.; Werner, A.P.; Koch, M.; Choi, A.; Narayanan, E.; Stewart-Jones, G.B.; Colpitts, T.; Bennett, H.; Boyoglu-Barnum, S.; Shi, W.; et al. Serum Neutralizing Activity Elicited by mRNA-1273 Vaccine. *N. Engl. J. Med.* **2021**, *384*, 1468–1470. [CrossRef]

104. Moore, J.P.; Offit, P.A. SARS-CoV-2 Vaccines and the Growing Threat of Viral Variants. *JAMA* **2021**, *325*, 821–822. [CrossRef]

105. IHME. COVID-19 Vaccine Efficacy Summary. Available online: http://www.healthdata.org/covid/covid-19-vaccine-efficacy-summary (accessed on 26 April 2021).

106. Lurie, N.; Sharfstein, J.M.; Goodman, J.L. The Development of COVID-19 Vaccines: Safeguards Needed. *JAMA* **2020**, *324*, 439–440. [CrossRef]

107. Madhi, S.A.; Baillie, V.; Cutland, C.L.; Voysey, M.; Koen, A.L.; Fairlie, L.; Padayachee, S.D.; Dheda, K.; Barnabas, S.L.; Bhorat, Q.E.; et al. Efficacy of the ChAdOx1 nCoV-19 COVID-19 Vaccine against the B.1.351 Variant. *N. Engl. J. Med.* **2021**, *384*, 1885–1898. [CrossRef] [PubMed]

108. Available online: https://www.sciencemag.org/news/2021/02/south-africa-suspends-use-astrazenecas-covid-19-vaccine-after-it-fails-clearly-stop (accessed on 8 February 2021).

109. Van Doremalen, N.; Lambe, T.; Spencer, A.; Belij-Rammerstorfer, S.; Purushotham, J.N.; Port, J.R.; Avanzato, V.A.; Bushmaker, T.; Flaxman, A.; Ulaszewska, M.; et al. ChAdOx1 nCoV-19 vaccine prevents SARS-CoV-2 pneumonia in rhesus macaques. *Nature* **2020**, *586*, 578–582. [CrossRef] [PubMed]

110. WHO World Health Organization. Buthan. The Current COVID-19 Situation. Available online: https://www.who.int/countries/btn/ (accessed on 11 May 2021).

111. Rossman, H.; Shilo, S.; Meir, T.; Gorfine, M.; Shalit, U.; Segal, E. COVID-19 dynamics after a national immunization program in Israel. *Nat. Med.* **2021**, *27*, 1055–1061. [CrossRef]

112. Vasileiou, E.; Simpson, C.R.; Robertson, C.; Shi, T.; Kerr, S.; Agrawal, U.; Akbari, A.; Bedston, S.; Beggs, J.; Bradley, D.; et al. Interim findings from first-dose mass COVID-19 vaccination roll-out and COVID-19 hospital admissions in Scotland: A national prospective cohort study. *Lancet* **2021**, *397*, 1646–1657. [CrossRef]

113. Benenson, S.; Oster, Y.; Cohen, M.J.; Nir-Paz, R. BNT162b2 mRNA COVID-19 Vaccine Effectiveness among Health Care Workers. *N. Engl. J. Med.* **2021**, *384*, 1775–1777. [CrossRef]

114. Available online: https://www.pfizer.com/news/press-release/press-release-detail/real-world-evidence-confirms-high-effectiveness-pfizer (accessed on 11 March 2021).

115. Available online: https://www.sciencemediacentre.org/expert-reaction-to-preprint-looking-at-the-pfizer-biontech-covid-19-vaccine-efficacy-in-israel/ (accessed on 23 February 2021).

116. Dougherty, K.; Mannell, M.; Naqvi, O.; Matson, D.; Stone, J. SARS-CoV-2 B.1.617.2 (Delta) Variant COVID-19 Outbreak Associat-ed with a Gymnastics Facility-Oklahoma, April–May 2021. *MMWR Morb. Mortal. Wkly. Rep.* **2021**, *70*, 1004–1007. [CrossRef]

117. Available online: https://www.pfizer.com/news/press-release/press-release-detail/pfizer-and-biontech-initiate-study-part-broad-development (accessed on 25 March 2021).

118. Available online: https://investors.modernatx.com/news-releases/news-release-details/moderna-announces-it-has-shipped-variant-specific-vaccine (accessed on 24 February 2021).

119. Shinde, V.; Bhikha, S.; Hoosain, Z.; Archary, M.; Bhorat, Q.; Fairlie, L.; Lalloo, U.; Masilela, M.S.; Moodley, D.; Hanley, S.; et al. Efficacy of NVX-CoV2373 COVID-19 Vaccine against the B.1.351 Variant. *N. Engl. J. Med.* **2021**, *384*, 1899–1909. [CrossRef] [PubMed]

120. Goldman, M.; Hermans, C. Thrombotic thrombocytopenia associated with COVID-19 infection or vaccination: Possible paths to platelet factor 4 autoimmunity. *PLoS Med.* **2021**, *18*, e1003648. [CrossRef]

121. Nguyen, T.-H.; Medvedev, N.; Delcea, M.; Greinacher, A. Anti-platelet factor 4/polyanion antibodies mediate a new mechanism of autoimmunity. *Nat. Commun.* **2017**, *8*, 14945. [CrossRef] [PubMed]

122. Vaxzevria (Previously COVID-19 Vaccine AstraZeneca). Available online: https://www.ema.europa.eu/en/medicines/human/EPAR/vaxzevria-previously-covid-19-vaccine-astrazeneca (accessed on 19 July 2021).

123. SARS-CoV-2 Variants of Concern as of 22 July 2021. Available online: https://www.ecdc.europa.eu/en/covid-19/variants-concern (accessed on 22 July 2021).

124. Greinacher, A.; Thiele, T.; Warkentin, T.E.; Weisser, K.; Kyrle, P.A.; Eichinger, S. Thrombotic Thrombocytopenia after ChAdOx1 nCov-19 Vaccination. *N. Engl. J. Med.* **2021**, *384*, 2092–2101. [CrossRef]

125. Schultz, N.H.; Sørvoll, I.H.; Michelsen, A.E.; Munthe, L.A.; Lund-Johansen, F.; Ahlen, M.T.; Wiedmann, M.; Aamodt, A.-H.; Skattør, T.H.; Tjønnfjord, G.E.; et al. Thrombosis and Thrombocytopenia after ChAdOx1 nCoV-19 Vaccination. *N. Engl. J. Med.* **2021**, *384*, 2124–2130. [CrossRef] [PubMed]
126. EMA. Meeting Highlights from the Pharmacovigilance Risk Assessment Committee (PRAC) 3–6 May 2021. Available online: https://www.ema.europa.eu/en/news/meeting-highlights-pharmacovigilance-risk-assessment-committee-prac-3-6-may-2021 (accessed on 5 July 2021).
127. EMA. Meeting Highlights from the Pharmacovigilance Risk Assessment Committee (PRAC) 5–8 July 2021. Available online: https://www.ema.europa.eu/en/news/meeting-highlights-pharmacovigilance-risk-assessment-committee-prac-5-8-july-2021 (accessed on 9 June 2021).
128. Pottegård, A.; Lund, L.C.; Karlstad, Ø.; Dahl, J.; Andersen, M.; Hallas, J.; Lidegaard, Ø.; Tapia, G.; Gulseth, H.L.; Ruiz, P.L.D.; et al. Arterial events, venous thromboembolism, thrombocytopenia, and bleeding after vaccination with Oxford-AstraZeneca ChAdOx1-S in Denmark and Norway: Population based cohort study. *BMJ* **2021**, *373*, n1114. [CrossRef] [PubMed]
129. Sadoff, J.; Davis, K.; Douoguih, M. Thrombotic Thrombocytopenia after Ad26.COV2.S Vaccination—Response from the Manufacturer. *N. Engl. J. Med.* **2021**, *384*, 1965–1966. [CrossRef]
130. Muir, K.-L.; Kallam, A.; Koepsell, S.A.; Gundabolu, K. Thrombotic Thrombocytopenia after Ad26.COV2.S Vaccination. *N. Engl. J. Med.* **2021**, *384*, 1964–1965. [CrossRef] [PubMed]
131. Wise, J. COVID-19: European countries suspend use of Oxford-AstraZeneca vaccine after reports of blood clots. *BMJ* **2021**, *372*, n699. [CrossRef] [PubMed]
132. European Medicines Agency (EMA). COVID-19 Vaccine AstraZeneca: PRAC Investigating Cases of Thromboembolic Events-Vaccine's Benefits Currently Still Outweigh Risks-Update. Available online: https://www.ema.europa.eu/en/news/covid-19-vaccine-astrazeneca-prac-investigating-cases-thromboembolic-events-vaccines-benefits (accessed on 11 March 2021).

 *biomedicines*

Review

H$_2$S as a Bridge Linking Inflammation, Oxidative Stress and Endothelial Biology: A Possible Defense in the Fight against SARS-CoV-2 Infection?

Francesca Gorini [1,*], Serena Del Turco [1,*], Laura Sabatino [1], Melania Gaggini [1] and Cristina Vassalle [2,*]

1   Institute of Clinical Physiology, National Research Council, 56124 Pisa, Italy; laura.sabatino@ifc.cnr.it (L.S.); melania.gaggini@ifc.cnr.it (M.G.)
2   Fondazione CNR-Regione Toscana G. Monasterio, 56124 Pisa, Italy
*   Correspondence: fgorini@ifc.cnr.it (F.G.); serena@ifc.cnr.it (S.D.T.); cristina.vassalle@ftgm.it (C.V.)

Abstract: The endothelium controls vascular homeostasis through a delicate balance between secretion of vasodilators and vasoconstrictors. The loss of physiological homeostasis leads to endothelial dysfunction, for which inflammatory events represent critical determinants. In this context, therapeutic approaches targeting inflammation-related vascular injury may help for the treatment of cardiovascular disease and a multitude of other conditions related to endothelium dysfunction, including COVID-19. In recent years, within the complexity of the inflammatory scenario related to loss of vessel integrity, hydrogen sulfide (H$_2$S) has aroused great interest due to its importance in different signaling pathways at the endothelial level. In this review, we discuss the effects of H$_2$S, a molecule which has been reported to demonstrate anti-inflammatory activity, in addition to many other biological functions related to endothelium and sulfur-drugs as new possible therapeutic options in diseases involving vascular pathobiology, such as in SARS-CoV-2 infection.

Keywords: endothelium; hydrogen sulfide; inflammation; therapeutic target; SARS-CoV-2; COVID-19

Citation: Gorini, F.; Del Turco, S.; Sabatino, L.; Gaggini, M.; Vassalle, C. H$_2$S as a Bridge Linking Inflammation, Oxidative Stress and Endothelial Biology: A Possible Defense in the Fight against SARS-CoV-2 Infection? *Biomedicines* **2021**, *9*, 1107. https://doi.org/10.3390/biomedicines9091107

Academic Editors: Byeong Hwa Jeon and Marika Cordaro

Received: 14 July 2021
Accepted: 26 August 2021
Published: 28 August 2021

Publisher's Note: MDPI stays neutral with regard to jurisdictional claims in published maps and institutional affiliations.

1. Introduction

The endothelium is the inner lining that covers all blood vessels with a very large spatial distribution, and consequently with potentially different characteristics depending on its position in the body. Not just an inert barrier, the endothelium is recognized as a biologically active tissue that regulates vascular tone and structure through autocrine, paracrine, and hormone-like mechanisms [1]. In fact, it may respond to various stimuli (e.g., shear stress on the endothelial surface) and release numerous vasoactive substances with vasodilator or vasoconstrictor properties [1]. Loss of physiological homeostasis leads to endothelial dysfunction, a condition underlying micro- and macrovascular diseases, characterized by reduced capacity for vasodilation, recruitment of neutrophils, enhanced inflammation and oxidative stress, prothrombotic properties, impaired cell growth and vessel permeability [1]. Inflammatory processes remain critical determinants which can provoke increases in prothrombotic and pro-oxidative events. Therefore, endothelial dysfunction has been observed associated with different clinical diseases, such as chronic kidney disease, liver failure, atherosclerosis, hypertension, dyslipidemia, diabetes, and obesity [2]. This implies that endothelial dysfunction can have important repercussions on health and on the onset and development of practically all diseases. In particular, endothelial dysfunction can be promoted or exacerbated by severe acute respiratory syndrome coronavirus-2 (SARS-CoV-2) infection, either by direct (interaction of SARS-CoV-2 virus with endothelial angiotensin-converting enzyme 2 (ACE2)) or indirect mechanisms (e.g., systemic inflammation, leukocyte recruitment, immune dysregulation, procoagulant state, impaired fibrinolysis, or activation of the complement system) [3–6]. Conversely, preexisting endothelial dysfunction underlies cardiovascular risk factors (e.g., hypertension, diabetes, obesity, and aging), all conditions which may favor the risk and severity of novel coronavirus disease 2019 (COVID-19) [4]. Increasing experimental, clinical, and translational findings suggest that many common drugs, such as lipid-lowering, antihypertensive, and antidiabetic drugs, and also antioxidants (e.g., vitamin C and E, N-acetylcysteine-NAC) and anti-inflammatories (e.g., cyclooxygenase-2 inhibitors, and glucocorticoids) may target the endothelium [2]. Consequently, further knowledge of endothelium pathophysiology may greatly help in patient care, given its enormous diagnostic and therapeutic potential.

In this context, the role of hydrogen sulfide (H_2S), defined as the third endogenous gaseous signaling molecule besides nitric oxide (NO) and carbon monoxide (CO), in recent years has aroused a great interest due to its importance on different signaling pathways at the endothelial level [5]. In particular, given the close interaction between reduced H_2S levels and endothelial dysfunction and inflammation, and the relationship between NO and H_2S, the therapeutic release of H_2S may represent a new and intriguing development in the prevention and treatment of inflammatory-related endothelial dysfunction conditions [6].

In this review, we discuss the effects of H_2S and the use of sulfur-drugs as a new possible therapeutic option in diseases involving vascular pathobiology, such as in the SARS-CoV-2 infection.

2. Hydrogen Sulfide: An Additive Key Factor in Vascular Homeostasis

In the last few years, the role of H_2S as an interesting novel mediator involved in inflammation and endothelial function has emerged [7], in terms of the relaxation of blood vessels, regulation of blood pressure, reduction of inflammatory response, and induction of antioxidant defense [8].

This gasotransmitter is mostly produced through the reverse trans-sulfuration pathway by three different enzymes. Two enzymes, cystathionine beta synthase (CBS) and cystathionine gamma lyase (CSE), use piridoxal $5'$ phosphate (PLP) as a cofactor, while 3-mercaptopiruvate sulfur transferase (3-MST), primarily located in the mitochondria, is not dependent on PLP [9,10].

The protective effect of H_2S is mediated by many different cellular and molecular mechanisms (Figure 1).

S-sulfhydration, a chemical modification on specific cysteine residues of target proteins to form a persulfide group (–SSH), is considered a primary mechanism through which H_2S alters the function of signaling proteins [11,12]. A striking example is the S-sulfhydration of endothelial nitric oxide synthase (eNOS), which promotes eNOS dimer stability, NO production, and consequent vasorelaxation [13]. Indeed, sulfhydration of the Kir 6.1 subunit of ATP-sensitive potassium (K_{ATP}) channels activates the channel, causing vascular endothelial and smooth muscle cell hyperpolarization and vasorelaxation [14]. Accordingly, exogenous administration of H_2S attenuates the rise of blood pressure in both spontaneously hypertensive rats [15] and in mice rendered hypertensive with angiotensin II (AngII) [16].

Although H_2S may directly inactivate reactive oxygen species (ROS), e.g., through the inhibition of peroxynitrite-mediated processes in vivo [17], it also protects cells via the upregulation of antioxidant defense systems [6,8]. Specifically, sodium hydrosulfide (NaHS, an H_2S donor) induces the sulfhydration of Kelch-like ECH-associating protein 1 (Keap1), a repressor of nuclear-factor-E2-related factor-2 (Nrf2), which is the main regulator of the antioxidant response. This action results in Keap1/Nrf2 disassociation, Nrf2 nuclear translocation, and increased mRNA expression of Nrf2-targeted downstream genes [18,19]. In human umbilical vein endothelial cells (HUVECs) exposed to H_2O_2, H_2S upregulates a wide range of enzymes attenuating oxidative stress, such as catalase (CAT), superoxide dismutase (SOD), glutathione peroxidase, and glutathione-S-transferase [20]. Moreover, in endothelial cells (ECs) and fibroblasts, H_2S-induced S-sulfhydration and activation of mitogen-activated extracellular signal-regulated kinase 1 are followed by phosphorylated ERK1/2 translocation into the nucleus to stimulate activity of PARP-1, a nuclear protein that exerts an important role in DNA damage repair [21]. H_2S is further implicated in the regulation of the other major cellular inflammatory signaling pathway, namely the nuclear factor-κB (NF-κB) pathway [22]. The multifunctional pro-inflammatory cytokine, tumor necrosis factor α (TNF-α), stimulates the transcription of CSE. Consequently, the H_2S-induced sulfhydration of the p65 subunit of NF-κB, promotes transcription of anti-apoptotic genes by enhancing its ability to bind the co-activator ribosomal protein S3 [23]. On the other hand, NaHS can exert an anti-inflammatory effect in ECs by inhibiting the expression of adhesion molecules (i.e., intercellular adhesion molecule-1 (ICAM-1), vascular cell adhesion molecule-1, P-selectin, and E-selectin), an early marker of endothelial activation and dysfunction, by upregulating the cytoprotective enzyme heme oxygenase-1 (HO-1), and decreasing TNF-α-induced NF-kB activation and ROS production [24]. Exogenous H_2S also attenuates the inflammation and cytotoxicity induced by AngII in HUVECs via inhibition of the NF-κB/endothelin-I signaling pathway [25]. The pretreatment of HUVECs with NaHS can inhibit high-glucose-induced ICAM-1 expression at both the protein and mRNA levels, leading to a reduction of ROS production and NF-kB activity [26]. Furthermore, the administration of NaHS to high-glucose-treated ECs attenuates both the apoptosis and impairment of the CAT and SOD expression and activities induced by type 2 diabetes [27]. Interestingly, both diabetic rats [27] and patients with type 2 diabetes display markedly decreased plasma H_2S levels [28,29].

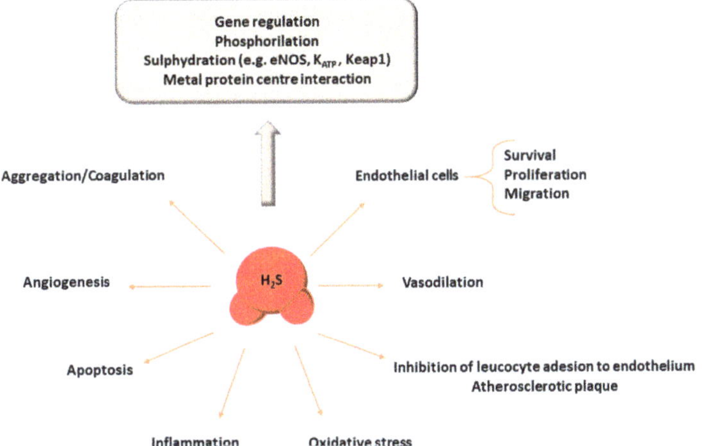

Figure 1. Main molecular, cellular, and systemic actions of hydrogen sulfide. H_2S attenuates apoptosis and decreases oxidative stress by a direct action, upregulating the cellular antioxidant system. H_2S also produces an anti-inflammatory response through the inhibition or induction of specific pathways. At the vascular level, in addition to promoting cell proliferation and migration, it induces angiogenesis, vasodilation, represses aggregation and coagulation, and reduces aortic atherosclerotic plaque formation. The mechanisms underlying the effects of H_2S include the phosphorylation and addition of cysteine residues to target proteins, interaction with the metal protein centers of proteins, and gene regulation (for more details see text). Abbreviations: eNOS: Endothelial nitric oxide synthase; H_2S: Hydrogen sulfide; K_{ATP}: ATP-sensitive potassium channels; Keap1: Kelch-like ECH-associating protein 1.

H_2S has also been ascertained to interact with the metal centers of proteins [30]. The interaction of H_2S with the oxygenated form of human hemoglobin and myoglobin produces a sulfheme protein complex that participates in the catabolism of H_2S [31]. Of interest is that H_2S affects the function of soluble guanylyl cyclase (sGC), a receptor for NO, through the reduction of the ferric sGC heme into a ferrous state, facilitating NO-dependent vasodilation and thus providing an additional level of cross-talk between NO and H_2S [32].

H_2S stimulates EC proliferation, adhesion and migration in vitro and in vivo [6,11]. The mechanism whereby H_2S regulates the key steps of angiogenesis can be driven by the increase in intracellular calcium levels in vascular ECs through the activation of many calcium-dependent signaling pathways and enzymes [9,33,34]. Also, H_2S donors activate the PI-3K/Akt axis and enhance the phosphorylation of components of the MAPK pathway (p38 and ERK1/2) in EC models in vitro, with a subsequent pro-angiogenic effect [35]. Other mechanisms in H_2S-induced angiogenesis in vascular ECs could also involve the opening of K_{ATP} channels by persulfidation of sulfonylurea receptor 1 subunit [36] and the activation of vascular endothelial growth factor receptor 2 (VEGFR2), a receptor tyrosine kinase that mediates most of the biological effects of the vascular endothelial growth factor through the breakage of a cysteine-cysteine disulfide bond within VEGFR2 [37].

Importantly, altered H_2S metabolism is likely to be involved in the initiation and progression of atherosclerosis [9]. The H_2S donors NaHS and GYY4137 may decrease aortic atherosclerotic plaque formation, macrophage infiltration, and aortic inflammation, and may partially restore endothelium-dependent relaxation in the aorta of apolipoprotein E (ApoE) gene-knockout mice [38,39]. In addition, CSE/H_2S treatment directly sulfhydrated sirtuin-1, a histone deacetylase with a crucial role in longevity, increasing its activity and stability, thereby reducing atherosclerotic plaque formation in the aorta of these animals [39].

A few studies finally have investigated the anti-aggregatory and anticoagulatory effects of H_2S [40]. In mice, GYY4137 seems to act as an anti-thrombotic and to regulate thrombogenesis by reducing platelet activation and adhesion molecule-mediated aggregation [41]. In mice treated with the H_2S donor sodium sulfide (Na_2S), thrombus formation induced using a phototoxic light/dye-injury model is significantly delayed compared to controls, due to the up-regulation of eNOS and inducible NOS (iNOS) [40]. Likewise, in an in vitro study on human whole blood, GYY4137 was observed to reduce platelet-leukocyte aggregation provoked by the thrombin-receptor activating peptide, and consequently facilitate microvascular thrombolysis [42].

It is important to note that H_2S, in biological systems, co-exists with the sulfane sulfur species, i.e., uncharged sulfur atoms carrying six valence electrons being able to attach reversibly to other sulfur atoms [43]. Sulfane sulfur does not exist in the free form and can be considered as a sort of H_2S storage, releasing H_2S under reducing conditions, following a physiological signal [44,45]. Inorganic hydrogen polysulfides (H_2S_n, $n \geq 2$), in particular, are endogenously produced through several enzymatic routes involving 3-MST [46], copper/zinc SOD [47], or the direct reaction between H_2S and NO [48]. H_2S_n is greatly reactive, and has recently emerged as a potential signaling molecule that immediately reacts with intracellular cysteine, glutathione (GSH), and protein cysteine residues [49].

2.1. Sulfur-Drugs as New Therapeutic Options in Endothelial Dysfunction

The use of sulfur moieties as therapeutic agents in a wide array of applications (e.g., arterial hypertension, atherosclerosis, myocardial hypertrophy, heart failure, ischemia-reperfusion, diabetic nephron- and retinopathy, and chronic inflammatory diseases), has attracted growing attention in the last few years [50,51] (Table 1).

2.1.1. H_2S Donors

Inorganic sulfide salts such as NaHS and Na_2S are cheap and readily available, and have been largely employed in vitro and in animal models with the main effects of protecting ECs from inflammation, oxidative stress, damage induced by hyperglycemia, and promoting vasorelaxation and neovascularization, as illustrated above [6,52]. Nevertheless, they are unsuitable for clinical use, owing to the fast increase in H_2S concentration to supraphysiological concentration, the change of intracellular pH if used in unbuffered solution, and possible toxicity characterized by pro-inflammatory effects [50,53]. In fact, in experimental studies, sulfide salts are frequently used at a very high concentration (100 μM to 10 mM), in great excess of the levels of H_2S measured in vivo [52].

Unlike sulfide salts, the phosphorodithioate GYY4137 belongs to the class of organic slow-release H_2S compounds, being able to release H_2S over 3–4 h after dissolving [50].

GYY4137 has been demonstrated to exert vasodilator, antihypertensive, anti-atherosclerotic, and anti-thrombotic activities [38,41]. Whereas NaHS increases the synthesis of the interleukins (IL)-1β and -6, NO and prostaglandin E2 at a high concentration, GYY4137 inhibits the release of these pro-inflammatory mediators in a dose-dependent manner and also promotes the production of the anti-inflammatory chemokine IL-10 in lipopolysaccharide (LPS)-treated macrophages [53].

Diphosphorothioates such as AP67 and AP105 derive from structural modifications of the phosphorodithioate core and, if compared to GYY4137, have the advantages of being able to be employed at a lower concentration, showing an even enhanced activity [52]. The novel mitochondria-targeted AP39 and AP123 were found to reduce the high-glucose-induced hyperpolarization of the mitochondrial membrane and inhibit ROS production in microvascular ECs, with a long-lasting effect suggesting their application in the treatment of diabetic vascular complications [54]. The H_2S prodrug sodium polysulthionate (SG1002) has been successfully used in both a swine model of acute limb ischemia [55] and in heart failure patients [56], in which it promotes an increase in circulating H_2S and NO and, consequently, coronary artery vasorelaxation.

Garlic, which has been associated with multiple health beneficial effects in folk medicine for centuries, is rich in organosulfur compounds considered responsible for most of its pharmacological activities [57]. In particular, garlic-derived organic polysulfides like diallyl disulfide and diallyl trisulfide, as well as their analogs, act as H_2S donors in the presence of GSH, and promote vasorelaxation (NO bioavailability) [57], lowering of arterial blood pressure [58], decreasing apoptosis and oxidative stress [59], and improved angiogenesis [60]. Similarly, the natural isothiocyanates commonly present in the Brassicaceae (e.g., broccoli, mustard, horseradish, rocket salad), can be considered as potential slow and long-lasting H_2S donors able to release H_2S in cell environments with high concentration of GSH and cysteine [61,62]. Sulforaphane and other isothiocyanates (i.e., benzyl isothiocyanate and phenethyl isocyanate) have anti-inflammatory properties mediated through the upregulation of HO-1 [63,64] and the glutamine cysteine ligase that plays a critical role in maintaining GSH homeostasis [63]. A diet based on sulforaphane-enriched foods was reported to significantly reduce oxidative stress related to improved endothelial-dependent relaxation of the aorta and lower blood pressure in spontaneously hypertensive rats [65].

Although L-cysteine is frequently employed in experimental studies to increase the production of endogenous H_2S, it is not suitable for clinical use due to its unstable nature, being metabolized in a number of pathways including GSH synthesis, taurine synthesis and oxidation to sulfate [51]. Conversely, NAC, a well-tolerated compound used in clinical settings to enhance cellular levels of GSH, could represent a promising compound able to generate H_2S [6], since the administration of NAC prevents the development of hypertension in rodents [66] and humans (Clinical Trial NCT01232257, 2011). Supplementation of taurine (2-aminoethanesulfonic acid), a metabolite of cysteine, has the ability to moderately lower blood pressure in prehypertensive subjects via endothelium-dependent and endothelium-independent vasodilation [67,68]. Like NAC, these antihypertensive effects are associated with increased expression of CSE and CBS, as observed both in the aorta of spontaneously hypertensive rats and in human vascular tissue cultures [67].

Synthetic cysteine derivatives such as S-propyl-cysteine, S-allyl-cysteine, and S-propargyl-cysteine (SPRC) are also enzymatically converted to H_2S by CSE and CBS [50]. SPRC, though not available for clinical use [68], can attenuate the LPS-induced inflammatory response in cardiac myocytes by reducing the mRNA expression of TNF-α, ICAM-1 and iNOS [69], increase H_2S levels through the upregulation of CSE, and protect HUVECs from TNF-α-induced dysfunction [70].

2.1.2. H_2S-Hybrid Drugs

In addition to H_2S donors, there is a group of compounds known as H_2S-hybrid drugs that, while able to release H_2S, have a mechanism of action independent of H_2S-mediated properties [61]. The group of ACE inhibitors represents one of the fundamental drug classes for the antihypertensive treatment [52]. Many of them, e.g., Omaprilat, Remikiren, Macitentan, Bosentan, Vardenafil, Sildenafil, have a sulfonil moiety in their structure [51]. Captocapril contains a thiol that, in plasma, can react with other thiol-containing compounds (cysteine, GSH) to form mixed disulfides, however it remains to be clarified whether the drug's effects depend on sulfide signaling [6,51]. On the other hand, the pro-angiogenic [71], anti-inflammatory [72], and anti-apoptotic [73] activities of Zofenopril in ECs are at least partially mediated by its ability to increase H_2S availability [72]. Of interest, Zofenopril, but not Enalapril (a non-thiol ACE inhibitor), also improves the vascular response to acetylcholine in spontaneously hypertensive rats, which is accompanied by increased H_2S concentration in the plasma and the vascular wall [74].

Non-steroidal anti-inflammatory drugs (NSAIDs) are among the most widely used classes of medicines [75]. H_2S-releasing derivatives of NSAIDs, synthetized by the conjugation of the parent NSAID with a dithiolethione moiety, show improved efficacy and reduced toxicity (i.e., gastrolesivity), which are mainly attributable to intracellular H_2S/GSH formation in comparison to the non-releasing H_2S compounds [75–77]. In particular, S-aspirin (ACS14), which exerts anti-platelet [78] and antithrombotic [79] activity in vivo and in vitro, prevents the formation and development of atherosclerosis in ApoE-deficient mice [80] and attenuates the oxidative stress caused by methylglyoxal (a chemically active metabolite of glucose and fructose) and high glucose in vascular smooth cultured cells, indicating a possible future use in the treatment of diabetic patients [81]. Likewise, S-diclofenac (ACS15) produces marked anti-inflammatory effects but significantly less gastric toxicity than diclofenac [82]. Notably, this drug can inhibit the smooth muscle cell growth that has been recognized as a fundamental event in vascular injury in diseases such as atherosclerosis [83].

Table 1. Summary of the main biological effects of sulfur drugs.

Sulfur Drugs	Effects	Reference
Inorganic sulfide salts NaHS and Na2S	Reduction of inflammation, oxidative stress and damage induced by hyperglycemia; promotion of vasorelaxation and neovascularization	[6,52]
Organic "slow-release" H_2S compounds GYY4137	Vasodilator, antihypertensive, anti-atherosclerotic, anti-thrombotic and anti-inflammatory effects	[38,41]
Diphosphorothioates AP67 and AP105	Promotion of high-glucose-induced hyperpolarization of the mitochondrial membrane and inhibition of ROS production in microvascular ECs,	[54]
H_2S prodrug sodium polysulthionate (SG1002)	Promotion of increase in circulating H_2S and NO and the consequent endothelial-dependent coronary artery vasorelaxation	[55,56]
Natural organosulfur compound Garlic Natural isothiocyanates	Promotion of vasorelaxation, lower arterial blood pressure, decreased apoptosis and oxidative stress, angiogenesis Anti-inflammatory and antioxidant effects	[57–60] [63–65]
N-acetyl-Cysteine (NAC) and taurine	Anti-hypertensive and anti-inflammatory effects	[66–68]
Synthetic cysteine derivatives (S-propyl-cysteine, S-allyl-cysteine and S-propargyl-cysteine)	Increase in H_2S levels, anti-inflammatory effects	[69,70]
H_2S-hybrid drug ACE inhibitors: Omaprilat, Remikiren, Macitentan, Bosentan, Vardenafil, Sildenafil	Pro-angiogenic, anti-inflammatory and anti-apoptotic activities Zofenopril: increase in H_2S concentration in plasma and vascular wall	[71–73] [74]
H_2S-releasing derivatives of NSAIDs	Anti-platelet, antithrombotic and antioxidant effects S-diclofenac (ACS15): anti-inflammatory and antiproliferative effects	[78,79,81] [82,83]

The classes of sulfur drugs are in bold.

2.2. H$_2$S-Producing Compounds: A Further Tool against COVID-19

A growing body of evidence supports the potential role of H$_2$S as an effective host defense factor against SARS-CoV-2 [84,85] (Figure 2).

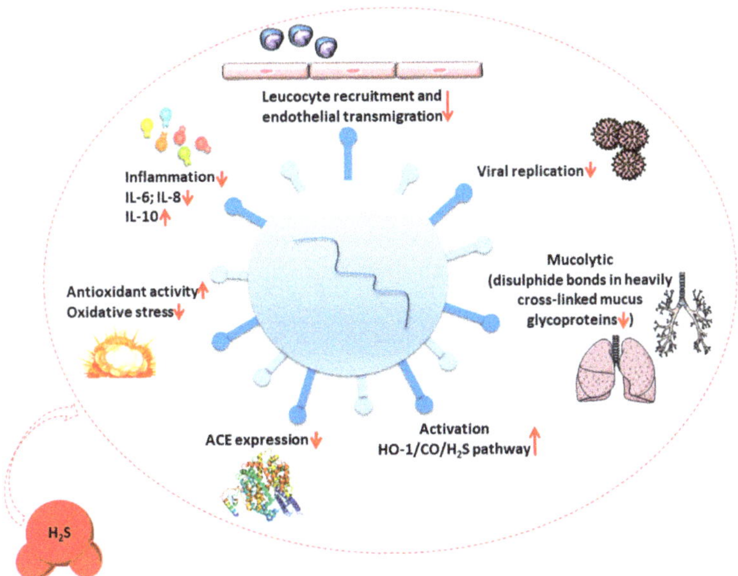

Figure 2. Primary biological mechanisms involving H$_2$S (donors) in the protective response against SARS-CoV-2. Following viral infection, H$_2$S counteracts inflammation by decreasing levels of IL-6 and IL-8, inducing an increase in IL-10 concentration and inhibiting the recruitment of leukocytes to the endothelium. It also represses ACE expression, which in turn leads to vasodilation, decreased inflammation and oxidative stress, and reduced fibrosis. H$_2$S donors enhance endogenous antioxidant defenses and play a mucolytic role. The activity of H$_2$S also results in the activation of the HO-1/CO/H$_2$S system and inhibition of virus replication. Abbreviations: ACE: Angiotensin-converting enzyme; CO: Carbon monoxide; H$_2$S: Hydrogen sulfide; HO-1: Heme oxygenase-1; IL: Interleukin.

In a SARS-CoV-2 infection, a pro-inflammatory cytokine storm is a primary event characterized by increases in IL-1β, IL-6, and TNF-α [85], and even moderately elevated IL-6 levels have been associated with a high risk of respiratory failure in COVID-19 patients [86]. Interestingly, the 4-day change in ratio of IL-6 to IL-10 (a cytokine with anti-inflammatory effects involved both in innate and adaptive immunity), named the Dublin–Boston score, has proven to be a more reliable tool than IL-6 alone in predicting clinical progression and poor outcome in COVID-19 patients [87,88].

H$_2$S can significantly downregulate the IL-6/STAT3 signaling pathway that is implicated in inflammatory responses and cell apoptosis [89]. Of note, serum H$_2$S levels were found to inversely correlate with IL-6 as well as with the severity and final outcome of pneumonia in a cohort of patients with COVID-19, suggesting that the reduction of H$_2$S bioavailability may be considered a biomarker of enhanced pro-inflammatory response, whereas exogenous administration of H$_2$S could represent a valuable strategy to counteract severe manifestations of the infection [90,91]. In addition to decreasing IL-6 and IL-8 levels and the infiltration of polymorphonuclear cells, NaHS increases plasma levels of IL-10 in an animal model of induced acute lung injury [92], and administering IL-10 to IL-10 deficient mice in turn restores H$_2$S production and homocysteine metabolism [93]. H$_2$S has also been reported to play a role in enhancing T cell activation [94] and, among the variety of cell types capable to produce IL-10, IL-10 generation by CD4(+) T cells is crucial to preventing early mortality caused by excessive inflammation [87].

The anti-inflammatory effects of H_2S further encompass the regulation of the ACE/ACE2 balance, although the underlying mechanisms remain elusive [95]. Specifically, NaHS may promote, dose-dependently, the expression of ACE2, a key enzymatic component of the renin-angiotensin-aldosterone system that is recognized to have vasodilating, anti-inflammatory, antioxidant, and antifibrotic effects, by catalyzing the generation of Ang (1–7) from AngII in ECs [96,97]. Conversely, NaHS treatment reduces ACE expression in spontaneously hypertensive rats [98], consistent with the recognized enhanced anti-hypertensive role of ACE-inhibitors containing a sulfur moiety in their chemical structure (see the previous section).

The role of H_2S in acute and chronic inflammatory pulmonary diseases has been extensively investigated [91]. Besides anti-inflammatory activities (e.g., reduced mRNA expression of NF-kB, macrophage inflammatory protein-2, and interferon regulatory factor 3 [99,100]), H_2S donors increase endogenous antioxidant defenses (e.g., SOD, glutathione peroxidase, and glutathione reductase) and inhibit leukocyte recruitment and transmigration through the inflamed endothelium [100,101]. Notably, there is evidence indicating that higher levels of GSH may improve an individual's responsiveness to viral infections, protecting host cells from oxidative damage of the lung [102]. De Flora et al. observed that a 6-month preventive administration of NAC, a known precursor of GSH, provides a significant attenuation of influenza and influenza-like episodes, especially in the elderly [103]. The addition of NAC to a conventional therapy ameliorates oxidative stress and inflammation parameters (i.e., a decrease of malondialdehyde and TNF-α, and an increase of total antioxidant capacity) in patients with pneumonia [104]. NAC can act both as a direct scavenger of free radicals [105] and as a mucolytic agent capable of reducing disulphide bonds in heavily cross-linked mucus glycoproteins [106]. Moreover, the efficacy of NAC in the treatment of patients with chronic bronchitis and chronic obstructive pulmonary disease has been documented in several clinical trials and meta-analyses [107].

The levels of GSH are also negatively related with COVID-19 severity, and patients with moderate and severe illnesses show increased levels of ROS and a higher ROS/GSH ratio than subjects with mild symptoms [102]. Hence, it is sensible to hypothesize that replenishing intracellular GSH could be a useful strategy against SARS-CoV-2, as shown in two cases in which a glutathione-based therapy (GSH and NAC) combined with antioxidants (alpha-lipoic acid and vitamin C) was immediately effective in relieving symptoms of dyspnea [108]. In another case report, despite treatment with antibiotics, antiviral, and antibacterial medications, a 64-year-old male COVID-19 patient developed respiratory failure on the 13th day of admission. The patient's conditions improved following NAC supplementation, and discharge occurred after 46 days of hospitalization [109]. In particular, on the basis of its mucolytic and antioxidant properties, it has been proposed that inclusion of 1200 mg/d oral NAC in the therapeutic schemes of patients with COVID-19 could be an effective measure to prevent a cytokine storm and the associated acute respiratory distress syndrome [110].

A case-control study found that both H_2S and NO were significantly higher in expired COVID-19 patients compared to those who survived, emphasizing the complex interaction between these two gasotransmitters and a more synergistic role in this context [111]. The reasons for this increase can hint a compensatory response of sicker patients to the detrimental effects of COVID-19 infection or, alternatively, an underutilization of NO and H_2S, which results in fatal outcomes [111].

Finally, it has recently been proposed that another mechanism of defense against COVID-19 could involve the activation of the HO-1/CO/H_2S system [112]. Indeed, HO-1 metabolizes the heme group of a variety of hemeproteins with the release of CO which, under endoplasmic reticulum stress conditions, inhibits CBS with decreased production of GSH, while the SH groups are enzymatically converted to H_2S by CSE [112,113]. Liu and Li have hypothesized that ORF8 and the surface glycoprotein of SARS-CoV-2 attack and destroy the hemes of heme proteins, resulting in the suppression of HO-1 activation and H_2S signaling [114,115]. Subjects with a long promoter for the HMOX1 gene (associated with decreased HO-1 anticoagulant activity) present an increased risk of recurrent venous thromboembolism [116], whilst HO-1 has been demonstrated to exert a significant antiviral activity against a wide variety of viruses [117]. Therefore, activation of the HO-1/CO/H_2S axis has the potential to improve the clinical manifestations of COVID-19, and pharmacological treatments based on H_2S delivery could once again represent an effective strategy in the treatment of patients with COVID-19 [112].

In addition to counteracting the inflammatory response in COVID-19, H_2S may interfere with viral replication [96]. In fact, there is some well-established evidence demonstrating that H_2S inhibits the replications of many other highly pathogenic RNA viruses in lungs [118], both by decreasing the expression of viral proteins and mRNA and by inhibiting syncytium formation and virus assembly/release [95,99].

Overall, a large array of data indicate that H_2S could be a potential target for attenuating viral replication, inflammation development and progression, and organ damage, which needs to be further explored in preclinical models of viral infections [95].

3. Discussion

Amid the complexity of events and effectors underlying processes leading to endothelial dysfunction, H_2S has emerged as one of the crucial determinants of endothelial homeostasis. Hence, sulfur drugs can represent advanced tools in the prevention and treatment of the numerous diseases involving endothelial dysfunction. For these reasons, several compounds are currently being investigated in clinical studies with promising results.

On the other hand, it has been recognized that a high H_2S concentration (i.e., >250 ppm/ ~350 mg/m^3) is harmful to health, and is associated with increased oxidative stress and inflammation [119]. Less well known are the long-term effects of chronic low-dose H_2S exposures in light of controversial results and due to differences between and limitations of the available studies [119].

In addition to endogenous production, it is also important to consider the intake of H_2S from exogenous sources. Indeed, H_2S may originate from a variety of natural sources, including volcanoes, sulfur springs, undersea vents, swamps, bogs, crude oil and natural gas, or from man-made activities, such as oil refineries, tanneries, natural gas, products petrochemicals, ovens, food processing plants, municipal sewage and wastewater treatment plants, fertilization processes, and paper mills. Occupational exposure to H_2S is therefore greater than that from environmental sources [119]. Furthermore, smokers have low serum H_2S levels, while chronic alcohol users have high levels of H_2S in their breath, suggesting that smoking and alcohol consumption may modulate endogenous H_2S concentration [120].

Dietary intake can also affect endogenous levels of H_2S. In addition to garlic, broccoli, mustard, etc., a high consumption of proteins and fats or a high carbohydrate content in the diet also appears to increase and reduce H_2S levels, respectively [121]. In particular, the intake of sulfur in the diet, together with the presence and different composition of sulfate-reducing bacteria in the gastrointestinal tract, represent the most significant modulators of H_2S production [121].

In this context, the discovery that H_2S can protect the mucus layer and reduce inflammation when produced at nanomolar to low micromolar levels, exerting adverse effects when released at a higher concentration (from high micromolar to millimolar) by the local microbiota, is interesting because it has highlighted the double face of H_2S in the balance between beneficial and harmful effects [122].

Although H_2S donors may represent valuable tools to protect the endothelium, further clinical studies targeting their effect on the endothelium in terms of reducing or slowing the progression of dysfunction in specific diseases (e.g., COVID-19) would be desirable. Furthermore, it is currently unknown whether any drugs added to endogenous levels resulting from environmental exposure, lifestyle habits (cigarette smoking or alcohol consumption and diet) and microbiota activity could reach toxic concentrations, and this requires further refinement work towards a more personalized and targeted therapy for each patient.

Author Contributions: Conceptualization, F.G., S.D.T. and C.V.; writing—original draft preparation, F.G., S.D.T. and C.V.; writing—review and editing, L.S. and M.G. All authors have read and agreed to the published version of the manuscript.

Funding: This research received no external funding.

Institutional Review Board Statement: Not applicable.

Informed Consent Statement: Not applicable.

Conflicts of Interest: The authors declare no conflict of interest.

References

1. Medina-Leyte, D.J.; Zepeda-García, O.; Domínguez-Pérez, M.; González-Garrido, A.; Villarreal-Molina, T.; Jacobo-Albavera, L. Endothelial Dysfunction, Inflammation and Coronary Artery Disease: Potential Biomarkers and Promising Therapeutical Approaches. *Int. J. Mol. Sci.* **2021**, *22*, 3850. [CrossRef]
2. Xu, S.; Ilyas, I.; Little, P.J.; Li, H.; Kamato, D.; Zheng, X.; Luo, S.; Li, Z.; Liu, P.; Han, J.; et al. Endothelial Dysfunction in Atherosclerotic Cardiovascular Diseases and Beyond: From Mechanism to Pharmacotherapies. *Pharmacol. Rev.* **2021**, *73*, 924–967. [CrossRef]
3. Dou, Q.; Wei, X.; Zhou, K.; Yang, S.; Jia, P. Cardiovascular Manifestations and Mechanisms in Patients with COVID-19. *Trends Endocrinol. Metab.* **2020**, *31*, 893–904. [CrossRef]
4. Guzik, T.J.; Mohiddin, S.A.; Dimarco, A.; Patel, V.; Savvatis, K.; Marelli-Berg, F.M.; Madhur, M.S.; Tomaszewski, M.; Maffia, P.; D'Acquisto, F.; et al. COVID-19 and the cardiovascular system: Implications for risk assessment, diagnosis, and treatment options. *Cardiovasc. Res.* **2020**, *116*, 1666–1687. [CrossRef] [PubMed]
5. Pan, L.L.; Qin, M.; Liu, X.H.; Zhu, Y.Z. The Role of Hydrogen Sulfide on Cardiovascular Homeostasis: An Overview with Update on Immunomodulation. *Front. Pharmacol.* **2017**, *8*, 686. [CrossRef] [PubMed]
6. Ciccone, V.; Genah, S.; Morbidelli, L. Endothelium as a Source and Target of H_2S to Improve Its Trophism and Function. *Antioxidants (Basel)* **2021**, *10*, 486. [CrossRef]
7. Wang, R.; Szabo, C.; Ichinose, F.; Ahmed, A.; Whiteman, M.; Papapetropoulos, A. The role of H2S bioavailability in endothelial dysfunction. *Trends Pharmacol. Sci.* **2015**, *36*, 568–578. [CrossRef]
8. Xiao, Q.; Ying, J.; Xiang, L.; Zhang, C. The biologic effect of hydrogen sulfide and its function in various diseases. *Medicine (Baltimore)* **2018**, *97*, e13065. [CrossRef]
9. Wallace, J.L.; Wang, R. Hydrogen sulfide-based therapeutics: Exploiting a unique but ubiquitous gasotransmitter. *Nat. Rev. Drug Discov.* **2015**, *14*, 329–345. [CrossRef]
10. Majtan, T.; Krijt, J.; Sokolová, J.; Křížková, M.; Ralat, M.A.; Kent, J.; Gregory, J.F., 3rd; Kožich, V.; Kraus, J.P. Biogenesis of Hydrogen Sulfide and Thioethers by Cystathionine Beta-Synthase. *Antioxid. Redox Signal.* **2018**, *8*, 311–323. [CrossRef]
11. Kanagy, N.L.; Szabo, C.; Papapetropoulos, A. Vascular biology of hydrogen sulfide. *Am. J. Physiol. Cell Physiol.* **2017**, *312*, C537–C549. [CrossRef]
12. Meng, G.; Zhao, S.; Xie, L.; Han, Y.; Ji, Y. Protein S-sulfhydration by hydrogen sulfide in cardiovascular system. *Br. J. Pharmacol.* **2018**, *175*, 1146–1156. [CrossRef] [PubMed]
13. Altaany, Z.; Ju, Y.; Yang, G.; Wang, R. The coordination of S-sulfhydration, S-nitrosylation, and phosphorylation of endothelial nitric oxide synthase by hydrogen sulfide. *Sci. Signal.* **2014**, *7*, ra87. [CrossRef] [PubMed]
14. Mustafa, A.K.; Sikka, G.; Gazi, S.K.; Steppan, J.; Jung, S.M.; Bhunia, A.K.; Barodka, V.M.; Gazi, F.K.; Barrow, R.K.; Wang, R.; et al. Hydrogen sulfide as endothelium-derived hyperpolarizing factor sulfhydrates potassium channels. *Circ. Res.* **2011**, *109*, 1259–1268. [CrossRef]
15. Yan, H.; Du, J.; Tang, C. The possible role of hydrogen sulfide on the pathogenesis of spontaneous hypertension in rats. *Biochem. Biophys. Res. Commun.* **2004**, *313*, 22–27. [CrossRef]
16. Al-Magableh, M.R.; Kemp-Harper, B.K.; Hart, J.L. Hydrogen sulfide treatment reduces blood pressure and oxidative stress in angiotensin II-induced hypertensive mice. *Hypertens Res.* **2015**, *38*, 13–20. [CrossRef] [PubMed]
17. Whiteman, M.; Armstrong, J.S.; Chu, S.H.; Jia-Ling, S.; Wong, B.S.; Cheung, N.S.; Halliwell, B.; Moore, P.K. The novel neuromodulator hydrogen sulfide: An endogenous peroxynitrite 'scavenger'? *J. Neurochem.* **2004**, *90*, 765–768. [CrossRef]
18. Yang, G.; Zhao, K.; Ju, Y.; Mani, S.; Cao, Q.; Puukila, S.; Khaper, N.; Wu, L.; Wang, R. Hydrogen sulfide protects against cellular senescence via S-sulfhydration of Keap1 and activation of Nrf2. *Antioxid. Redox Signal.* **2013**, *18*, 1906–1919. [CrossRef] [PubMed]
19. Guo, C.; Liang, F.; Shah Masood, W.; Yan, X. Hydrogen sulfide protected gastric epithelial cell from ischemia/reperfusion injury by Keap1 s-sulfhydration, MAPK dependent anti-apoptosis and NF-κB dependent anti-inflammation pathway. *Eur. J. Pharmacol.* **2014**, *725*, 70–78. [CrossRef]
20. Wen, Y.D.; Wang, H.; Kho, S.H.; Rinkiko, S.; Sheng, X.; Shen, H.M.; Zhu, Y.Z. Hydrogen sulfide protects HUVECs against hydrogen peroxide induced mitochondrial dysfunction and oxidative stress. *PLoS ONE* **2013**, *8*, e53147. [CrossRef] [PubMed]

21. Zhao, K.; Ju, Y.; Li, S.; Altaany, Z.; Wang, R.; Yang, G. S-sulfhydration of MEK1 leads to PARP-1 activation and DNA damage repair. *EMBO Rep.* **2014**, *15*, 792–800. [CrossRef] [PubMed]

22. Li, L.; Rose, P.; Moore, P.K. Hydrogen sulfide and cell signaling. *Annu. Rev. Pharmacol. Toxicol.* **2011**, *51*, 169–187. [CrossRef]

23. Sen, N.; Paul, B.D.; Gadalla, M.M.; Mustafa, A.K.; Sen, T.; Xu, R.; Kim, S.; Snyder, S.H. Hydrogen sulfide-linked sulfhydration of NF-κB mediates its antiapoptotic actions. *Mol. Cell.* **2012**, *45*, 13–24. [CrossRef] [PubMed]

24. Pan, L.L.; Liu, X.H.; Gong, Q.H.; Wu, D.; Zhu, Y.Z. Hydrogen sulfide attenuated tumor necrosis factor-α-induced inflammatory signaling and dysfunction in vascular endothelial cells. *PLoS ONE* **2011**, *6*, e19766. [CrossRef]

25. Hu, H.J.; Jiang, Z.S.; Zhou, S.H.; Liu, Q.M. Hydrogen sulfide suppresses angiotensin II-stimulated endothelin-1 generation and subsequent cytotoxicity-induced endoplasmic reticulum stress in endothelial cells via NF-κB. *Mol. Med. Rep.* **2016**, *14*, 4729–4740. [CrossRef]

26. Guan, Q.; Wang, X.; Gao, L.; Chen, J.; Liu, Y.; Yu, C.; Zhang, N.; Zhang, X.; Zhao, J. Hydrogen sulfide suppresses high glucose-induced expression of intercellular adhesion molecule-1 in endothelial cells. *J. Cardiovasc. Pharmacol.* **2013**, *62*, 278–284. [CrossRef] [PubMed]

27. Liu, J.; Wu, J.; Sun, A.; Sun, Y.; Yu, X.; Liu, N.; Dong, S.; Yang, F.; Zhang, L.; Zhong, X.; et al. Hydrogen sulfide decreases high glucose/palmitate-induced autophagy in endothelial cells by the Nrf2-ROS-AMPK signaling pathway. *Cell Biosci.* **2016**, *6*, 33. [CrossRef] [PubMed]

28. Jain, S.K.; Bull, R.; Rains, J.L.; Bass, P.F.; Levine, S.N.; Reddy, S.; McVie, R.; Bocchini, J.A. Low levels of hydrogen sulfide in the blood of diabetes patients and streptozotocin-treated rats causes vascular inflammation? *Antioxid. Redox Signal.* **2010**, *12*, 1333–1337. [CrossRef]

29. Suzuki, K.; Sagara, M.; Aoki, C.; Tanaka, S.; Aso, Y. Clinical Implication of Plasma Hydrogen Sulfide Levels in Japanese Patients with Type 2 Diabetes. *Intern. Med.* **2017**, *56*, 17–21. [CrossRef]

30. Boubeta, F.M.; Bieza, S.A.; Bringas, M.; Palermo, J.C.; Boechi, L.; Estrin, D.A.; Bari, S.E. Hemeproteins as Targets for Sulfide Species. *Antioxid. Redox Signal.* **2020**, *32*, 247–257. [CrossRef]

31. Ríos-González, B.B.; Román-Morales, E.M.; Pietri, R.; López-Garriga, J. Hydrogen sulfide activation in hemeproteins: The sulfheme scenario. *J. Inorg. Biochem.* **2014**, *133*, 78–86. [CrossRef] [PubMed]

32. Zhou, Z.; Martin, E.; Sharina, I.; Esposito, I.; Szabo, C.; Bucci, M.; Cirino, G.; Papapetropoulos, A. Regulation of soluble guanylyl cyclase redox state by hydrogen sulfide. *Pharmacol. Res.* **2016**, *111*, 556–562. [CrossRef] [PubMed]

33. Bauer, C.C.; Boyle, J.P.; Porter, K.E.; Peers, C. Modulation of Ca^{2+} signalling in human vascular endothelial cells by hydrogen sulfide. *Atherosclerosis* **2010**, *209*, 374–380. [CrossRef] [PubMed]

34. Moccia, F.; Bertoni, G.; Pla, A.F.; Dragoni, S.; Pupo, E.; Merlino, A.; Mancardi, D.; Munaron, L.; Tanzi, F. Hydrogen sulfide regulates intracellular Ca^{2+} concentration in endothelial cells from excised rat aorta. *Curr. Pharm. Biotechnol.* **2011**, *12*, 1416–1426. [CrossRef] [PubMed]

35. Szabó, C.; Papapetropoulos, A. Hydrogen sulphide and angiogenesis: Mechanisms and applications. *Br. J. Pharmacol.* **2011**, *164*, 853–865. [CrossRef] [PubMed]

36. Jiang, B.; Tang, G.; Cao, K.; Wu, L.; Wang, R. Molecular mechanism for H_2S-induced activation of K(ATP) channels. *Antioxid. Redox Signal.* **2010**, *12*, 1167–1178. [CrossRef]

37. Tao, H.; Shi, K.H.; Yang, J.J. Vascular endothelial growth factor: A novel potential therapeutic target for hypertension. *J. Clin. Hypertens. (Greenwich)* **2013**, *15*, 514. [CrossRef]

38. Liu, Z.; Han, Y.; Li, L.; Lu, H.; Meng, G.; Li, X.; Shirhan, M.; Peh, M.T.; Xie, L.; Zhou, S.; et al. The hydrogen sulfide donor, GYY4137, exhibits anti-atherosclerotic activity in high fat fed apolipoprotein $E^{(-/-)}$ mice. *Br. J. Pharmacol.* **2013**, *169*, 1795–1809. [CrossRef]

39. Du, C.; Lin, X.; Xu, W.; Zheng, F.; Cai, J.; Yang, J.; Cui, Q.; Tang, C.; Cai, J.; Xu, G.; et al. Sulfhydrated Sirtuin-1 Increasing Its Deacetylation Activity Is an Essential Epigenetics Mechanism of Anti-Atherogenesis by Hydrogen Sulfide. *Antioxid. Redox Signal.* **2019**, *30*, 184–197. [CrossRef]

40. Kram, L.; Grambow, E.; Mueller-Graf, F.; Sorg, H.; Vollmar, B. The anti-thrombotic effect of hydrogen sulfide is partly mediated by an upregulation of nitric oxide synthases. *Thromb. Res.* **2013**, *132*, e112–e117. [CrossRef]

41. Grambow, E.; Mueller-Graf, F.; Delyagina, E.; Frank, M.; Kuhla, A.; Vollmar, B. Effect of the hydrogen sulfide donor GYY4137 on platelet activation and microvascular thrombus formation in mice. *Platelets* **2014**, *25*, 166–174. [CrossRef]

42. Grambow, E.; Leppin, C.; Leppin, K.; Kundt, G.; Klar, E.; Frank, M.; Vollmar, B. The effects of hydrogen sulfide on platelet-leukocyte aggregation and microvascular thrombolysis. *Platelets* **2017**, *28*, 509–517. [CrossRef]

43. Kharma, A.; Grman, M.; Misak, A.; Domínguez-Álvarez, E.; Nasim, M.J.; Ondrias, K.; Chovanec, M.; Jacob, C. Inorganic Polysulfides and Related Reactive Sulfur–Selenium Species from the Perspective of Chemistry. *Molecules* **2019**, *24*, 1359. [CrossRef]

44. Kimura, H. Hydrogen sulfide and polysulfides as signaling molecules. *Proc. Jpn. Acad. Ser. B Phys. Biol. Sci.* **2015**, *91*, 131–159. [CrossRef]

45. Toohey, J.I. Sulfur signaling: Is the agent sulfide or sulfane? *Anal. Biochem.* **2011**, *413*, 1–7. [CrossRef]

46. Kimura, Y.; Toyofuku, Y.; Koike, S.; Shibuya, N.; Nagahara, N.; Lefer, D.; Ogasawara, Y.; Kimura, H. Identification of H2S3 and H2S produced by 3-mercaptopyruvate sulfurtransferase in the brain. *Sci. Rep.* **2015**, *5*, 14774. [CrossRef] [PubMed]

47. Olson, K.R.; Gao, Y.; Arif, F.; Arora, K.; Patel, S.; DeLeon, E.R.; Sutton, T.R.; Feelisch, M.; Cortese-Krott, M.M.; Straub, K.D. Metabolism of hydrogen sulfide (H_2S) and Production of Reactive Sulfur Species (RSS) by superoxide dismutase. *Redox. Biol.* **2018**, *15*, 74–85. [CrossRef] [PubMed]

48. Kimura, H. Hydrogen polysulfide (H2Sn) signaling along with hydrogen sulfide (H2S) and nitric oxide (NO). *J. Neural Transm. (Vienna)* **2016**, *123*, 1235–1245. [CrossRef] [PubMed]

49. Kimura, Y.; Koike, S.; Shibuya, N.; Lefer, D.; Ogasawara, Y.; Kimura, H. 3-Mercaptopyruvate sulfurtransferase produces potential redox regulators cysteine- and glutathione-persulfide (Cys-SSH and GSSH) together with signaling molecules H2S2, H2S3 and H2S. *Sci. Rep.* **2017**, *7*, 10459. [CrossRef] [PubMed]

50. Bełtowski, J. Hydrogen sulfide in pharmacology and medicine—An update. *Pharmacol. Rep.* **2015**, *67*, 647–658. [CrossRef] [PubMed]

51. Zaorska, E.; Tomasova, L.; Koszelewski, D.; Ostaszewski, R.; Ufnal, M. Hydrogen Sulfide in Pharmacotherapy, Beyond the Hydrogen Sulfide-Donors. *Biomolecules* **2020**, *10*, 323. [CrossRef]

52. Whiteman, M.; Perry, A.; Zhou, Z.; Bucci, M.; Papapetropoulos, A.; Cirino, G.; Wood, M.E. Phosphinodithioate and Phosphoramidodithioate Hydrogen Sulfide Donors. *Handb. Exp. Pharmacol.* **2015**, *230*, 337–363. [CrossRef]

53. Whiteman, M.; Li, L.; Rose, P.; Tan, C.H.; Parkinson, D.B.; Moore, P.K. The effect of hydrogen sulfide donors on lipopolysaccharide-induced formation of inflammatory mediators in macrophages. *Antioxid. Redox Signal.* **2010**, *12*, 1147–1154. [CrossRef] [PubMed]

54. Gerő, D.; Torregrossa, R.; Perry, A.; Waters, A.; Le-Trionnaire, S.; Whatmore, J.L.; Wood, M.; Whiteman, M. The novel mitochondria-targeted hydrogen sulfide (H2S) donors AP123 and AP39 protect against hyperglycemic injury in microvascular endothelial cells in vitro. *Pharmacol. Res.* **2016**, *113 Pt A*, 186–198. [CrossRef]

55. Rushing, A.M.; Donnarumma, E.; Polhemus, D.J.; Au, K.R.; Victoria, S.E.; Schumacher, J.D.; Li, Z.; Jenkins, J.S.; Lefer, D.J.; Goodchild, T.T. Effects of a novel hydrogen sulfide prodrug in a porcine model of acute limb ischemia. *J. Vasc. Surg.* **2019**, *69*, 1924–1935. [CrossRef]

56. Polhemus, D.J.; Li, Z.; Pattillo, C.B.; Gojon, G., Sr.; Gojon, G., Jr.; Giordano, T.; Krum, H. A novel hydrogen sulfide prodrug, SG1002, promotes hydrogen sulfide and nitric oxide bioavailability in heart failure patients. *Cardiovasc. Ther.* **2015**, *33*, 216–226. [CrossRef] [PubMed]

57. Benavides, G.A.; Squadrito, G.L.; Mills, R.W.; Patel, H.D.; Isbell, T.S.; Patel, R.P.; Darley-Usmar, V.M.; Doeller, J.E.; Kraus, D.W. Hydrogen sulfide mediates the vasoactivity of garlic. *Proc. Natl. Acad. Sci. USA* **2007**, *104*, 17977–17982. [CrossRef] [PubMed]

58. Sharma, D.K.; Manral, A.; Saini, V.; Singh, A.; Srinivasan, B.P.; Tiwari, M. Novel diallyldisulfide analogs ameliorate cardiovascular remodeling in rats with L-NAME-induced hypertension. *Eur. J. Pharmacol.* **2012**, *691*, 198–208. [CrossRef]

59. Hayashida, R.; Kondo, K.; Morita, S.; Unno, K.; Shintani, S.; Shimizu, Y.; Calvert, J.W.; Shibata, R.; Murohara, T. Diallyl Trisulfide Augments Ischemia-Induced Angiogenesis via an Endothelial Nitric Oxide Synthase-Dependent Mechanism. *Circ. J.* **2017**, *81*, 870–878. [CrossRef]

60. Yang, H.B.; Liu, H.M.; Yan, J.C.; Lu, Z.Y. Effect of Diallyl Trisulfide on Ischemic Tissue Injury and Revascularization in a Diabetic Mouse Model. *J. Cardiovasc. Pharmacol.* **2018**, *71*, 367–374. [CrossRef]

61. Calderone, V.; Martelli, A.; Testai, L.; Citi, V.; Breschi, M.C. Using hydrogen sulfide to design and develop drugs. *Expert Opin. Drug. Discov.* **2016**, *11*, 163–175. [CrossRef]

62. Martelli, A.; Piragine, E.; Citi, V.; Testai, L.; Pagnotta, E.; Ugolini, L.; Lazzeri, L.; Di Cesare Mannelli, L.; Manzo, O.L.; Bucci, M.; et al. Erucin exhibits vasorelaxing effects and antihypertensive activity by H2 S-releasing properties. *Br. J. Pharmacol.* **2020**, *177*, 824–835. [CrossRef] [PubMed]

63. Huang, C.S.; Lin, A.H.; Liu, C.T.; Tsai, C.W.; Chang, I.S.; Chen, H.W.; Lii, C.K. Isothiocyanates protect against oxidized LDL-induced endothelial dysfunction by upregulating Nrf2-dependent antioxidation and suppressing NFκB activation. *Mol. Nutr. Food Res.* **2013**, *57*, 1918–1930. [CrossRef]

64. Kashfi, K.; Olson, K.R. Biology and therapeutic potential of hydrogen sulfide and hydrogen sulfide-releasing chimeras. *Biochem. Pharmacol.* **2013**, *85*, 689–703. [CrossRef]

65. Wu, L.; Noyan Ashraf, M.H.; Facci, M.; Wang, R.; Paterson, P.G.; Ferrie, A.; Juurlink, B.H. Dietary approach to attenuate oxidative stress, hypertension, and inflammation in the cardiovascular system. *Proc. Natl. Acad. Sci. USA* **2004**, *101*, 7094–7099. [CrossRef] [PubMed]

66. Nagababu, E.; Steppan, J.; Santhanam, L.; Frank, S.M.; Berkowitz, D.E. Effect of Nitrite and N-acetylcysteine Treatment on Blood Pressure, Arterial Stiffness and Vascular Function in Spontaneously Hypertensive Rats. *Free Radic. Biol. Med.* **2017**, *112*, 118–119. [CrossRef]

67. Sun, Q.; Wang, B.; Li, Y.; Sun, F.; Li, P.; Xia, W.; Zhou, X.; Li, Q.; Wang, X.; Chen, J.; et al. Taurine Supplementation Lowers Blood Pressure and Improves Vascular Function in Prehypertension: Randomized, Double-Blind, Placebo-Controlled Study. *Hypertension* **2016**, *67*, 541–549. [CrossRef] [PubMed]

68. DiNicolantonio, J.J.; OKeefe, J.H.; McCarty, M.F. Boosting endogenous production of vasoprotective hydrogen sulfide via supplementation with taurine and N-acetylcysteine: A novel way to promote cardiovascular health. *Open Heart* **2017**, *4*, e000600. [CrossRef] [PubMed]

69. Pan, L.L.; Liu, X.H.; Gong, Q.H.; Zhu, Y.Z. S-Propargyl-cysteine (SPRC) attenuated lipopolysaccharide-induced inflammatory response in H9c2 cells involved in a hydrogen sulfide-dependent mechanism. *Amino Acids* **2011**, *41*, 205–215. [CrossRef] [PubMed]

70. Pan, L.L.; Liu, X.H.; Zheng, H.M.; Yang, H.B.; Gong, Q.H.; Zhu, Y.Z. S-propargyl-cysteine, a novel hydrogen sulfide-modulated agent, attenuated tumor necrosis factor-α-induced inflammatory signaling and dysfunction in endothelial cells. *Int. J. Cardiol.* **2012**, *155*, 327–332. [CrossRef]

71. Terzuoli, E.; Monti, M.; Vellecco, V.; Bucci, M.; Cirino, G.; Ziche, M.; Morbidelli, L. Characterization of zofenoprilat as an inducer of functional angiogenesis through increased H2 S availability. *Br. J. Pharmacol.* **2015**, *172*, 2961–2973. [CrossRef] [PubMed]

72. Monti, M.; Terzuoli, E.; Ziche, M.; Morbidelli, L. H$_2$S dependent and independent anti-inflammatory activity of zofenoprilat in cells of the vascular wall. *Pharmacol. Res.* **2016**, *113*, 426–437. [CrossRef]

73. Monti, M.; Terzuoli, E.; Ziche, M.; Morbidelli, L. The sulphydryl containing ACE inhibitor Zofenoprilat protects coronary endothelium from Doxorubicin-induced apoptosis. *Pharmacol. Res.* **2013**, *76*, 171–181. [CrossRef]

74. Bucci, M.; Vellecco, V.; Cantalupo, A.; Brancaleone, V.; Zhou, Z.; Evangelista, S.; Calderone, V.; Papapetropoulos, A.; Cirino, G. Hydrogen sulfide accounts for the peripheral vascular effects of zofenopril independently of ACE inhibition. *Cardiovasc. Res.* **2014**, *102*, 138–147. [CrossRef] [PubMed]

75. Wallace, J.L. Hydrogen sulfide-releasing anti-inflammatory drugs. *Trends Pharmacol. Sci.* **2007**, *28*, 501–505. [CrossRef]

76. Sparatore, A.; Perrino, E.; Tazzari, V.; Giustarini, D.; Rossi, R.; Rossoni, G.; Erdmann, K.; Schröder, H.; Del Soldato, P. Pharmacological profile of a novel H$_2$S-releasing aspirin. *Free Radic. Biol. Med.* **2009**, *46*, 586–592. [CrossRef]

77. Liu, L.; Cui, J.; Song, C.J.; Bian, J.S.; Sparatore, A.; Soldato, P.D.; Wang, X.Y.; Yan, C.D. H$_2$S-releasing aspirin protects against aspirin-induced gastric injury via reducing oxidative stress. *PLoS ONE* **2012**, *7*, e46301. [CrossRef]

78. Gao, L.; Cheng, C.; Sparatore, A.; Zhang, H.; Wang, C. Hydrogen sulfide inhibits human platelet aggregation in vitro in part by interfering gap junction channels: Effects of ACS14, a hydrogen sulfide-releasing aspirin. *Heart Lung Circ.* **2015**, *24*, 77–85. [CrossRef] [PubMed]

79. Pircher, J.; Fochler, F.; Czermak, T.; Mannell, H.; Kraemer, B.F.; Wörnle, M.; Sparatore, A.; Del Soldato, P.; Pohl, U.; Krötz, F. Hydrogen sulfide-releasing aspirin derivative ACS14 exerts strong antithrombotic effects in vitro and in vivo. *Arterioscler. Thromb. Vasc. Biol.* **2012**, *32*, 2884–2891. [CrossRef] [PubMed]

80. Zhang, H.; Guo, C.; Zhang, A.; Fan, Y.; Gu, T.; Wu, D.; Sparatore, A.; Wang, C. Effect of S-aspirin, a novel hydrogen-sulfide-releasing aspirin (ACS14), on atherosclerosis in apoE-deficient mice. *Eur. J. Pharmacol.* **2012**, *697*, 106–116. [CrossRef]

81. Huang, Q.; Sparatore, A.; Del Soldato, P.; Wu, L.; Desai, K. Hydrogen sulfide releasing aspirin, ACS14, attenuates high glucose-induced increased methylglyoxal and oxidative stress in cultured vascular smooth muscle cells. *PLoS ONE* **2014**, *9*, e97315. [CrossRef] [PubMed]

82. Li, L.; Rossoni, G.; Sparatore, A.; Lee, L.C.; Del Soldato, P.; Moore, P.K. Anti-inflammatory and gastrointestinal effects of a novel diclofenac derivative. *Free Radic. Biol. Med.* **2007**, *42*, 706–719. [CrossRef] [PubMed]

83. Baskar, R.; Sparatore, A.; Del Soldato, P.; Moore, P.K. Effect of S-diclofenac, a novel hydrogen sulfide releasing derivative inhibit rat vascular smooth muscle cell proliferation. *Eur. J. Pharmacol.* **2008**, *594*, 1–8. [CrossRef] [PubMed]

84. Zhu, N.; Zhang, D.; Wang, W.; Li, X.; Yang, B.; Song, J.; Zhao, X.; Huang, B.; Shi, W.; Lu, R.; et al. A novel coronavirus from patients with pneumonia in China, 2019. *N. Engl. J. Med.* **2020**, *382*, 727–733. [CrossRef]

85. Bourgonje, A.R.; Offringa, A.K.; van Eijk, L.E.; Abdulle, A.E.; Hillebrands, J.L.; van der Voort, P.H.J.; van Goor, H.; van Hezik, E.J. N-Acetylcysteine and Hydrogen Sulfide in Coronavirus Disease 2019. *Antioxid. Redox Signal.* **2021**. [CrossRef] [PubMed]

86. Herold, T.; Jurinovic, V.; Arnreich, C.; Hellmuth, J.C.; von Bergwelt-Baildon, M.; Klein, M.; Weinberger, T. Level of IL-6 Predicts Respiratory Failure in Hospitalized Symptomatic COVID-19 Patients. *medRxiv* **2020**. [CrossRef]

87. Trinchieri, G. Interleukin-10 production by effector T cells: Th1 cells show self control. *J. Exp. Med.* **2007**, *204*, 239–243. [CrossRef]

88. McElvaney, O.J.; Hobbs, B.D.; Qiao, D.; McElvaney, O.F.; Moll, M.; McEvoy, N.L.; Clarke, J.; O'Connor, E.; Walsh, S.; Cho, M.H.; et al. A linear prognostic score based on the ratio of interleukin-6 to interleukin-10 predicts outcomes in COVID-19. *EBioMedicine* **2020**, *61*, 103026. [CrossRef]

89. Fouad, A.A.; Hafez, H.M.; Hamouda, A. Hydrogen sulfide modulates IL-6/STAT3 pathway and inhibits oxidative stress, inflammation, and apoptosis in rat model of methotrexate hepatotoxicity. *Hum. Exp. Toxicol.* **2020**, *39*, 77–85. [CrossRef]

90. Renieris, G.; Katrini, K.; Damoulari, C.; Akinosoglou, K.; Psarrakis, C.; Kyriakopoulou, M.; Dimopoulos, G.; Lada, M.; Koufargyris, P.; Giamarellos-Bourboulis, E.J. Serum Hydrogen Sulfide and Outcome Association in Pneumonia by the SARS-CoV-2 Coronavirus. *Shock* **2020**, *54*, 633–637. [CrossRef]

91. Citi, V.; Martelli, A.; Gorica, E.; Brogi, S.; Testai, L.; Calderone, V. Role of hydrogen sulfide in endothelial dysfunction: Pathophysiology and therapeutic approaches. *J. Adv. Res.* **2020**, *27*, 99–113. [CrossRef] [PubMed]

92. Li, T.; Zhao, B.; Wang, C.; Wang, H.; Liu, Z.; Li, W.; Jin, H.; Tang, C.; Du, J. Regulatory effects of hydrogen sulfide on IL-6, IL-8 and IL-10 levels in the plasma and pulmonary tissue of rats with acute lung injury. *Exp. Biol. Med. (Maywood)* **2008**, *233*, 1081–1087. [CrossRef] [PubMed]

93. Flannigan, K.L.; Agbor, T.A.; Blackler, R.W.; Kim, J.J.; Khan, W.I.; Verdu, E.F.; Ferraz, J.G.; Wallace, J.L. Impaired hydrogen sulfide synthesis and IL-10 signaling underlie hyperhomocysteinemia-associated exacerbation of colitis. *Proc. Natl. Acad. Sci. USA* **2014**, *111*, 13559–13564. [CrossRef]

94. Miller, T.W.; Wang, E.A.; Gould, S.; Stein, E.V.; Kaur, S.; Lim, L.; Amarnath, S.; Fowler, D.H.; Roberts, D.D. Hydrogen sulfide is an endogenous potentiator of T cell activation. *J. Biol. Chem.* **2012**, *287*, 4211–4221. [CrossRef]

95. Yang, G. H$_2$S as a potential defense against COVID-19? *Am. J. Physiol. Cell Physiol.* **2020**, *319*, C244–C249. [CrossRef] [PubMed]

96. Danser, A.H.J.; Epstein, M.; Batlle, D. Renin-Angiotensin System Blockers and the COVID-19 Pandemic: At Present There Is No Evidence to Abandon Renin-Angiotensin System Blockers. *Hypertension* **2020**, *75*, 1382–1385. [CrossRef] [PubMed]

97. Lin, Y.; Zeng, H.; Gao, L.; Gu, T.; Wang, C.; Zhang, H. Hydrogen Sulfide Attenuates Atherosclerosis in a Partially Ligated Carotid Artery Mouse model via Regulating Angiotensin Converting Enzyme 2 Expression. *Front. Physiol.* **2017**, *8*, 782. [CrossRef]

98. Tain, Y.L.; Hsu, C.N.; Lu, P.C. Early short-term treatment with exogenous hydrogen sulfide postpones the transition from prehypertension to hypertension in spontaneously hypertensive rat. *Clin. Exp. Hypertens.* **2018**, *40*, 58–64. [CrossRef]

99. Li, H.; Ma, Y.; Escaffre, O.; Ivanciuc, T.; Komaravelli, N.; Kelley, J.P.; Coletta, C.; Szabo, C.; Rockx, B.; Garofalo, R.P.; et al. Role of hydrogen sulfide in paramyxovirus infections. *J. Virol.* **2015**, *89*, 5557–5568. [CrossRef] [PubMed]

100. Faller, S.; Hausler, F.; Goeft, A.; von Itter, M.A.; Gyllenram, V.; Hoetzel, A.; Spassov, S.G. Hydrogen sulfide limits neutrophil transmigration, inflammation, and oxidative burst in lipopolysaccharide-induced acute lung injury. *Sci. Rep.* **2018**, *8*, 14676. [CrossRef]

101. Benetti, L.R.; Campos, D.; Gurgueira, S.A.; Vercesi, A.E.; Guedes, C.E.; Santos, K.L.; Wallace, J.L.; Teixeira, S.A.; Florenzano, J.; Costa, S.K.; et al. Hydrogen sulfide inhibits oxidative stress in lungs from allergic mice in vivo. *Eur. J. Pharmacol.* **2013**, *698*, 463–469. [CrossRef]

102. Polonikov, A. Endogenous Deficiency of Glutathione as the Most Likely Cause of Serious Manifestations and Death in COVID-19 Patients. *ACS Infect. Dis.* **2020**, *6*, 1558–1562. [CrossRef] [PubMed]

103. De Flora, S.; Grassi, C.; Carati, L. Attenuation of influenza-like symptomatology and improvement of cell-mediated immunity with long-term N-acetylcysteine treatment. *Eur. Respir. J.* **1997**, *10*, 1535–1541. [CrossRef] [PubMed]

104. Zhang, Q.; Ju, Y.; Ma, Y.; Wang, T. N-acetylcysteine improves oxidative stress and inflammatory response in patients with community acquired pneumonia: A randomized controlled trial. *Medicine (Baltimore)* **2018**, *97*, e13087. [CrossRef] [PubMed]

105. Benrahmoune, M.; Thérond, P.; Abedinzadeh, Z. The reaction of superoxide radical with N-acetylcysteine. *Free Radic. Biol. Med.* **2000**, *29*, 775–782. [CrossRef]

106. Aldini, G.; Altomare, A.; Baron, G.; Vistoli, G.; Carini, M.; Borsani, L.; Sergio, F. N-Acetylcysteine as an antioxidant and disulphide breaking agent: The reasons why. *Free Radic. Res.* **2018**, *52*, 751–762. [CrossRef]

107. Santus, P.; Corsico, A.; Solidoro, P.; Braido, F.; Di Marco, F.; Scichilone, N. Oxidative stress and respiratory system: Pharmacological and clinical reappraisal of N-acetylcysteine. *COPD* **2014**, *11*, 705–717. [CrossRef]

108. Horowitz, R.I.; Freeman, P.R.; Bruzzese, J. Efficacy of glutathione therapy in relieving dyspnea associated with COVID-19 pneumonia: A report of 2 cases. *Respir. Med. Case Rep.* **2020**, *30*, 101063. [CrossRef]

109. Liu, Y.; Wang, M.; Luo, G.; Qian, X.; Wu, C.; Zhang, Y.; Chen, B.; Leung, E.L.; Tang, Y. Experience of N-acetylcysteine airway management in the successful treatment of one case of critical condition with COVID-19: A case report. *Medicine (Baltimore)* **2020**, *99*, e22577. [CrossRef]

110. Assimakopoulos, S.F.; Marangos, M. N-acetyl-cysteine may prevent COVID-19-associated cytokine storm and acute respiratory distress syndrome. *Med. Hypotheses* **2020**, *140*, 109778. [CrossRef]

111. Dominic, P.; Ahmad, J.; Bhandari, R.; Pardue, S.; Solorzano, J.; Jaisingh, K.; Watts, M.; Bailey, S.R.; Orr, A.W.; Kevil, C.G.; et al. Decreased availability of nitric oxide and hydrogen sulfide is a hallmark of COVID-19. *Redox Biol.* **2021**, *43*, 101982. [CrossRef]

112. Dattilo, M. The role of host defences in Covid 19 and treatments thereof. *Mol. Med.* **2020**, *26*, 90. [CrossRef]

113. Kabil, O.; Yadav, V.; Banerjee, R. Heme-dependent Metabolite Switching Regulates H_2S Synthesis in Response to Endoplasmic Reticulum (ER) Stress. *J. Biol. Chem.* **2016**, *291*, 16418–16423. [CrossRef]

114. Liu, W.; Li, H. COVID-19 Disease: ORF8 and Surface Glycoprotein Inhibit Heme Metabolism by Binding to Porphyrin. *ChemRxiv Preprint* **2020**. [CrossRef]

115. Liu, W.; Li, H. COVID-19: Attacks the 1-Beta chain of hemoglobin and captures the Porphyrin to inhibit human Heme metabolism. *ChemRxiv Preprint* **2020**. [CrossRef]

116. Mustafa, S.; Weltermann, A.; Fritsche, R.; Marsik, C.; Wagner, O.; Kyrle, P.A.; Eichinger, S. Genetic variation in heme oxygenase 1 (HMOX1) and the risk of recurrent venous thromboembolism. *J. Vasc. Surg.* **2008**, *47*, 566–570. [CrossRef] [PubMed]

117. Espinoza, J.A.; González, P.A.; Kalergis, A.M. Modulation of Antiviral Immunity by Heme Oxygenase-1. *Am. J. Pathol.* **2017**, *187*, 487–493. [CrossRef] [PubMed]

118. Bazhanov, N.; Escaffre, O.; Freiberg, A.N.; Garofalo, R.P.; Casola, A. Broad-Range Antiviral Activity of Hydrogen Sulfide Against Highly Pathogenic RNA Viruses. *Sci. Rep.* **2017**, *7*, 41029. [CrossRef]

119. Gorini, F.; Bustaffa, E.; Chatzianagnostou, K.; Bianchi, F.; Vassalle, C. Hydrogen sulfide and cardiovascular disease: Doubts, clues, and interpretation difficulties from studies in geothermal areas. *Sci. Total Environ.* **2020**, *743*, 140818. [CrossRef]

120. Read, E.; Zhu, J.; Yang, G. Disrupted H_mS Signaling by Cigarette Smoking and Alcohol Drinking: Evidence from Cellular, Animal, and Clinical Studies. *Antioxidants (Basel)* **2021**, *10*, 49. [CrossRef]

121. Teigen, L.M.; Geng, Z.; Sadowsky, M.J.; Vaughn, B.P.; Hamilton, M.J.; Khoruts, A. Dietary Factors in Sulfur Metabolism and Pathogenesis of Ulcerative Colitis. *Nutrients* **2019**, *11*, 931. [CrossRef] [PubMed]

122. Blachier, F.; Beaumont, M.; Kim, E. Cysteine-derived hydrogen sulfide and gut health: A matter of endogenous or bacterial origin. *Curr. Opin. Clin. Nutr. Metab. Care* **2019**, *22*, 68–75. [CrossRef] [PubMed]

 biomedicines

 MDPI

Review

Lupus Vasculitis: An Overview

Patrizia Leone [1,†], **Marcella Prete** [1,†], **Eleonora Malerba** [1], **Antonella Bray** [1], **Nicola Susca** [1], **Giuseppe Ingravallo** [2] **and Vito Racanelli** [1,*]

[1] Department of Biomedical Sciences and Human Oncology, "Aldo Moro" University of Bari Medical School, 70124 Bari, Italy; patrizia.leone@uniba.it (P.L.); marcella.prete@uniba.it (M.P.); ele94.malerba@gmail.com (E.M.); antonella.bray94@gmail.com (A.B.); susnic@libero.it (N.S.)

[2] Section of Pathology, Department of Emergency and Organ Transplantation, "Aldo Moro" University of Bari Medical School, 70124 Bari, Italy; giuseppe.ingravallo@uniba.it

[*] Correspondence: vito.racanelli1@uniba.it

[†] Patrizia Leone and Marcella Prete have contributed equally to this work.

Abstract: Lupus vasculitis (LV) is one of the secondary vasculitides occurring in the setting of systemic lupus erythematosus (SLE) in approximately 50% of patients. It is most commonly associated with small vessels, but medium-sized vessels can also be affected, whereas large vessel involvement is very rare. LV may involve different organ systems and present in a wide variety of clinical manifestations according to the size and site of the vessels involved. LV usually portends a poor prognosis, and a prompt diagnosis is fundamental for a good outcome. The spectrum of involvement ranges from a relatively mild disease affecting small vessels or a single organ to a multiorgan system disease with life-threatening manifestations, such as mesenteric vasculitis, pulmonary hemorrhage, or mononeuritis multiplex. Treatment depends upon the organs involved and the severity of the vasculitis process. In this review, we provide an overview of the different forms of LV, describing their clinical impact and focusing on the available treatment strategies.

Keywords: vasculitis; systemic lupus erythematosus; lupus vasculitis; small vessel vasculitis

Citation: Leone, P.; Prete, M.; Malerba, E.; Bray, A.; Susca, N.; Ingravallo, G.; Racanelli, V. Lupus Vasculitis: An Overview. *Biomedicines* **2021**, *9*, 1626. https://doi.org/10.3390/biomedicines9111626

Academic Editor: Byeong Hwa Jeon

Received: 28 September 2021
Accepted: 3 November 2021
Published: 5 November 2021

Publisher's Note: MDPI stays neutral with regard to jurisdictional claims in published maps and institutional affiliations.

1. Introduction

Systemic lupus erythematosus (SLE) is a multifactorial systemic autoimmune disease caused by a loss of tolerance to self-antigens, mainly nuclear antigens (DNA, RNA, and nuclear proteins). It is characterized by aberrant T- and B-cell responses, autoantibody production, and immune complex deposition in tissues with complement activation, inflammation, and irreversible organ damage [1]. SLE can affect any organ system, resulting in a wide range of clinical presentations. One of these is vasculitis, which can occur in approximately 50% of SLE patients and principally involve small vessels; medium-sized vessels can also be affected, whereas large vessel involvement is very rare [2,3]. Lupus vasculitis (LV), also known as lupus vasculopathy, may take many clinical forms dependent on the size of the affected vessels and the sites involved, with prognoses that range from mild to life-threatening [3]. Ninety percent of cases affect the skin. The kidneys, gastrointestinal tract, nervous system (central and peripheral), lungs, and heart can be involved with lower frequency [3–5]. LV usually appears during an active disease associated with general inflammatory symptoms (fever, fatigue, and weight loss) and laboratory abnormalities (anemia, a high erythrocyte sedimentation rate, and elevated inflammatory markers) [5].

LV can be associated with the antiphospholipid syndrome (APS) characterized by antiphospholipid antibody positivity (lupus anticoagulant, anticardiolipin antibodies, and/or anti-β2-glycoprotein−1 antibodies) [4,6].

In this paper, we provide an overview of the main clinical, diagnostic, and immunological features of LV, with special attention to their impact on SLE severity and outcome. Moreover, we describe the therapeutic strategies commonly used by clinicians, focusing mainly on the available evidence.

2. Pathogenesis

The pathogenesis of vasculitis remains poorly understood, but certainly, complex interactions among the vascular endothelium, inflammatory cells, cytokines, and the autoantibodies and immune complexes play crucial roles. LV is a secondary vasculitis that can manifest as an acute or subacute lupus symptom due to an inflammatory process triggered by the deposition of immune complexes in blood vessel walls [7]. Alternatively, it may develop as an associated comorbidity (steroid-related atherosclerosis) or represent the synergistic pathogenetic consequence of enhanced atherosclerosis in a proinflammatory environment [8]. As an autoimmune disease, SLE is characterized by a loss of tolerance to self-antigens, altered T- and B-cell responses, and the production of autoantibodies.

The binding of autoantibodies to antigens generates soluble antigen–antibody complexes (immune complexes) that, because of a defective clearance by the reticuloendothelial system and an increase in vascular permeability, are deposited in blood vessel walls. The increased vascular permeability results from the action of platelet-derived vasoactive amines and IgE-mediated reactions [9]. After a complex deposition, the complement system is activated, leading to inflammation and a complement protein consumption [10]. Some complement components act as chemotactic factors for polymorphonuclear leukocytes, which diffusely infiltrate the vessel wall and release their lysosomal enzymes, principally collagenase and elastase, causing damage and necrosis of the vessel wall (Figure 1). This may be associated with thrombosis, occlusion, hemorrhage, and ischemic changes in the surrounding tissue [9]. In addition to the localized increase in vascular permeability, other factors, including the hydrostatic pressure and blood flow turbulence at bifurcations, favor the localization of immune complexes in specific sites of the vascular tree and, thus, the distribution of vascular lesions [11].

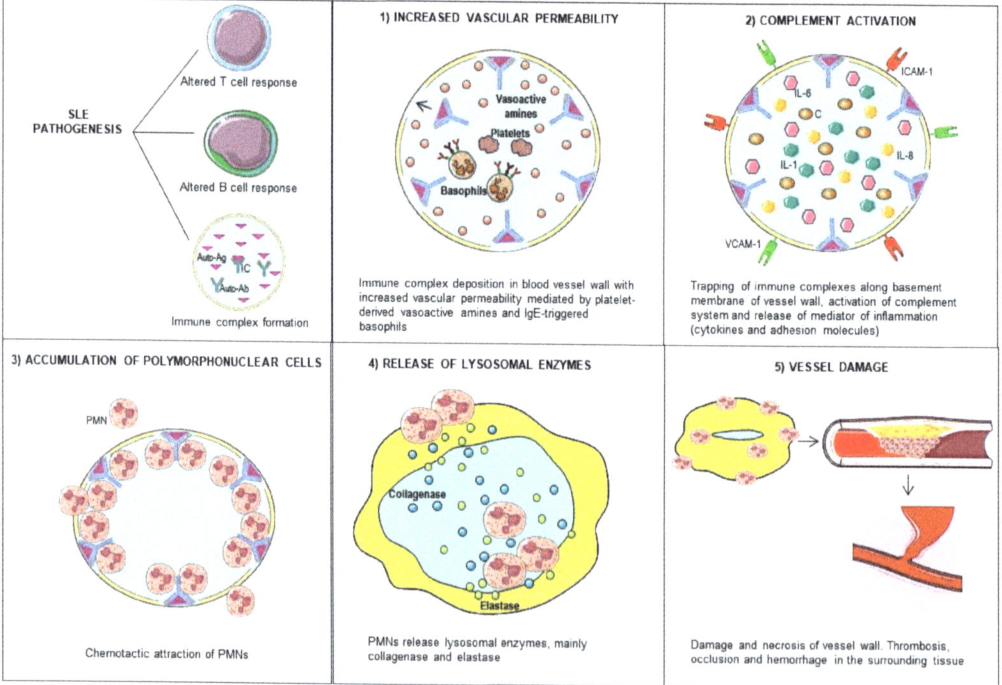

Figure 1. Pathogenesis of lupus vasculitis. Abbreviations: SLE = systemic lupus erythematosus, Ag = antigen, Ab = antibody, IC = immune complex, ICAM-1 = intercellular adhesion molecule 1, VCAM-1 = vascular cell adhesion molecule 1, IL = interleukin, and PMN = polymorphonuclear cell.

Among autoantibodies, anti-endothelial cell antibodies (AECA) are the main cause of endothelial damage. Their binding can induce endothelial cell activation with the upregulation of adhesion molecules (E-selectin, intercellular adhesion molecule 1 (ICAM-1), vascular cell adhesion molecule 1 (VCAM-1)) and the release of cytokines and chemokines, leading to inflammation (interleukin-1 (IL-1), IL-6, and IL-8) [12]; the production of tissue factors favoring coagulation; and the activation of endothelial cell cytotoxicity [13,14]. One study has identified that AECA was positive in more than 80% of the SLE patients [14].

Other types of autoantibodies that may be involved in the pathogenesis of LV are antineutrophil cytoplasmic antibodies (ANCA), anti-phospholipids antibodies (aPL), and anti-double strain DNA (Anti-dsDNA) [15]. ANCA are mainly associated with primary systemic vasculitis, but they also occur in secondary vasculitis linked to systemic connective tissue disorders, including SLE, and may be positive in 15–20% of SLE patients [16,17]. ANCA form immune complexes with proteinase 3 (PR3) or myeloperoxidase (MPO) antigens, leading to neutrophil adhesion to endothelial cells, with a consequent extravascular permeation, vessel damage, and endothelial cell apoptosis [16–18]. Focal necrosis of capillaries, venules, and sometimes arterioles occur due to the sequestration of activated neutrophils and monocytes in the microcirculation [18,19]. aPL bind to endothelial cell phospholipids exposed after endothelial damage, causing further endothelial cell damage and activation. Moreover, these antibodies have a role in the complement system activation, resulting in pro-adhesive, proinflammatory, and prothrombotic effects on endothelial cells, and activate endothelial cell thrombin formation by binding to platelet membrane phospholipids [20,21]. Anti-dsDNA induces endothelial cell activation by the stimulation of IL-6 and IL-8 release [22]. They have an anti-endothelial activity, even if a direct cytotoxic effect on endothelial cells has not been established [15].

Changes in cell death pathways, such as apoptosis and the neutrophil-specific kind of death called NETosis, are important contributing factors in the pathogenesis of SLE and LV. An imbalance between the production of apoptotic cells and clearance of apoptotic cells and DNA-containing neutrophil extracellular traps (NETs) can represent a potential source of autoantigens involved in immune complex formation. Immune complexes are cleared by bloodborne macrophages and dendritic cells, resulting in proinflammatory cytokine secretion and the perpetuation of inflammation and tissue damage in SLE patients [23]. NETs are extracellular networks of DNA scaffolds composed of cytosolic and granule proteins that can vary depending on the pathophysiologic context of each disease [24,25]. For instance, the expression of the tissue factor on NETs promotes thromboinflammation in sepsis [26], SLE [27], and vasculitis [28]. In SLE, rheumatoid arthritis and ANCA-associated vasculitis NETs expose immunostimulatory proteins and autoantigens [29].

Some patients with SLE, particularly those with central nervous system involvement, can manifest an inflammatory complement-mediated vascular injury in the absence of immune complex deposition (the Shwartzman phenomenon) [30].

Other forms of vasculitis in SLE patients are drug-induced vasculitis and infection-induced vasculitis [7]. Some drugs, including penicillin, allopurinol, thiazides, pyrazolones, retinoids, cytokines, monoclonal antibodies, chinolons, hydantoin, carbamazepine, and other anticonvulsants, may act as a hapten that, when binding to autoantigens, elicit an immune response, resulting in inflammatory vascular lesions [31,32].

Infection-induced SLE vasculitis may be the result of a direct attack by microbes of the blood vessel wall, followed by an infectious inflammatory process or endothelial cell invasion and activation by certain viruses, e.g., the cytomegalovirus, or the deposition of immune complexes consisting of microbial antigens and their corresponding antibodies [32,33]. In this respect, the association of hepatitis C virus with cryoglobulinemia is remarkable [33,34].

3. Cutaneous Vasculitis

Cutaneous vasculitis is the most frequent type of vasculitis among patients with SLE [35]. It is reported in 17–28% of patients with SLE [35,36] and in 89% of the cases of vasculitis in this disease [5]. High levels of anti-Ro and aPL [4,37] and positivity for cryoglobulins [38] are associated with a major risk of developing cutaneous lupus vasculitis. The clinical presentation is heterogeneous and includes palpable purpura, petechiae, papulonodular lesions, urticaria lesions, or bullous lesions of the extremities, livedo reticularis, cutaneous infarction, erythematous plaques or macules, erythema with necrosis, panniculitis, splinter hemorrhages, and superficial ulcerations [4,5] (Figures 2 and 3).

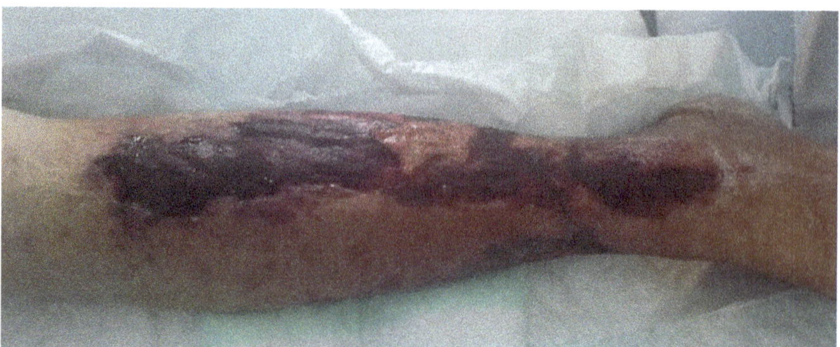

Figure 2. Necrotic purpuric plaques with ulcerations.

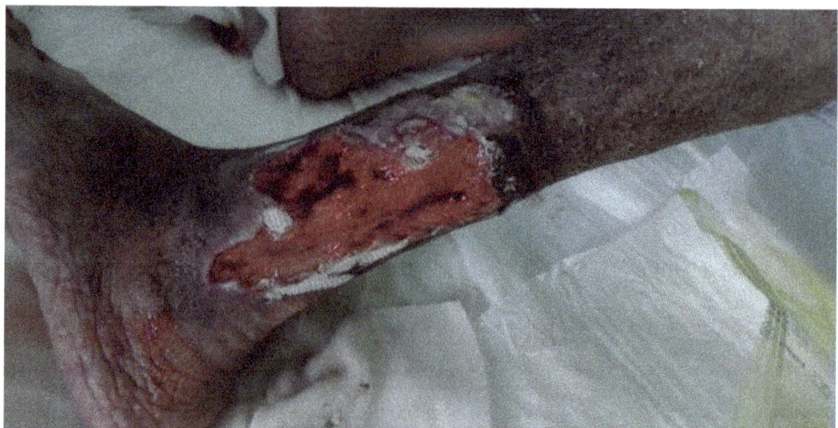

Figure 3. Deep leg ulcer.

Most skin lesions consist of discoid erythematosus lesions usually located on the fingertips, erythema of the hand dorsum, and nodular lesions [5].

Small vessels, principally post-capillary venules, are involved in most cases. Medium-vessel vasculitis is less frequent and appears as subcutaneous nodules or ischemic ulcers [3]. Myositis and hematological manifestations such as anemia, Coombs' positivity, leucopenia, anti-Smith, and anti-RNP (ribonucleoprotein) may predict cutaneous vasculitis developments [35].

Skin biopsies from patients with lupus cutaneous vasculitis displayed fragmentated neutrophilic nuclei (a leukocytoclastic variant), dermal chronic inflammation infiltrates, variable fibrinoid necrosis of the vessel walls, and secondary changes in the overlying epidermis and sweat glands (Figure 4).

In two large cohort studies in patients with vasculitis and SLE, the most frequent type of vasculitis was leukocytoclastic vasculitis (60%), followed by cryoglobulinemic vasculitis (25–30%) and urticarial vasculitis (7%) [4,5].

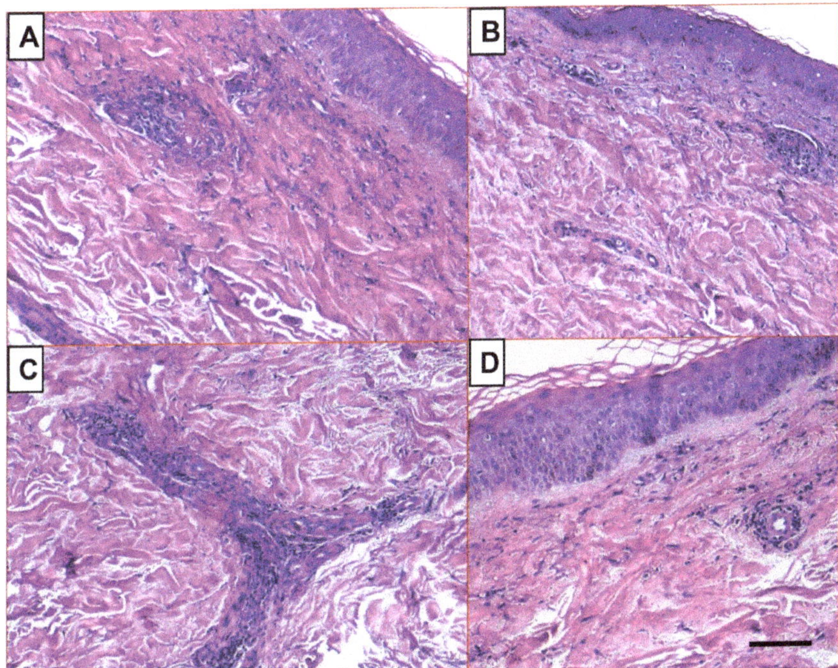

Figure 4. Skin biopsies showing leukocytoclastic vasculitis in SLE patients. The inflammatory lesions are evident in the perivascular derma, associated with nuclear dust and subepithelial myxoid degeneration. (**A**,**B**) original magnification: 40×, scale bar, 50 µm and (**C**,**D**) original magnification: 100×, scale bar, 25 µm.

4. Nervous System Vasculitis

The nervous system involvement is one of the most complex and heterogeneous features of SLE and occurs more frequently in patients with a high SLE disease activity index. The wide prevalence of neuropsychiatric manifestations ranging between 37% and 95% is due to the difficultly in discerning among the forms due to the disease or to another concomitant process. Noninflammatory microangiopathy in association with brain microinfarctions and thrombosis are common pathogenetic features, whereas inflammatory vasculitis is rare and usually affects the microvasculature [39,40]. Nervous system vasculitis can involve both the peripheral and central nervous systems in SLE patients.

At the peripheral level, the most common clinical manifestation is the mononeuritis multiplex histologically characterized by chronic axonal degeneration, necrotizing, or occlusive vasculitis of the vasa nervorum and demyelination. It affects the individual nerves focally or multifocally rather than many nerves diffusely. The clinical features include pain; weakness; sensory loss; and asymmetrical, progressive, and asynchronous sensory and motor peripheral neuropathy involving at least two separate nerve areas. With disease progression, contiguous nerves become affected, producing a syndrome that mimics a generalized polyneuropathy [41,42]. A mild-to-moderately severe peripheral symmetric sensory polyneuropathy can also develop in SLE patients [43]. Recently, an observational cross-sectional study evaluated the peripheral nerve disease in SLE patients. It found that polyneuropathy was the most frequent manifestation, with an independent statistically significant association with older age at SLE diagnosis and the absence of hematologic involvement as the cumulative SLE manifestation. Three out of nine patients who had undergone a peripheral nerve biopsy showed non-necrotizing small vessel vasculitis [44].

At the central level, cognitive dysfunction, demyelinating syndrome, cerebrovascular disease, and seizure disorders are the most frequently reported clinical features that can appear together or separately in the course of the disease [45]. The diagnosis of central nervous vasculitis is a challenge for clinicians. A brain biopsy is the gold standard for the diagnosis, although it is a highly invasive procedure with a limited sensitivity due to the segmental nature of the vascular lesions. A combination of neuroimaging with clinical features and appropriate diagnostic studies often allows reaching an early diagnosis without a brain biopsy [45]. Actually, the most sensitive noninvasive image study for cerebral SLE-related vasculitis is magnetic resonance imaging (MRI) [45]. However, this technique can reveal wall-thickening and intramural contrast material uptake in vasculitis affecting large brain arteries but does not detect small-vessel involvement [46,47]. Contrast-enhanced MR angiography at 3.0 T and intracranial vessel wall imaging (VWI) modality are also often key supports for a more accurate diagnosis and better differentiation between vasculopathies and intracranial atherosclerotic disease (ICAD). Angiographic imaging can provide information about the vessel lumen and only indirect evidence of vessel wall-thickening. It can show segmental stenosis and dilatation in multiple vascular territories, but these findings are also common in atherosclerotic disease. Moreover, small-vessel disease is beyond angiography resolutions [48]. A cerebrospinal fluid examination does not allow a direct diagnosis of lupus vasculitis but may be the most useful in excluding infections caused by bacterial, viral, and parasitic pathogens [49], as well as autoimmune causes such as multiple sclerosis that can mimic vasculitis [50].

5. Gastrointestinal Vasculitis

SLE-related vasculitis of the gastrointestinal tract (also named lupus enteritis) is uncommon; the estimated prevalence varies between 0.2 and 14.2% among all SLE patients [51,52].

A common manifestation of gastrointestinal vasculitis is lupus mesenteric vasculitis (LMV) [53]. It is one of the most devastating complications of SLE, with a mortality rate of 50%, when severe, occlusive damage progresses to bowel ischemia and potential necrosis of the small or large bowel, which may evolve to perforation and hemorrhage [53,54]. In 80–85% of the cases, the superior mesenteric artery is involved, and the structures affected are the ileum and the jejunum; involvement of the large bowel and the rectum is less frequent [55]. LMV often occurs in patients with high disease activity, demonstrated by higher scores on disease activity measurements such as the British Isles Lupus Assessment Group (BILAG) or SLE Disease Activity Index (SLEDAI) [55,56]; altered laboratory data (thrombocytopenia, lymphopenia, hypoalbuminemia, and elevated serum amylase are associated with adverse outcomes); and other coexistent organ involvements, mostly of the skin, kidneys, heart, joints, serositis, lungs, and central nervous system [57]. The main symptoms include acute abdominal pain, nausea, vomiting, diarrhea, melena, hematemesis, and bloating [58]. Urinary symptoms (lupus cystitis and dysuria) can also be associated with LMV in 22.7% of cases [57,59].

The gold standard for diagnosis is computed tomography (CT), which allows visualizing both the bowel wall and the abdominal vasculature. The typical tomographic findings are bowel wall edema, target signs, the dilatation of intestinal segments, prominent mesenteric vessels, increased attenuation of mesenteric fat, and ascites [57,60]. The timely diagnosis of lupus enteritis is crucial to prevent life-threatening complications such as gastrointestinal perforation, hemorrhage, and sepsis [61]. Relapses can occur in 31.7% of cases [62].

Within the abdominal cavity, vasculitis may also affect the liver and pancreas. Necrotizing arteritis of the liver has been reported in 18–20% of SLE autopsy cases [63,64]. Alazani et al. described a hepatic vasculitis mimicking multiple liver abscesses in a patient with SLE, which showed clinical improvement after steroid therapy [65]. Spontaneous hepatic rupture due to small- and medium-sized vessel vasculitis is an unusual complication [66,67]. Although SLE-related acute pancreatitis is uncommon, it is more severe and frequently fatal and should be suspected in SLE patients with abdominal pain [68]. Vasculitis has been related to pancreatitis in a subset of SLE patients [69].

6. Renal Vasculitis

Five pathological types of renal microvascular lesions have been described in patients with lupus nephritis (Table 1) [70]. So far, the attention has been mainly focused on glomerular pathology, and renal vascular lesions have been overlooked.

Table 1. Pathological types of renal microvascular lesions.

Uncomplicated vascular immune deposits	Immune deposits in the wall of small renal arteries without inflammation, necrosis, or thrombosis are more commonly associated with active glomerular proliferative forms of lupus nephritis. By the light microscopy examination of renal biopsy specimens, the normal histology is assessed. By immunofluorescence microscopy, staining for IgG, IgA, IgM, and various complement components (often C1q or C3) can be observed in the vessel wall. By electron microscopy, the deposits are electron dense, with a granular texture, and are most commonly observed below an intact vascular endothelium or within the basement membranes.
Arteriosclerosis	It is characterized by an increased arterial wall thickness and reduction of the vascular lumen due to fibrotic intimal thickening and replication of the internal elastic lamina.
Noninflammatory necrotizing vasculopathy	It may be considered a complication of more severe forms of immune complex deposition. The immune complex deposits can cause luminal narrowing or occlusion and are accompanied by necrotizing damage, frequently found in preglomerular arterioles and less in interlobular arteries. Abundant glassy eosinophilic materials may occupy the lumen and intima and, sometimes, may extend into the media. The endothelium is usually swollen or denuded, and the elastic membrane is often disrupted. The inflammatory infiltrate is rare. IgG, IgM, and IgA positivity can be detected by immunofluorescence microscopy in the vessel wall and in the lumen, as well as complement components and fibrin-related antigens. By electron microscopy, swelling or loss of the endothelium can be seen along with abundant intraluminal and mural deposits of granular electron-dense materials.
Thrombotic microangiopathy	It is most frequent in SLE patients with thrombotic thrombocytopenic purpura or anticardiolipin syndrome. In the early phase, there is swelling of the endothelial cells and subendothelial space. During the acute phase, a severe narrowing or total occlusion of the arteriolar lumen may be found. Fibrinoid necrosis may also be detect. The chronic phase presents swelling of the intima of the interlobular arteries associated with mucoid intimal edema and/or "onion skin" pattern lesions as result of the cellular intimal proliferation. By immunofluorescence microscopy, fibrinogen or fibrin in the walls of arterioles and small arteries can be observed, as well as IgM, IgG, IgA, C3, and C1q positivity. Electron microscopy may highlight the swelling and detachment of the endothelium from the underlying structures and an expanded intima.
True renal vasculitis	It is the least common renal lupus vascular lesion that usually involves small arteries, most commonly intralobular arteries. Histologically, it is indistinguishable from the polyarteritis nodosa. Morphologically, these lesions are characterized by neutrophils and mononuclear leukocytes that eccentrically or circumferentially infiltrate the intima and media. In the acute phase, this infiltration is often associated with fibrinoid necrosis and rupture of the elastic lamellae. Immunofluorescence reveals strong staining for fibrin-related antigens, with weak and variable staining for immunoglobulin and the complement.

A Chinese study analyzed 341 patients with lupus nephritis and found 279 (81.8%) patients with renal vascular lesions, including 74.2% with immune complex deposition, 24.0% with nonspecific arteriosclerosis, 17.6% with thrombotic microangiopathy, 3.8% with noninflammatory necrotizing vasculopathy, and 0.6% with true renal vasculitis. Approximately 40% of the cases presented with more than two types of vascular lesions [71].

True renal vasculitis is the least frequent renal vascular injury found in lupus nephritis and has been infrequently reported in the literature. It was found retrospectively in 2.8% [72], 2.4% [73], and 0.6% of renal biopsies [71]. True renal vasculitis can be morphologically differentiated from the other, more frequent forms of vascular renal lesions in SLE given that it is the only form yet described in which there is true inflammatory infiltration of the intima and media [71].

More attention should be dedicated to the patterns of renal microvascular lesions. The presence of vascular in a on lupus nephritis biopsy is associated with a worse prognosis and the risk of end-stage renal disease (ESRD) but not independent of the serum creatinine and nephritis class [71,74–76]. Histological lesions can present as glomerular, tubulointerstitial, and microvascular lesions and can affect small- and medium-sized arteries, most commonly intralobular arteries [69]. Mural inflammation with prominent inflammatory cell infiltrates and fibrinoid necrosis may be found [77]. Although clinical presentations vary with different types of vascular lesions, in general, SLE patients with renal vasculitis manifest with glomerular lesions, severe hypertension that likely worsens the vascular changes, anemia, hematuria, severe renal insufficiently with rapid progression to renal failure, and high SLEDAI scores [70].

A diffuse proliferative glomerulonephritis is considered by some authors as the most frequent form of renal vasculitis involving glomerular capillaries. However, the general agreement is that these lesions should be classified as proliferative lupus glomerulonephritis [78].

7. Retinal Vasculitis

Retinal vasculitis is a very uncommon complication documented in few case reports. Although the exact pathogenesis is unclear, it is thought that immune complex deposition, complement activation and aPL play a role [79]. Typically, the precapillary superficial arterial vasculature is involved [80]. Retinal vasculitis can present as asymptomatic or with painless blurred vision, decreased vision, or even permanent visual loss. A funduscopic examination reveals retinal vessel sheathing, cotton wool spots, retinal hemorrhage, and vascular occlusion [81]. Retinal imaging, including fluorescein angiography and optical coherence tomography, can be helpful in the identification and characterization of retinal vasculitis [81,82]. Using multimodal imaging techniques and electrophysiology, Chin et al. recently described a rare case of severe bilateral lupus retinal vasculitis associated with paracentral acute middle maculopathy. Both the superficial and deep retinal capillary vasculature was involved, resulting in marked generalized retinal dysfunction. The combination of immunomodulatory therapy with localized pan-retinal laser photocoagulation has led to an improvement in vision, the prevention of neovascularization, and remission of SLE [80].

8. Coronary Vasculitis

Coronary vasculitis is a rare condition with few case reports published in the literature. There is no strong association between SLE clinical activity and coronary arteritis. It often manifests in the absence of clinical SLE flare and laboratory evidence of active SLE [39]. The diagnosis is usually made by serial coronary angiographic studies that disclose arterial aneurysms, tapered stenoses, and/or rapidly developing arterial occlusions [69]. Histopathologically, coronary artery thrombosis or immune complex deposits, with an infiltration of lymphocytes and neutrophils, and fibrinoid necrosis can be detected [83,84]. Rare examples of cardiac valve dysfunction and myocardial dysfunction due to small vessel vasculitis have also been reported [69].

9. Pulmonary Vasculitis

The most common clinical manifestation of lupus pulmonary vasculitis is diffuse alveolar hemorrhage (DAH) due to the access of red blood cells within the alveolar spaces as a result of the widespread damage of the pulmonary vessels with disruption of the alveolar-capillary basement membrane [85]. The imaging studies often describe classical bilateral alveolar interstitial infiltrates (Figure 5).

It usually occurs in the context of high disease activity and could be a severe complication of SLE with a mortality rate of 35.3% [86]. Symptoms may include cough, progressive and severe dyspnea, fever, chest pain, and hemoptysis in over 60% of cases. Some patients may be asymptomatic [87]. The diagnosis is based on CT, a chest radiography, bronchoscopy, and bronchial lavage. Histopathologically, capillaritis and mononuclear infiltrates, alveolar necrosis, and immune complex deposits of IgG and C3 can be found [39].

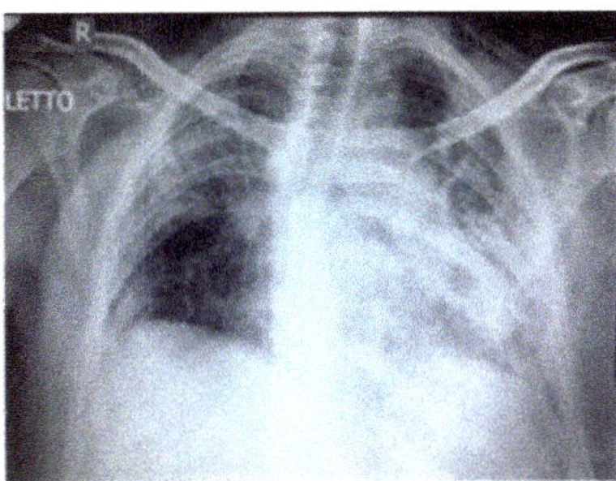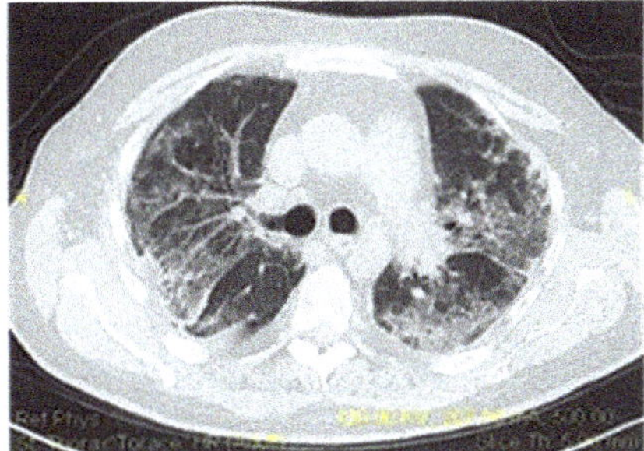

Figure 5. Diffuse alveolar hemorrhage. A chest radiography shows bilateral widespread infiltrates in both medium lower fields. A chest computed tomography shows left pulmonary embolism and massive diffuse infiltration of both lung fields. R means right.

10. Association between Lupus Vasculitis and Antiphospholipid Syndrome

SLE is the most common disease with which APS occurs [20], and within SLE patients, there is an association between LV and APS. Vascular injury and APS are often present simultaneously in SLE patients [88–90] and are closely connected with each other given that aPL can contribute to the damage of vascular endothelium during the vasculitic process [20,21].

Specifically, aPL plays a pathogenetic role in some forms of LV, including retinal vasculitis [79], DAH [91–93], and renal vasculitis [70]. The concomitant presence of vasculitis and APS is associated with a poor outcome. In particular, a comparison of the renal disease severity and outcome in patients with primary APS, APS secondary to SLE, and SLE alone revealed that APS worsens the prognosis of lupus nephritis [94].

11. Treatment

The treatment of LV is extremely varied and requires a timely diagnostic framework to limit the potentially severe consequences and life-threatening manifestations. The wide spectrum of clinical manifestations as a result of the inflammatory involvement of different-sized vessels and different organs is the main limiting factor in the management of these patients. Mesenteric vasculitis with bowel ischemia [52]; nervous systemic vasculitis—specifically, multiple mononeuropathy, seizures, and transverse myelitis [95]—and pulmonary vasculitis as diffuse alveolar hemorrhage [96] are the main severe complications in SLE that need timely aggressive therapy.

Unfortunately, no robust body of literature is available to guide their management, and therapeutic recommendations are commonly based upon other autoimmune conditions or case reports, case series, and expert opinions [97] (Table 2).

Table 2. The treatments of different vascular manifestations in SLE.

Vascular Manifestation	Type of Study	No. of Patients	Treatment	% Response (or Remission)	Adverse Events	Ref.
Cutaneous	Case report	1	Hydroxychloroquine (200–400 mg/day)	100	N.R.	[98]
	Case series	13	Colchicine (0.5–0.6 mg twice daily)	69	Mild (adominal cramping and diarrhea)	[99]
	Clinical experience	10		70	Mild	[100]
	Prospective randomized controlled trial	41		29	Mild	[101]
	Multicenter study	69	Thalidomide (50 mg/day)	100	Mild (drowsiness and constipation)	[102]

Table 2. Cont.

Vascular Manifestation	Type of Study	No. of Patients	Treatment	% Response (or Remission)	Adverse Events	Ref.
Cutaneous	Case series	6	Azathioprine (2 mg/kg/day)	33	Leukopenia, hepatic injury, hypersensitivity reaction, and infections	[103]
	Clinical trial	12	Immunoglobulin (1 g/kg for 2 consecutive days followed by 400 mg/kg monthly)	>75	N.R	[104]
	Case report	1			Patient died from septic shock	[105]
	Case series	2	Rituximab	100	N.R.	[106]
Gastrointestinal	Retrospective cohort study	97	Cyclophosphamide (500–1000 mg/m^2) and prednisone (>30 mg) Mycophenolate mofetil (2 g/day) Corticosteroid only	84.5	Severe adverse events occurred in 15 patients	[57]
	Retrospective study	38	Methylprednisolone (1 mg/kg/day)	100	N.R.	[58]
	Case series	5	Cyclophosphamide and corticosteroids	80	N.R.	[107]
			Rituximab	20		
	Case series	3	Methylprednisolone (20 mg/kg/day for 5 days) and cyclophosphamide (1 g/m^2)	100	N.R.	[108]
	Case series	19	Methylprednisolone pulse therapy (1 g/day for 3 days), followed by cyclophosphamide (1 g/m^2 intravenously) in 4 cases.	90	N.R.	[109]
Nervous system	Controlled clinical trial	32	Cyclophosphamide (0.75 g/m^2 monthly for 1 year and then every 3 months for another year). Methylprednisolone (1 g daily for 3 days, monthly for 4 months, then bimonthly for 6 months and subsequently every 3 months for 1 year) Oral prednisone on the fourth day of treatment (1 mg/kg/day)	75	Infections of the gastrointestinal, urinary and upper respiratory tract. Herpes zoster	[95]
	Case report and review of 34 cases	35	Rituximab and concomitant treatment with corticosteroids, methylprednisolone, cyclophosphamide or azathioprine	50	Herpes zoster Infections	[110]
Renal	Meta-analysis (18 studies)	1989	Mycophenolate mofetil and cyclophosphamide	N.R.	Infections	[111]
	Meta-analysis (74 studies)	5175	High-dose steroids with cyclophosphamide or mycophenolate mofetil as induction therapy, and low-dose steroids combined with varying regimens of azathioprine or mycophenolate mofetil for the maintenance phase	N.R.	Diarrhea	[112]
	Retrospective study	61	Cyclophosphamide or mycophenolate mofetil in combination with glucocorticoids for the induction phase. Mycophenolate mofetil or azathioprine combined with low-dose glucocorticoid regimens for the maintence phase.	N.R.	No severe adverse events	[113]
		9	Plasmapheresis and baseline immunosuppressive therapy	33	None	
	Systematic review (31 studies)	1259	Rituximab alone or in combination with cyclophosphamide or mycophenolate mofetil	77 Caucasian 38 East-Asian 28 Hispanic	N.R.	[114]

Table 2. *Cont.*

Vascular Manifestation	Type of Study	No. of Patients	Treatment	% Response (or Remission)	Adverse Events	Ref.
Renal	Systematic review (15 studies) and case report	20	Eculizumab	85	N.R.	[115]
Pulmonary	Case series	16	High-dose corticosteroid, followed by pulse methylprednisolone, plasmapheresis, pulse cyclophosphamide, and rituximab	N.R.	Infections	[86]
	Case series	34	High dose of methylprednisolone (>3 g) and cyclophosphamide	N.R.	N.R.	[96]
	Retrospective clinical trials	40 SLE (11 DHA)	Therapeutic plasma exchange	N.R.	Mild (bleeding)	[116]
	Case control study	22	Various combinations of corticosteroids, plasmapheresis, cyclophosphamide, rituximab, and mycophenolate mofetil.	N.R.	N.R.	[92]
	Case report	1	rFVIIa 75 µg/kg	N.R.	None	[117]
	Case report	1	Extracorporeal Membrane Oxygenation	N.R.	None	[118]
	Retrospective study	4	Umbilical cord-derived mesenchymal stem cell transplantation	N.R.	None	[119]
Retinal	Case report	1	Plasmapheresis and the bilateral administration of intravitreal ranibizumab (0.5 mg) and rituximab	N.R.	None	[120]
	Case report	1	Plasmapheresis, followed by rituximab and mycophenolate mofetil	100	None	[121]
	Case report	1	Rituximab (1 g) and cyclophosphamide (10 mg/kg)	100	None	[122]
	Case series	2	Plasmapheresis, followed by a single intravenous infusion of cyclophosphamide (750 mg) Plasmapheresis and methotrexate (15 mg weekly)	N.R.	None	[123]
	Case series	2	Panretinal photocoagulation, rituximab (750 mg/m^2 × 2 weeks), and cyclophosphamide (750 mg/m2 per dose) with a concurrent pulse of methylprednisolone (1000 mg) Methylprednisolone (1000 mg) plus rituximab (750 mg/m^2 × 2 weeks) and monthly cyclophosphamide (750 mg/m^2 × 7 months).	N.R.	N.R.	[124]
Coronary	Case report	1	Prednisone (60 mg once a day) and cyclophosphamide (1300 mg monthly)	100	N.R.	[83]
	Case report	1	Methylprednisolone (1 g) and cyclophosphamide (1 g)	100	N.R.	[125]
	Case report	1	Methylprednisolone 1000 mg daily for 3 days, followed by 1-mg/kg/day prednisone in addition to a single dose of intravenous cyclophosphamide 860 mg and oral hydroxychloroquine 400 mg daily. Orthotopic heart transplant	0	N.R.	[126]

Abbreviations: N.R. = not reported; DHA = diffuse alveolar hemorrhage; rFVIIa = recombinant activated coagulation factor VII.

In this context, the updated EULAR (European League Against Rheumatism) recommendations for SLE are a clinical reference point, although LV treatment should be tailored to the severity of the disease and its associated symptoms [127]. Mild-to-moderate manifestations are usually handled with oral corticosteroids and immunosuppressants such as methotrexate, azathioprine, and mycophenolate mofetil. A more aggressive therapy with intravenous high-dose corticosteroids, cyclophosphamide, rituximab, intravenous immunoglobulin, and/or plasmapheresis is considered for the severe and life-threatening forms [69].

Cutaneous vasculitis often requires anti-malarias; hydroxychloroquine (200–400 mg/day) has been considered in some patients with success, primarily those with hypocomplementemia urticarial vasculitis [98,128]. It is generally well-tolerated, though its potential ophthalmological toxicity is well-known and needs regular monitoring. In skin-limited vasculitis, the responses to colchicine (0.6 mg twice daily) have been described in several open-label case series [99–101], although relapses have been reported after colchicine therapy interruption. In the case of poor efficacy or contraindications for the drug, thalidomide and dapsone (50–200 mg/day) can be used, with good results [97,102,128]. In a multicenter Chinese study, 69 patients with cutaneous lesions of SLE were treated for 8 weeks with a starting dose of thalidomide at 25 mg daily and gradually increased. The maximum ratio of an effective and maintenance dose of thalidomide were both at 50 mg daily, and the rate of total remission rose to 100% at the eighth week [102]. Oral glucocorticoids (methylprednisolone > 15 mg/day) may be required for a short period of time for painful, ulcerative, or otherwise severe diseases in order to speed up the resolution [1,97,129]. Among Conventional Disease-Modifying Antirheumatic Drugs (cDMARDs), azathioprine (2 mg/kg/day) has been successfully used in the treatment of various types of systemic vasculitis, including severe lupus cutaneous vasculitis resistant to conventional therapy, with some adverse events such as leukopenia, hepatic injury, hypersensitivity reaction, and infections [97,100,103]. For resistant cases of lupus cutaneous vasculitis, immunoglobulin could be an off-label option. The usual starting dose is 1 g/kg for 2 consecutive days, followed by 400 mg/kg monthly until disease resolution or for 6 months [104,105]. The anti-CD20 antibody rituximab is a safe, effective treatment for refractory chronic cutaneous small vessel vasculitis that is nonresponsive to traditional therapies [106]. Although there are no data about the effectiveness of JAK inhibitors in LV, the high efficacy of these drugs in many skin manifestations, including atopic dermatitis [130], psoriasis [131], and graft-versus-host disease [132], suggests that they may represent a new effective weapon for treating LV.

The current treatment for gastrointestinal LV includes a high dose of corticosteroids, intravenous infusions of methylprednisolone with subsequent tapering, and for patients with recurrent disease or that do not respond to intravenous prednisolone alone, intravenous cyclophosphamide should be considered [57,58,107–109,133]. The successful use of rituximab has also been reported in case series [107]. When a rapid response is not achieved, surgical options should be considered.

In a controlled long-term clinical trial, multiplex mononeuropathy, seizures, and transverse myelitis were successfully treated with a frontline therapy based on methylprednisolone 3 g daily for 3 days as the induction treatment, followed by cyclophosphamide for 2 years at 0.75 g/m^2 of the body surface, monthly for 1 year, and then every 3 months for another year [95]. Several studies have supported the off-label use of rituximab in cases of severe refractory neuropsychiatric SLE, but a relapse after rituximab treatment was observed in 45% of the cases [110].

The treatment of renal microvascular lesions in lupus nephritis remains undefined, and the current therapeutic strategies are based on glomerular pathology. The gold standard for inducing remission in systemic necrotizing vasculitis and severe lupus nephritis is the combination of high-dose steroids with cyclophosphamide or mycophenolate mofetil as the induction therapy and low-dose steroids combined with varying regimens of azathioprine or mycophenolate mofetil for the maintenance phase [111,112,134]. An update of a Cochrane review first published in 2004 showed that a mycophenolate mofetil treatment can result in increased complete disease remission compared with cyclophosphamide, with acceptable toxicity, although with low certainty evidence. Calcineurin inhibitors, alone and in combination with mycophenolate mofetil, may have comparable or improved rates of disease remission compared with cyclophosphamide and a lower toxicity but uncertain effects. Maintenance therapy based on azathioprine may increase the disease relapse compared with mycophenolate mofetil [112]. A retrospective study assessed the efficacy of plasmapheresis in patients with lupus nephritis combined with thrombotic microangiopathy, highlighting the improvements in recovery and renal outcome in patients who received plasmapheresis associated with corticosteroid and immunosuppressive drugs compared with those treated with corticosteroid and immunosuppressive drugs alone [113]. The role of rituximab in the induction therapy has not been clearly established for lupus nephritis. In a recent systematic review, an analysis of 31 studies of rituximab for class I-VI lupus nephritis revealed the heterogeneous efficacy of rituximab alone or in combination with cyclophosphamide or mycophenolate mofetil among patients of different ethnic and racial backgrounds, lupus nephritis classes, time courses of the disease, ages, and prior immunosuppressive uses [114]. For cases resistant to conventional therapy, alternative strategies may be used. Eculizumab, a fully humanized monoclonal antibody that inhibits the human C5 complement component, might be an alternative treatment of severe refractory lupus renal vasculitis [115,135]. Recently, new drugs, including Obinutuzumab (anti-CD20 monoclonal antibody) for B-cell depletion or belimumab (anti-"B-cell activating factor" monoclonal antibody neutralization) for B-cell neutralization and Voclosporin (a calcineurin inhibitor with a low profile of renal and systemic toxicity) have shown promising results regarding an improvement in the renal response in addition to the standard therapy in patients with lupus nephritis and might represent new potential therapeutic strategies for lupus renal vasculitis [136–138].

In pulmonary vasculitis with DAH, a high dose of methylprednisolone (4–8 g, above the conventional dosage of 3 g) [96] and cyclophosphamide remain the most commonly used therapies [139,140]. Plasmapheresis [116] and rituximab [86] are the other beneficial treatment options in refractory cases [92]. Experimental strategies such as Intrapulmonary Factor VII therapy [117], extracorporeal membrane oxygenation [118], and umbilical cord-derived mesenchymal stem cell transplantation [119] are limited to selective severe cases.

The mainstay of the treatment of retinal vasculitis in lupus is systemic immunosuppression with high-dose oral corticosteroids. Once disease control is reached, low-dose systemic glucocorticoids and hydroxychloroquine are often used long term to prevent disease flares. Cases of severe retinal vasculitis that are refractory to steroid therapy are treated with plasmapheresis; rituximab; or a combination of plasmapheresis, rituximab, and intravitreal ranibizumab (a monoclonal antibody against vascular endothelial growth factor A) [120–123]. In pediatric patients with SLE and occlusive retinopathy, an early intervention with a combination of B-cell depletion therapy and a traditional cytotoxic agent such as cyclophosphamide should be considered [124]. Pan-retinal photocoagulation and intravitreal anti-vascular endothelial growth factor injections can induce the regression of macular edema and retinal neovascularization [80,124].

There is no established therapy for SLE coronary vasculitis, although few case reports have reported clinical benefits with intravenous pulse dose methylprednisolone associated with intravenous cyclophosphamide in active SLE patients [83,125,141]. When SLE coronary vasculitis is refractory to immunosuppressant therapy, an orthotropic heart transplant should be performed [126].

12. Conclusions

Vasculitis occurs frequently in SLE patients with an active disease and poor prognosis. Lupus vasculitis is characterized by varying manifestations, given that it can affect any organ system. A prompt diagnosis and adequate treatment are essential. The modalities for treatment are tailored according to the presentation and severity of the disease. Due to a lack of randomized controlled trials specific for LV, therapeutic decisions are based on experiences treating other autoimmune conditions, other forms of vasculitis syndromes, or case reports.

Author Contributions: Conceptualization, P.L., M.P. and V.R.; data curation, E.M., A.B., N.S. and G.I.; writing, P.L., M.P. and E.M.; and supervision, V.R. All of the authors reviewed the manuscript, approved the draft submission, and accept responsibility for all aspects of this study. All authors have read and agreed to the published version of the manuscript.

Funding: The Emergency Medicine Residency Program General Fund.

Institutional Review Board Statement: Not applicable.

Informed Consent Statement: Not applicable.

Acknowledgments: The authors acknowledge Smart Servier Medical Art (http://smart.servier.com, accessed on 15 October 2021) for providing comprehensive medical and biological figures and datasets that are fruitful for the international scientific community.

Conflicts of Interest: The authors declare no conflict of interest.

References

1. Rahman, A.; Isenberg, D.A. Systemic lupus erythematosus. *N. Engl. J. Med.* **2008**, *358*, 929–939. [CrossRef]
2. Doyle, M.K. Vasculitis associated with connective tissue disorders. *Curr. Rheumatol. Rep.* **2006**, *8*, 312–316. [CrossRef]
3. Leone, P.; Cicco, S.; Vacca, A.; Dammacco, F.; Racanelli, V. Vasculitis in connective tissue diseases. In *Systemic Vasculitides: Current Status and Perspectives*; Dammcco, F., Ribatti, D., Vacca, A., Eds.; Springer International Publishing: Cham, Switzerland, 2016; pp. 345–359.
4. Drenkard, C.; Villa, A.R.; Reyes, E.; Abello, M.; Alarcon-Segovia, D. Vasculitis in systemic lupus erythematosus. *Lupus* **1997**, *6*, 235–242. [CrossRef] [PubMed]
5. Ramos-Casals, M.; Nardi, N.; Lagrutta, M.; Brito-Zeron, P.; Bove, A.; Delgado, G.; Cervera, R.; Ingelmo, M.; Font, J. Vasculitis in systemic lupus erythematosus: Prevalence and clinical characteristics in 670 patients. *Medicine (Baltimore)* **2006**, *85*, 95–104. [CrossRef] [PubMed]
6. Alarcón-Segovia, D.; Drenkard, C. Vasculitis and the antiphospholipid syndrome. *Rheumatology (Oxford)* **2000**, *8*, 922–923. [CrossRef]
7. Kallenberg, C.G.; Heeringa, P. Pathogenesis of vasculitis. *Lupus* **1998**, *7*, 280–284. [CrossRef] [PubMed]
8. Pyrpasopoulou, A.; Chatzimichailidou, S.; Aslanidis, S. Vascular disease in systemic lupus erythematosus. *Autoimmune Dis.* **2012**, *2012*, 876456. [CrossRef]
9. Fauci, A.S.; Haynes, B.; Katz, P. The spectrum of vasculitis: Clinical, pathologic, immunologic and therapeutic considerations. *Ann. Intern. Med.* **1978**, *89*, 660–676. [CrossRef]
10. Manderson, A.P.; Botto, M.; Walport, M.J. The role of complement in the development of systemic lupus erythematosus. *Annu Rev. Immunol.* **2004**, *22*, 431–456. [CrossRef]
11. Jennette, J.C. Implications for pathogenesis of patterns of injury in small- and medium-sized-vessel vasculitis. *Cleve. Clin. J. Med.* **2002**, *69* (Suppl. 2), SII33-8. [CrossRef]
12. Belmont, H.M.; Abramson, S.B.; Lie, J.T. Pathology and pathogenesis of vascular injury in systemic lupus erythematosus. Interactions of inflammatory cells and activated endothelium. *Arthritis Rheum* **1996**, *39*, 9–22. [CrossRef]
13. Cid, M.C. Endothelial cell biology, perivascular inflammation, and vasculitis. *Cleve. Clin. J. Med.* **2002**, *69* (Suppl. 2), SII45–SII49. [CrossRef] [PubMed]
14. Renaudineau, Y.; Dugue, C.; Dueymes, M.; Youinou, P. Antiendothelial cell antibodies in systemic lupus erythematosus. *Autoimmun Rev.* **2002**, *1*, 365–372. [CrossRef]
15. Cieslik, P.; Hrycek, A.; Klucinski, P. Vasculopathy and vasculitis in systemic lupus erythematosus. *Pol. Arch. Med. Wewn.* **2008**, *118*, 57–63. [CrossRef] [PubMed]
16. Manolova, I.; Dancheva, M.; Halacheva, K. Antineutrophil cytoplasmic antibodies in patients with systemic lupus erythematosus: Prevalence, antigen specificity, and clinical associations. *Rheumatol. Int.* **2001**, *20*, 197–204. [CrossRef]
17. Sen, D.; Isenberg, D.A. Antineutrophil cytoplasmic autoantibodies in systemic lupus erythematosus. *Lupus* **2003**, *12*, 651–658. [CrossRef] [PubMed]
18. Jennette, J.C.; Xiao, H.; Falk, R.J. Pathogenesis of vascular inflammation by anti-neutrophil cytoplasmic antibodies. *J. Am. Soc. Nephrol.* **2006**, *17*, 1235–1242. [CrossRef]
19. Wiik, A. What you should know about PR3-ANCA. An introduction. *Arthritis Res.* **2000**, *2*, 252–254. [CrossRef]
20. Levine, J.S.; Branch, D.W.; Rauch, J. The antiphospholipid syndrome. *N. Engl. J. Med.* **2002**, *346*, 752–763. [CrossRef]
21. Meroni, P.L.; Raschi, E.; Camera, M.; Testoni, C.; Nicoletti, F.; Tincani, A.; Khamashta, M.A.; Balestrieri, G.; Tremoli, E.; Hess, D.C. Endothelial activation by aPL: A potential pathogenetic mechanism for the clinical manifestations of the syndrome. *J. Autoimmun.* **2000**, *15*, 237–240. [CrossRef] [PubMed]
22. Praprotnik, S.; Blank, M.; Meroni, P.L.; Rozman, B.; Eldor, A.; Shoenfeld, Y. Classification of anti-endothelial cell antibodies into antibodies against microvascular and macrovascular endothelial cells: The pathogenic and diagnostic implications. *Arthritis Rheum.* **2001**, *44*, 1484–1494. [CrossRef]

23. Mahajan, A.; Herrmann, M.; Munoz, L.E. Clearance Deficiency and Cell Death Pathways: A Model for the Pathogenesis of SLE. *Front. Immunol.* **2016**, *7*, 35. [CrossRef] [PubMed]

24. Brinkmann, V.; Reichard, U.; Goosmann, C.; Fauler, B.; Uhlemann, Y.; Weiss, D.S.; Weinrauch, Y.; Zychlinsky, A. Neutrophil extracellular traps kill bacteria. *Science* **2004**, *303*, 1532–1535. [CrossRef]

25. Fuchs, T.A.; Abed, U.; Goosmann, C.; Hurwitz, R.; Schulze, I.; Wahn, V.; Weinrauch, Y.; Brinkmann, V.; Zychlinsky, A. Novel cell death program leads to neutrophil extracellular traps. *J. Cell Biol.* **2007**, *176*, 231–241. [CrossRef]

26. Kambas, K.; Mitroulis, I.; Apostolidou, E.; Girod, A.; Chrysanthopoulou, A.; Pneumatikos, I.; Skendros, P.; Kourtzelis, I.; Koffa, M.; Kotsianidis, I.; et al. Autophagy mediates the delivery of thrombogenic tissue factor to neutrophil extracellular traps in human sepsis. *PLoS ONE* **2012**, *7*, e45427. [CrossRef]

27. Frangou, E.; Vassilopoulos, D.; Boletis, J.; Boumpas, D.T. An emerging role of neutrophils and NETosis in chronic inflammation and fibrosis in systemic lupus erythematosus (SLE) and ANCA-associated vasculitides (AAV): Implications for the pathogenesis and treatment. *Autoimmun. Rev.* **2019**, *18*, 751–760. [CrossRef]

28. Kambas, K.; Chrysanthopoulou, A.; Vassilopoulos, D.; Apostolidou, E.; Skendros, P.; Girod, A.; Arelaki, S.; Froudarakis, M.; Nakopoulou, L.; Giatromanolaki, A.; et al. Tissue factor expression in neutrophil extracellular traps and neutrophil derived microparticles in antineutrophil cytoplasmic antibody associated vasculitis may promote thromboinflammation and the thrombophilic state associated with the disease. *Ann. Rheum. Dis.* **2014**, *73*, 1854–1863. [CrossRef]

29. Villanueva, E.; Yalavarthi, S.; Berthier, C.C.; Hodgin, J.B.; Khandpur, R.; Lin, A.M.; Rubin, C.J.; Zhao, W.; Olsen, S.H.; Klinker, M.; et al. Netting neutrophils induce endothelial damage, infiltrate tissues, and expose immunostimulatory molecules in systemic lupus erythematosus. *J. Immunol.* **2011**, *187*, 538–552. [CrossRef]

30. Riemekasten, G.; Ziemer, S.; Haupl, T.; Melzer, C.; Loddenkemper, K.; Hauptmann, S.; Burmester, G.R.; Hiepe, F. Shwartzman phenomenon in a patient with active systemic lupus erythematosus preceding fatal disseminated intravascular coagulation. *Lupus* **2002**, *11*, 204–207. [CrossRef] [PubMed]

31. D'Cruz, D. Vasculitis in systemic lupus erythematosus. *Lupus* **1998**, *7*, 270–274. [CrossRef] [PubMed]

32. Radic, M.; Martinovic Kaliterna, D.; Radic, J. Drug-induced vasculitis: A clinical and pathological review. *Neth. J. Med.* **2012**, *70*, 12–17. [PubMed]

33. Guillevin, L.; Lhote, F.; Gherardi, R. The spectrum and treatment of virus-associated vasculitides. *Curr. Opin. Rheumatol.* **1997**, *9*, 31–36. [CrossRef]

34. Agnello, V.; Chung, R.T.; Kaplan, L.M. A role for hepatitis C virus infection in type II cryoglobulinemia. *N. Engl. J. Med.* **1992**, *327*, 1490–1495. [CrossRef]

35. Kallas, R.; Goldman, D.; Petri, M.A. Cutaneous vasculitis in SLE. *Lupus Sci. Med.* **2020**, *7*, e000411. [CrossRef]

36. Sharma, A.; Dhooria, A.; Aggarwal, A.; Rathi, M.; Chandran, V. Connective Tissue Disorder-Associated Vasculitis. *Curr. Rheumatol. Rep.* **2016**, *18*, 31. [CrossRef] [PubMed]

37. Fukuda, M.V.; Lo, S.C.; de Almeida, C.S.; Shinjo, S.K. Anti-Ro antibody and cutaneous vasculitis in systemic lupus erythematosus. *Clin. Rheumatol.* **2009**, *28*, 301–304. [CrossRef]

38. Garcia-Carrasco, M.; Ramos-Casals, M.; Cervera, R.; Trejo, O.; Yague, J.; Siso, A.; Jimenez, S.; de La Red, G.; Font, J.; Ingelmo, M. Cryoglobulinemia in systemic lupus erythematosus: Prevalence and clinical characteristics in a series of 122 patients. *Semin. Arthritis Rheum.* **2001**, *30*, 366–373. [CrossRef] [PubMed]

39. Barile-Fabris, L.; Hernandez-Cabrera, M.F.; Barragan-Garfias, J.A. Vasculitis in systemic lupus erythematosus. *Curr. Rheumatol. Rep.* **2014**, *16*, 440. [CrossRef]

40. Jafri, K.; Patterson, S.L.; Lanata, C. Central Nervous System Manifestations of Systemic Lupus Erythematosus. *Rheum. Dis. Clin. N. Am.* **2017**, *43*, 531–545. [CrossRef]

41. Bortoluzzi, A.; Silvagni, E.; Furini, F.; Piga, M.; Govoni, M. Peripheral nervous system involvement in systemic lupus erythematosus: A review of the evidence. *Clin. Exp. Rheumatol.* **2019**, *37*, 146–155.

42. Gorson, K.C. Vasculitic neuropathies: An update. *Neurologist* **2007**, *13*, 12–19. [CrossRef]

43. Florica, B.; Aghdassi, E.; Su, J.; Gladman, D.D.; Urowitz, M.B.; Fortin, P.R. Peripheral neuropathy in patients with systemic lupus erythematosus. *Semin. Arthritis Rheum.* **2011**, *41*, 203–211. [CrossRef] [PubMed]

44. Toledano, P.; Orueta, R.; Rodriguez-Pinto, I.; Valls-Sole, J.; Cervera, R.; Espinosa, G. Peripheral nervous system involvement in systemic lupus erythematosus: Prevalence, clinical and immunological characteristics, treatment and outcome of a large cohort from a single centre. *Autoimmun. Rev.* **2017**, *16*, 750–755. [CrossRef] [PubMed]

45. Rodrigues, M.; Galego, O.; Costa, C.; Jesus, D.; Carvalho, P.; Santiago, M.; Malcata, A.; Ines, L. Central nervous system vasculitis in systemic lupus erythematosus: A case series report in a tertiary referral centre. *Lupus* **2017**, *26*, 1440–1447. [CrossRef]

46. Abdel Razek, A.A.; Alvarez, H.; Bagg, S.; Refaat, S.; Castillo, M. Imaging spectrum of CNS vasculitis. *Radiographics* **2014**, *34*, 873–894. [CrossRef]

47. Schwartz, N.; Stock, A.D.; Putterman, C. Neuropsychiatric lupus: New mechanistic insights and future treatment directions. *Nat. Rev. Rheumatol.* **2019**, *15*, 137–152. [CrossRef] [PubMed]

48. Calle-Botero, E.; Abril, A. Lupus Vasculitis. *Curr. Rheumatol. Rep.* **2020**, *22*, 71. [CrossRef]

49. Younger, D.S.; Coyle, P.K. Central Nervous System Vasculitis due to Infection. *Neurol. Clin.* **2019**, *37*, 441–463. [CrossRef]

50. Govoni, M.; Bortoluzzi, A.; Padovan, M.; Silvagni, E.; Borrelli, M.; Donelli, F.; Ceruti, S.; Trotta, F. The diagnosis and clinical management of the neuropsychiatric manifestations of lupus. *J. Autoimmun.* **2016**, *74*, 41–72. [CrossRef]

51. Kroner, P.T.; Tolaymat, O.A.; Bowman, A.W.; Abril, A.; Lacy, B.E. Gastrointestinal Manifestations of Rheumatological Diseases. *Am. J. Gastroenterol.* **2019**, *114*, 1441–1454. [CrossRef]

52. Maruyama, A.; Nagashima, T.; Iwamoto, M.; Minota, S. Clinical characteristics of lupus enteritis in Japanese patients: The large intestine-dominant type has features of intestinal pseudo-obstruction. *Lupus* **2018**, *27*, 1661–1669. [CrossRef]

53. Ju, J.H.; Min, J.K.; Jung, C.K.; Oh, S.N.; Kwok, S.K.; Kang, K.Y.; Park, K.S.; Ko, H.J.; Yoon, C.H.; Park, S.H.; et al. Lupus mesenteric vasculitis can cause acute abdominal pain in patients with SLE. *Nat. Rev. Rheumatol.* **2009**, *5*, 273–281. [CrossRef]

54. Janssens, P.; Arnaud, L.; Galicier, L.; Mathian, A.; Hie, M.; Sene, D.; Haroche, J.; Veyssier-Belot, C.; Huynh-Charlier, I.; Grenier, P.A.; et al. Lupus enteritis: From clinical findings to therapeutic management. *Orphanet J. Rare Dis.* **2013**, *8*, 67. [CrossRef] [PubMed]

55. Calamia, K.T.; Balabanova, M. Vasculitis in systemic lupus erythematosis. *Clin. Dermatol.* **2004**, *22*, 148–156. [CrossRef]

56. Brewer, B.N.; Kamen, D.L. Gastrointestinal and Hepatic Disease in Systemic Lupus Erythematosus. *Rheum. Dis. Clin. N. Am.* **2018**, *44*, 165–175. [CrossRef]

57. Yuan, S.; Ye, Y.; Chen, D.; Qiu, Q.; Zhan, Z.; Lian, F.; Li, H.; Liang, L.; Xu, H.; Yang, X. Lupus mesenteric vasculitis: Clinical features and associated factors for the recurrence and prognosis of disease. *Semin. Arthritis Rheum.* **2014**, *43*, 759–766. [CrossRef]

58. Lee, C.K.; Ahn, M.S.; Lee, E.Y.; Shin, J.H.; Cho, Y.S.; Ha, H.K.; Yoo, B.; Moon, H.B. Acute abdominal pain in systemic lupus erythematosus: Focus on lupus enteritis (gastrointestinal vasculitis). *Ann. Rheum. Dis.* **2002**, *61*, 547–550. [CrossRef]

59. Xu, D.; Lin, J. Urinary tract involvement in systemic lupus erythematosus: Coexistence with lupus mesenteric vasculitis or intestinal pseudo-obstruction? *Semin. Arthritis Rheum.* **2015**, *44*, e9. [CrossRef] [PubMed]

60. Koo, B.S.; Hong, S.; Kim, Y.J.; Kim, Y.G.; Lee, C.K.; Yoo, B. Lupus enteritis: Clinical characteristics and predictive factors for recurrence. *Lupus* **2015**, *24*, 628–632. [CrossRef] [PubMed]

61. Wang, Y.S.; Huang, I.F.; Feng, W.B.; Chiou, Y.H. Recurrent lupus mesenteric vasculitis leading to gastrointestinal perforation and sepsis. *Kaohsiung J. Med. Sci.* **2015**, *31*, 440–441. [CrossRef]

62. Kwok, S.K.; Seo, S.H.; Ju, J.H.; Park, K.S.; Yoon, C.H.; Kim, W.U.; Min, J.K.; Park, S.H.; Cho, C.S.; Kim, H.Y. Lupus enteritis: Clinical characteristics, risk factor for relapse and association with anti-endothelial cell antibody. *Lupus* **2007**, *16*, 803–809. [CrossRef] [PubMed]

63. Matsumoto, T.; Kobayashi, S.; Shimizu, H.; Nakajima, M.; Watanabe, S.; Kitami, N.; Sato, N.; Abe, H.; Aoki, Y.; Hoshi, T.; et al. The liver in collagen diseases: Pathologic study of 160 cases with particular reference to hepatic arteritis, primary biliary cirrhosis, autoimmune hepatitis and nodular regenerative hyperplasia of the liver. *Liver* **2000**, *20*, 366–373. [CrossRef]

64. Matsumoto, T.; Yoshimine, T.; Shimouchi, K.; Shiotu, H.; Kuwabara, N.; Fukuda, Y.; Hoshi, T. The liver in systemic lupus erythematosus: Pathologic analysis of 52 cases and review of Japanese Autopsy Registry Data. *Hum. Pathol.* **1992**, *23*, 1151–1158. [CrossRef]

65. Alanazi, T.; Alqahtani, M.; Al Duraihim, H.; Al Khathlan, K.; Al Ahmari, B.; Makanjuola, D.; Afzal, M. Hepatic vasculitis mimicking liver abscesses in a patient with systemic lupus erythematosus. *Ann. Saudi Med.* **2009**, *29*, 474–477. [CrossRef] [PubMed]

66. Levitin, P.M.; Sweet, D.; Brunner, C.M.; Katholi, R.E.; Bolton, W.K. Spontaneous rupture of the liver. An unusual complication of SLE. *Arthritis Rheum.* **1977**, *20*, 748–750. [CrossRef]

67. Trambert, J.; Reinitz, E.; Buchbinder, S. Ruptured hepatic artery aneurysms in a patient with systemic lupus erythematosus: Case report. *Cardiovasc. Interv. Radiol.* **1989**, *12*, 32–34. [CrossRef]

68. Nesher, G.; Breuer, G.S.; Temprano, K.; Moore, T.L.; Dahan, D.; Baer, A.; Alberton, J.; Izbicki, G.; Hersch, M. Lupus-associated pancreatitis. *Semin. Arthritis Rheum.* **2006**, *35*, 260–267. [CrossRef]

69. Fessler, B.J.; Hoffman, G.S. Vasculitis. In *Systemic Lupus Erythematosus*, 5th ed.; Lahita, R.G., Ed.; Academic Press: Cambridge, MA, USA, 2011; pp. 833–845.

70. Ding, Y.; Tan, Y.; Qu, Z.; Yu, F. Renal microvascular lesions in lupus nephritis. *Ren. Fail.* **2020**, *42*, 19–29. [CrossRef] [PubMed]

71. Wu, L.H.; Yu, F.; Tan, Y.; Qu, Z.; Chen, M.H.; Wang, S.X.; Liu, G.; Zhao, M.H. Inclusion of renal vascular lesions in the 2003 ISN/RPS system for classifying lupus nephritis improves renal outcome predictions. *Kidney Int.* **2013**, *83*, 715–723. [CrossRef]

72. Banfi, G.; Bertani, T.; Boeri, V.; Faraggiana, T.; Mazzucco, G.; Monga, G.; Sacchi, G. Renal vascular lesions as a marker of poor prognosis in patients with lupus nephritis. Gruppo Italiano per lo Studio della Nefrite Lupica (GISNEL). *Am. J. Kidney Dis.* **1991**, *2*, 240–248. [CrossRef]

73. Descombes, E.; Droz, D.; Drouet, L.; Grünfeld, J.P.; Lesavre, P. Renal vascular lesions in lupus nephritis. *Medicine (Baltimore)* **1997**, *5*, 355–368. [CrossRef] [PubMed]

74. Broder, A.; Mowrey, W.B.; Khan, H.N.; Jovanovic, B.; Londono-Jimenez, A.; Izmirly, P.; Putterman, C. Tubulointerstitial damage predicts end stage renal disease in lupus nephritis with preserved to moderately impaired renal function: A retrospective cohort study. *Semin. Arthritis Rheum.* **2018**, *47*, 545–551. [CrossRef]

75. Huang, J.; Han, S.S.; Qin, D.D.; Wu, L.H.; Song, Y.; Yu, F.; Wang, S.X.; Liu, G.; Zhao, M.H. Renal Interstitial Arteriosclerotic Lesions in Lupus Nephritis Patients: A Cohort Study from China. *PLoS ONE* **2015**, *10*, e0141547. [CrossRef]

76. Leatherwood, C.; Speyer, C.B.; Feldman, C.H.; D'Silva, K.; Gomez-Puerta, J.A.; Hoover, P.J.; Waikar, S.S.; McMahon, G.M.; Rennke, H.G.; Costenbader, K.H. Clinical characteristics and renal prognosis associated with interstitial fibrosis and tubular atrophy (IFTA) and vascular injury in lupus nephritis biopsies. *Semin. Arthritis Rheum.* **2019**, *49*, 396–404. [CrossRef] [PubMed]

77. Barber, C.; Herzenberg, A.; Aghdassi, E.; Su, J.; Lou, W.; Qian, G.; Yip, J.; Nasr, S.H.; Thomas, D.; Scholey, J.W.; et al. Evaluation of clinical outcomes and renal vascular pathology among patients with lupus. *Clin. J. Am. Soc. Nephrol.* **2012**, *7*, 757–764. [CrossRef]

78. Abdellatif, A.A.; Waris, S.; Lakhani, A.; Kadikoy, H.; Haque, W.; Truong, L.D. True vasculitis in lupus nephritis. *Clin. Nephrol.* **2010**, *74*, 106–112. [CrossRef] [PubMed]

79. Nag, T.C.; Wadhwa, S. Vascular changes of the retina and choroid in systemic lupus erythematosus: Pathology and pathogenesis. *Curr. Neurovasc. Res.* **2006**, *2*, 159–168. [CrossRef] [PubMed]

80. Chin, D.; Gan, N.Y.; Holder, G.E.; Tien, M.; Agrawal, R.; Manghani, M. Severe retinal vasculitis in systemic lupus erythematosus leading to vision threatening paracentral acute middle maculopathy. *Mod. Rheumatol. Case Rep.* **2021**, *5*, 265–271. [CrossRef]

81. Butendieck, R.R.; Parikh, K.; Stewart, M.; Davidge-Pitts, C.; Abril, A. Systemic lupus erythematosus-associated retinal vasculitis. *J. Rheumatol.* **2012**, *39*, 1095–1096. [CrossRef]

82. Androudi, S.; Dastiridou, A.; Symeonidis, C.; Kump, L.; Praidou, A.; Brazitikos, P.; Kurup, S.K. Retinal vasculitis in rheumatic diseases: An unseen burden. *Clin. Rheumatol.* **2013**, *32*, 7–13. [CrossRef]

83. Caracciolo, E.A.; Marcu, C.B.; Ghantous, A.; Donohue, T.J.; Hutchinson, G. Coronary vasculitis with acute myocardial infarction in a young woman with systemic lupus erythematosus. *J. Clin. Rheumatol.* **2004**, *10*, 66–68. [CrossRef]

84. Sokalski, D.G.; Copsey Spring, T.R.; Roberts, W.N. Large artery inflammation in systemic lupus erythematosus. *Lupus* **2013**, *22*, 953–956. [CrossRef]

85. Nasser, M.; Cottin, V. Alveolar Hemorrhage in Vasculitis (Primary and Secondary). *Semin. Respir. Crit. Care Med.* **2018**, *39*, 482–493. [CrossRef]

86. Wang, C.R.; Liu, M.F.; Weng, C.T.; Lin, W.C.; Li, W.T.; Tsai, H.W. Systemic lupus erythematosus-associated diffuse alveolar haemorrhage: A single-centre experience in Han Chinese patients. *Scand. J. Rheumatol.* **2018**, *47*, 392–399. [CrossRef]

87. Schwab, E.P.; Schumacher, H.R., Jr.; Freundlich, B.; Callegari, P.E. Pulmonary alveolar hemorrhage in systemic lupus erythematosus. *Semin. Arthritis Rheum.* **1993**, *23*, 8–15. [CrossRef]

88. Rocca, P.V.; Siegel, L.B.; Cupps, T.R. The concomitant expression of vasculitis and coagulopathy: Synergy for marked tissue ischemia. *J. Rheumatol.* **1994**, *3*, 556–560.

89. Alarcón-Segovia, D.; Pérez-Vázquez, M.E.; Villa, A.R.; Drenkard, C.; Cabiedes, J. Preliminary classification criteria for the antiphospholipid syndrome within systemic lupus erythematosus. *Semin. Arthritis Rheum.* **1992**, *5*, 275–286. [CrossRef]

90. Cervera, R.; Piette, J.; Font, J.; Khamashta, M.A.; Shoenfeld, Y.; Camps, M.T.; Jacobsen, S.; Lakos, G.; Tincani, A.; Ingelmo, M.; et al. Antiphospholipid syndrome: Clinical and immunologic manifestations and patterns of disease expression in a cohort of 1000 patients. *Arthritis Rheum.* **2002**, *4*, 1019–1027. [CrossRef] [PubMed]

91. Nguyen, V.A.; Gotwald, T.; Prior, C.; Oberrnoser, G.; Sepp, N. Acute pulmonary edema, capillaritis and alveolar hemorrhage: Pulmonary manifestations coexistent in antiphospholipid syndrome and systemic lupus erythematosus? *Lupus* **2005**, *14*, 557–560. [CrossRef] [PubMed]

92. Kazzaz, N.M.; Coit, P.; Lewis, E.E.; McCune, W.J.; Sawalha, A.H.; Knight, J.S. Systemic lupus erythematosus complicated by diffuse alveolar haemorrhage: Risk factors, therapy and survival. *Lupus Sci. Med.* **2015**, *2*, e000117. [CrossRef] [PubMed]

93. Andrade, C.; Mendonça, T.; Farinha, F.; Correia, J.; Marinho, A.; Almeida, I.; Vasconcelos, C. Alveolar hemorrhage in systemic lupus erythematosus: A cohort review. *Lupus* **2016**, *25*, 75–80. [CrossRef]

94. Moss, K.E.; Isenberg, D.A. Comparison of renal disease severity and outcome in patients with primary antiphospholipid syndrome, antiphospholipid syndrome secondary to systemic lupus erythematosus (SLE) and SLE alone. *Rheumatology (Oxford)* **2001**, *40*, 863–867. [CrossRef]

95. Barile-Fabris, L.; Ariza-Andraca, R.; Olguin-Ortega, L.; Jara, L.J.; Fraga-Mouret, A.; Miranda-Limon, J.M.; Fuentes de la Mata, J.; Clark, P.; Vargas, F.; Alocer-Varela, J. Controlled clinical trial of IV cyclophosphamide versus IV methylprednisolone in severe neurological manifestations in systemic lupus erythematosus. *Ann. Rheum. Dis.* **2005**, *64*, 620–625. [CrossRef] [PubMed]

96. Barile, L.A.; Jara, L.J.; Medina-Rodriguez, F.; Garcia-Figueroa, J.L.; Miranda-Limon, J.M. Pulmonary hemorrhage in systemic lupus erythematosus. *Lupus* **1997**, *6*, 445–448. [CrossRef]

97. Micheletti, R.G.; Pagnoux, C. Management of cutaneous vasculitis. *Presse Med.* **2020**, *49*, 104033. [CrossRef] [PubMed]

98. Lopez, L.R.; Davis, K.C.; Kohler, P.F.; Schocket, A.L. The hypocomplementemic urticarial-vasculitis syndrome: Therapeutic response to hydroxychloroquine. *J. Allergy Clin. Immunol.* **1984**, *73*, 600–603. [CrossRef]

99. Callen, J.P. Colchicine is effective in controlling chronic cutaneous leukocytoclastic vasculitis. *J. Am. Acad. Dermatol.* **1985**, *13*, 193–200. [CrossRef]

100. Callen, J.P.; af Ekenstam, E. Cutaneous leukocytoclastic vasculitis: Clinical experience in 44 patients. *South Med. J.* **1987**, *80*, 848–851. [CrossRef] [PubMed]

101. Sais, G.; Vidaller, A.; Jucgla, A.; Gallardo, F.; Peyri, J. Colchicine in the treatment of cutaneous leukocytoclastic vasculitis. Results of a prospective, randomized controlled trial. *Arch. Dermatol.* **1995**, *131*, 1399–1402. [CrossRef] [PubMed]

102. Wang, D.; Chen, H.; Wang, S.; Zou, Y.; Li, J.; Pan, J.; Wang, X.; Ren, T.; Zhang, Y.; Chen, Z.; et al. Thalidomide treatment in cutaneous lesions of systemic lupus erythematosus: A multicenter study in China. *Clin. Rheumatol.* **2016**, *35*, 1521–1527. [CrossRef]

103. Callen, J.P.; Spencer, L.V.; Burruss, J.B.; Holtman, J. Azathioprine. An effective, corticosteroid-sparing therapy for patients with recalcitrant cutaneous lupus erythematosus or with recalcitrant cutaneous leukocytoclastic vasculitis. *Arch. Dermatol.* **1991**, *127*, 515–522. [CrossRef] [PubMed]

104. Goodfield, M.; Davison, K.; Bowden, K. Intravenous immunoglobulin (IVIg) for therapy-resistant cutaneous lupus erythematosus (LE). *J. Dermatolog. Treat* **2004**, *15*, 46–50. [CrossRef] [PubMed]

105. Stumpf, M.A.M.; Quintino, C.R.; Rodrigues, M.; de Campos, F.P.F.; Maruta, C.W. Cutaneous vasculitis in lupus treated with IV immunoglobulin. *Clin. Rheumatol.* **2021**, *40*, 3023–3024. [CrossRef]

106. Chung, L.; Funke, A.A.; Chakravarty, E.F.; Callen, J.P.; Fiorentino, D.F. Successful use of rituximab for cutaneous vasculitis. *Arch. Dermatol.* **2006**, *142*, 1407–1410. [CrossRef] [PubMed]

107. Fotis, L.; Baszis, K.W.; French, A.R.; Cooper, M.A.; White, A.J. Mesenteric vasculitis in children with systemic lupus erythematosus. *Clin. Rheumatol.* **2016**, *35*, 785–793. [CrossRef]

108. Liu, Y.; Zhu, J.; Lai, J.M.; Sun, X.F.; Hou, J.; Zhou, Z.X.; Yuan, X.Y. Reports of three cases with the initial presentation of mesenteric vasculitis in children with system lupus erythematous. *Clin. Rheumatol.* **2018**, *37*, 277–283. [CrossRef]

109. Medina, F.; Ayala, A.; Jara, L.J.; Becerra, M.; Miranda, J.M.; Fraga, A. Acute abdomen in systemic lupus erythematosus: The importance of early laparotomy. *Am. J. Med.* **1997**, *103*, 100–105. [CrossRef]

110. Narvaez, J.; Rios-Rodriguez, V.; de la Fuente, D.; Estrada, P.; Lopez-Vives, L.; Gomez-Vaquero, C.; Nolla, J.M. Rituximab therapy in refractory neuropsychiatric lupus: Current clinical evidence. *Semin. Arthritis Rheum.* **2011**, *41*, 364–372. [CrossRef]

111. Jiang, Y.P.; Zhao, X.X.; Chen, R.R.; Xu, Z.H.; Wen, C.P.; Yu, J. Comparative efficacy and safety of mycophenolate mofetil and cyclophosphamide in the induction treatment of lupus nephritis: A systematic review and meta-analysis. *Medicine (Baltimore)* **2020**, *99*, e22328. [CrossRef]

112. Tunnicliffe, D.J.; Palmer, S.C.; Henderson, L.; Masson, P.; Craig, J.C.; Tong, A.; Singh-Grewal, D.; Flanc, R.S.; Roberts, M.A.; Webster, A.C.; et al. Immunosuppressive treatment for proliferative lupus nephritis. *Cochrane Database Syst. Rev.* **2018**, *6*, CD002922. [CrossRef]

113. Li, Q.Y.; Yu, F.; Zhou, F.D.; Zhao, M.H. Plasmapheresis Is Associated With Better Renal Outcomes in Lupus Nephritis Patients With Thrombotic Microangiopathy: A Case Series Study. *Medicine (Baltimore)* **2016**, *95*, e3595. [CrossRef]

114. Stolyar, L.; Lahita, R.G.; Panush, R.S. Rituximab use as induction therapy for lupus nephritis: A systematic review. *Lupus* **2020**, *29*, 892–912. [CrossRef]

115. De Holanda, M.I.; Porto, L.C.; Wagner, T.; Christiani, L.F.; Palma, L.M.P. Use of eculizumab in a systemic lupus erythemathosus patient presenting thrombotic microangiopathy and heterozygous deletion in CFHR1-CFHR3. A case report and systematic review. *Clin. Rheumatol.* **2017**, *36*, 2859–2867. [CrossRef] [PubMed]

116. Aguirre-Valencia, D.; Naranjo-Escobar, J.; Posso-Osorio, I.; Macia-Mejia, M.C.; Nieto-Aristizabal, I.; Barrera, T.; Obando, M.A.; Tobon, G.J. Therapeutic Plasma Exchange as Management of Complicated Systemic Lupus Erythematosus and Other Autoimmune Diseases. *Autoimmune Dis.* **2019**, *2019*, 5350960. [CrossRef] [PubMed]

117. Alabed, I.B. Treatment of diffuse alveolar hemorrhage in systemic lupus erythematosus patient with local pulmonary administration of factor VIIa (rFVIIa): A case report. *Medicine (Baltimore)* **2014**, *93*, e72. [CrossRef]

118. Pais, F.; Fayed, M.; Evans, T. The Successful Use of Extracorporeal Membrane Oxygenation in Systemic Lupus Erythematosus-Induced Diffuse Alveolar Haemorrhage. *Eur. J. Case Rep. Intern. Med.* **2017**, *4*, 000515. [CrossRef]

119. Shi, D.; Wang, D.; Li, X.; Zhang, H.; Che, N.; Lu, Z.; Sun, L. Allogeneic transplantation of umbilical cord-derived mesenchymal stem cells for diffuse alveolar hemorrhage in systemic lupus erythematosus. *Clin. Rheumatol.* **2012**, *31*, 841–846. [CrossRef]

120. Ali Dhirani, N.; Ahluwalia, V.; Somani, S. Case of combination therapy to treat lupus retinal vasculitis refractory to steroids. *Can. J. Ophthalmol.* **2017**, *52*, e13–e15. [CrossRef] [PubMed]

121. Damato, E.; Chilov, M.; Lee, R.; Singh, A.; Harper, S.; Dick, A. Plasma exchange and rituximab in the management of acute occlusive retinal vasculopathy secondary to systemic lupus erythematosus. *Ocul. Immunol. Inflamm.* **2011**, *19*, 379–381. [CrossRef]

122. Hickman, R.A.; Denniston, A.K.; Yee, C.S.; Toescu, V.; Murray, P.I.; Gordon, C. Bilateral retinal vasculitis in a patient with systemic lupus erythematosus and its remission with rituximab therapy. *Lupus* **2010**, *19*, 327–329. [CrossRef] [PubMed]

123. Papadaki, T.G.; Zacharopoulos, I.P.; Papaliodis, G.; Iaccheri, B.; Fiore, T.; Foster, C.S. Plasmapheresis for lupus retinal vasculitis. *Arch. Ophthalmol.* **2006**, *124*, 1654–1656. [CrossRef]

124. Donnithorne, K.J.; Read, R.W.; Lowe, R.; Weiser, P.; Cron, R.Q.; Beukelman, T. Retinal vasculitis in two pediatric patients with systemic lupus erythematosus: A case report. *Pediatr. Rheumatol. Online J.* **2013**, *11*, 25. [CrossRef]

125. Shriki, J.; Shinbane, J.S.; Azadi, N.; Su, T.I.; Hirschbein, J.; Quismorio, F.P., Jr.; Bhargava, P. Systemic lupus erythematosus coronary vasculitis demonstrated on cardiac computed tomography. *Curr. Probl. Diagn. Radiol.* **2014**, *43*, 294–297. [CrossRef] [PubMed]

126. Nandkeolyar, S.; Kim, H.B.; Doctorian, T.; Stoletniy, L.N.; Sandhu, V.K.; Yu, M.; Zuppan, C.W.; Razzouk, A.; Hilliard, A.; Parwani, P. A case report of heart transplant for ischaemic cardiomyopathy from lupus coronary vasculitis. *Eur. Heart J. Case Rep.* **2019**, *3*, 1–7. [CrossRef]

127. Fanouriakis, A.; Kostopoulou, M.; Alunno, A.; Aringer, M.; Bajema, I.; Boletis, J.N.; Cervera, R.; Doria, A.; Gordon, C.; Govoni, M.; et al. 2019 update of the EULAR recommendations for the management of systemic lupus erythematosus. *Ann. Rheum. Dis.* **2019**, *78*, 736–745. [CrossRef] [PubMed]

128. Uthman, I. Pharmacological therapy of vasculitis: An update. *Curr. Opin. Pharmacol.* **2004**, *4*, 177–182. [CrossRef]

129. Yang, L.; Zeng, Y.P. Cutaneous Vasculitis in Systemic Lupus Erythematosus. *JAMA Dermatol.* **2021**, *157*, 991. [CrossRef] [PubMed]

130. Guttman-Yassky, E.; Brunner, P.M.; Neumann, A.U.; Khattri, S.; Pavel, A.B.; Malik, K.; Singer, G.K.; Baum, D.; Gilleaudeau, P.; Sullivan-Whalen, M.; et al. Efficacy and safety of fezakinumab (an IL-22 monoclonal antibody) in adults with moderate-to-severe atopic dermatitis inadequately controlled by conventional treatments: A randomized, double-blind, phase 2a trial. *J. Am. Acad. Dermatol.* **2018**, *78*, 872–881 e876. [CrossRef] [PubMed]
131. Papp, K.A.; Menter, M.A.; Raman, M.; Disch, D.; Schlichting, D.E.; Gaich, C.; Macias, W.; Zhang, X.; Janes, J.M. A randomized phase 2b trial of baricitinib, an oral Janus kinase (JAK) 1/JAK2 inhibitor, in patients with moderate-to-severe psoriasis. *Br. J. Dermatol.* **2016**, *174*, 1266–1276. [CrossRef]
132. MacDonald, K.P.A.; Betts, B.C.; Couriel, D. Reprint of: Emerging Therapeutics for the Control of Chronic Graft-versus-Host Disease. *Biol. Blood Marrow Transpl.* **2018**, *24*, S7–S14. [CrossRef]
133. Smith, E.M.D.; Lythgoe, H.; Hedrich, C.M. Vasculitis in Juvenile-Onset Systemic Lupus Erythematosus. *Front. Pediatr.* **2019**, *7*, 149. [CrossRef] [PubMed]
134. Fontana, F.; Alfano, G.; Cappelli, G. The treatment of lupus nephritis, between consolidated strategies and new therapeutic options: A narrative review. *G Ital. Nefrol.* **2021**, *38*.
135. El-Husseini, A.; Hannan, S.; Awad, A.; Jennings, S.; Cornea, V.; Sawaya, B.P. Thrombotic microangiopathy in systemic lupus erythematosus: Efficacy of eculizumab. *Am. J. Kidney Dis.* **2015**, *65*, 127–130. [CrossRef]
136. Dooley, M.A.; Houssiau, F.; Aranow, C.; D'Cruz, D.P.; Askanase, A.; Roth, D.A.; Zhong, Z.J.; Cooper, S.; Freimuth, W.W.; Ginzler, E.M.; et al. Effect of belimumab treatment on renal outcomes: Results from the phase 3 belimumab clinical trials in patients with SLE. *Lupus* **2013**, *22*, 63–72. [CrossRef]
137. Lei, Y.; Loutan, J.; Anders, H.J. B-cell depletion or belimumab or voclosporin for lupus nephritis? *Curr. Opin. Nephrol. Hypertens.* **2021**, *30*, 237–244. [CrossRef]
138. Rovin, B.H.; Teng, Y.K.O.; Ginzler, E.M.; Arriens, C.; Caster, D.J.; Romero-Diaz, J.; Gibson, K.; Kaplan, J.; Lisk, L.; Navarra, S.; et al. Efficacy and safety of voclosporin versus placebo for lupus nephritis (AURORA 1): A double-blind, randomised, multicentre, placebo-controlled, phase 3 trial. *Lancet* **2021**, *397*, 2070–2080. [CrossRef]
139. Al-Adhoubi, N.K.; Bystrom, J. Systemic lupus erythematosus and diffuse alveolar hemorrhage, etiology and novel treatment strategies. *Lupus* **2020**, *29*, 355–363. [CrossRef]
140. Kim, D.; Choi, J.; Cho, S.K.; Choi, C.B.; Kim, T.H.; Jun, J.B.; Yoo, D.H.; Bae, S.C.; Sung, Y.K. Clinical characteristics and outcomes of diffuse alveolar hemorrhage in patients with systemic lupus erythematosus. *Semin. Arthritis Rheum.* **2017**, *46*, 782–787. [CrossRef] [PubMed]
141. Luqmani, R.A. State of the art in the treatment of systemic vasculitides. *Front. Immunol.* **2014**, *5*, 471. [CrossRef]

Article

Perioperative Vascular Biomarker Profiling in Elective Surgery Patients Developing Postoperative Delirium: A Prospective Cohort Study

Jan Menzenbach [1], Stilla Frede [1], Janine Petras [1], Vera Guttenthaler [1], Andrea Kirfel [1], Claudia Neumann [1], Andreas Mayr [2], Maria Wittmann [1], Mark Coburn [1], Sven Klaschik [1,†] and Tobias Hilbert [1,*,†]

[1] Department of Anesthesiology and Intensive Care Medicine, University Hospital Bonn, Venusberg-Campus 1, 53127 Bonn, Germany; jan.menzenbach@ukbonn.de (J.M.); stilla.frede@ukbonn.de (S.F.); s4japetr@uni-bonn.de (J.P.); vera.guttenthaler@ukbonn.de (V.G.); andrea.kirfel@ukbonn.de (A.K.); claudia.neumann@ukbonn.de (C.N.); maria.wittmann@ukbonn.de (M.W.); mark.coburn@ukbonn.de (M.C.); sven.klaschik@ukbonn.de (S.K.)
[2] Institute of Medical Biometrics, Informatics and Epidemiology (IMBIE), University Hospital Bonn, Venusberg-Campus 1, 53127 Bonn, Germany; amayr@uni-bonn.de
* Correspondence: thilbert@uni-bonn.de; Tel.: +49-228-287-14114; Fax: +49-228-287-14115
† These authors contributed equally to this work.

Citation: Menzenbach, J.; Frede, S.; Petras, J.; Guttenthaler, V.; Kirfel, A.; Neumann, C.; Mayr, A.; Wittmann, M.; Coburn, M.; Klaschik, S.; et al. Perioperative Vascular Biomarker Profiling in Elective Surgery Patients Developing Postoperative Delirium: A Prospective Cohort Study. *Biomedicines* 2021, 9, 553. https://doi.org/10.3390/biomedicines9050553

Academic Editor: Byeong Hwa Jeon

Received: 18 April 2021
Accepted: 13 May 2021
Published: 15 May 2021

Publisher's Note: MDPI stays neutral with regard to jurisdictional claims in published maps and institutional affiliations.

Abstract: Background: Postoperative delirium (POD) ranks among the most common complications in surgical patients. Blood-based biomarkers might help identify the patient at risk. This study aimed to assess how serum biomarkers with specificity for vascular and endothelial function and for inflammation are altered, prior to or following surgery in patients who subsequently develop POD. Methods: This was a study on a subcohort of consecutively recruited elective non-cardiac as well as cardiac surgery patients (age > 60 years) of the single-center PROPDESC trial at a German tertiary care hospital. Serum was sampled prior to and following surgery, and the samples were subjected to bead-based multiplex analysis of 17 serum proteins (IL-3, IL-8, IL-10, Cripto, CCL2, RAGE, Resistin, ANGPT2, TIE2, Thrombomodulin, Syndecan-1, E-Selectin, VCAM-1, ICAM-1, CXCL5, NSE, and uPAR). Development of POD was assessed during the first five days after surgery, using the Confusion Assessment Method for ICU (CAM-ICU), the CAM, the 4-'A's test (4AT), and the Delirium Observation Scale (DOS). Patients were considered positive if POD was detected at least once during the visitation period by any of the applied methods. Non-parametric testing, as well as propensity score matching were used for statistical analysis. Results: A total of 118 patients were included in the final analysis; 69% underwent non-cardiac surgery, median overall patient age was 71 years, and 59% of patients were male. In the whole cohort, incidence of POD was 28%. The male gender was significantly associated with the development of POD ($p = 0.0004$), as well as a higher ASA status III ($p = 0.04$). Incidence of POD was furthermore significantly increased in cardiac surgery patients ($p = 0.002$). Surgery induced highly significant changes in serum levels of almost all biomarkers except uPAR. In preoperative serum samples, none of the analyzed parameters was significantly altered in subsequent POD patients. In postoperative samples, CCL2 was significantly increased by a factor of 1.75 in POD patients ($p = 0.03$), as compared to the no-POD cohort. Following propensity score matching, CCL2 remained the only biomarker that showed significant differences in postoperative values ($p = 0.01$). In cardiac surgery patients, postoperative CCL2 serum levels were more than 3.5 times higher than those following non-cardiac surgery ($p < 0.0001$). Moreover, after cardiac surgery, Syndecan-1 serum levels were significantly increased in POD patients, as compared to no-POD cardiac surgery patients ($p = 0.04$). Conclusions: In a mixed cohort of elective non-cardiac as well as cardiac surgery patients, preoperative serum biomarker profiling with specificity for vascular dysfunction and for systemic inflammation was not indicative of subsequent POD development. Surgery-induced systemic inflammation—as evidenced by the significant increase in CCL2 release—was associated with POD, particularly following cardiac surgery. In those patients, postoperative glycocalyx injury might furthermore contribute to POD development.

Keywords: postoperative delirium; biomarker; vascular inflammation; elective surgery; cardiac surgery; glycocalyx; CCL2; MCP-1

1. Introduction

Postoperative delirium (POD) ranks among the most common perioperative complications in surgical patients. Since the prevalence of POD is known to increase with patient age, and the average age of the general surgical patient population likewise increases, so does the incidence of POD. Comorbidities and the type of surgery do have a further impact, resulting in a statistical risk for POD ranging from 2.5% in the general surgical population up to 70% in patients requiring emergency femur fracture repair [1]. POD is characterized as cognitive impairment with an acute and fluctuating disturbance in awareness and attention, appearing in either hyper- or hypoactive manifestation or a combination of both [2]. Commonly occurring within the first five postoperative days, POD was shown to increase the length of stay (LOS) in the intensive care unit (ICU) and in hospital [3]. It is also associated with an up to 10 times increased 30-day mortality, as compared to individuals without POD [4]. Furthermore, the risk for long-term care dependency is likewise increased by up to three times [5]. Not least for these reasons, POD is associated with significantly increased health care costs and thus represents a substantial economic factor [3].

As strategies exist to reduce the risk for POD, especially in the vulnerable population, there is a particular need to identify those patients who are prone to develop POD. To this end, scores were developed and validated to determine the risk by including clinical parameters such as age, comorbidities, preoperative mental status, or details on surgery [6]. Blood-based biomarkers might provide additional information. It was shown that stress-induced mediators such as cortisol, C-reactive protein (CRP), or IL-6 might be associated with subsequent POD development [7]. The significant increase in the prevalence of delirium in septic patients provides further evidence for the role of systemic inflammation in delirogenesis [8].

Systemic inflammation and sepsis are furthermore associated with altered microvascular function, resulting in blood-brain barrier disruption, for example [9,10]. Meanwhile, biomarkers with specificity for vascular and endothelial injury, such as soluble adhesion molecules (Selectins and CAMs (Cellular Adhesion Molecules)), Angiopoietins, and soluble TIE2 receptor, or glycocalyx components (Syndecan-1) play a substantial role in estimating the prognosis of septic patients [11]. Moreover, clinical risk factors for vascular damage, including nicotine abuse, diabetes, or atherosclerosis as well as intraoperative vascular disturbances were shown to be associated with an increased risk for POD, underscoring the significance of impaired vascular and endothelial integrity for the pathogenesis of POD [12,13]. However, whether biomarkers with specificity for vascular function might be altered prior to or following surgery in patients that subsequently develop POD is not yet systemically investigated. Therefore, we assessed a panel of 17 biomarkers characteristic for vascular activation and permeability and for systemic inflammation in serum samples from elective adult patients across all relevant types of surgery. Analyses were performed on a subcohort of the single-center PROPDESC trial [14].

2. Materials and Methods

2.1. Study Design and Patient Population

The PRe-Operative Prediction of postoperative DElirium by appropriate SCreening (PROPDESC) trial is an investigator-initiated prospective monocentric observational study at the Department of Anesthesiology and Intensive Care Medicine, University Hospital Bonn. Details of the study design were previously described [14]. The trial was conducted in accordance with the Declaration of Helsinki and after approval by the institutional review board (IRB) of the University of Bonn (protocol number 255/17, date of approval

18 September 2017; Chairman—Professor K. Racké). The study was registered in the German Clinical Trials Register (protocol number DRKS00015715). Analyses of the herein presented study were performed on a subcohort of participants consecutively recruited between July and September 2019. Inpatients from various surgical specialties, including general, orthopedic, and trauma, cardiac, thoracic, vascular, ear–nose–throat, urologic, and plastic surgery were prospectively screened for eligibility to the study. Inclusion criteria were—elective surgery scheduled for a duration of at least 60 minutes, patient age >60 years, willingness and ability to provide written informed consent. Exclusion criteria were—emergency procedures, substantial language barriers, and a lack of patients' compliance with the study protocol, which was determined by the respective physician. A checklist according to the STROBE statement can be found in Supplementary Materials.

Recorded preoperative baseline characteristics included patient age, sex, body mass index (BMI), American Society of Anesthesiologists (ASA) Physical Status Classification System, surgical risk, surgical specialty, and preoperative routine laboratory values (hemoglobin (Hgb), HbA_{1c}, leukocyte count, sodium, potassium, creatinine, CRP, total protein, high-sensitive cardiac troponin T (hs-TnT), and NT-proBNP). For surgical risk classification, the 5-level Johns Hopkins classification of intervention risk was transformed into a 3-level modified classification (low, intermediate, high risk (adapted from Glance et al. [15]).

2.2. Serum Sampling and Biomarker Profiling

All patients recruited to PROPDESC during the above-mentioned period were subjected to an additional serum biomarker profiling and included in the study. Ten milliliter of blood were drawn prior to and following surgery. The coagulated samples were centrifuged (3.000 rpm, 4 °C, 10 min), and serum aliquots were stored at −80 °C for subsequent analysis. In serum samples, the following 17 proteins were assessed using custom-made Luminex™ multiplex arrays (RnD Systems, Minneapolis, MN, USA):

Interleukin-3 (IL-3), IL-8, IL-10, Cripto, CC-chemokine Ligand 2 (CCL2), soluble Receptor for Advanced Glycation Endproducts (RAGE), Resistin, Angiopoietin-2 (ANGPT2), soluble Tyrosine Kinase with Immunoglobulin-like and EGF-like domains 2 (TIE2), Thrombomodulin (THBD), Syndecan-1 (SDC1), E-Selectin, soluble Vascular Cell Adhesion Protein 1 (VCAM-1), soluble Intercellular Adhesion Molecule 1 (ICAM-1), C-X-C Motif Chemokine 5 (CXCL5), Neuron-specific Enolase (NSE), and Urokinase Plasminogen Activator Surface Receptor (uPAR).

All analyses were performed according to the manufacturer's protocol. Bead-based multiplex arrays such as the Luminex™ system are described in the work from Zhang et al. [16]. Arrays were analyzed using a MAGPIX™ reader (Luminex Corp., Austin, TX, USA). Results are given in pg/mL serum. Values below the assay's lower detection limit (undetected values) were set to half the lower detection limit when the number of undetected values did not exceed 15% of all values of the respective protein in the whole cohort. When number of undetected values exceeded 15% of all values of the respective protein in the whole pre- as well as postoperative cohort, these analytes were totally excluded from further analysis. All experimental tests were performed in duplicates. The mean value was calculated from these results and used for further statistical analysis. All personnel performing the serum analyses were blinded for intra- and postoperative patient data.

2.3. Assessment of POD

POD assessment and postoperative data recording were implemented through regular patient rounds in the ICU and peripheral wards. Primary endpoint was development of POD within the first five days following surgery. POD assessments were performed every morning by trained study personnel on each of the first five days after surgery or after ending of sedation (RASS (Richmond Agitation and Sedation Scale) level $\geq$ −3), respectively. The Confusion Assessment Method for ICU (CAM-ICU) was used for ICU patients [17], while the Confusion Assessment Method (CAM) [18] and the 4-'*A*'s test (4AT) [19] were used for the patients in the peripheral ward. In order not to miss any POD-positive patients due to spot examination, the Delirium Observation Scale (DOS) [20] was additionally used. A 4AT score of 4 points and a DOS score of 3 points onwards, respectively, was considered positive. Primary endpoint was considered to be achieved if POD was detected at least once during the 5-day visitation period by any of the applied assessment methods (CAM-ICU, CAM, 4AT, or DOS). If patients died during the 5-day visitation period, they were classified as delirious for subsequent analysis if they showed a positive test result prior to death. If not, they were excluded from later analysis.

Additionally recorded data included details on anesthesia and surgery (general or regional or combined anesthesia, duration of surgery, duration of mechanical ventilation, postoperative admission to ICU, and LOS in hospital).

2.4. Statistical and Bioinformatical Analysis

Structured patient data and results of POD assessment were entered into the electronic database "*REDCap*", which is administered by the Institute for Medical Biometrics, Informatics and Epidemiology (IMBIE) of the University Hospital Bonn. All data including results from biomarker assessment were then transferred into MS Excel 2019 (Microsoft Corp., Redmond, CA, USA). Statistical analysis and visualization were performed using GraphPad PRISM 8.4.3 (La Jolla, CA, USA) and the statistical computing environment R 3.5.1 (Vienna, Austria).

All data are presented as percentage values or as median values with 25th and 75th percentile, respectively. To assess differences between groups (non-paired samples), non-parametric Mann-Whitney U test was used; in case of comparisons of pre- and postoperative measurements (paired samples), Wilcoxon rank-sum tests were applied. Binary variables were compared using Fisher's exact test. p values < 0.05 (two-sided) were considered statistically significant.

For the analysis of serum samples, a nearest neighbor-based propensity score matching [21] was performed to generate a control group with similar baseline characteristics to the POD group. Variables considered for the logistic regression model to estimate the propensity score were surgical risk, CRP, leukocyte count, BMI, patient age, duration of surgery, and patient gender. For the comparison of the matched groups, the corresponding Wilcoxon rank-sum test for paired samples was applied.

The datasets generated and analyzed during the current study are available from the corresponding author on request.

3. Results

A total of 123 consecutive participants of the PROPDESC trial were prospectively recruited to this subcohort study. Five patients were excluded from later analysis as multiplex array analysis of postoperative serum samples failed, resulting in a total of 118 assessed patients with complete data, as well as serum sample sets. A total of 81 patients (69%) underwent non-cardiac surgery. All patients but one, who received conscious sedation, received general anesthesia. Overall median patient age was 71 years (66–78), and 70 patients (59%) were male. Three patients (2.5%) were classified ASA I, 35 patients (29.7%) ASA II, 70 patients (59.3%) ASA III, and 10 patients (8.5%) were classified as ASA IV. Risk of surgery was considered low in 11 cases (9.3%), intermediate in 47 cases (39.8%), while it was considered high in 60 cases (50.9%). Sixty-six patients (55%) were immediately admitted to ICU (including Intermediate Care Unit (IMC)), while the others were transferred to the peripheral ward from the recovery room. Details on further patient characteristics as well as on surgical procedures are given in Table 1.

Table 1. Patient and procedural details.

Parameter	Median (25th–75th Percentile)
Patient details:	
n	118
Age (years)	71 (66–78)
Male gender (n [%])	70 (59)
Body mass index (kg/m^2)	27.3 (24.4–30.2)
ASA status:	
I (n [%])	3 (2.5)
II (n [%])	35 (29.7)
III (n [%])	70 (59.3)
IV (n [%])	10 (8.5)
Preoperative routine laboratory values:	
Hemoglobin (g/dL)	13.4 (12.2–14.5)
HbA$_{1C}$ (%)	5.6 (5.4–6.2)
Leukocyte count (G/L)	6.9 (5.9–8.8)
Sodium (mmol/L)	140 (138–142)
Potassium (mmol/L)	4.4 (4.1–4.7)
Creatinine (mg/dL)	0.9 (0.78–1.05)
C-reactive protein (mg/L)	3.1 (1.1–8.8)
Total protein (g/L)	69 (65–73)
High-sensitive cardiac troponin T (ng/L)	12.9 (8.1–20.3)
NT-proBNP (pg/mL)	235 (114–796)
Procedural and anesthesia details:	
Surgical risk:	
Low (n [%])	11 (9.3)
Intermediate (n [%])	47 (39.8)
High (n [%])	60 (50.9)
Surgical specialty:	
General (n [%])	18 (15)
Orthopedic and trauma (n [%])	35 (30)
Cardiac (n [%])	37 (31)
Thoracic (n [%])	2 (2)
Vascular (n [%])	4 (3)
Ear-nose-throat (n [%])	11 (9)
Urologic (n [%])	9 (8)
Plastic (n [%])	2 (2)
Placement of epidural catheter (n [%])	6 (5)
Duration of surgery (min)	219 (136–298)
Duration of mechanical ventilation (h)	6.0 (3.6–14.0)
Postoperative admission to ICU (n [%])	66 (55)
Length of hospital stay (days)	13 (10–23)

Data are given as percentage values or as median values with 25th and 75th percentile, respectively. ASA = American Society of Anesthesiologists, NT-proBNP = N-terminal prohormone of brain natriuretic peptide, and ICU = Intensive Care Unit.

Biomarker profiling was performed in pre- and in postoperative serum samples. In total, 236 serum samples were analyzed. For IL-3, IL-10, and Cripto, the number of undetected values exceeded 15% of all values of the respective protein in the whole pre- as well as postoperative cohort, therefore, these proteins were totally excluded from further analysis, resulting in 14 serum proteins being analyzed. As shown in Figure 1 (and in Supplemental Table S1), according to the Wilcoxon rank-sum test, surgery induced highly significant changes in serum levels of almost all biomarkers for vascular activation, permeability, and inflammation, with the exception of uPAR. While there was a considerable interindividual spread in pre- as well as postoperative markers, some of them were decreased in postoperative samples (E-Selectin, ICAM-1, THBD, TIE2, CXCL5), while others were increased following surgery, as compared to preoperative levels. Percentage decrease of median levels was most pronounced in TIE2 (reduction to 70.3%), and percentage increase was most pronounced in IL-8 (increase to 203%).

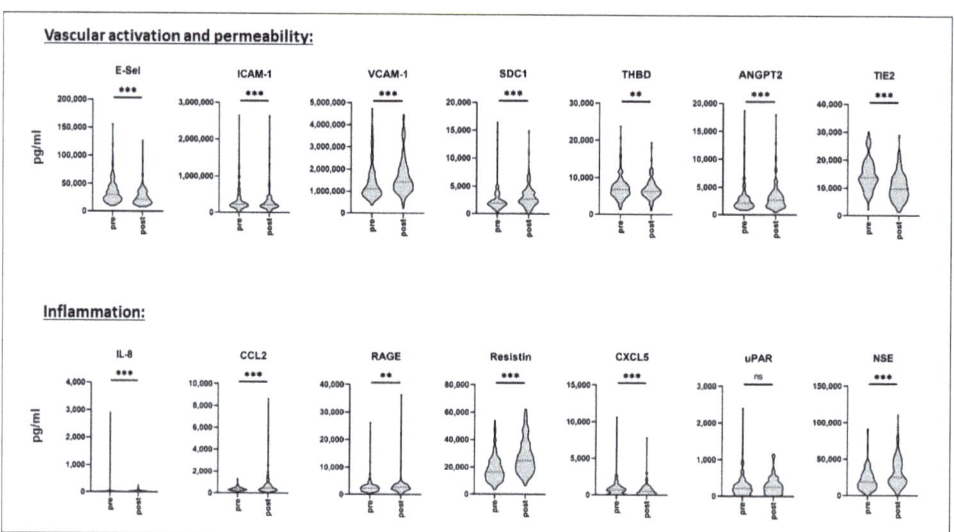

Figure 1. Pre- and postoperative serum biomarker profiling. Serum was sampled prior to and following surgery from a mixed cohort of consecutively recruited elective non-cardiac as well as cardiac surgery patients. Biomarker profiling was performed using the multiplex array technique. Data are given as median values (black dashed line) with 25th and 75th percentile (white dashed lines) and were compared using the Wilcoxon rank-sum test. $n = 118$. ns = not significant, ** $p < 0.01$, *** $p < 0.005$. E-Sel = E-Selectin, ICAM-1 = Intercellular Adhesion Molecule 1, VCAM-1 = Vascular Cell Adhesion Protein 1, SDC1 = Syndecan-1, THBD = Thrombomodulin, ANGPT2 = Angiopoietin-2, TIE2 = Tyrosine Kinase with Immunoglobulin-like and EGF-like domains 2, IL-8 = Interleukin-8, CCL2 = CC-chemokine Ligand 2, RAGE = Receptor for Advanced Glycation Endproducts, CXCL5 = C-X-C Motif Chemokine 5, uPAR = Urokinase Plasminogen Activator Surface Receptor, and NSE = Neuron.

Development of POD was assessed during the first five days after surgery or after ending of sedation, respectively. In the whole cohort, incidence of POD was 28% (33 patients). While 18 patients were considered POD-positive for one day, POD remained for two days in 9 patients and for three or more days in 6 patients. A total of 20 patients developed POD on postoperative day 1 or 2 (early-onset), while 13 developed POD on day 3 to 5 (late-onset). According to the Fisher's exact test, the male sex was significantly associated with the development of POD (28 male vs. 5 female patients, $p = 0.0004$), as was higher ASA status III ($p = 0.04$). In the POD subcohort, overall duration of mechanical ventilation was significantly increased (11.0 (4.6–26.1) vs. 5.1 (3.4–10.8) h), and immediate postoperative admission to ICU was more common ($p = 0.008$). In contrast, there was no difference in POD incidence between low to intermediate and high-risk surgery ($p = 0.1$) or concerning the preoperative routine laboratory parameters ($p = 0.07$). Duration of surgery was likewise not different between the POD and the no-POD group (249 (168–336) vs. 204 (132–278) min, $p = 0.13$). Details on patient characteristics as well as on surgical procedures in both subcohorts are given in Supplementary Table S2.

The results of the pre- and the postoperative multiplex biomarker assessment for the POD and the no-POD cohort, respectively, are shown in Figure 2 and in Supplemental Table S3. According to the Mann-Whitney test, in preoperative serum samples, none of the analyzed parameters was significantly altered in the POD patients, as compared to the no-POD cohort (Figure 2A). As for the whole cohort, there was a considerable interindividual spread. This also applied to the postoperative serum sample values. However, as shown in Figure 2B, values for CCL2 following surgery were significantly increased by a factor of 1.75 in POD patients ($p = 0.03$), as compared to the no-POD cohort.

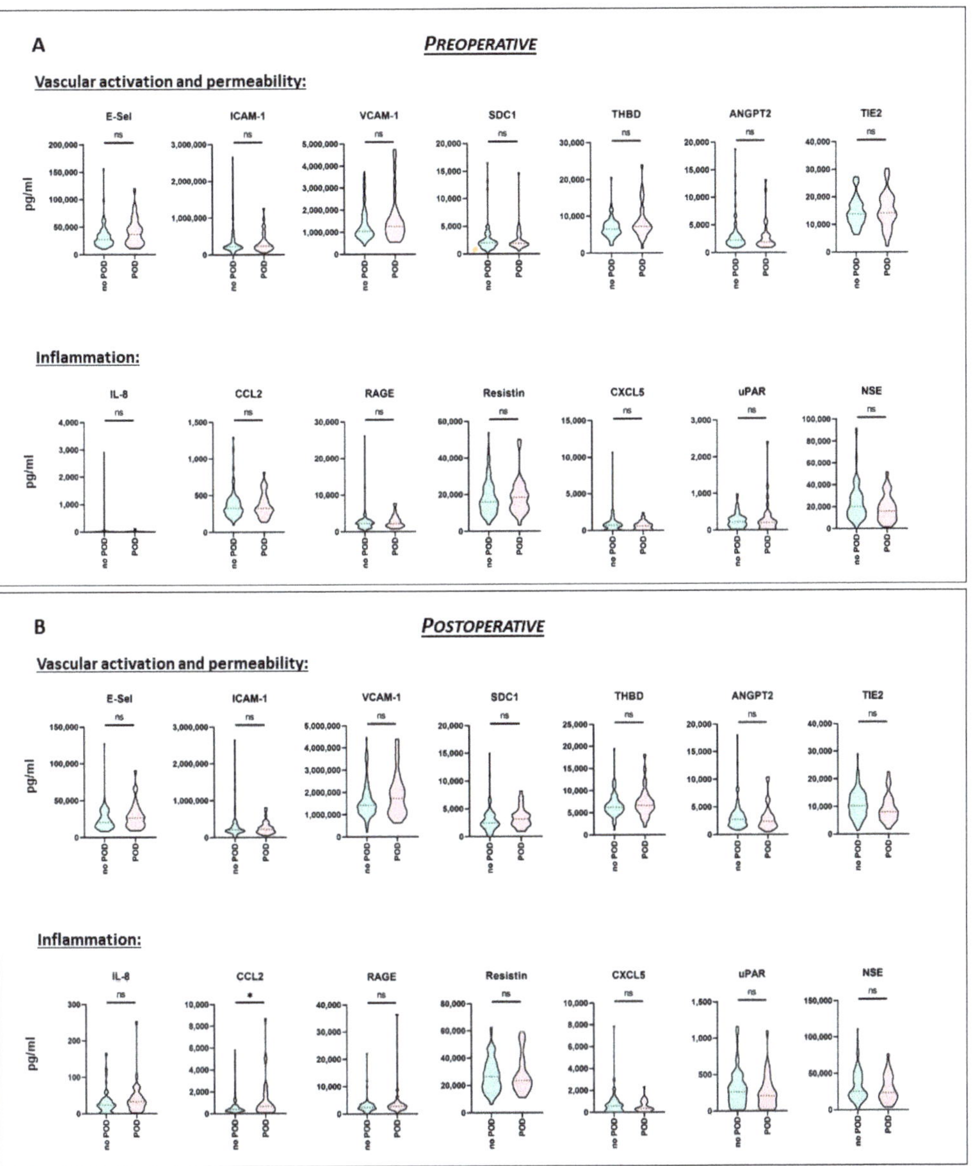

Figure 2. Pre- and postoperative serum biomarker profiling (no-POD and POD). Serum was sampled prior to (**A**) and following surgery (**B**) from a mixed cohort of consecutively recruited elective non-cardiac, as well as cardiac surgery patients. Biomarker profiling was performed using the multiplex array technique. The whole cohort was divided according to development of postoperative delirium (POD). Data are given as median values (black dashed line) with 25th and 75th percentile (white dashed lines) and were compared using the Mann-Whitney U test. no POD: $n = 85$, POD: $n = 33$. ns = not significant, * $p < 0.05$. E-Sel = E-Selectin, ICAM-1 = Intercellular Adhesion Molecule 1, VCAM-1 = Vascular Cell Adhesion Protein 1, SDC1 = Syndecan-1, THBD = Thrombomodulin, ANGPT2 = Angiopoietin-2, TIE2 = Tyrosine Kinase with Immunoglobulin-like and EGF-like domains 2, IL-8 = Interleukin-8, CCL2 = CC-chemokine Ligand 2, RAGE = Receptor for Advanced Glycation Endproducts, CXCL5 = C-X-C Motif Chemokine 5, uPAR = Urokinase Plasminogen Activator Surface Receptor, and NSE = Neuron-specific Enolase.

Since some of the analyzed (particularly postoperative) biomarkers tended to be different in the POD cohort but without actually reaching statistical significance when being compared to the control cohort, a propensity score matching was performed in order to correct for the potential confounders and to increase statistical power. However, CCL2 remained the only biomarker showing significant differences in postoperative absolute ($p = 0.01$) as well as fold-change values ($p = 0.03$), while the preoperative samples showed no difference between the two subcohorts (Figure 3). Complete results of the matched-pairs serum analysis are given in Supplementary Table S4.

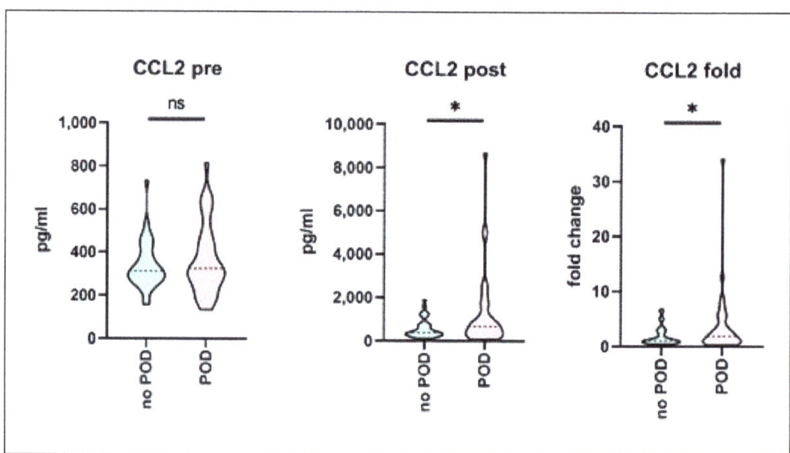

Figure 3. Pre- and postoperative serum biomarker profiling (no-POD and POD) following propensity score matching. Serum was sampled prior to and following surgery from a mixed cohort of consecutively recruited elective non-cardiac as well as cardiac surgery patients. Biomarker profiling was performed using the multiplex array technique. Development of postoperative delirium (POD) was assessed, and POD-positive patients were matched with respect to no-POD controls using propensity score matching. Figure shows results for the CCL2 (CC-chemokine Ligand 2) serum levels. Data are given as median values (black dashed line) with 25th and 75th percentile (white dashed lines) and were compared using Wilcoxon rank-sum test. no POD: $n = 33$, POD: $n = 33$. ns = not significant, * $p <0.05$.

According to Fisher's exact test, the incidence of POD was significantly increased in cardiac surgery patients, as compared to non-cardiac surgery (18 out of 37 (49%) vs. 15 out of 81 patients (19%), $p = 0.002$; Figure 4A). In cardiac surgery patients, CCL2 was the only one among all analyzed biomarkers that showed a highly significant difference in postoperative absolute (1117 (670.8–1749) vs. 315.9 (240.1–481.6) pg/mL), as well as fold-change increase (3.21 (2.51–6.17) vs. 0.93 (0.71–1.29)), irrespective of POD development, as compared to the non-cardiac surgery patients ($p < 0.0001$) (Figure 4B). Postoperative CCL2 serum levels were more than 3.5 times higher following cardiac surgery than following non-cardiac surgery. When the subcohort of cardiac surgery patients was furthermore divided according to subsequent POD development, only postoperative SDC1 serum levels were significantly increased in POD patients, as compared to no-POD cardiac surgery patients (3985 (3005–5871) vs. 2951 (2451–3959) pg/mL, $p = 0.04$) (Figure 4C).

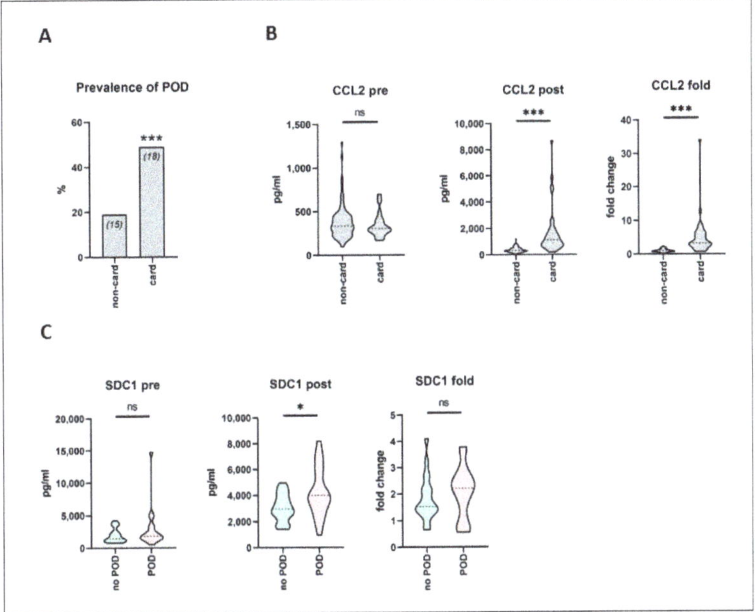

Figure 4. Pre- and postoperative serum biomarker profiling in cardiac and non-cardiac surgery patients. Serum was sampled prior to and following surgery from a mixed cohort of consecutively recruited elective non-cardiac as well as cardiac surgery patients. Biomarker profiling was performed using multiplex array technique. Development of postoperative delirium (POD) was assessed. (**A**) Prevalence of POD was significantly increased in cardiac surgery patients, as compared to non-cardiac surgery (18 out of 37 (49%) vs. 15 out of 81 patients (19%)). Fisher's exact test. *** $p < 0.005$. (**B**) In cardiac surgery patients, CCL2 (CC-chemokine Ligand 2) showed a significant difference in postoperative absolute as well as fold-change increase, irrespective of POD development, as compared to the non-cardiac surgery patients. non-cardiac: $n = 81$, cardiac: $n = 37$. (**C**) In cardiac surgery patients, postoperative SDC1 (Syndecan-1) serum levels were significantly increased in patients that subsequently developed POD, as compared to the no-POD cardiac surgery patients. no POD: n = 19, POD: $n = 18$. (**B,C**) Data are given as median values (black dashed line) with 25th and 75th percentile (white dashed lines) and were compared using the Mann-Whitney U test. ns = not significant, * $p < 0.05$, *** $p < 0.005$.

4. Discussion

Vascular injury as well as clinically evident systemic inflammation were both associated with increased risk of POD [7,8,12]. Our herein presented results of the analysis of 17 serum biomarkers with specificity for vascular activation and permeability and for inflammation, revealed that almost all markers were significantly altered following surgery, as compared to the preoperative values. However, with the exception of CCL2, none of the other biomarkers analyzed was significantly different in pre- or in postoperative samples in patients that later developed POD, as compared to the no-POD subcohort. A paired analysis following propensity score matching confirmed the results. Incidence of POD was furthermore significantly increased following cardiac surgery, as compared to non-cardiac surgery patients. Among all biomarkers, CCL2 was the only one that showed a highly significant difference in postoperative increase, following cardiac surgery, as compared to the non-cardiac surgery patients, suggesting a critical role of proinflammatory activation in the pathogenesis of POD in this patient population. Furthermore, postoperative SDC1 serum levels were significantly increased in cardiac surgery patients who developed POD, as compared to those without subsequent POD.

POD is supposed to be mediated by neuroinflammation and might therefore be designated as an inflammatory "state of mind" [22]. Although still elusive, the underlying factors in the pathogenesis of POD appear to be numerous. Patient age and gender, preexisting cognitive decline, comorbidities including (among numerous others) alcohol abuse, diabetes, and hypertension, and emergency surgery were identified as predisposing factors that might not or at least might not be significantly influenced [1]. In contrast, precipitating factors such as perioperative pharmacological and anesthesiological management, extent and invasiveness of surgical measures, and preservation of fluid and temperature homeostasis can be actively controlled, thereby offering the chance to reduce the risk for POD development. The rationale of any preoperative screening instrument, whether score- or biomarker-based, is to identify those patients that might either benefit from further and thorough preoperative evaluation, from an adjustment in perioperative management, or from prolonged or more intense postoperative monitoring. Therefore, the first step in a successful reduction of the risk for POD is to identify the patient at risk.

In preoperative risk stratification, biomarkers are gaining more and more importance. Brain natriuretic peptide (BNP) or its N-terminal prohormone, for example, were shown to be validly associated with the incidence of myocardial injury, as well as with major adverse cardiac events after non-cardiac surgery [23,24]. Therefore, preoperative BNP assessment for perioperative cardiac risk stratification is recommended by recent national guidelines [25,26]. It was demonstrated that the predictive power of score-based screening tools for perioperative risk assessment was significantly improved when combined with biomarkers [27].

Myocardial injury, besides sepsis and bleeding, is still one of the most important determinants of postoperative mortality, therefore, the use of biomarkers with specificity for cardiac morbidity is most validated and established in preoperative risk assessment. However, the primary goal of perioperative medicine should not only be survival but rather a favorable treatment outcome in terms of physical, as well as mental performance, particularly avoiding cognitive decline to preserve a satisfying quality of life. The single-center PROPDESC trial was designed to identify score-based parameters that allow estimation of the patients' individual risk for POD development from preoperative routine data in a cohort of elective surgery patients aged >60 years [14]. The results of the herein presented analysis were obtained from a large subcohort of consecutively enrolled PROPDESC patients who underwent additional biomarker profiling to identify potential candidate markers that might help improve score-based screening [27]. This subcohort comprised patients from relevant types of non-cardiac surgery as well as from cardiac surgery, and covered low- to high-risk procedures, thereby forming a representative sample from the population of elderly surgical patients at a German tertiary care hospital. The 14 assessed biomarkers were grouped into either vascular-specific or markers representing systemic inflammation. Surgery resulted in significant changes of postoperative serum levels in almost all biomarkers, as compared to preoperative levels. Increase in proinflammatory mediators following surgery was well described previously, and was among others, associated with the invasiveness of procedures [7,28,29]. In contrast, markers for vascular activation showed a more heterogenous kinetic. In line with results from others, some of them like E-Selectin, ICAM-1, or TIE2 were decreased in postoperative serum samples [29,30], while others (Syndecan-1, ANGPT2) were increased, indicating surgery-induced vascular and endothelial activation or injury [30,31].

Development of POD in our study was assessed using three different methodologies, each validated for the detection of hypo- or hyperactive or combined POD [17–20]. In our whole study cohort, 28% of patients were rated POD-positive during the first five days following surgery. Incidence of POD in cardiac surgery patients was significantly increased, as compared to non-cardiac surgery. These results were in line with previously reported epidemiological data [3–5,28]. When the whole study cohort was divided according to POD diagnosis, neither vascular activation nor inflammatory biomarkers significantly differed between the two groups prior to surgery, suggesting no association with subsequent POD development. Although only few authors assessed preoperative blood-based biomarkers in POD patients, this is at least partly in line with the results of others. In a recent meta-analysis, Liu et al. could show that preoperative serum levels of IL-8 and NSE were not altered in patients developing POD [7]. The same was demonstrated by Vasunilashorn et al. for IL-8 [28].

Given the impact of surgery on postoperative levels of almost all biomarkers, we also expected to see an association with subsequent POD development. However, only the serum levels of CCL2 following surgery were significantly increased in POD patients, as compared to the no-POD controls. This was a stable result of our analyses even after controlling for potential confounders by performing a propensity score matching, which additionally revealed a significant intergroup-difference in CCL2 fold-change levels. CCL2, also referred to as MCP-1 (Monocyte Chemoattractant Protein 1), is a chemokine primarily secreted by monocytes and macrophages in states of inflammation, exhibiting chemotactic properties on monocytes and subsets of granulocytes. Postoperative CCL2 release is directly correlated with the severity of surgical stress [32]. Systemic inflammation was shown to be associated with the risk for delirium [8]. In line with this, postoperative serum or plasma levels of CCL2 were found to be increased in elderly orthopedic patients developing POD, as compared to no-POD patients [33,34]. Increased CCL2 activity might furthermore induce acute neuroinflammation and thereby contribute to POD pathogenesis, as demonstrated by animal studies [22,35]. Appropriately, the upcoming INTUIT study will shed light on the role of neuroinflammation in POD pathogenesis, explicitly focusing on surgery-induced CCL2 release [36].

As cardiac surgery was demonstrated to be associated with the risk for POD [3,37], incidence was significantly increased in those patients in our study, as compared to non-cardiac surgery. Dividing the whole cohort into a cardiac and a non-cardiac surgery subcohort, this revealed that CCL2 serum levels were the only ones among all analyzed biomarkers of which the postoperative increase was markedly and significantly greater, following cardiac surgery, as compared to the other subcohort. While cardiac surgery induced a more than three-fold increase in non-cardiac surgery patients, postoperative CCL2 serum levels were further decreased rather than increased. This, on the one hand, again underlines the impact of surgical stress on systemic inflammation and CCL2 release, which furthermore was shown to explicitly follow cardiac surgery [32,38]. On the other hand, this might provide a mechanistic explanation for increased prevalence of POD in cardiac surgery patients, given the pathogenetic role of CCL2 for neuroinflammation and POD development [22,35].

When we focused our analysis on the subcohort of cardiac surgery patients, this furthermore revealed SDC1 as the only biomarker that showed a significant postoperative increase in POD patients, as compared to the no-POD group. SDC1 is a transmembrane heparan sulfate proteoglycan and as such a key component of the endothelial glycocalyx layer. Its serum levels are increased following surgery, suggesting endothelial damage induced by surgical trauma [30]. Particularly during cardiac surgery with cardiopulmonary bypass, vascular disturbances were associated with POD development [13], and markers of endothelial activation and damage including SCD1 were shown to be markedly increased [31,39]. Impairment of the glycocalyx with SDC1 release was furthermore demonstrated to be associated with delayed neuroinflammation, following cerebral hemorrhage [40]. A specific relation between SDC1 and POD development was not demonstrated so far but it seems likely, since vascular and endothelial injury were shown to be associated with the pathogenesis of delirium in septic patients [9–11]. Therefore, postoperative SDC1 release might serve as specific marker for increased risk for POD in cardiac surgery patients.

Our study has several strengths, but also specific limitations. First, unlike others, we assessed pre- as well as postoperative serum samples. This is justified by our aim to identify, on one hand, biomarkers associated with subsequent POD development that are altered already prior to surgery and thus might serve as 'true' predictors. On the other hand, with the assessment of perioperative change of biomarker serum levels (pre- vs.

postoperative), we sought to shed further light on the pathogenesis of POD. However, our study lacks the inclusion of additional time-points of serum sampling, and therefore, later changes in biomarker profile remain undetected. Second, we drew a random sample of consecutive patients from the whole PROPDESC cohort in order to obtain representative and generalizable results for non-cardiac as well as cardiac surgery. The longitudinal design of the study results in little loss of follow-up. Patients aged < 60 years as well as neurosurgery patients were not included. This was done on one hand in order to increase overall POD prevalence in the whole study population (which is known to be associated with increasing age). On the other hand, we wanted to exclude the effect of intracranial surgical trauma on POD development. However, this might limit the transferability of our data to other patient populations. Consequently, our results should be confirmed in a larger multi-center approach. Third, the strength of our study was the use of multiple validated and complementary POD detection methods to ensure that not POD-positive cases were missed. We performed additional propensity score matching, which allowed for greater efficiency with the analysis of fewer samples in total, while controlling for potential confounders. We also examined 17 biomarkers with specificity for vascular and endothelial function and for inflammation, rather than assessing only one or a limited number of cytokines (as done in other studies). However, our selection missed some important candidate cytokines, which could also have served as internal methodological control, since, e.g., TNF-alpha was previously shown to be associated with POD [41]. Moreover, the use of multiplex arrays that allow for the measurement of a substantial number of biomarkers at once, while using only small serum sample sizes, could also have resulted in less accurate detection of some proteins over others. Therefore, when reproducing our results, the use of enzyme-linked immunosorbent assay as the 'gold standard' for cytokine assessment would be preferable. Last, as our data indicate, we did not succeeded in identifying preoperative candidate serum biomarkers that allow for the identification of the patient at risk, and this could also be seen as a shortcoming of our study.

5. Conclusions

In a mixed, representative cohort of elective non-cardiac as well as cardiac surgery patients, preoperative profiling including 17 serum biomarkers with specificity for vascular dysfunction and for systemic inflammation was not indicative of subsequent POD development. Surgery-induced systemic inflammation, evidenced by a significant increase in CCL2 release, was associated with POD, particularly following cardiac surgery. In those patients, postoperative glycocalyx injury might further contribute to POD development.

Supplementary Materials: The following are available online at https://www.mdpi.com/article/10.3390/biomedicines9050553/s1. STROBE checklist. Supplementary Table S1: Pre- and postoperative serum biomarker profiling. Supplementary. Table S2: Patient and procedural details (no-POD and POD). Supplementary Table S3: Pre- and postoperative serum biomarker profiling (no-POD and POD). Supplementary Table S4: Pre- and postoperative serum biomarker profiling (no-POD and POD) following propensity score matching.

Author Contributions: Conceptualization: J.M., M.W. and T.H. Methodology: J.M., S.F., A.M., M.W. and T.H. Software: A.M. Validation: J.M., A.M. and T.H. Formal Analysis: A.M. and T.H. Investigation: J.M., J.P., V.G., A.K. and C.N. Resources: M.W. and M.C. Data Curation: J.M., S.F., J.P., V.G., A.K., C.N., S.K. and T.H. Writing—Original Draft Preparation: T.H. Writing—Review & Editing: J.M., A.M., M.C. and S.K. Visualization: T.H. Supervision: M.W. and M.C. Project Administration: M.W., M.C. and T.H. Funding Acquisition: J.M. and M.W. All authors have read and agreed to the published version of the manuscript.

Funding: This research was funded by the Clinical Studies Funding Program (FKS) of the Study Center (SZB) at University Hospital Bonn (UKB) (application number: 2018-FKS-01, grant number: O-417.0002). SZB had no impact on the design of the study, on collection, analysis, and interpretation of data and on writing the manuscript.

Institutional Review Board Statement: The study was conducted according to the guidelines of the Declaration of Helsinki, and approved by the Institutional Review Board (or Ethics Committee) of the University Hospital Bonn, Germany (protocol code 255/17, date of approval 18 September 2017).

Informed Consent Statement: Informed consent was obtained from all subjects involved in the study.

Data Availability Statement: All datasets generated and analyzed during the current study are available from the corresponding author on request.

Acknowledgments: The authors wish to thank Frank Splettstoesser for providing technical support.

Conflicts of Interest: The authors declare no conflict of interest.

References

1. Jin, Z.; Hu, J.; Ma, D. Postoperative delirium: Perioperative assessment, risk reduction, and management. *Br. J. Anaesth.* **2020**, *125*, 492–504. [CrossRef]
2. American Psychiatric Association. *Diagnostic and Statistical Manual of Mental Disorders*, 5th ed.; American Psychiatric Association: Arlington, VA, USA, 2013; ISBN 978-0-89042-555-8.
3. Brown, C.H.; Laflam, A.; Max, L.; Lymar, D.; Neufeld, K.J.; Tian, J.; Shah, A.S.; Whitman, G.J.; Hogue, C.W. The Impact of Delirium After Cardiac Surgical Procedures on Postoperative Resource Use. *Ann. Thorac. Surg.* **2016**, *101*, 1663–1669. [CrossRef]
4. Raats, J.W.; Van Eijsden, W.A.; Crolla, R.; Steyerberg, E.; Van Der Laan, L. Risk Factors and Outcomes for Postoperative Delirium after Major Surgery in Elderly Patients. *PLoS ONE* **2015**, *10*, e0136071. [CrossRef]
5. Gleason, L.J.; Schmitt, E.M.; Kosar, C.M.; Tabloski, P.; Saczynski, J.S.; Robinson, T.N.; Cooper, Z.; Rogers, S.O.; Jones, R.N.; Marcantonio, E.R.; et al. Effect of Delirium and Other Major Complications on Outcomes After Elective Surgery in Older Adults. *JAMA Surg.* **2015**, *150*, 1134–1140. [CrossRef] [PubMed]
6. Van Meenen, L.C.C.; Van Meenen, D.M.P.; De Rooij, S.E.; Ter Riet, G. Risk Prediction Models for Postoperative Delirium: A Systematic Review and Meta-Analysis. *J. Am. Geriatr. Soc.* **2014**, *62*, 2383–2390. [CrossRef] [PubMed]
7. Liu, X.; Yu, Y.; Zhu, S. Inflammatory markers in postoperative delirium (POD) and cognitive dysfunction (POCD): A meta-analysis of observational studies. *PLoS ONE* **2018**, *13*, e0195659. [CrossRef]
8. Atterton, B.; Paulino, M.C.; Povoa, P.; Martin-Loeches, I. Sepsis Associated Delirium. *Intensive Care Med.* **2020**, *56*, 240. [CrossRef]
9. Lee, W.L.; Liles, W.C. Endothelial activation, dysfunction and permeability during severe infections. *Curr. Opin. Hematol.* **2011**, *18*, 191–196. [CrossRef] [PubMed]
10. Hughes, C.G.; Pandharipande, P.P.; Thompson, J.L.; Chandrasekhar, R.; Ware, L.B.; Ely, E.W.; Girard, T.D. Endothelial Activation and Blood-Brain Barrier Injury as Risk Factors for Delirium in Critically Ill Patients. *Crit. Care Med.* **2016**, *44*, e809–e817. [CrossRef] [PubMed]
11. Page, A.V.; Liles, W.C. Biomarkers of endothelial activation/dysfunction in infectious diseases. *Virulence* **2013**, *4*, 507–516. [CrossRef]
12. Rudolph, J.L.; Jones, R.N.; Rasmussen, L.S.; Silverstein, J.H.; Inouye, S.K.; Marcantonio, E.R. Independent Vascular and Cognitive Risk Factors for Postoperative Delirium. *Am. J. Med.* **2007**, *120*, 807–813. [CrossRef] [PubMed]
13. Thudium, M.; Ellerkmann, R.K.; Heinze, I.; Hilbert, T. Relative cerebral hyperperfusion during cardiopulmonary bypass is associated with risk for postoperative delirium: A cross-sectional cohort study. *BMC Anesthesiol.* **2019**, *19*, 35. [CrossRef] [PubMed]
14. Menzenbach, J.; Guttenthaler, V.; Kirfel, A.; Ricchiuto, A.; Neumann, C.; Adler, L.; Kieback, M.; Velten, L.; Fimmers, R.; Mayr, A.; et al. Estimating patients' risk for postoperative delirium from preoperative routine data—Trial design of the PRe-Operative prediction of postoperative DElirium by appropriate SCreening (PROPDESC) study—A monocentre prospective observational trial. *Contemp. Clin. Trials Commun.* **2020**, *17*, 100501. [CrossRef]
15. Glance, L.G.; Lustik, S.J.; Hannan, E.L.; Osler, T.M.; Mukamel, D.B.; Qian, F.; Dick, A.W. The Surgical Mortality Probability Model: Derivation and Validation of a Simple Risk Prediction Rule for Noncardiac Surgery. *Ann. Surg.* **2012**, *255*, 696–702. [CrossRef] [PubMed]
16. Zhang, Y.; Birru, R.; Di, Y.P. Analysis of Clinical and Biological Samples Using Microsphere-Based Multiplexing Luminex System. *Mol. Toxicol. Protoc.* **2014**, *1105*, 43–57. [CrossRef]
17. Guenther, U.; Popp, J.; Koecher, L.; Muders, T.; Wrigge, H.; Ely, E.W.; Putensen, C. Validity and Reliability of the CAM-ICU Flowsheet to diagnose delirium in surgical ICU patients. *J. Crit. Care* **2010**, *25*, 144–151. [CrossRef]
18. Inouye, S.K.; Van Dyck, C.H.; Alessi, C.A.; Balkin, S.; Siegal, A.P.; Horwitz, R.I. Clarifying Confusion: The Confusion Assessment Method. A New Method for Detection of Delirium. *Ann. Intern. Med.* **1990**, *113*, 941–948. [CrossRef]
19. Bellelli, G.; Morandi, A.; Davis, D.H.; Mazzola, P.; Turco, R.; Gentile, S.; Ryan, T.; Cash, H.; Guerini, F.; Torpilliesi, T.; et al. Validation of the 4AT, a new instrument for rapid delirium screening: A study in 234 hospitalised older people. *Age Ageing* **2014**, *43*, 496–502. [CrossRef]
20. Schuurmans, M.J.; Shortridge-Baggett, L.M.; Duursma, S.A. The Delirium Observation Screening Scale: A Screening Instrument for Delirium. *Res. Theory Nurs. Pract.* **2003**, *17*, 31–50. [CrossRef]

21. Ho, D.E.; Imai, K.; King, G.; Stuart, E.A. Matching as Nonparametric Preprocessing for Reducing Model Dependence in Parametric Causal Inference. *Political Anal.* **2007**, *15*, 199–236. [CrossRef]
22. Alam, A.; Hana, Z.; Jin, Z.; Suen, K.C.; Ma, D. Surgery, neuroinflammation and cognitive impairment. *EBioMedicine* **2018**, *37*, 547–556. [CrossRef] [PubMed]
23. Khan, A.; Johnson, D.K.; Carlson, S.; Hocum-Stone, L.; Kelly, R.F.; Gravely, A.A.; Mbai, M.; Green, D.L.; Santilli, S.; Garcia, S.; et al. NT-Pro BNP Predicts Myocardial Injury Post-vascular Surgery and is Reduced with CoQ10: A Randomized Double-Blind Trial. *Ann. Vasc. Surg.* **2020**, *64*, 292–302. [CrossRef] [PubMed]
24. Rodseth, R.N.; Biccard, B.M.; Chu, R.; Lurati Buse, G.A.; Thabane, L.; Bakhai, A.; Bolliger, D.; Cagini, L.; Cahill, T.J.; Cardinale, D.; et al. Postoperative B-Type Natriuretic Peptide for Prediction of Major Cardiac Events in Patients Undergoing Noncardiac Surgery: Systematic Review and Individual Patient Meta-Analysis. *Anesthesiology* **2013**, *119*, 270–283. [CrossRef]
25. Raslau, D.; Bierle, D.M.; Stephenson, C.R.; Mikhail, M.A.; Kebede, E.B.; Mauck, K.F. Preoperative Cardiac Risk Assessment. *Mayo Clin. Proc.* **2020**, *95*, 1064–1079. [CrossRef]
26. De Hert, S.; Staender, S.; Fritsch, G.; Hinkelbein, J.; Afshari, A.; Bettelli, G.; Bock, M.; Chew, M.S.; Coburn, M.; De Robertis, E.; et al. Pre-operative evaluation of adults undergoing elective noncardiac surgery: Updated guideline from the european society of anaesthesiology. *Eur. J. Anaesthesiol.* **2018**, *35*, 407–465. [CrossRef]
27. Marković, D.Z.; Jevtović-Stoimenov, T.; Ćosić, V.; Stošić, B.; Živković, B.M.; Janković, R.J. Addition of biomarker panel improves prediction performance of American College of Surgeons National Surgical Quality Improvement Program (ACS NSQIP) calculator for cardiac risk assessment of elderly patients preparing for major non-cardiac surgery: A pilot study. *Aging Clin. Exp. Res.* **2018**, *30*, 419–431. [CrossRef]
28. Vasunilashorn, S.M.; Ngo, L.; Inouye, S.K.; Libermann, T.A.; Jones, R.N.; Alsop, D.C.; Guess, J.; Jastrzebski, S.; McElhaney, J.E.; Kuchel, G.; et al. Cytokines and Postoperative Delirium in Older Patients Undergoing Major Elective Surgery. *J. Gerontol. Ser. A Boil. Sci. Med. Sci.* **2015**, *70*, 1289–1295. [CrossRef]
29. Lin, M.-T.; Yeh, S.-L.; Wu, M.-S.; Lin, J.-T.; Lee, P.-H.; Liaw, K.-Y.; Chang, K.-J.; Chen, W.-J. Impact of surgery on local and systemic responses of cytokines and adhesion molecules. *Hepatogastroenterology* **2009**, *56*, 1341–1345. [PubMed]
30. Klaschik, S.; Gehlen, J.; Neumann, C.; Keyver-Paik, M.-D.; Soehle, M.; Frede, S.; Velten, M.; Hoeft, A.; Hilbert, T. Network of Mediators for Vascular Inflammation and Leakage Is Dysbalanced during Cytoreductive Surgery for Late-Stage Ovarian Cancer. *Mediat. Inflamm.* **2019**, *2019*, 5263717. [CrossRef]
31. Hilbert, T.; Duerr, G.D.; Hamiko, M.; Frede, S.; Rogers, L.; Baumgarten, G.; Hoeft, A.; Velten, M. Endothelial permeability following coronary artery bypass grafting: An observational study on the possible role of angiopoietin imbalance. *Crit. Care Lond. Engl.* **2016**, *20*, 51. [CrossRef]
32. Shibasaki, H.; Furukawa, K.; Yamamori, H.; Kimura, F.; Tashiro, T.; Miyazaki, M. The Post-Operative Level of Serum Monocyte Chemoattractant Protein-1 and Its Correlation with the Severity of Surgical Stress. *J. Surg. Res.* **2006**, *136*, 314–319. [CrossRef] [PubMed]
33. Skrede, K.; Wyller, T.B.; Watne, L.O.; Seljeflot, I.; Juliebø, V. Is there a role for monocyte chemoattractant protein-1 in delirium? Novel observations in elderly hip fracture patients. *BMC Res. Notes* **2015**, *8*, 186. [CrossRef] [PubMed]
34. Hirsch, J.; Vacas, S.; Terrando, N.; Yuan, M.; Sands, L.P.; Kramer, J.; Bozic, K.; Maze, M.M.; Leung, J.M. Perioperative cerebrospinal fluid and plasma inflammatory markers after orthopedic surgery. *J. Neuroinflamm.* **2016**, *13*, 1–12. [CrossRef] [PubMed]
35. Thompson, W.L.; Karpus, W.J.; Van Eldik, L.J. MCP-1-deficient mice show reduced neuroinflammatory responses and increased peripheral inflammatory responses to peripheral endotoxin insult. *J. Neuroinflamm.* **2008**, *5*, 35. [CrossRef]
36. Berger, M.; Oyeyemi, D.; Olurinde, M.O.; Whitson, H.E.; Weinhold, K.J.; Woldorff, M.G.; Lipsitz, L.A.; Moretti, E.; Giattino, C.M.; Roberts, K.C.; et al. The INTUIT Study: Investigating Neuroinflammation Underlying Postoperative Cognitive Dysfunction. *J. Am. Geriatr. Soc.* **2019**, *67*, 794–798. [CrossRef]
37. Cereghetti, C.; Siegemund, M.; Schaedelin, S.; Fassl, J.; Seeberger, M.D.; Eckstein, F.S.; Steiner, L.A.; Goettel, N. Independent Predictors of the Duration and Overall Burden of Postoperative Delirium After Cardiac Surgery in Adults: An Observational Cohort Study. *J. Cardiothorac. Vasc. Anesth.* **2017**, *31*, 1966–1973. [CrossRef]
38. Moledina, D.; Isguven, S.; McArthur, E.; Thiessen-Philbrook, H.; Garg, A.X.; Shlipak, M.; Whitlock, R.; Kavsak, P.A.; Coca, S.G.; Parikh, C.R.; et al. Plasma Monocyte Chemotactic Protein-1 Is Associated with Acute Kidney Injury and Death After Cardiac Operations. *Ann. Thorac. Surg.* **2017**, *104*, 613–620. [CrossRef]
39. Pesonen, E.; Passov, A.; Andersson, S.; Suojaranta, R.; Niemi, T.; Raivio, P.; Salmenperä, M.; Schramko, A. Glycocalyx Degradation and Inflammation in Cardiac Surgery. *J. Cardiothorac. Vasc. Anesth.* **2019**, *33*, 341–345. [CrossRef]
40. Bell, J.D.; Rhind, S.G.; Di Battista, A.P.; Macdonald, R.L.; Baker, A.J. Biomarkers of Glycocalyx Injury are Associated with Delayed Cerebral Ischemia Following Aneurysmal Subarachnoid Hemorrhage: A Case Series Supporting a New Hypothesis. *Neurocrit. Care* **2016**, *26*, 339–347. [CrossRef] [PubMed]
41. Kazmierski, J.; Banys, A.; Latek, J.; Bourke, J.; Jaszewski, R. Raised IL-2 and TNF-α concentrations are associated with postoperative delirium in patients undergoing coronary-artery bypass graft surgery. *Int. Psychogeriatr.* **2014**, *26*, 845–855. [CrossRef]

biomedicines

Article

17β-Estradiol Increases APE1/Ref-1 Secretion in Vascular Endothelial Cells and Ovariectomized Mice: Involvement of Calcium-Dependent Exosome Pathway

Yu-Ran Lee [1,2], Hee-Kyoung Joo [1,2], Eun-Ok Lee [1,2], Sungmin Kim [1,2], Hao Jin [1,2], Yeon-Hee Choi [1,2], Cuk-Seong Kim [1,2] and Byeong-Hwa Jeon [1,2,*]

[1] Research Institute for Medical Sciences, College of Medicine, Chungnam National University, 266 Munhwa-ro, Jung-gu, Daejeon 35015, Korea; lyr0913@cnu.ac.kr (Y.-R.L.); hkjoo79@cnu.ac.kr (H.-K.J.); y21c486@naver.com (E.-O.L.); s13845@naver.com (S.K.); jinhao0508@gmail.com (H.J.); yeonhee970@gmail.com (Y.-H.C.); cskim@cnu.ac.kr (C.-S.K.)
[2] Department of Physiology, College of Medicine, Chungnam National University, 266 Munhwa-ro, Jung-gu, Daejeon 35015, Korea
* Correspondence: bhjeon@cnu.ac.kr; Tel.: +82-42-580-8214

Citation: Lee, Y.-R.; Joo, H.-K.; Lee, E.-O.; Kim, S.; Jin, H.; Choi, Y.-H.; Kim, C.-S.; Jeon, B.-H. 17β-Estradiol Increases APE1/Ref-1 Secretion in Vascular Endothelial Cells and Ovariectomized Mice: Involvement of Calcium-Dependent Exosome Pathway. *Biomedicines* **2021**, *9*, 1040. https://doi.org/10.3390/biomedicines9081040

Academic Editor: Andreas Weber

Received: 11 June 2021
Accepted: 16 August 2021
Published: 18 August 2021

Publisher's Note: MDPI stays neutral with regard to jurisdictional claims in published maps and institutional affiliations.

Abstract: Apurinic/apyrimidinic endonuclease-1/redox factor-1 (APE1/Ref-1) is a multifunctional protein that can be secreted, and recently suggested as new biomarker for vascular inflammation. However, the endogenous hormones for APE1/Ref-1 secretion and its underlying mechanisms are not defined. Here, the effect of twelve endogenous hormones on APE1/Ref-1 secretion was screened in cultured vascular endothelial cells. The endogenous hormones that significantly increased APE1/Ref-1 secretion was 17β-estradiol (E2), 5α-dihydrotestosterone, progesterone, insulin, and insulin-like growth factor. The most potent hormone inducing APE1/Ref-1 secretion was E2, which in cultured endothelial cells, E2 for 24 h increased APE1/Ref-1 secretion level of 4.56 ± 1.16 ng/mL, compared to a basal secretion level of 0.09 ± 0.02 ng/mL. Among the estrogens, only E2 increased APE1/Ref-1 secretion, not estrone and estriol. Blood APE1/Ref-1 concentrations decreased in ovariectomized (OVX) mice but were significantly increased by the replacement of E2 (0.39 ± 0.09 ng/mL for OVX vs. 4.67 ± 0.53 ng/mL for OVX + E2). E2-induced APE1/Ref-1 secretion was remarkably suppressed by the estrogen receptor (ER) blocker fulvestrant and intracellular Ca^{2+} chelator 1,2-Bis(2-aminophenoxy)ethane-N,N,N′,N′-tetraacetic acid tetrakis (acetoxymethyl ester) (BAPTA-AM), suggesting E2-induced APE1/Ref-1 secretion was dependent on ER and intracellular calcium. E2-induced APE1/Ref-1 secretion was significantly inhibited by exosome inhibitor GW4869. Furthermore, APE1/Ref-1 level in CD63-positive exosome were increased by E2. Finally, fluorescence imaging data showed that APE1/Ref-1 co-localized with CD63-labled exosome in the cytoplasm of cells upon E2 treatment. Taken together, E2 was the most potent hormone for APE1/Ref-1 secretion, which appeared to occur through exosomes that were dependent on ER and intracellular Ca^{2+}. Furthermore, hormonal effects should be considered when analyzing biomarkers for vascular inflammation.

Keywords: apurinic/apyrimidinic endonuclease-1/redox factor-1; 17β-estradiol; estrogen receptor; endothelial cells; calcium; exosome

1. Introduction

Apurinic/apyrimidinic endonuclease 1/redox factor-1 (APE1/Ref-1) is a multifunctional protein suggested as a new biomarker of vascular inflammation [1,2]. Basically, APE1/Ref-1 plays roles in transcriptional regulation through redox modification and base excision repair [3]. Since it was first reported in 2013 that APE1/Ref-1 secretion is increased by intracellular hyperacetylation [4], it has subsequently been reported that its secretion is increased in lipopolysaccharide-induced endotoxemic animal models [5], apolipoprotein

E-deficient mice fed Western-type diets [1], and patients with coronary artery disease [6], suggesting its usefulness as a new biomarker for vascular inflammation.

Cell signaling are affected by a variety of stimulators that are responsible for the secretion of proteins in cells [7]. Hormone secretion is involved in maintaining homeostasis in vivo. Changes in specific protein secretion can play an important role in the diagnosis or prognosis of various diseases, such as systemic inflammation [8]. The presence of APE1/Ref-1 in the extracellular environment as a biomarker has been suggested as it is actively secreted in specific diseases, and it may also be used as evidence of non-specific tissue damage since it can be released from cells upon cell death. Identifying substances that increase the secretion of proteins without causing cell death could help in the development of new biomarkers for specific conditions or diseases. The secreted substances in the response to hormone have been hypothesized to play an important role in the regulation of vascular inflammation.

The biological functions of secreted APE1/Ref-1 are uncovering. Extracellular APE1/Ref-1 is known to have its own function and can affect surrounding and distant cells. Enzymatically active APE1/Ref-1 protein that functions as an endonuclease is secreted in response to genotoxic stress [9]. Secretory APE1/Ref-1 has been reported to have a role in the inhibition of vascular inflammation via thiol-disulfide exchange in the tumor necrosis factor (TNF) receptor [10]. Therefore, the secreted APE1/Ref-1 might have anti-inflammatory properties and the redox activity of its cysteine residue has recently been associated with its anti-inflammatory activity in vivo in animal models [11].

The subcellular location of APE1/Ref-1 is regulated by post-transcriptional modifications, including acetylation. The histone deacetylase inhibitor trichostatin A induces APE1/Ref-1 secretion in human embryonic kidney (HEK) 293T cells and mutations in APE1/Ref-1 at lysine residues 6 and 7 (K6R/K7R) markedly diminish its secretion [4]. Secretion of APE1/Ref-1 is observed in response to hyperacetylation in triple-negative breast cancer cell lines, resulting in significantly decreased cell viability and the induction of apoptosis [12]. In addition, nitrosylation selectively induces cytosolic translocation of APE1/Ref-1 with Cys93 and Cys310 being critical for the nitrosylation-mediated cytosolic translocation [13].

Protein secretion may occur through either the classical or non-classical secretory pathway. Proteins secreted through the classical pathway typically contain N-terminal signal peptides that direct the proteins to the translocation apparatus of the endoplasmic reticulum [14]. The mechanisms of APE1/Ref-1 secretion have not yet been fully elucidated. Some reports on the pathway of APE1/Ref-1 secretion suggested that it may be secreted via exosomes [9,12,15] or the ATP-binding cassette transporter [16].

Finding out which endogenous hormone can affect the secretion of APE1/Ref-1 can be utilized to understand the pathophysiology of hormone imbalance in vascular inflammation. Therefore, the current study aimed to identify potential endogenous hormones that increase APE1/Ref-1 secretion in vascular endothelial cells under conditions that do not induce cell death; it also aimed to reveal the underlying secretion mechanism.

2. Materials and Methods

2.1. Cell Culture and Reagents

Human umbilical vein endothelial cells (HUVECs) (C2517A, Lonza, Walkersville, MD, USA) were cultured in endothelial growth medium (EGM-2) purchased from Lonza Bioscience (Walkersville, MD, USA). The HUVECs were maintained in a humidified atmosphere of 95% air and 5% CO_2 at 37 °C. Norepinephrine, acetylcholine, triiodothyronine (T3), thyroxine (T4), insulin-like growth factor (IGF), cortisol, aldosterone, insulin, glucagon, 5α-dihydrotestosterone (DHT), 17β-estradiol (E2), progesterone (P4), estrone (E1), estriol (E3), fulvestrant, N(ω)-nitro-L-arginine methyl ester (L-NAME), 1,2-Bis(2-aminophenoxy)ethane-N,N,N′,N′-tetraacetic acid tetrakis(acetoxymethyl ester) (BAPTA-AM), GW4869, and dimethyl sulfoxide (DMSO) were purchased from Sigma Aldrich (St. Louis, MO, USA). An ExoQuick-Tc exosome isolation kit was purchased from System

Biosciences (Palo Alto, CA, USA). The polyclonal antibody against APE1/Ref-1 was obtained from MediRedox, Inc. (Daejeon, Korea) and, the anti-CD63 antibody was obtained from Biobyt (Cambridge, UK). The monoclonal antibody against anti-heat shock protein-70 (HSP70, C92F3A-5) was obtained from Enzo Life Science (Farmingdale, NY, USA), the anti-CD9 (C-4) antibody was obtained from Santa Cruz Biotechnology (Dallas, TX, USA) and anti-ALG-2 interacting protein X (Alix, 3A9) antibody was obtained from Thermo-Fisher Scientific Inc (Waltham, MA, USA).

2.2. Cell Viability Assay Using Reducing Potentials of Cells

The viability and cytotoxicity of HUVECs were analyzed using a RealTime-Glo™ MT cell viability assay kit (Promega, Madison, WI, USA), according to the manufacturer's instructions. Briefly, HUVECs were plated into 96-well white cell culture plates at a density of 5×10^3 cells/well. After 24 h of incubation, the pro-substrate and luciferase were added at the same time as that of the hormones to continuously monitor the viability of the HUVECs in real-time. Luminescence intensity at the desired time points was measured using a Glo-Max™ multimode reader (Promega, Madison, WI, USA).

2.3. Quantification of Secretory APE1/Ref-1

Mouse plasma and cell culture media were centrifuged at 3000 rpm for 10 min to obtain cell-free samples as previously reported [11]. Secreted APE1/Ref-1 levels were determined using a APE1/Ref-1 sandwich enzyme-linked immunosorbent assay (ELISA) kit (MediRedox, Inc., Daejeon, Korea) according to the manufacturer's instructions. Secreted APE1/Ref-1 levels (ng/mL for plasma and ng/10^5 cells for supernatants) were calculated against a standard curve generated using recombinant human APE1/Ref-1 protein (MediRedox, Inc., Daejeon, Korea).

2.4. Isolation of Exosome in Cell Culture Media

To isolate the exosome from the cell culture medium of HUVECs, the medium was changed to medium without fetal bovine serum. The HUVECs were then treated with E2 for 24 h and the cell culture medium was collected and centrifuged at $500 \times g$ for 10 min. The clarified supernatant was collected, and debris and vesicle over 0.22 μm in diameter was removed through filtration using a 0.22 μm syringe filter (Millipore, Billerica, MA). exosomes were isolated using an Exoquick-TC isolation kit (System Biosciences, Palo Alto, CA, USA) that precipitates exosomes based on polyethylene glycol precipitation as recommended by the manufacturer. The isolated exosomes were confirmed using immunoblotting for CD63, CD9, HSP70, and Alix.

2.5. Establishment of an Ovariectomized Mice Model

Animal experiments were performed using female C57BL/6J mice, 7–8 weeks of age (DooYeol Biotech, Seoul, Korea). The animal protocol was approved by the Ethics Committee of Animal Experimentation of Chungnam National University Hospital (CNUH-017-A0025) and all experiments were performed in accordance with the Guide for the Care of Use of Laboratory Animals published by the US National Institutes of Health (NIH Publication, 8th edition, 2011). All surgeries and pump implantations were performed using aseptic procedures.

All mice in the ovariectomized (OVX) group underwent bilateral ovariectomy [17] using a single ventral approach. The sham group suffered the same surgery except that their ovaries were preserved. The OVX and sham mice were randomly divided into two groups for implantation of Alzet osmotic minipumps (model number 1002, for 14-day delivery at 0.26 μL/hour, Durect Corp, Cupertino, CA, USA). The osmotic pumps with E2 were implanted subcutaneously between the scapulae via a small incision. E2 was dissolved in 10% dimethyl sulfoxide (DMSO; Sigma-Aldrich, St. Louis, MO, USA). The concentration of the experimental agent was 35 μg/mL, which resulted in a delivery rate of approximately 0.25 μg/day.

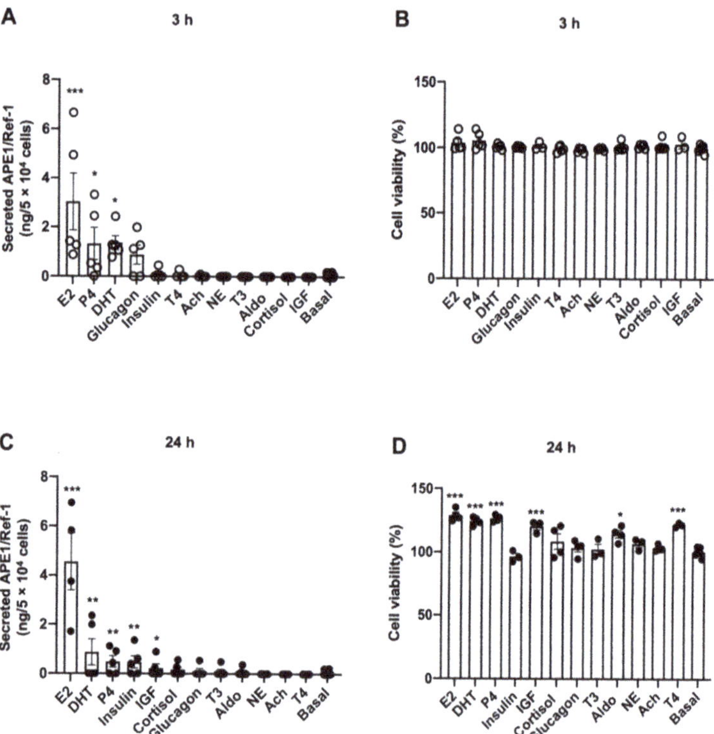

Figure 1. Identification of endogenous hormone for induction of apurinic/apyrimidinic endonuclease-1/redox factor-1 (APE1/Ref-1) secretion. (**A**) Effect of short-term (3 h) treatment of hormones on APE1/Ref-1 levels in the culture media of vascular endothelial cells. (**B**) Effect of short-term (3 h) treatment of hormones on cell viability of vascular endothelial cells. (**C**) Effect of long-term (24 h) treatment of hormones on APE1/Ref-1 levels in the culture media of vascular endothelial cells. (**D**) Effect of long-term (24 h) treatment of hormones on cell viability of vascular endothelial cells. Vascular endothelial cells were treated with each hormone for 3 h or 24 h. APE1/Ref-1 levels in culture supernatant were measured using an APE1/Ref-1 sandwich ELISA assay as describe Material and Methods. Columns, mean (n = 4–5); dot plot, SE. *** $p < 0.001$, ** $p < 0.01$, * $p < 0.05$ indicates a significant difference compared to the control cells (Basal) according to an unpaired *t*-test.

3.2. 17β-Estradiol Induced APE1/Ref-1 Secretion in HUVECs

Estrogen is known to exist as three main types: estrone (E1), 17β-estradiol (E2), and estriol (E3). Next, we investigated which type(s) of estrogen was able to induce APE/1Ref-1 secretion in the endothelial cells. To evaluate the effect of three types of estrogen on APE1/Ref-1 secretion, APE1/Ref-1 levels were measured after 3 h and 24 h of treatment in various dosage of the estrogens. E1-treated or E3-treated HUVECs did not demonstrate the ability to induce APE1/Ref-1 secretion (Figure 2A,E). Treatment with E1 and E3 for 3 h did not affect cell viability of the vascular endothelial cells (Figure 2B,F). However, when HUVECs were treated with E2, APE1/Ref-1 secretion was increased in a dose-dependent manner (Figure 2C). E2 treatment for 24 h induced an increase in cell viability, suggesting cell growth (Figure 2D). Therefore, we confirmed that E2 is a hormone to increase APE1/Ref-1 secretion without causing cell death.

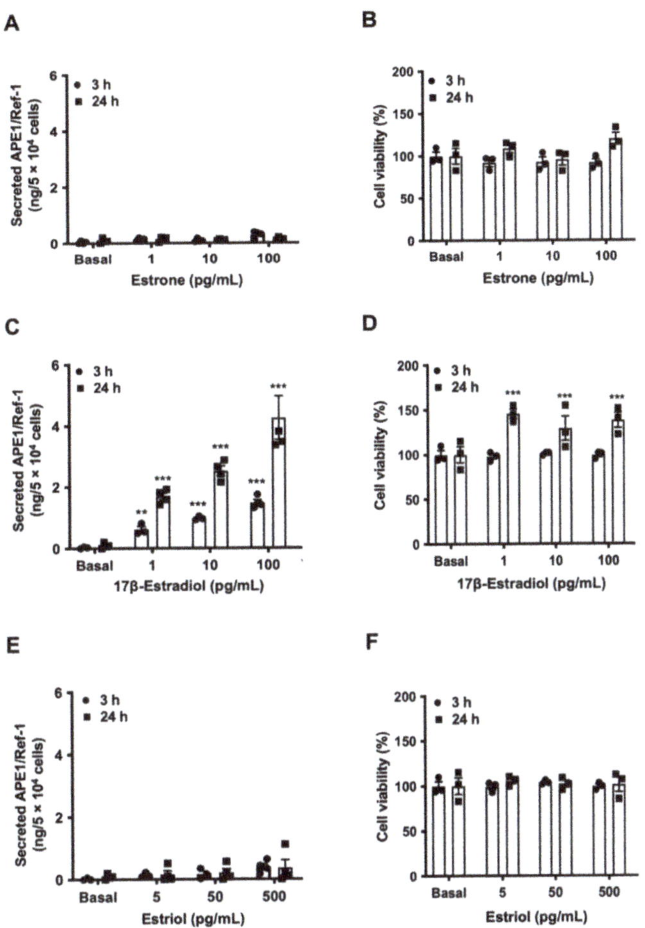

Figure 2. Differential regulation of apurinic/apyrimidinic endonuclease-1/redox factor-1 (APE1/Ref-1) secretion by estrogen. The effect of estrone (**A**), 17β-estradiol (**C**), and estriol (**E**) on APE1/Ref-1 levels in the culture media of vascular endothelial cells. The effect of estrone (**B**), 17β -estradiol (**D**), and estriol (**F**) on cell viability of vascular endothelial cells. Columns, mean (n s= 3–4); dot plot, SE. *** $p < 0.001$, ** $p < 0.01$, indicates a significant difference compared to that of control cells (Basal) according to an unpaired *t*-test.

3.3. 17β-Estradiol Increased APE1/Ref-1 Secretion in Ovariectomized (OVX) Mice

To extend the novel concept of 17β-estradiol (E2) as an inducer of APE1/Ref-1 secretion in cultured endothelial cells, we attempted to determine whether it could increase the levels of plasma APE1/Ref-1 in OVX mice. Figure 3A shows the experimental design for the evaluation of the role of estradiol in OVX mice. The effect of E2 on APE1/Ref-1 concentration in the blood was evaluated 14 days after insertion of the E2-containing osmotic pump. As shown in Figure 3B, plasma APE1/Ref-1 levels in basal conditions and OVX mice were 1.98 ± 0.17 ng/mL and 0.39 ± 0.09 ng/mL, respectively. These results are shown that the plasma APE1/Ref-1 levels were significantly reduced by the removal of ovary, suggesting a physiological role of the ovary in regulating blood APE1/Ref-1 concentrations in vivo. The replacement of E2 for 14 days increased plasma APE1/Ref-1 levels both normal and OVX mice (4.51 ± 0.41 ng/mL and 4.67 ± 0.53 ng/mL, respectively), compared with OVX mice (0.39 ± 0.09 ng/mL). Taken together, these findings confirmed that estrogen in ovary is important hormone for the in vivo regulation of blood APE1/Ref-1, and administration of E2 can increases the concentration of APE1/Ref-1 in the blood.

A

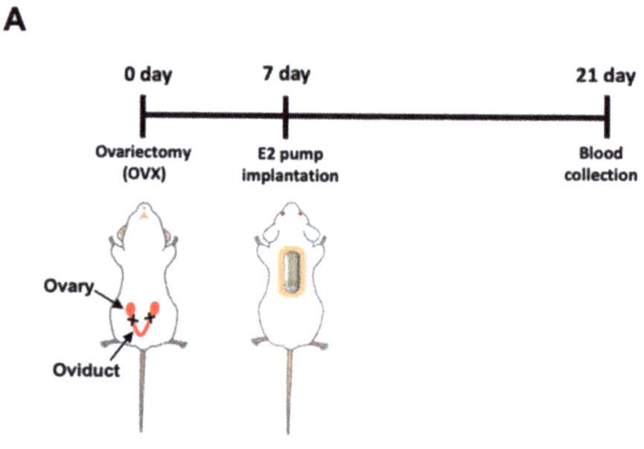

B

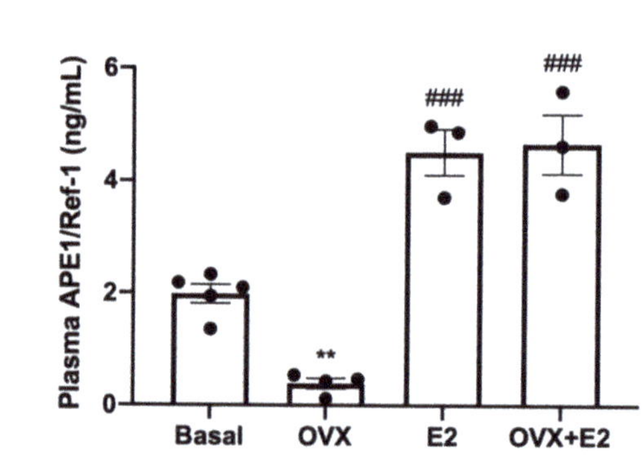

Figure 3. 17β-Estradiol (E2) increases plasma apurinic/apyrimidinic endonuclease-1/redox factor-1 (APE1/Ref-1) levels in ovariectomized (OVX) mice. (**A**) Experimental schedule for 17β-estradiol treatment and performing the APE1/Ref-1 assay. Ovariectomy (OVX) was performed 7 d prior to the implantation of the E2 pump in mice. APE1/Ref-1 blood levels were analyzed using an APE1/Ref-1 sandwich enzyme-linked immunosorbent assay (ELISA) 14 d after implantation of the E2 pump. (**B**) Effect of E2 implantation on plasma APE1/Ref-1 levels in ovariectomized mice. Basal group, sham OVX surgery with 10% dimethyl sulfoxide (DMSO) in the osmotic pump; OVX group, ovariectomized mice with 10% DMSO in the osmotic pump; E2 group, sham OVX surgery with E2 in the osmotic pump (10 μg/kg/day); OVX + E2 group, ovariectomy with E2 in the osmotic pump (10 μg/kg/day). Columns, mean (n = 3–4 animals per group.); dot plot, SE. ** $p < 0.05$ vs. the Basal group and ### $p < 0.001$ vs. the OVX group based on one-way ANOVA analysis followed by Bonferroni's multiple comparison test.

3.4. 17β-Estradiol-Induced APE1/Ref-1 Secretion Depend on the Binding of ER and Intracellular Calcium

The biological action of estrogen is mediated by estrogen receptor (ER) binding and its activation. Therefore, we evaluated whether E2 increased APE1/Ref-1 secretion via the ER (Figure 4A). Fulvestrant, a selective ER inhibitor, significantly reduced by about 80% of E2-induced APE1/Ref-1 secretion, compared to untreated cells (1.48 ± 0.09 ng/mL for E2 vs. 0.3 ± 0.09 ng/mL at 3 h for fulvestrant +E2, 4.14 ng/mL ± 0.42 ng/mL for E2 vs. 0.7 ± 0.27 ng/mL at 24 h for fulvestrant +E2). ER activation in vascular cells results in increased endothelial nitric oxide synthase (eNOS) activity [38]. As shown in Figure 4B, we attempted to determine whether E2-induced APE1/Ref-1 secretion was due to eNOS activity by comparing the results to those in the presence of the eNOS inhibitor L-NAME. However, pretreatment with L-NAME did not affect E2-induced increase of APE1/Ref-1 secretion in cultured endothelial cells.

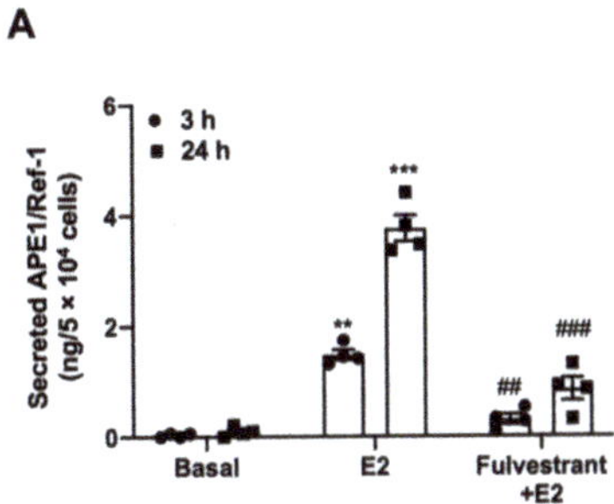

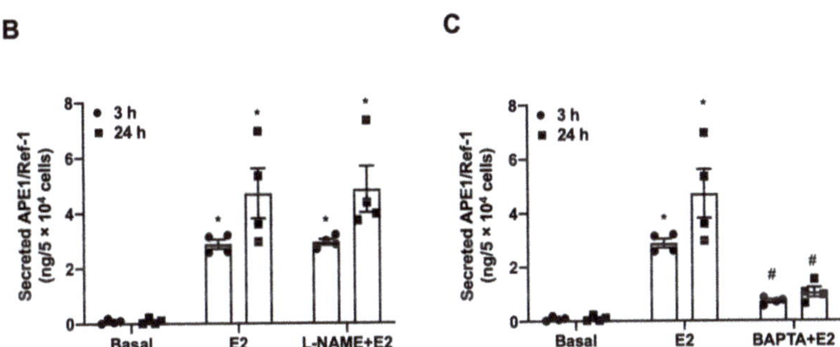

Figure 4. 17β-estradiol (E2)-induced apurinic/apyrimidinic endonuclease-1/redox factor-1 (APE1/Ref-1) secretion depends on estrogen receptor binding and intracellular calcium. (**A**) Effect of the competitive estrogen receptor (ER) inhibitor fulvestrant (100 nM for 2 h) on E2-induced APE1/Ref-1 secretion in cultured endothelial cells. (**B**) Effect of the endothelial nitric oxide synthase inhibitor N(ω)-nitro-L-arginine methyl ester (L-NAME) (10 mM for 1 h) on 17β-estradiol (E2)-induced APE1/Ref-1 secretion in cultured endothelial cells. (**C**) Effect of the cell permeable calcium chelator 1,2-Bis(2-aminophenoxy)ethane-N,N,N′,N′-tetraacetic acid tetrakis (acetoxymethyl ester) (BAPTA-AM) (10 μM for 30 min) on E2-induced APE1/Ref-1 secretion in cultured endothelial cells. Columns, mean (n = 4); dot plot, SE. *** $p < 0.001$, ** $p < 0.01$, and * $p < 0.05$ vs. Basal; ### $p < 0.001$, ## $p < 0.01$, and # $p < 0.05$ vs. E2 treated based on one-way ANOVA analysis followed by Bonferroni's multiple comparison test.

It is also known that 17β-estradiol is involved in intracellular Ca^{2+} homeostasis in human endothelial cells [39]. To determine whether the APE1/Ref-1 secretion induced by E2 was caused by an increase in intracellular Ca^{2+} concentration, the cells were pretreated with the intracellular Ca^{2+} chelator BAPTA-AM and then evaluated. Interestingly, as shown in Figure 4C, pretreatment of vascular endothelial cells with BAPTA-AM significantly inhibited the APE1/Ref-1 secretion induced by E2 (2.87 ± 0.17 ng/mL for E2 vs. 0.7 ± 0.07 ng/mL at 3 h for BAPTA-AM + E2, 4.71 ± 0.9 ng/mL for E2 vs. 1.0 ± 0.18 ng/mL at 24 h for BAPTA-AM + E2). These findings suggest the binding of E2 to ER and increase in intracellular Ca^{2+} were major signaling processes required for E2-induced APE1/Ref-1 secretion in the cultured endothelial cells.

3.5. 17β-Estradiol-Induced APE1/Ref-1 Secretion Was Mediated through Exosomes

Many cellular proteins lacking a signal peptide can be secreted through unconventional secretion processes, such as vesicle transport. Accordingly, we next investigated whether APE1/Ref-1 secretion induced by E2 occurred through the exosome pathway. The effect of the exosome inhibitor GW4869 was analyzed to determine whether E2-induced APE1/Ref-1 secretion is mediated with exosome pathway. As shown in Figure 5A, pretreatment of HUVECs with GW4869 significantly inhibited E2-induced APE1/Ref-1 secretion, suggested it was mediated through the exosome in endothelial cells.

We next investigated whether APE1/Ref-1 existed in exosome following E2 treatment. The exosome in the cell culture media was isolated as described with Material and Methods, and its experimental step was shown in Figure 5B. The successful exosome isolation from the culture media could be confirmed by the existence of exosome markers [40]. As shown in Figure 5C, exosome-specific markers such as CD63, CD9, HSP70, and Alix were expressed in isolated exosomes of cultured endothelial cells. A small amount of APE1/Ref-1 in basal condition was detected in exosomes, however, the exposure of E2 for 24 h in cultured endothelial cells was significantly increased approximately three-fold of APE1/Ref-1 in exosomes (Figure 5C,D). Interestingly, the expressions of exosome markers such as CD63, CD9, HSP70, and Alix were not changed by the exposure of E2.

Finally, we evaluated whether APE1/Ref-1 co-localized in vascular endothelial cell vesicles using immunofluorescence staining. As shown in Figure 5E, APE1/Ref-1 (green signal) was present in the nuclei of the cells while the exosome marker CD63 (red signal) was mainly present in the cytoplasm. However, when the cells were treated with E2 (100 pg/mL) for 3 h, the cytoplasmic expression of APE1/Ref-1 increased and co-localized with the CD63. The cytoplasmic APE1/Ref-1 and CD63 signals are merged to exhibit an orange color in the enlarged region of Figure 5E. Taken together, these data suggest that APE1/Ref-1 in response to E2 treatment was co-localized with CD63-positive exosome, suggesting APE/Ref-1 can be secreted through the exosome pathway in vascular endothelial cells.

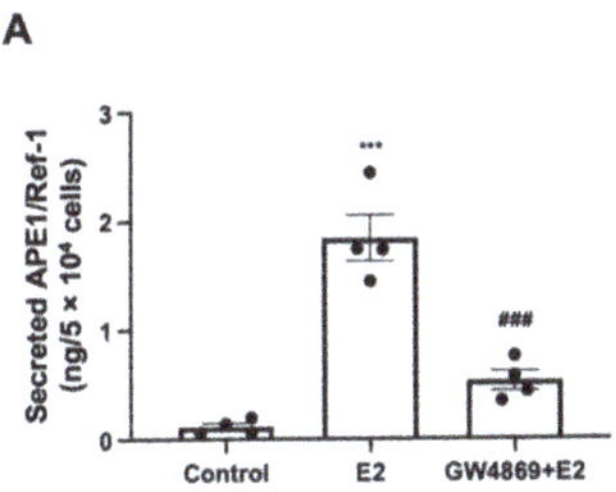

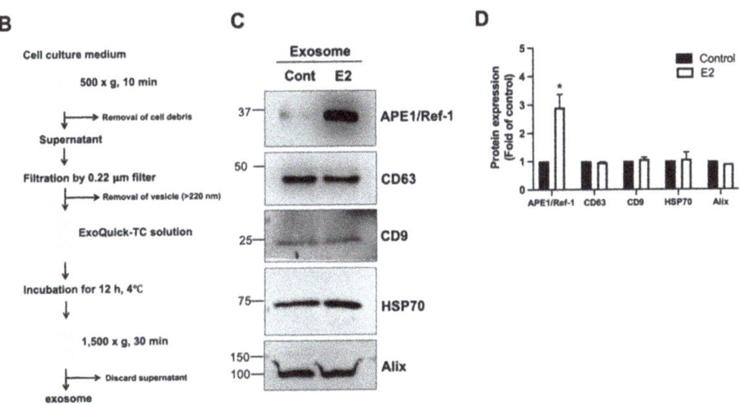

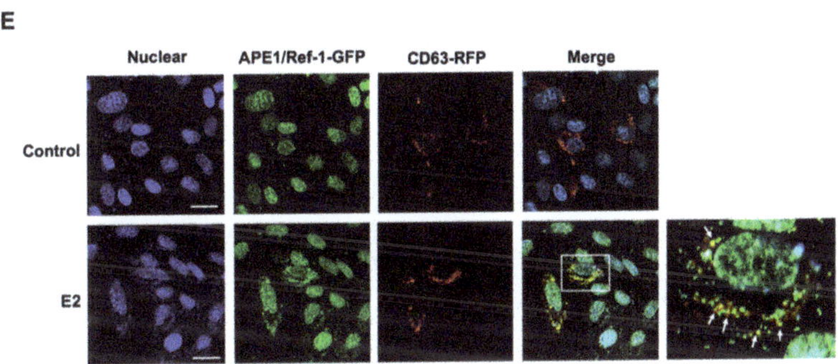

Figure 5. 17β-estradiol(E2)-induced apurinic/apyrimidinic endonuclease-1/redox factor-1 (APE1/Ref-1) secretion is mediated through exosome pathway. (**A**) 17β-estradiol-induced APE1/Ref-1 secretion is inhibited by the exosome inhibitor GW4869 (5 μM). Columns, mean (n = 4); dot plot, SE. *** $p < 0.001$ and * $p < 0.05$ vs. control; ### $p < 0.001$ vs. E2-treated. (**B**) Schematic experimental steps for exosome isolation from the cultured medium of HUVECs. (**C**) 17β-estradiol (E2) increases APE1/Ref-1 expression in exosomes. Exosome isolation from culture media was confirmed with the presence of exosome-specific markers such as CD63, CD9, heat shock protein-70 (HSP70), and ALG-2 interacting protein X (Alix). (**D**) Summarized data of APE1/Ref-1 or exosome markers expression in exosomes. Columns, mean (n = 4); bar, SE. * $p < 0.05$ vs. control. Protein expressions are expressed as relative fold of control bands. (**E**) Immunofluorescence image in cultured endothelial cells transfected with plasmid *pAPE1/Ref-1-GFP* or *pCD63-RFP*. The images illustrate green fluorescent protein (GFP) fluorescence from APE1/Ref-1 and red fluorescent protein (RFP) fluorescence from CD63. The bottom right photo shows a 2.5× magnified image of the white box displayed in the merged image. Note, the cytoplasmic APE1/Ref-1 signals (green) are merged with the CD63 signals (red). Arrows indicate typical merged signal (orange). Scale bar showed 100 μm in length (white line).

4. Discussion

A significant finding of this study was the identification of an endogenous hormone that promotes APE1/Ref-1 secretion in vascular endothelial cells. We demonstrated that among the 12 hormones used, estrogen was the most potent in promoting the APE1/Ref-1 secretion.

APE1/Ref-1 has been detected in various biological solutions, including blood and urine [5,10,12,41–43]. In 2013, we reported that trichostatin A, a histone deacetylase inhibitor, increased the secretion of APE1/Ref-1 in HEK293 cells through intracellular hyperacetylation [4]. However, trichostatin A has been known to inhibit cell growth in cancer cells by inducing apoptosis [44], and even minimal cell death can alter the concentration of a specific protein in culture media, leading to doubts regarding the actual secretion of APE1/Ref-1. In the present study, we designed to exclude the possibility of passive APE1/Ref-1 release due to cell death, and to identify endogenous hormone that may actively increase APE1/Ref-1 secretion. Interestingly, our results showed that the endogenous hormones such as estrogen increased APE1/Ref-1 secretion without cell death as shown in Figure 1. Considering these results, we confirmed that actual secretion of APE1/Ref-1 in response to hormone.

In vascular cells, E2 binds to the ER, activates phosphoinositide 3-kinases (PI3K)/AKT signaling, stimulates eNOS, and consequently produces nitric oxide signaling [45,46]. In the present study, L-NAME and eNOS inhibitors did not directly affect E2-induced APE1/Ref-1 secretion in cultured endothelial cells. eNOS is constitutively expressed in vascular endothelial cells, and produces low nanomolar level of nitric oxide (NO) [47]. Therefore, low nanomolar level of NO did not affect APE1/Ref-1 secretion in endothelial cells. However, it is difficult to conclude that all nitric oxide concentrations or nitrosylation would not affect APE1/Ref-1 secretion. Previous study showed that plasma APE1/Ref-1 levels are increased in lipopolysaccharide-treated experimental animals [5]. Certain oxidative nitrogen-donating agents, such as S-nitrosoglutathione (GSNO) and S-nitroso-N-acetylpenicillamine (SNAP), promote nitrosylation by transferring their nitroxyl group to the protein thiol residue [48]. Previous reports have shown that GSNO selectively induces cytosolic translocation of APE1/Ref-1, where Cys93 and Cys310 are critical for nitrosylation-mediated cytosolic translocation [13].

Fulvestrant is a 7α-alkylsulphinyl analog of the 17β-estradiol structure and the first of the new type of ER antagonist that downregulates the receptor [49]. Fulvestrant competitively inhibits the binding of E2 to the ER and has a binding capacity of 89% compared to that of 17β-estradiol [50]. In the present study, fulvestrant effectively inhibited E2-induced APE1/Ref-1 secretion, suggesting that ER binding of estrogen is an important signal for APE1/Ref-1 secretion. In general, secretion pathways can be influenced by intracellular calcium, which is an important ion that regulates exocytosis and exosome release [39,51]. Cell membrane swelling and binding to other cell membranes are required during exosome formation and exocytosis, and Ca^{2+} is required for this process; therefore, it is likely that calcium may be required for the fusion events involved in exosome generation [51,52]. As APE1/Ref-1 secretion was completely inhibited by the calcium chelator BAPTA-AM, our findings indicate that APE1/Ref-1 secretion induced by E2 is dependent on intracellular calcium.

Recent studies have also demonstrated a great interest in exosomes as an important potential source of biomarkers as they participate in intercellular communication in cardiovascular disorders and tumors by transporting various proteins [53,54]. Proteins secreted via the classical secretory pathway require a secretory signal peptide or leader peptide [55,56]. However, computer-based analysis using SecretomeP predicts non-classical secretion of APE1/Ref-1 and indicates the absence of a putative N-terminal signal peptide [57]. Moreover, APE1/Ref-1 secretion is not blocked by brefeldin A, a typical inhibitor of the classical secretory pathway, again suggesting a non-classical pathway [16]. Secreted membrane-enclosed vesicles are collectively called extracellular vesicles and include exosomes and microvesicles.

Exosomes are typically 30–150 nm in diameter, and the size of the microvesicle range from 100 to 1000 nm in diameter [58]. In the present study, in order to minimize the unwanted mixing of microvesicles during exosome separation, pure exosomes were isolated using an Exoquick isolation kit that precipitates exosomes based on polyethylene glycol precipitation after removing microvesicle over 200 nm in diameter at 0.22 μm filter through filtration [59]. Exosome purification was confirmed by the expression of exosome-specific markers, CD63, CD9, HSP70, and Alix. Based on tetraspanins content, CD9 and CD63 were primarily used as classical exosome marker [60]. Exosome formation can be regulated by endosomal sorting complex required for transport proteins and its accessary protein such as Alix or heat shock protein-90 (HSP90) which is also used as exosome markers [40]. GW4869 is a neutral sphingomyelinase inhibitor and a commonly used pharmacological agent that inhibits exosome generation [61]. It blocks ceramide-mediated inward budding of multivesicular bodies and the release of mature exosomes from multivesicular bodies [62,63]. Our data showed that exosome inhibitor GW4869 significantly inhibited E2-induced APE1/Ref-1 secretion. APE1/Ref-1 secretion in CD63-positive exosomes was significantly increased by E2. These results suggest that APE1/Ref-1 secretion in response to E2 was mediated via exosomes in cultured endothelial cells.

Estrogen is a sex hormone responsible for the development of female reproductive systems and is produced primarily by the ovaries. At menopause, the ovarian follicles degenerate and circulating estrogen decreased to levels of castration. In postmenopausal women, a marked reduction of estrogen is risk factor of cardiovascular disease or osteoporosis [64–66]. In the present study, we also confirmed that reduced plasma APE1/Ref-1 levels in ovariectomized mice, administration of estrogen induced APE1/Ref-1 secretion in mice. These results suggested that low level of APE1/Ref-1 might be related with increased risks of cardiovascular disorders. Considering another aspect, abnormal production of female hormones or hormone-producing tumors can affect the level of APE1/Ref-1 in the blood. Therefore, when analyzing biomarkers such as APE1/Ref-1 in vascular inflammatory diseases, hormonal changes should also be considered.

Author Contributions: Conceptualization, Y.-R.L. and B.-H.J.; Data curation, Y.-R.L., E.-O.L., H.J. and Y.-H.C.; Formal analysis, Y.-R.L., H.-K.J. and S.K.; Writing—original draft writing, B.-H.J.; Supervision, C.-S.K.; Writing—review and editing, B.-H.J. All authors have read and agreed to the published version of the manuscript.

Funding: This research was supported by grants from the Basic Science Research Program through the National Research Foundation of Korea (NRF) funded by the Ministry of Education (NRF-2014R1A6A1029617 to B.-H.J. and 2020R1I1A1A01072327 to Y.-R.L.) and Ministry of Science, ICT & Future Planning (2020R1C1C1014490 to H.-K.J.)

Institutional Review Board Statement: The animal study was approved by the Ethics Committee of Animal Experimentation of Chungnam National University Hospital (CNUH-017-A0025).

Informed Consent Statement: Not applicable.

Conflicts of Interest: The authors declare no conflict of interest. The funders played no role in the design of the study; in the collection, analyses, or interpretation of data; in the writing of the manuscript; or in the decision to publish the results.

References

1. Lee, Y.R.; Joo, H.K.; Lee, E.O.; Park, M.S.; Cho, H.S.; Kim, S.; Jin, H.; Jeong, J.O.; Kim, C.S.; Jeon, B.H. Plasma APE1/Ref-1 Correlates with Atherosclerotic Inflammation in ApoE(-/-) Mice. *Biomedicines* **2020**, *8*, 366. [CrossRef]
2. Lee, Y.R.; Joo, H.K.; Jeon, B.H. The Biological Role of Apurinic/Apyrimidinic Endonuclease1/Redox Factor-1 as a Therapeutic Target for Vascular Inflammation and as a Serologic Biomarker. *Biomedicines* **2020**, *8*, 57. [CrossRef] [PubMed]
3. Jeon, B.H.; Irani, K. APE1/Ref-1: Versatility in progress. *Antioxid. Redox Signal.* **2009**, *11*, 571–574. [CrossRef]
4. Choi, S.; Lee, Y.R.; Park, M.S.; Joo, H.K.; Cho, E.J.; Kim, H.S.; Kim, C.S.; Park, J.B.; Irani, K.; Jeon, B.H. Histone deacetylases inhibitor trichostatin A modulates the extracellular release of APE1/Ref-1. *Biochem. Biophys. Res. Commun.* **2013**, *435*, 403–407. [CrossRef] [PubMed]

5. Park, M.S.; Lee, Y.R.; Choi, S.; Joo, H.K.; Cho, E.J.; Kim, C.S.; Park, J.B.; Jo, E.K.; Jeon, B.H. Identification of plasma APE1/Ref-1 in lipopolysaccharide-induced endotoxemic rats: Implication of serological biomarker for an endotoxemia. *Biochem. Biophys. Res. Commun.* **2013**, *435*, 621–626. [CrossRef]

6. Jin, S.A.; Lim, B.K.; Seo, H.J.; Kim, S.K.; Ahn, K.T.; Jeon, B.H.; Jeong, J.O. Elevation of Serum APE1/Ref-1 in Experimental Murine Myocarditis. *Int. J. Mol. Sci.* **2017**, *18*, 2664. [CrossRef]

7. Habener, J.F.; Powell, D.; Murray, T.M.; Mayer, G.P.; Potts, J.T., Jr. Parathyroid hormone: Secretion and metabolism in vivo. *Proc. Natl. Acad. Sci. USA* **1971**, *68*, 2986–2991. [CrossRef]

8. Tanaka, T.; Narazaki, M.; Kishimoto, T. IL-6 in inflammation, immunity, and disease. *Cold Spring Harb. Perspect. Biol.* **2014**, *6*, a016295. [CrossRef] [PubMed]

9. Mangiapane, G.; Parolini, I.; Conte, K.; Malfatti, M.C.; Corsi, J.; Sanchez, M.; Pietrantoni, A.; D'Agostino, V.G.; Tell, G. Enzymatically active apurinic/apyrimidinic endodeoxyribonuclease 1 is released by mammalian cells through exosomes. *J. Biol. Chem.* **2021**, *296*, 100569–100584. [CrossRef]

10. Park, M.S.; Choi, S.; Lee, Y.R.; Joo, H.K.; Kang, G.; Kim, C.S.; Kim, S.J.; Lee, S.D.; Jeon, B.H. Secreted APE1/Ref-1 inhibits TNF-alpha-stimulated endothelial inflammation via thiol-disulfide exchange in TNF receptor. *Sci. Rep.* **2016**, *6*, 23015. [CrossRef] [PubMed]

11. Joo, H.K.; Lee, Y.R.; Lee, E.O.; Park, M.S.; Choi, S.; Kim, C.S.; Park, J.B.; Jeon, B.H. The extracellular role of Ref-1 as anti-inflammatory function in lipopolysaccharide-induced septic mice. *Free Radic. Biol. Med.* **2019**, *139*, 16–23. [CrossRef]

12. Lee, Y.R.; Kim, K.M.; Jeon, B.H.; Choi, S. Extracellularly secreted APE1/Ref-1 triggers apoptosis in triple-negative breast cancer cells via RAGE binding, which is mediated through acetylation. *Oncotarget* **2015**, *6*, 23383–23398. [CrossRef]

13. Qu, J.; Liu, G.H.; Huang, B.; Chen, C. Nitric oxide controls nuclear export of APE1/Ref-1 through S-nitrosation of cysteines 93 and 310. *Nucleic Acids Res.* **2007**, *35*, 2522–2532. [CrossRef] [PubMed]

14. Muesch, A.; Hartmann, E.; Rohde, K.; Rubartelli, A.; Sitia, R.; Rapoport, T.A. A novel pathway for secretory proteins? *Trends Biochem. Sci.* **1990**, *15*, 86–88. [CrossRef]

15. Nath, S.; Roychoudhury, S.; Kling, M.J.; Song, H.; Biswas, P.; Shukla, A.; Band, H.; Joshi, S.; Bhakat, K.K. The extracellular role of DNA damage repair protein APE1 in regulation of IL-6 expression. *Cell Signal.* **2017**, *39*, 18–31. [CrossRef] [PubMed]

16. Lee, Y.R.; Joo, H.K.; Lee, E.O.; Cho, H.S.; Choi, S.; Kim, C.S.; Jeon, B.H. ATP Binding Cassette Transporter A1 is Involved in Extracellular Secretion of Acetylated APE1/Ref-1. *Int. J. Mol. Sci.* **2019**, *20*, 3178. [CrossRef] [PubMed]

17. Sophocleous, A.; Idris, A.I. Rodent models of osteoporosis. *Bonekey Rep.* **2014**, *3*, 614–623. [CrossRef]

18. Park, M.S.; Kim, C.S.; Joo, H.K.; Lee, Y.R.; Kang, G.; Kim, S.J.; Choi, S.; Lee, S.D.; Park, J.B.; Jeon, B.H. Cytoplasmic localization and redox cysteine residue of APE1/Ref-1 are associated with its anti-inflammatory activity in cultured endothelial cells. *Mol. Cells* **2013**, *36*, 439–445. [CrossRef] [PubMed]

19. Xiu, F.; Stanojcic, M.; Jeschke, M.G. Norepinephrine inhibits macrophage migration by decreasing CCR2 expression. *PLoS ONE* **2013**, *8*, e69167. [CrossRef] [PubMed]

20. Chavez-Noriega, L.E.; Gillespie, A.; Stauderman, K.A.; Crona, J.H.; Claeps, B.O.; Elliott, K.J.; Reid, R.T.; Rao, T.S.; Velicelebi, G.; Harpold, M.M.; et al. Characterization of the recombinant human neuronal nicotinic acetylcholine receptors alpha3beta2 and alpha4beta2 stably expressed in HEK293 cells. *Neuropharmacology* **2000**, *39*, 2543–2560. [CrossRef]

21. Chopra, I.J. An assessment of daily production and significance of thyroidal secretion of 3, 3′, 5′-triiodothyronine (reverse T3) in man. *J. Clin. Investig.* **1976**, *58*, 32–40. [CrossRef]

22. Hosur, M.B.; Puranik, R.S.; Vanaki, S.; Puranik, S.R. Study of thyroid hormones free triiodothyronine (FT3), free thyroxine (FT4) and thyroid stimulating hormone (TSH) in subjects with dental fluorosis. *Eur. J. Dent.* **2012**, *6*, 184–190. [CrossRef]

23. Keating, G.M. Mecasermin. *BioDrugs* **2008**, *22*, 177–188. [CrossRef] [PubMed]

24. Liu, Y.; Mladinov, D.; Pietrusz, J.L.; Usa, K.; Liang, M. Glucocorticoid response elements and 11 beta-hydroxysteroid dehydrogenases in the regulation of endothelial nitric oxide synthase expression. *Cardiovasc. Res.* **2009**, *81*, 140–147. [CrossRef]

25. Pratt, R.E.; Flynn, J.A.; Hobart, P.M.; Paul, M.; Dzau, V.J. Different secretory pathways of renin from mouse cells transfected with the human renin gene. *J. Biol. Chem.* **1988**, *263*, 3137–3141. [CrossRef]

26. Mahalle, N.P.; Garg, M.K.; Kulkarni, M.V.; Naik, S.S. Differences in traditional and non-traditional risk factors with special reference to nutritional factors in patients with coronary artery disease with or without diabetes mellitus. *Indian J. Endocrinol. Metab.* **2013**, *17*, 844–850. [CrossRef]

27. Van Beek, A.P.; de Haas, E.R.; van Vloten, W.A.; Lips, C.J.; Roijers, J.F.; Canninga-van Dijk, M.R. The glucagonoma syndrome and necrolytic migratory erythema: A clinical review. *Eur. J. Endocrinol.* **2004**, *151*, 531–537. [CrossRef] [PubMed]

28. Yassin, A.A.; Saad, F. Plasma levels of dihydrotestosterone remain in the normal range in men treated with long-acting parenteral testosterone undecanoate. *Andrologia* **2007**, *39*, 181–184. [CrossRef] [PubMed]

29. Urysiak-Czubatka, I.; Kmiec, M.L.; Broniarczyk-Dyla, G. Assessment of the usefulness of dihydrotestosterone in the diagnostics of patients with androgenetic alopecia. *Postepy Derm. Alergol.* **2014**, *31*, 207–215. [CrossRef] [PubMed]

30. Stricker, R.; Eberhart, R.; Chevailler, M.C.; Quinn, F.A.; Bischof, P.; Stricker, R. Establishment of detailed reference values for luteinizing hormone, follicle stimulating hormone, estradiol, and progesterone during different phases of the menstrual cycle on the Abbott ARCHITECT analyzer. *Clin. Chem. Lab. Med.* **2006**, *44*, 883–887. [CrossRef] [PubMed]

31. Tulic, L.; Tulic, I.; Bila, J.; Nikolic, L.; Dotlic, J.; Lazarevic-Suntov, M.; Kalezic, I. Correlation of progesterone levels on the day of oocyte retrieval with basal hormonal status and the outcome of ART. *Sci. Rep.* **2020**, *10*, 22291. [CrossRef] [PubMed]

32. Mazer, N.A. A novel spreadsheet method for calculating the free serum concentrations of testosterone, dihydrotestosterone, estradiol, estrone and cortisol: With illustrative examples from male and female populations. *Steroids* **2009**, *74*, 512–519. [CrossRef]

33. Reed, M.J.; Cheng, R.W.; Noel, C.T.; Dudley, H.A.; James, V.H. Plasma levels of estrone, estrone sulfate, and estradiol and the percentage of unbound estradiol in postmenopausal women with and without breast disease. *Cancer Res.* **1983**, *43*, 3940–3943.

34. Wu, C.H.; Motohashi, T.; Abdel-Rahman, H.A.; Flickinger, G.L.; Mikhail, G. Free and protein-bound plasma estradiol-17 beta during the menstrual cycle. *J. Clin. Endocrinol. Metab.* **1976**, *43*, 436–445. [CrossRef] [PubMed]

35. Dawood, M.Y.; Brown, J.B.; Newnam, K.L. Serum free estriol and estriol glucuronide fractions in hydatidiform mole measured by radioimmunoassay. *Obs. Gynecol.* **1977**, *49*, 303–307.

36. Cleary, R.E.; Young, P.C. Serum unconjugated estriol in normal and abnormal pregnancy. *Am. J. Obs. Gynecol.* **1974**, *118*, 18–24. [CrossRef]

37. Rotti, K.; Stevens, J.; Watson, D.; Longcope, C. Estriol concentrations in plasma of normal, non-pregnant women. *Steroids* **1975**, *25*, 807–816. [CrossRef]

38. MacRitchie, A.N.; Jun, S.S.; Chen, Z.; German, Z.; Yuhanna, I.S.; Sherman, T.S.; Shaul, P.W. Estrogen upregulates endothelial nitric oxide synthase gene expression in fetal pulmonary artery endothelium. *Circ. Res.* **1997**, *81*, 355–362. [CrossRef] [PubMed]

39. Thor, D.; Uchizono, J.A.; Lin-Cereghino, G.P.; Rahimian, R. The effect of 17 beta-estradiol on intracellular calcium homeostasis in human endothelial cells. *Eur. J. Pharm.* **2010**, *630*, 92–99. [CrossRef]

40. Kalluri, R.; LeBleu, V.S. The biology, function, and biomedical applications of exosomes. *Science* **2020**, *367*, eaau6977. [CrossRef]

41. Choi, S.; Shin, J.H.; Lee, Y.R.; Joo, H.K.; Song, K.H.; Na, Y.G.; Chang, S.J.; Lim, J.S.; Jeon, B.H. Urinary APE1/Ref-1: A Potential Bladder Cancer Biomarker. *Dis. Markers* **2016**, *2016*, 7276502. [CrossRef]

42. Jin, S.A.; Seo, H.J.; Kim, S.K.; Lee, Y.R.; Choi, S.; Ahn, K.T.; Kim, J.H.; Park, J.H.; Lee, J.H.; Choi, S.W.; et al. Elevation of the Serum Apurinic/Apyrimidinic Endonuclease 1/Redox Factor-1 in Coronary Artery Disease. *Korean Circ. J.* **2015**, *45*, 364–371. [CrossRef]

43. Lee, Y.R.; Lim, J.S.; Shin, J.H.; Choi, S.; Joo, H.K.; Jeon, B.H. Altered Secretory Activity of APE1/Ref-1 D148E Variants Identified in Human Patients with Bladder Cancer. *Int. Neurourol. J.* **2016**, *20*, S30–S37. [CrossRef]

44. Galfi, P.; Neogrady, Z.; Csordas, A. Apoptosis sensitivity is not correlated with sensitivity to proliferation inhibition by the histone deacetylase inhibitors butyrate and TSA. *Cancer Lett.* **2002**, *188*, 141–152. [CrossRef]

45. Chambliss, K.L.; Yuhanna, I.S.; Mineo, C.; Liu, P.; German, Z.; Sherman, T.S.; Mendelsohn, M.E.; Anderson, R.G.; Shaul, P.W. Estrogen receptor alpha and endothelial nitric oxide synthase are organized into a functional signaling module in caveolae. *Circ. Res.* **2000**, *87*, E44–E52. [CrossRef]

46. Haynes, M.P.; Li, L.; Sinha, D.; Russell, K.S.; Hisamoto, K.; Baron, R.; Collinge, M.; Sessa, W.C.; Bender, J.R. Src kinase mediates phosphatidylinositol 3-kinase/Akt-dependent rapid endothelial nitric-oxide synthase activation by estrogen. *J. Biol. Chem.* **2003**, *278*, 2118–2123. [CrossRef] [PubMed]

47. Hall, C.N.; Garthwaite, J. What is the real physiological NO concentration in vivo? *Nitric Oxide* **2009**, *21*, 92–103. [CrossRef]

48. Arnelle, D.R.; Stamler, J.S. NO+, NO, and NO- donation by S-nitrosothiols: Implications for regulation of physiological functions by S-nitrosylation and acceleration of disulfide formation. *Arch. Biochem. Biophys.* **1995**, *318*, 279–285. [CrossRef] [PubMed]

49. Osborne, C.K.; Wakeling, A.; Nicholson, R.I. Fulvestrant: An oestrogen receptor antagonist with a novel mechanism of action. *Br. J. Cancer* **2004**, *90* (Suppl. 1), S2–S6. [CrossRef]

50. Wakeling, A.E.; Bowler, J. Steroidal pure antioestrogens. *J. Endocrinol.* **1987**, *112*, R7–R10. [CrossRef]

51. Barclay, J.W.; Morgan, A.; Burgoyne, R.D. Calcium-dependent regulation of exocytosis. *Cell Calcium* **2005**, *38*, 343–353. [CrossRef]

52. Savina, A.; Furlan, M.; Vidal, M.; Colombo, M.I. Exosome release is regulated by a calcium-dependent mechanism in K562 cells. *J. Biol. Chem.* **2003**, *278*, 20083–20090. [CrossRef]

53. Zhou, B.; Xu, K.; Zheng, X.; Chen, T.; Wang, J.; Song, Y.; Shao, Y.; Zheng, S. Application of exosomes as liquid biopsy in clinical diagnosis. *Signal. Transduct. Target. Ther.* **2020**, *5*, 144. [CrossRef] [PubMed]

54. Logozzi, M.; Mizzoni, D.; Di Raimo, R.; Fais, S. Exosomes: A Source for New and Old Biomarkers in Cancer. *Cancers* **2020**, *12*, 2566. [CrossRef] [PubMed]

55. Eder, C. Mechanisms of interleukin-1beta release. *Immunobiology* **2009**, *214*, 543–553. [CrossRef] [PubMed]

56. Morrison, W.H.; Lou, M.F.; Hamilton, P.B. The determination of hexoses and pentoses by anion-exchange chromatography: A method of high sensitivity. *Anal. Biochem.* **1976**, *71*, 415–425. [CrossRef]

57. Choi, S.; Joo, H.K.; Jeon, B.H. Dynamic Regulation of APE1/Ref-1 as a Therapeutic Target Protein. *Chonnam Med. J.* **2016**, *52*, 75–80. [CrossRef]

58. Zaborowski, M.P.; Balaj, L.; Breakefield, X.O.; Lai, C.P. Extracellular Vesicles: Composition, Biological Relevance, and Methods of Study. *Bioscience* **2015**, *65*, 783–797. [CrossRef] [PubMed]

59. Doyle, L.M.; Wang, M.Z. Overview of Extracellular Vesicles, Their Origin, Composition, Purpose, and Methods for Exosome Isolation and Analysis. *Cells* **2019**, *8*, 727. [CrossRef]

60. Andreu, Z.; Yanez-Mo, M. Tetraspanins in extracellular vesicle formation and function. *Front. Immunol* **2014**, *5*, 442. [CrossRef]

61. Chen, L.; Brigstock, D.R. Integrins and heparan sulfate proteoglycans on hepatic stellate cells (HSC) are novel receptors for HSC-derived exosomes. *FEBS Lett.* **2016**, *590*, 4263–4274. [CrossRef]

62. Kosaka, N.; Iguchi, H.; Yoshioka, Y.; Takeshita, F.; Matsuki, Y.; Ochiya, T. Secretory mechanisms and intercellular transfer of microRNAs in living cells. *J. Biol. Chem.* **2010**, *285*, 17442–17452. [CrossRef] [PubMed]

63. Aydin, Y.; Koksal, A.R.; Reddy, V.; Lin, D.; Osman, H.; Heidari, Z.; Rhadhi, S.M.; Wimley, W.C.; Parsi, M.A.; Dash, S. Extracellular Vesicle Release Promotes Viral Replication during Persistent HCV Infection. *Cells* **2021**, *10*, 984. [CrossRef] [PubMed]
64. Hodis, H.N.; Mack, W.J.; Azen, S.P.; Lobo, R.A.; Shoupe, D.; Mahrer, P.R.; Faxon, D.P.; Cashin-Hemphill, L.; Sanmarco, M.E.; French, W.J.; et al. Hormone therapy and the progression of coronary-artery atherosclerosis in postmenopausal women. *N. Engl. J. Med.* **2003**, *349*, 535–545. [CrossRef] [PubMed]
65. Vehkavaara, S.; Hakala-Ala-Pietila, T.; Virkamaki, A.; Bergholm, R.; Ehnholm, C.; Hovatta, O.; Taskinen, M.R.; Yki-Jarvinen, H. Differential effects of oral and transdermal estrogen replacement therapy on endothelial function in postmenopausal women. *Circulation* **2000**, *102*, 2687–2693. [CrossRef] [PubMed]
66. Ji, M.X.; Yu, Q. Primary osteoporosis in postmenopausal women. *Chronic Dis. Transl. Med.* **2015**, *1*, 9–13.

 *biomedicines*

Article

Elevated Plasma Apurinic/Apyrimidinic Endonuclease 1/Redox Effector Factor-1 Levels in Refractory Kawasaki Disease

Yu-Ran Lee [1], Eun Young Bae [2], Hong Ryang Kil [2], Byeong-Hwa Jeon [1,*] and Geena Kim [3,*]

1   Department of Physiology, College of Medicine, Chungnam National University, Daejeon 35015, Korea; lyr0913@gmail.com
2   Department of Pediatrics, College of Medicine, Chungnam National University Hospital, Chungnam National University, Daejeon 35015, Korea; pebble1217@hanmail.net (E.Y.B.); gilhongr@gmail.com (H.R.K.)
3   Department of Pediatrics, College of Medicine, Chungnam National University Sejong Hospital, Chungnam National University, Sejong 30099, Korea
*   Correspondence: bhjeon@cnu.ac.kr (B.-H.J.); drgnkim@naver.com (G.K.); Tel.: +82-42-580-8214 (B.-H.J.); +82-44-995-4761 (G.K.)

Abstract: Kawasaki disease (KD) refers to systemic vasculitis of medium-sized vessels accompanied by fever. The multifunctional protein apurinic/apyrimidinic endonuclease-1/redox factor-1 (APE1/Ref-1) is a new biomarker for vascular inflammation. Here, we investigated the association between APE1/Ref-1 and KD. Three groups, including 32 patients with KD (KD group), 33 patients with fever (Fever group), and 19 healthy individuals (Healthy group), were prospectively analyzed. APE1/Ref-1 levels were measured, and the clinical characteristics of KD were evaluated. The mean age of all patients was 2.7 ± 1.8 years, but the Healthy group participants were older than the other participants. Fever duration was longer in the KD group than in the fever group. APE1/Ref-1 levels were significantly higher in the KD group ($p = 0.004$) than in the other two groups, but there was no difference between the healthy and fever groups. APE1/Ref-1 levels did not differ according to fever duration or coronary arterial lesion but were higher in refractory KD cases than in non-refractory cases. APE1/Ref-1 levels were significantly higher during the acute phase of KD. We propose that APE1/Ref-1 could be a beneficial biological marker for the diagnosis and prognosis of KD, especially in refractory KD.

Keywords: apurinic/apyrimidinic endonuclease-1/redox factor-1; mucocutaneous lymph node syndrome; vasculitis

Citation: Lee, Y.-R.; Bae, E.Y.; Kil, H.R.; Jeon, B.-H.; Kim, G. Elevated Plasma Apurinic/Apyrimidinic Endonuclease 1/Redox Effector Factor-1 Levels in Refractory Kawasaki Disease. *Biomedicines* **2022**, *10*, 190. https://doi.org/10.3390/biomedicines10010190

Academic Editor: Estefania Nuñez

Received: 30 October 2021
Accepted: 13 January 2022
Published: 17 January 2022

Publisher's Note: MDPI stays neutral with regard to jurisdictional claims in published maps and institutional affiliations.

1. Introduction

Kawasaki disease (KD) is an acute febrile disease diagnosed in young children or infants aged less than 6 years [1]. Kawasaki and Naoe reported KD in 1967 [2]. It is characterized by a persistent fever above 38 °C over 5 days, and its clinical symptoms include nonsuppurative bilateral conjunctival injection, red lips, strawberry tongue, atypical exanthema, cervical lymph node enlargement, swelling and erythema of hands and feet, and membranous desquamation [3]. Patients who meet the criteria according to principal clinical findings are considered as complete KD cases, and those who do not meet the criteria are diagnosed with incomplete KD [3]. Most KD cases are manageable by treatment, without any complications. Intravenous immune globulin is the standard treatment for the acute stage of KD [3]. However, this disease may be accompanied by diverse cardiovascular complications such as coronary aneurysms, heart failure in the acute complication stage, and myocardial infarction in patients with large coronary aneurysms, and these complications are more likely associated with refractory KD. Refractory KD means that the patients are resistant to the standard immunoglobulin treatment for KD. [4]. KD is a type of systemic vasculitis primarily involving the arterial wall [5]. Various KD biomarkers have been suggested for definitive KD diagnosis [6,7]. Apurinic/apyrimidinic endonuclease-1/redox factor-1 (APE1/Ref-1) is a multifunctional protein that plays roles in transcriptional regulation through redox modification and base excision repair [8]. It has been suggested that alterations in APE1/Ref-1 are associated with various diseases, including cancer, neurodegenerative disease, coronary arterial disease, murine myocarditis, and hypertension [9–12]. APE1/Ref-1 acts as a reductive activator of various transcription factors for controlling cell apoptosis, inflammation, and proliferation [12]. We hypothesized that APE1/Ref-1 could be a diagnostic biomarker for KD patients and may distinguish other patients with fever. The main objective of this study was to investigate the association between APE1/Ref-1 and KD and further analyze APE1/Ref-1 according to KD characteristics.

2. Materials and Methods

2.1. Patients and Data Collection

Patients with KD and fever and the healthy controls were prospectively enrolled from January 2020 to February 2021 at the pediatric department of Chungnam National University Hospital. The study was approved by the Chungnam National University Hospital Institutional Review Board, and informed consent forms were signed by the parents/legal guardians of the patients. The inclusion criteria for the KD group were admission and treatment of children with KD, for the fever group were admission as patients with fever, and for the healthy group were healthy patients from a pediatric outpatient clinic who visited for simple cardiac murmur or short stature. All the enrolled participants were >1 month and <8 years of age. Exclusion criteria included neonates, a severe infection that required intensive care, or a malignant condition accompanied by congenital heart disease. KD patients met the diagnostic criteria for KD established in 2017 by the American Heart Association [3]. Age at diagnosis, sex, weight, fever duration, presence of complete KD, findings from blood examination, responsiveness to immunoglobulin, the Kobayashi score, Egami score, Sano score, and echocardiography results were examined for children in the KD group [13]. All patients with KD underwent echocardiography at diagnosis, at 2 weeks, and at 2 months. The acute-phase coronary arterial lesion was determined based on echocardiography findings conducted 2 weeks after admission [3]. Patients with KD were treated with 2 g/kg intravenous immunoglobulin (IVIG) administered as a single infusion and medium-dose aspirin (50 mg/kg/day). Immunoglobulin was re-administered if persistent fever occurred at least 36 h after the IVIG infusion. A methylprednisolone pulse of 30 mg/kg and infliximab were sequentially administered when there was no response to immunoglobulin re-administration. Three days after the initial aspirin administration, the dose was reduced to 3–5 mg/kg/day. The fever group participants were evaluated for age, sex, weight, fever duration, and blood examination. Some patients were diagnosed with viral infections such as rhinovirus, adenovirus, acute pharyngitis, acute otitis media, acute gastroenteritis, acute bronchitis, or pneumonia in the fever group; they were given supportive care and antibiotics treatment according to diagnosis.

2.2. Measurement of APE1/Ref-1 Levels

All blood samples were collected in vacuum tubes during KD and fever diagnoses. Blood samples in the fever group were obtained at admission due to fever. The plasma was centrifuged at $3000 \times g$ rpm for 10 min to obtain cell-free samples. APE1/Ref-1 levels were determined using an APE1/Ref-1 sandwich enzyme-linked immunosorbent assay (ELISA) kit (MediRedox, Daejeon, Korea) according to the manufacturer's instructions. Secreted APE1/Ref-1 levels in plasma (ng/mL) were calculated against a standard curve generated using recombinant human APE1/Ref-1 protein (MediRedox).

2.3. Statistical Analysis

SPSS Statistical Package version 21.0 (IBM Corp., Armonk, NY, USA) was used for all statistical analyses. Data are presented as mean ± standard deviation (SD), number (%), or mean value ± standard deviation for normally distributed data. Independent *t*-tests were used to compare continuous variables, while chi-square tests were used for comparing categorical variables. Variables between two groups were compared using an unpaired Student's *t*-test, and the three groups were compared using a one-way ANOVA. The correlation between APE1/Ref-1 levels and other laboratory data was analyzed using Pearson's correlation analysis. Statistical significance was set at $p < 0.05$.

3. Results

3.1. Patient Characteristics

Patient baseline characteristics are presented in Table 1. The mean age was 2.6 ± 1.6 years for the KD group, 2.3 ± 1.9 years for the fever group, and 3.6 ± 1.7 years for the healthy group participants. The age of the healthy group was significantly higher than that of the other two groups ($p = 0.029$). No difference was found in sex distribution or bodyweight among the three groups. Fever duration in the KD group was longer than that in the fever group, at 5.0 ± 1.6 d and 3.0 ± 2.8 d, respectively.

Table 1. Patient baseline characteristics.

Variables	KD (n = 32)	Fever (n = 33)	Healthy Control (n = 19)	*p* Value
	n (%) or Mean $\pm$ SD	n (%) or Mean $\pm$ SD	n (%) or Mean $\pm$ SD	
Sex (male/female)	20 (62)	17 (51)	13 (68)	0.445
Age (years)	2.6 ± 1.6	2.3 ± 1.9	3.6 ± 1.7	0.029
Bodyweight (kg)	14.6 ± 5.4	12.7 ± 5.2	14.9 ± 4.5	0.235
Fever duration (days)	5.0 ± 1.6	3.0 ± 2.8		0.001

3.2. Laboratory Results

Laboratory results for the KD, fever, and healthy groups are shown in Table 2. White blood cell (WBC) counts, hemoglobin levels, platelet counts, number of segment neutrophils, and levels of aspartate aminotransferase (AST), alanine aminotransferase (ALT), protein, albumin, bilirubin, blood urea nitrogen (BUN), creatinine, sodium, and C-reactive protein (CRP) were analyzed. Only the KD group was evaluated for N-terminal pro-b-type natriuretic peptide (NT-proBNP) levels. As shown in Table 2, platelet count and total protein, bilirubin, BUN, and creatinine levels showed no significant difference among the three groups. WBC counts and the percentage of segment neutrophils were significantly higher in the KD and fever groups than in the healthy group ($p < 0.001$, $p < 0.001$, respectively). Hemoglobin levels were significantly different among the three groups ($p = 0.003$). The mean $\pm$ 2 SD of hemoglobin levels in the KD group was 11.0 ± 1.2 g/dL, which was lower than that in the fever group at 11.5 ± 1.0 g/dL ($p = 0.041$). The mean $\pm$ 2 SD of CRP level was higher in the KD group than in the fever group (8.5 ± 6.0 mg/dL vs. 3.4 ± 5.0 mg/dL, $p < 0.001$). Albumin showed differences among the three groups ($p < 0.001$). The mean $\pm$ 2 SD of albumin levels in the KD group was lower than that in the fever group (3.4 ± 0.4 g/dL vs. 4.0 ± 0.4 g/dL; $p < 0.001$). AST and ALT levels showed differences among the three groups ($p = 0.039$ and $p < 0.001$, respectively). AST levels were higher in the KD group than in the healthy group. ALT levels were higher in the KD group than in the fever group (139 ± 177 U/L vs. 27 ± 46 U/L; $p = 0.001$). Sodium levels differed among the three groups ($p < 0.001$), and the mean $\pm$ 2 SD of sodium levels in the KD group was higher than that in the fever group (135 ± 1 mEq/L, 137 ± 2 mEq/L, $p = 0.007$). The mean $\pm$ 2 SD of NT-proBNP in the KD group was 1315 ± 3345 pg/mL. We could not obtain the related data in the other two groups. However, a study that investigated the normal reference value of NT-proBNP in normal children reported that the median of NT-proBNP in 2–6-year-old children were 70 pg/mL (range 5–391 pg/mL) [14]. The mean $\pm$ 2 SD of APE1/Ref-1 levels was significantly higher in the KD group (0.654 ± 0.265 ng/mL than in the fever group (0.459 ± 0.290 ng/mL) and healthy group (0.442 ± 0.199 ng/mL), and the differences were significant ($p = 0.004$). Figure 1 shows the difference in APE1/Ref-1 levels among the three groups; the levels were higher in the KD group than in the fever group ($p = 0.019$) and the healthy group ($p = 0.007$). No difference was found in APE1/Ref-1 levels between groups 2 and 3. Additionally, the ROC curve of APE1/Ref-1 comparing KD and fever showed a cutoff level of 0.542 ng/mL for predicting KD, with an area under the curve of 0.682, a sensitivity of 60.6%, and specificity of 62.5%.

Table 2. Laboratory data.

Variables	KD Group (n = 32)	Fever Group (n = 33)	Healthy Group (n = 19)	p Value
	Mean ± SD	Mean ± SD	Mean ± SD	
WBC (/μL)	13,600 ± 3690	12,644 ±7201	7260 ± 1319	<0.001
Seg (%)	63 ±12	59 ± 17	40 ± 15	<0.001
Hb (g/dL)	11.0 ± 1.2	11.5 ± 1.0	12.1 ± 1.1	0.003
Platelet ($\times 10^3$/μL)	336 ± 83	302 ± 113	331 ± 61	0.312
CRP (mg/dL)	8.5 ± 6.0	3.4 ± 5.0		<0.001
Total protein	6.4 ± 0.6	6.7 ± 0.6	6.5 ± 0.3	0.110
Albumin (g/dL)	3.4 ± 0.4	4.0 ± 0.4	4.2 ± 0.1	<0.001
AST (U/L)	98 ± 166	41 ± 23	30 ± 5	0.039
ALT (U/L)	139 ± 177	27 ± 46	15 ± 5	<0.001
Bilirubin (mg/dL)	0.7 ± 0.8	0.4 ± 0.2	0.5 ± 0.5	0.091
BUN (mg/dL)	10.0 ± 3.0	9.7 ±3.8	11.5 ±2.7	0.178
Creatinine (mg/dL)	0.25 ± 0.05	0.25 ± 0.09	0.30 ± 0.06	0.103
Sodium (mEq/L)	135 ± 1	137 ± 2	138 ± 1	<0.001
NT-proBNP (pg/mL)	1315 ± 3345			
APE1/Ref-1 (ng/mL)	0.654 ± 0.265	0.459 ± 0.290	0.442 ± 0.199	0.004

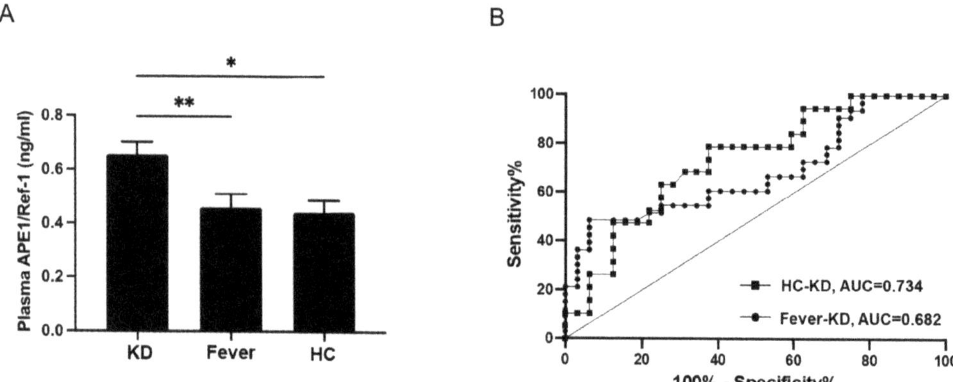

Figure 1. APE1/Ref-1 levels in the three groups, i.e., KD, fever, and healthy control groups. (**A**) APE1/Ref-1 levels were higher in the KD group than in the fever group (** *p* = 0.019) and the healthy control group (* *p* = 0.007). (**B**) The ROC curve of APE1/Ref-1 predicting KD, the cutoff value of APE1/Ref-1 predicting KD compared with that in the fever group was 0.542 ng/mL (AUC = 0.682; sensitivity of 60.6%; specificity of 62.5%), and the cutoff value of APE1/Ref-1 predicting KD compared with that in the healthy group was 0.482 ng/mL (AUC = 0.734; sensitivity of 68.4%; specificity of 68.7%), KD, Kawasaki disease; HC, healthy control.

Next, we analyzed the correlation between APE1/Ref-1 levels and fever duration. No correlation was found. Additionally, there was no correlation between APE1/Ref-1 levels and other laboratory results, including hemoglobin, CRP, albumin, alanine aminotransferase, bilirubin, sodium, and NT-proBNP (Figure S1).

3.3. APE1/Ref-1 According to Characteristics of KD

Complete KD was observed in 13 of the 32 (40.6%) patients with KD, but there was no difference in APE1/Ref-1 levels relative to complete KD or incomplete KD. Four of the KD patients (12.5%) had coronary arterial lesions. Three patients with coronary arterial lesions were diagnosed at the time of diagnosis. One patient with coronary arterial lesion was diagnosed at 2 weeks of echocardiography. Yet, there was no difference in APE/Ref-1 levels based on the presence or absence of coronary arterial lesions. Predictive scores, i.e., the Kobayashi, Egami, and Sano scores, for responsiveness to immunoglobulins in KD were 2.4 ± 2.0, 1.8 ± 1.3, and 0.8 ± 0.8 (mean $\pm$ standard deviation), respectively. We further analyzed APE1/Ref-1 by dividing the patients into high-score and low-score groups and found no difference in the subgroups. Refractory KD that was nonresponsive to immunoglobulin was observed in 9 of the 32 cases of KD (28.1%). All the patients with refractory KD received methylprednisolone treatment, and none received infliximab. As shown in Figure 2, the mean level of APE1/Ref-1 in the refractory KD group was 0.803 ± 0.111 ng/mL compared to that of 0.459 ± 0.290 ng/mL in the fever group ($p = 0.011$) and 0.596 ± 0.044 ng/mL in non-refractory KD ($p = 0.046$). The ROC curve of APE1/Ref-1 comparing refractory vs. non-refractory and fever vs. refractory is presented in Figure 2.

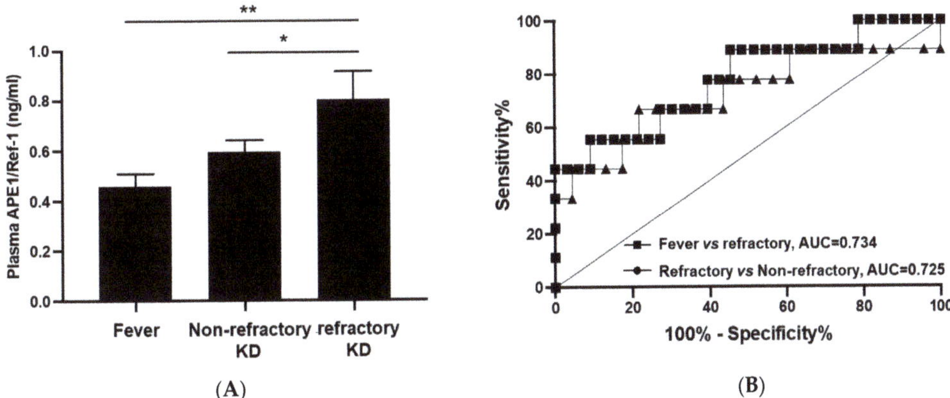

(A) (B)

Figure 2. APE1/Ref-1 levels between the fever, non-refractory KD, and refractory KD groups. (A) APE1/Ref-1 levels between the fever and refractory KD groups showed a significant difference (** $p = 0.011$) and those between the non-refractory and refractory KD also showed a significant difference (* $p = 0.046$) (B) The ROC curve of APE1/Ref-1 predicting refractory KD, (Fever vs. refractory KD, AUC = 0.734, Refractory vs. non-refractory, AUC = 0.725).

4. Discussion

This study revealed the elevation of APE1/Ref-1 levels in patients with KD compared to that in patients with fever and healthy individuals. Patients with fever showed no difference in APE1/Ref-1 levels compared to healthy children. No correlation was found between APE1/Ref-1 levels and CRP or NT-proBNP, which are the established serum biomarkers of KD. Moreover, APE1/Ref-1 levels were higher in the refractory KD group than in the non-refractory KD and the fever group.

Many conditions, including cancer, hypertension, atherosclerosis, and particularly cardiovascular diseases, such as myocarditis and coronary artery disease, can lead to an elevation in APE1/Ref-1 levels [9–12,15,16]. APE1/Ref-1 levels are also elevated in response to radiation, reactive oxygen species (ROS), ischemia/reperfusion, and hypoxia. We found an association between the acute stage of KD and the point at which APE1/Ref-1 levels were elevated in conditions of vascular inflammation.

According to a study by Jin et al., the elevation in APE1/Ref-1 levels is associated with chronic inflammation, ROS, and ischemia/reperfusion injury in coronary artery disease and myocarditis [10]. These investigators also suggested that APE1/Ref-1 elevation is not a non-specific result due solely to inflammation, as they found no correlation with hsCRP. This is consistent with the findings of our current study in that our data also showed that APE1/Ref-1 was not correlated with CRP or NT-proBNP levels (Figure S1). The protective effect of APE1/Ref-1 is different from that of other inflammatory biomarkers or other cardiac biomarkers. The repair role of APE1/Ref-a is also suggested from findings of a murine myocarditis model in that APE1/Ref-1 is elevated at a later time point, in contrast to that of other markers, such as IL-1β, IFN-β, and IL-6 expression, which disappear with a decline in virus titer [11]. We believe these protective and repairable effects of APE1/Ref-1 were influenced by the prolonged elevation of APE1/Ref-1 after the acute stage of KD.

We hypothesize several mechanisms underlying APE1/Ref-1 elevation in KD. During the acute stage of KD, monocyte/macrophage-dominant inflammatory cell infiltration is observed throughout the entire arterial wall [17]. Activated inflammatory cells are highly stimulatory and orchestrate various reactions, including ROS production. Increased inducible nitric oxide synthase (iNOS) levels are observed in the infiltrating and accumulating inflammatory cells and the vascular smooth muscle cells, producing a large amount of nitric oxide (NO). The NO of iNOS origin is an unstable radical and forms peroxynitrite under oxidative stress, which leads to vascular damage [18].

In animal studies, plasma APE1/Ref-1 secretion might be strongly correlated with plasma NO levels induced by TNF-α or lipopolysaccharides [19]. In addition, recombinant human APE1/Ref-1 treatment suppresses TNF-α-induced VCAM-1 expression in endothelial cells, suggesting a functional role for extracellular APE1/Ref-1 [19]. TNF-α is an inflammatory mediator that is a potent activator of endothelial cells and is a key mediator in KD. An early study showed that TNF-α induces endothelial cell apoptosis in the blood serum of children with KD [20]. Furthermore, serum TNF-α levels are significantly elevated in children with acute KD, and they correlate with the incidence of coronary artery aneurysms [21]. In an animal model of KD induced by *Lactobacillus casei* cell wall extract, the process of coronary arteritis and aneurysms can be ablated by blocking the TNF receptor, suggesting that TNF-α can directly induce coronary artery lesions [22]. Various studies have also reported the clinical efficacy of blocking TNF-α production in children with refractory KD resistant to standard immunoglobulin and aspirin treatments. The fact that TNF-α is a key mediator of KD may be associated with the observation in the current study of APE1/Ref-1 being higher in refractory KD compared to that of non-refractory KD. We could not determine if the action of TNF-α induced the secretion of APE1/Ref-1 in refractory KD.

A major concern regarding KD is the diagnosis and treatment of refractory KD. Refractory KD presents with prolonged fever, eventually leading to a coronary arterial aneurysm [23]. New sensitive and specific biomarkers are needed for the diagnosis and prognosis of refractory KD. High APE1/Ref-1 levels in refractory KD observed in this study represent meaningful data. Although this study revealed that APE1/Ref-1 levels in patients with KD were higher than those in patients with fever and the healthy individuals, the non-refractory KD group did not show a significant difference in APE1/Ref-1 levels when compared with the fever group. However, considering that it is difficult to predict during diagnosis whether KD is refractory or non-refractory, and refractory KD cannot be determined in an earlier phase, the elevation of APE1/Ref-1 levels in KD compared to that in the fever group could be a helpful and important finding, which can be applied in the clinical setting. Elevated APE1/Ref-1 levels could be used to predict refractory KD and to distinguish between non-refractory and refractory KD.

The levels of APE1/Ref-1 in our study did not correlate with the presence of coronary arterial lesions, a KD complication. We obtained a sample during diagnosis. Although coronary arterial complications often begin during the first two weeks after fever onset, they can continue for months to years in a small subset of patients [3]. The small number of study participants with coronary arterial lesions among all patients is notable, and further study is needed in the future.

Our study did have several limitations. For instance, the study included a small number of patients, and there was a lack of data regarding serial time points of follow-up in KD patients. Thus, this was a cross-sectional measurement of KD at the point of diagnosis. It will be necessary to analyze serial data of APE1/Ref-1 in the long-term aspects of KD. However, follow-up research on KD may be complicated by aspirin being one of the treatments for KD as it could influence the APE1/Ref-1 levels. It is necessary to conduct a follow-up study. Using an animal model of KD would allow the exception of aspirin, providing insight into its effect. Another limitation is that there is no standard value for APE1/Ref-1 in children based on sex or other parameters.

In conclusion, APE1/Ref-1 levels were significantly higher in patients with KD during the acute phase of disease than in patients with fever and healthy children. Moreover, APE1/Ref-1 levels were significantly higher in patients with refractory KD. We suggest that APE1/Ref-1 could help diagnose KD, especially patients with refractory KD, in addition to improving its prognosis.

Supplementary Materials: The following supporting information can be downloaded at: https://www.mdpi.com/article/10.3390/biomedicines10010190/s1, Figure S1: Correlation between plasma APE1/Ref-1 levels and laboratory data.

Author Contributions: Conceptualization, H.R.K., G.K. and B.-H.J.; methodology, B.-H.J. and G.K.; software, G.K.; validation, Y.-R.L. and G.K.; formal analysis, G.K. and Y.-R.L.; investigation, E.Y.B.; resources, G.K.; data curation, G.K. and Y.-R.L.; writing—original draft preparation, G.K.; writing—review and editing, G.K., Y.-R.L. and B.-H.J.; visualization, G.K. and Y.-R.L.; supervision, G.K. and Y.-R.L.; project administration, G.K. and Y.-R.L.; funding acquisition, G.K and B.-H.J. All authors have read and agreed to the published version of the manuscript.

Funding: This research was supported by Chungnam National University Hospital 2019 research support fund (2019-1568-01). This research was supported by grants from the Basic Science Research Program through the National Research Foundation of Korea (NRF) funded by the Ministry of Education (NRF-2014R1A6A1029617 to B.-H.J.).

Institutional Review Board Statement: The study was conducted according to the guidelines of the Declaration of Helsinki and approved by the Institutional Review Board of Chungnam National University Hospital (2019-11-056).

Informed Consent Statement: Informed consent was obtained from all subjects involved in the study.

Conflicts of Interest: The authors declare no conflict of interest. The funders had no role in the design of the study; in the collection, analyses, or interpretation of data; in the writing of the manuscript, or in the decision to publish the results.

References

1. Sosa, T.; Brower, L.; Divanovic, A. Diagnosis and management of kawasaki disease. *JAMA Pediatr.* **2019**, *173*, 278–279. [CrossRef]
2. Kawasaki, T.; Naoe, S. History of Kawasaki disease. *Clin. Exp. Nephrol.* **2014**, *18*, 301–304. [CrossRef] [PubMed]
3. McCrindle, B.W.; Rowley, A.H.; Newburger, J.W.; Burns, J.C.; Bolger, A.F.; Gewitz, M.; Tani, L.Y.; Burns, J.C.; Shulman, S.T.; Bolger, A.F.; et al. Diagnosis, treatment, and long-term management of kawasaki disease: A scientific statement for health professionals from the american heart association. *Circulation* **2017**, *135*, e927–e999. [CrossRef] [PubMed]
4. McCandless, R.T.; Minich, L.L.; Wilkinson, S.E.; McFadden, M.L.; Tani, L.Y.; Menon, S.C. Myocardial strain and strain rate in Kawasaki disease. *Eur. Heart J. Cardiovasc. Imaging* **2013**, *14*, 1061–1068. [CrossRef]
5. Gordon, J.B.; Kahn, A.M.; Burns, J.C. When children with Kawasaki disease grow up: Myocardial and vascular complications in adulthood. *J. Am. Coll. Cardiol.* **2009**, *54*, 1911–1920. [CrossRef] [PubMed]
6. Sato, Y.Z.; Molkara, D.P.; Daniels, L.B.; Tremoulet, A.H.; Shimizu, C.; Kanegaye, J.T.; Best, B.M.; Snider, J.V.; Frazer, J.R.; Maisel, A.; et al. Cardiovascular biomarkers in acute Kawasaki disease. *Int. J. Cardiol.* **2013**, *164*, 58–63. [CrossRef] [PubMed]
7. Parthasarathy, P.; Agarwal, A.; Chawla, K.; Tofighi, T.; Mondal, T.K. Upcoming biomarkers for the diagnosis of Kawasaki disease: A review. *Clin. Biochem.* **2015**, *48*, 1188–1194. [CrossRef]
8. Jeon, B.H.; Irani, K. APE1/Ref-1: Versatility in progress. *Antioxid. Redox Signal.* **2009**, *11*, 571–574. [CrossRef]
9. Choi, S.; Shin, J.H.; Lee, Y.R.; Joo, H.K.; Song, K.H.; Na, Y.G.; Chang, S.J.; Lim, J.S.; Jeon, B.H. Urinary APE1/Ref-1: A potential bladder cancer biomarker. *Dis. Markers* **2016**, *2016*, 7276502. [CrossRef]
10. Jin, S.A.; Seo, H.J.; Kim, S.K.; Lee, Y.R.; Choi, S.; Ahn, K.T.; Kim, J.H.; Park, J.H.; Lee, J.H.; Choi, S.W.; et al. Elevation of the serum apurinic/apyrimidinic endonuclease 1/redox factor-1 in coronary artery disease. *Korean Circ. J.* **2015**, *45*, 364–371. [CrossRef]
11. Jin, S.A.; Lim, B.K.; Seo, H.J.; Kim, S.K.; Ahn, K.T.; Jeon, B.H.; Jeong, J.O. Elevation of serum APE1/Ref-1 in experimental murine myocarditis. *Int. J. Mol. Sci.* **2017**, *18*, 2664. [CrossRef]
12. Thakur, S.; Sarkar, B.; Cholia, R.P.; Gautam, N.; Dhiman, M.; Mantha, A.K. APE1/Ref-1 as an emerging therapeutic target for various human diseases: Phytochemical modulation of its functions. *Exp. Mol. Med.* **2014**, *46*, e106. [CrossRef] [PubMed]

13. Fukazawa, R.; Kobayashi, J.; Ayusawa, M.; Hamada, H.; Miura, M.; Mitani, Y.; Tsuda, E.; Nakajima, H.; Matsuura, H.; Ikeda, K.; et al. JCS/JSCS 2020 guideline on diagnosis and management of cardiovascular sequelae in Kawasaki Disease. *Circ. J.* **2020**, *84*, 1348–1407. [CrossRef] [PubMed]

14. Nir, A.; Lindinger, A.; Rauh, M.; Bar-Oz, B.; Laer, S.; Schwachtgen, L.; Koch, A.; Falkenberg, J.; Mir, T.S. NT-pro-B-type natriuretic peptide in infants and children: Reference values based on combined data from four studies. *Pediatric Cardiol.* **2009**, *30*, 3–8. [CrossRef] [PubMed]

15. Jeon, B.H.; Gupta, G.; Park, Y.C.; Qi, B.; Haile, A.; Khanday, F.A.; Liu, Y.-X.; Kim, J.-M.; Ozaki, M.; White, A.R.; et al. Apurinic/apyrimidinic endonuclease 1 regulates endothelial NO production and vascular tone. *Circ. Res.* **2004**, *95*, 902–910. [CrossRef] [PubMed]

16. Shin, J.H.; Choi, S.; Lee, Y.R.; Park, M.S.; Na, Y.G.; Irani, K.; Do Lee, S.; Park, J.B.; Kim, J.M.; Lim, J.S.; et al. APE1/Ref-1 as a serological biomarker for the detection of bladder cancer. *Cancer Res. Treat.* **2015**, *47*, 823–833. [CrossRef]

17. Yahata, T.; Hamaoka, K. Oxidative stress and Kawasaki disease: How is oxidative stress involved from the acute stage to the chronic stage? *Rheumatology* **2017**, *56*, 6–13. [CrossRef]

18. Buttery, L.D.; Springall, D.R.; Chester, A.H.; Evans, T.J.; Standfield, E.N.; Parums, D.V.; Yacoub, M.H.; Polak, J.M. Inducible nitric oxide synthase is present within human atherosclerotic lesions and promotes the formation and activity of peroxynitrite. *Lab. Investig. A J. Tech. Methods Pathol.* **1996**, *75*, 77–85.

19. Park, M.S.; Lee, Y.R.; Choi, S.; Joo, H.K.; Cho, E.J.; Kim, C.S.; Park, J.B.; Jo, E.-K.; Jeon, B.H. Identification of plasma APE1/Ref-1 in lipopolysaccharide-induced endotoxemic rats: Implication of serological biomarker for an endotoxemia. *Biochem. Biophys. Res. Commun.* **2013**, *435*, 621–626. [CrossRef]

20. Leung, D.Y.; Geha, R.S.; Newburger, J.W.; Burns, J.C.; Fiers, W.; Lapierre, L.A.; Pober, J.S. Two monokines, interleukin 1 and tumor necrosis factor, render cultured vascular endothelial cells susceptible to lysis by antibodies circulating during Kawasaki syndrome. *J. Exp. Med.* **1986**, *164*, 1958–1972. [CrossRef]

21. Furukawa, S.; Matsubara, T.; Jujoh, K.; Yone, K.; Sugawara, T.; Sasai, K.; Kato, H.; Yabuta, K. Peripheral blood monocyte/macrophages and serum tumor necrosis factor in Kawasaki disease. *Clin. Immunol. Immunopathol.* **1988**, *48*, 247–251. [CrossRef]

22. Hui-Yuen, J.S.; Duong, T.T.; Yeung, R.S. TNF-alpha is necessary for induction of coronary artery inflammation and aneurysm formation in an animal model of Kawasaki disease. *J. Immunol.* **2006**, *176*, 6294–6301. [CrossRef] [PubMed]

23. Yamaji, N.; da Silva Lopes, K.; Shoda, T.; Ishitsuka, K.; Kobayashi, T.; Ota, E.; Mori, R. TNF-α blockers for the treatment of Kawasaki disease in children. *Cochrane Database Syst. Rev.* **2019**, *8*, Cd012448. [CrossRef] [PubMed]

Article

Transcriptomic Analysis Identifies Differentially Expressed Genes Associated with Vascular Cuffing and Chronic Inflammation Mediating Early Thrombosis in Arteriovenous Fistula

Vikrant Rai and Devendra K. Agrawal *

Department of Translational Research, Western University of Health Sciences, Pomona, CA 91766, USA; vrai@westernu.edu
* Correspondence: dagrawal@westernu.edu; Tel.: +1-909-469-7040; Fax: +1-909-469-5577

Abstract: Arteriovenous fistula (AVF) is vascular access created for hemodialysis in end-stage renal disease patients. AVF creation causes increased blood flow in the outflow vein with increased pressure. Increased blood flow, blood volume, and shear stress causes outward remodeling so that the outflow vein can withstand the increased pressure. Outward remodeling of the vein involved in AVF is necessary for AVF maturation, however, inward remodeling due to excessive neointimal hyperplasia (NIH) and chronic inflammation may end up with vessel thrombosis and AVF maturation failure. Early thrombosis of the vessel may be due to the luminal factors including NIH and chronic inflammation or due to chronic inflammation of the adventitial due to perivascular cuffing. Inflammation may either be due to an immune response to the vascular injury during AVF creation or injury to the surrounding muscles and fascia. Several studies have discussed the role of inflammation in vascular thrombosis due to intimal injury during AVF creation, but there is limited information on the role of inflammation due to surrounding factors like a muscle injury. The concept of perivascular cuffing has been reported in the nervous system, but there is no study of perivascular cuffing in AVF early thrombosis. We performed the bulk RNA sequencing of the femoral arterial tissue and contralateral arteries as we found thrombosed arteries after AVF creation. RNA sequencing revealed several significantly differentially expressed genes (DEGs) related to chronic inflammation and perivascular cuffing, including tripartite motif-containing protein 55 (TRIM55). Additionally, DEGs like myoblast determination protein 1 (MYOD1) increased after muscle injury and relates to skeletal muscle differentiation, and network analysis revealed regulation of various genes regulating inflammation via MYOD1. The findings of this study revealed multiple genes with increased expression in the AVF femoral artery and may provide potential therapeutic targets or biomarkers of early thrombosis in AVF maturation failure. Thus, not only the luminal factors but also the surrounding factors mediating vascular cuffing contribute to vessel thrombosis and AVF failure via early thrombosis, and targeting the key regulatory factors may have therapeutic potential.

Keywords: adventitial inflammation; arteriovenous fistula; chronic inflammation; early thrombosis; fibrosis; maturation; maturation failure; perivascular cuffing

Citation: Rai, V.; Agrawal, D.K. Transcriptomic Analysis Identifies Differentially Expressed Genes Associated with Vascular Cuffing and Chronic Inflammation Mediating Early Thrombosis in Arteriovenous Fistula. *Biomedicines* **2022**, *10*, 433. https://doi.org/10.3390/biomedicines10020433

Academic Editor: Byeong Hwa Jeon

Received: 11 January 2022
Accepted: 11 February 2022
Published: 13 February 2022

Publisher's Note: MDPI stays neutral with regard to jurisdictional claims in published maps and institutional affiliations.

1. Introduction

Arteriovenous fistula (AVF), an abnormal connection between an artery and a vein, is vascular access created for long-term hemodialysis in end-stage renal disease (ESRD) patients. In AVF, blood flows directly from an artery into a vein bypassing capillaries and subjecting the outflow vein to increased blood pressure and shear stress [1]. The creation of AVF is associated with acute inflammation which is necessary for the wound healing and resolution phase but chronicity of inflammation results in thrombosis of the vessels involved in AVF and leads to AVF maturation failure. Thrombosed AVF with chronic inflammation is characterized by increased C-reactive protein (CRP), infiltration of the immune cells including neutrophils and macrophages, increased expression of vascular cell adhesion protein (VCAM)-1, interleukin (IL)-6, and tumor necrosis factor (TNF)-α, neoangiogenesis, neointimal hyperplasia (NIH), and atheromatous plaque formation [2–5]. Intimal injury while creating AVF may cause endothelial dysfunction and chronic inflammation leading to AVF maturation failure [6]. Additionally, inflammation within the vicinity of AVF either due to vessel injury increasing vascular leakage or due to inflammation in the surrounding tissues including muscles causing vascular cuffing may also contribute to vessel thrombosis and AVF maturation failure. Muscle injury, either mechanical or contusion, precipitates inflammation associated with increased infiltration of immune cells including neutrophils, macrophages, natural killer (NK) cells, B- and T-lymphocytes; the secretion of pro-inflammatory cytokines including TNF-α, IL-6, IL-8, IL-1β, IL-1α, macrophage inflammatory protein 1 alpha (MIP-1α), and monocyte chemoattractant protein (MCP)-1 and growth factors including granulocyte colony-stimulating factor (G-CSF), macrophage colony-stimulating factor (M-CSF), vascular endothelial growth factor (VEGF), hepatocyte growth factor (HGF), fibroblast growth factor (FGF), and platelet-derived growth factor (PDGF).

Acute inflammation is necessary for the resolution and remodeling phase of inflammation; however, chronicity of inflammation leads to fibrosis [7–9]. Creation of AVF involving the femoral artery and vein induces muscle injury and non-resolving inflammation may lead to the fibrosis of the muscle and the surrounding tissue. After creation of AVF between the femoral artery and vein, the tissues are sutured in multiple layers within the muscles, subcutaneous tissues, and skin. Post-surgery, dissected tissues and surgical wound heal may induce acute inflammation, but chronicity of inflammation may cause fibrosis of the tissue surrounding AVF and thrombosis of the vessels. Based on this, we hypothesize that the chronicity of inflammation in the surrounding tissue precipitating vascular cuffing mediate thrombosis, stenosis, and fibrosis of the vessels [10] and the area surrounding AVF. This will lead to early thrombosis of the vessels and early AVF maturation failure. Thus, to investigate the effects of muscle injury during AVF creation, the presence of chronic inflammation, perivascular cuffing, and thrombosis and its association with early thrombosis of the vessels and to compare the transcriptomic profile of the contralateral control femoral artery (contralateral FA), untreated and treated femoral artery involved in AVF (AVF FA), we performed the bulk RNA sequencing of the control femoral artery and AVF tissue femoral artery treated with the inhibitors of triggering receptor expressed on myeloid cells-1 (TREM-1) and toll-like receptor-4 (TLR-4). The tissues were collected from the miniswine being used for another ongoing study in the lab to investigate the effect of TREM-1 and TLR-4 inhibition on AVF maturation. The effects of TREM-1 and TLR-4 inhibition on early vessel thrombosis was under investigation because pro-inflammatory mediators TREM-1 and TLR-4 play a critical role in atheromatous plaque formation, atherosclerosis, and vessel stenosis [11–14]. The aim was to compare the gene expression profile between the groups to elucidate differentially expressed genes (DEGs) related to chronic inflammation.

2. Materials and Methods

2.1. Animal Model, AVF Creation, and Tissue Collection

For this study, female Yucatan miniswine, four to seven months old and weighing between 20–30 kg, purchased from Premier Bio-resources (Cotati Ramona, CA, USA) randomly divided into two experimental groups were used to create AVF fistula involving the right femoral artery (FA) and femoral vein (FV). The contralateral FA and FV were used as biological controls. Yucatan miniswine were housed in the vivarium of Western University of Health Sciences, Pomona, CA with 12 h light and dark cycle at a temperature range of 72–74 °F and fed with the Mini-Pig Grower Diet (Test Diet # 5801) and allowed to drink water ad libitum. Female pigs were used because of their less aggressive behavior and ease of handling compared to the males. All experiments involving the animals were performed as per National Institutes of Health and USDA guidelines for the care and use of experimental animals. The Institutional Animal Care and Use Committee (IACUC) at Western University of Health Sciences approved protocol No. R20IACUC038 for this study.

For AVF creation, there were two experimental groups: (i) animals treated with LR-12 + TAK242, and (ii) animals treated with scrambled peptide and 30% ethanol (the vehicle for TAK-242). These animals were also used for another ongoing study in the lab to investigate the effect of inhibiting the triggering receptor expressed on myeloid cells-1 (TREM-1) and toll-like receptor-4 (TLR-4) on early thrombosis of the artery and fistula after creating AVF. Each experimental group consisted of three to four animals with similar body weights. For creating AVF, minipigs were first given a preanesthetic sedative injection of 2.5–5 mg/kg Telazol (a combination of tiletamine and zolazepam) and 1–2 mg/kg xylazine subcutaneously. After sedation, the animals were moved to the operating suite and intubated with an appropriately sized endotracheal tube, maintained on inhaled isoflurane in oxygen 1–3% and mechanical ventilation. After starting an intravenous ringer lactate (5 mL/kg/h) using the ear vein, the AVF was surgically created between FA and FV after preparing them for side-to-side anastomosis. For AVF creation, a 1 cm incision was made on the medial side of FA and FV opposing each other. AVF was created using a 6–0 proline to join FA and FV. LR-12 (>10^9 particles in 1 mL), bolus TAK-242 (3 mg/kg dissolved in 30% ethanol), scrambled peptide, and vehicle control (30% ethanol) was injected at the AVF side in the lumen of FA and FV, and we waited for 5–10 min before closing the AVF. The wound was closed in layers suturing muscles, subcutaneous tissues, and skin in layers using 3–0 vicryl sutures. The swine were given a maintenance dose of TAK-242 once daily 0.1 mg/kg i.v. for six days and then weekly once for four weeks. During surgery, the level of sedation, percentage of isoflurane received, oxygen flow rate, heart rate, respiratory rate, mucous membrane color, presence or absence of withdrawal reflex, and body temperature were maintained every 15 min by a veterinary technician. Post-surgery, the swine were given 1 g of cefazolin prophylactically and buprenorphine for pain. The sedation was reversed with flumazenil 0.01 mg/kg. The swine were monitored post-surgically until they were on their feet, and recovery parameters including consciousness level, recumbency, respiratory rate and character, and mucous membrane color were monitored every 1 h until the swine started walking.

After completing 12 weeks of AVF creation, the swine were sedated, and radiological assessments of FA, AVF, and FV using ultrasonogram (USG), angiography, and optical coherence tomography (OCT) were done. This was followed by euthanasia using intravenous administration of a single dose of euthanasia solution consisting of pentobarbital sodium (85 mg/kg) and phenytoin sodium (11 mg/kg) while the pigs were under anesthesia. Swine were observed for the absence of heartbeats and respiration for at least 10 min before tissue harvest. The tissue of interest including AVF involved FA and FV, tissue around the fistula, and contralateral FA and FV were harvested after dissecting the groin area. The tissues were harvested for histomorphological studies in 10% formalin, for RT-PCR and sequencing studies in RNA later, and for protein isolation at 4 °C and stored at −80 °C. For histomorphological studies, the harvested tissues were processed in a tissue processor and paraffin embedded. Five μm thin sections using a tungsten carbide knife (LeicaTM, Germany) in a Leica RM2265 rotary microtome (LeicaTM, Germany) were sectioned and attached to glass slides for histology.

2.2. Radiological Assessment of the AVF

To assess the diameter of the FA and FV, flow velocities and flow volume in the FA at baseline, a preoperative color Doppler ultrasound (Phillips EPIQ-7 US system) was done before creating AVF. After 12 weeks, to evaluate the AVF patency, FA flow and diameter, outflow vein diameter, and the flow velocity in the vein, a postoperative USG of the AVF, FA, and FV at the AV anastomosis site was done before sacrificing the miniswine. Each AVF site was evaluated for the peak systolic and end-diastolic velocity and blood flow at different locations in the artery and vein of the AVF, including the anastomosis site. Color Doppler ultrasound was followed by femoral angiography using a 7F (Concierge) guide catheter from Merit Medical USA through the carotid artery to assess the patency and blood flow in the fistula on the anastomosis site and the contralateral side FA for comparison.

Briefly, angiography was performed by percutaneous needle puncture in the common carotid artery under ultrasound guidance and an angiography catheter was advanced from the carotid artery into the descending aorta, external iliac arteries to the FA of the AVF side as well as to the contralateral side. A contrast dye (Iopromide; Ultravist) was injected into the catheter while taking X-rays of the area of interest to visualize the AVF and contralateral femoral artery patency. Angiography was followed by optical coherence tomography (OPTIS OCT from St. Jude Medical) of the anastomosis and contralateral side via carotid approach to measure the inside diameter and the cross-sectional area of the vessels to delineate the vessel's wall anatomy (intima, media, and adventitia), the presence as well as characterization of neointimal hyperplasia, thrombosis, neo-vascularization, luminal diameter, and percent diameter stenosis. For OCT, a 0.014-inch guidewire was positioned in the proximal FA and the OCT catheter (Dragonfly™ DUO imaging catheter; Abbott, Illinois, USA) was advanced over the guidewire. A nonionic, low-osmolar iodinated contrast agent, Ultravist, was simultaneously injected during OCT pullback, and the entire region of interest was scanned and images were analyzed using Light Lab OCT imaging proprietary software (Light Lab Imaging/Abbott). A comparative analysis between the AVF side femoral artery and contralateral side femoral artery was done for USG, angiography, and OCT results.

2.3. Histomorphology

Hematoxylin and Eosin (H&E) and Movat Pentachrome staining were done as per the standard protocol in our lab to assess the inflammation, fibrosis, vessel stenosis, and thrombosis. For H&E staining, after deparaffinization and rehydration through a series of xylene, alcohol, and distilled water, the tissue sections were stained with hematoxylin (45 s) followed by eosin (8–10 dips). The stained slides were mounted with xylene-based mounting media. For Movat Pentachrome staining, the tissue sections were deparaffinized and rehydrated and the sections were stained using a modified Russell-Movat Pentachrome kit following the manufacturer's protocol (Cat no. KTRMPPT from American MasterTech scientific laboratory supplies). Stained tissue sections were scanned at 100 μm using a light microscope (Leica DM6). All the scanned images were blindly reviewed by two independent observers.

2.4. Bulk RNA Sequencing and DEGs for Inflammation

Total RNA was extracted from the collected samples (three AVF sides and three contralateral sides) using TRIZOL (Trizol reagent, Sigma, Cat# T9424, St Louis, MO, USA) following the manufacturers' guidelines. The yields of total RNA were measured using a NanoDrop 2000 (Thermo Scientific, Waltham, MA, USA) and 1 ug of total RNA was sent (Genewiz LLC, South Plainfield, NJ, USA) for bulk RNA sequencing. The RNA samples with RIN > 6 were subjected to sequencing. The RNA samples were quantified using a Qubit 2.0 Fluorometer (ThermoFisher Scientific, Waltham, MA, USA) and RNA integrity was checked using TapeStation (Agilent Technologies, Palo Alto, CA, USA). The RNA sequencing libraries were prepared using the NEBNext Ultra II RNA Library Prep Kit for Illumina according to the manufacturer's instructions (New England Biolabs, Ipswich, MA, USA). Briefly, mRNAs were initially enriched with Oligod(T) beads. Enriched mRNAs were fragmented for 15 min at 94 °C. First-strand and second-strand cDNA were subsequently synthesized. cDNA fragments were end-repaired and adenylated at 3′ends, and universal adapters were ligated to cDNA fragments, followed by index addition and library enrichment by PCR with limited cycles. The sequencing libraries were validated on the Agilent TapeStation (Agilent Technologies, Palo Alto, CA, USA), and quantified by using Qubit 2.0 Fluorometer (ThermoFisher Scientific, Waltham, MA, USA) as well as by quantitative PCR (KAPA Biosystems, Wilmington, MA, USA).

The sequencing libraries were multiplexed and clustered onto a flowcell. After clustering, the flowcell was loaded onto the Illumina HiSeq instrument according to the manufacturer's instructions. The samples were sequenced using a 2 × 150 bp Paired-End (PE) configuration. Image analysis and base calling were conducted by the HiSeq Control Software (HCS). Raw sequence data (.bcl files) generated from Illumina HiSeq was converted into fastq files and de-multiplexed using Illumina bcl2fastq 2.17 software. One mismatch was allowed for index sequence identification. After investigating the quality of the raw data, sequence reads were trimmed to remove possible adapter sequences and nucleotides with poor quality using a Trimmomatic v.0.36. The trimmed reads were mapped to the Sus scrofa reference genome available on ENSEMBL using the STAR aligner v.2.5.2b. BAM files were generated because of this step. Unique gene hit counts were calculated by using feature counts from the Subread package v.1.5.2. Only the unique reads that fell within exon regions were counted. After extraction of gene hit counts, the gene hit counts table was used for downstream differential expression analysis. Using DESeq2, a comparison of gene expression between the groups of samples was performed. The Wald test was used to generate p values and Log$_2$ fold changes. Genes with adjusted p values < 0.05 and absolute log$_2$ fold changes >1 were called differentially expressed genes for each comparison.

2.5. Quantitative Real-Time PCR

To examine the gene expression of a few selected genes, qRT-PCR was done after preparing cDNA from isolated mRNA. An iScript cDNA Synthesis Kit (BioRad # 1708891) was used to prepare cDNA and the prepared cDNA was subjected to qRT-PCR in triplicates with SYBR green (BioRad # 1725122) using a CFX96 Touch Real-Time PCR Detection System. All the primers used in this study (Table 1) were designed using NCBI (assessed on 13 December 2021) and purchased from Integrated DNA Technologies (Coralville, IA, USA). The PCR cycling conditions were 5 min at 95 °C for initial denaturation, 40 cycles of the 30 s each at 95 °C (denaturation), 30 s at 55–60 °C (depending on primer annealing temperature), and 30 s at 72 °C (extension) followed by melting curve analysis. The folds change in mRNA expression relative to controls was analyzed using $2^{-\Delta\Delta ct}$ after normalization with 18S as a housekeeping gene.

Table 1. The nucleotide sequence (5′–3′) of the primers used for a real-time quantitative polymerase chain reaction. Interleukin (IL), C-C Motif Chemokine Ligand 2 (CCL2), CCAAT Enhancer Binding Protein Alpha (CEBPA), Lactotransferrin (LTF), Galectin 12 (LGALS12), Dual Oxidase 2 (DUOX2), Vanin 2 (VNN2), Myoblast Determination Protein 1 (MYOD1), Tripartite Motif Containing 55 (TRIM55), Pentraxin 3 (PTX3), Matrix metalloproteinase 25 (MMP25).

Gene Name	Forward Primer	Reverse Primer
IL-8	5′-GACCCCAAGGAAAAGTGGGT-3′	5′-TGACCAGCACAGGAATGAGG-3′
IL-18	5′-ATGGCTGCTGAACCGGAAG-3′	5′-GGTCTTCATCGTTTTCAGCTACA-3′
MYOD1	5′-GCTCCGCGACGTAGATTTGA-3′	5′-GGAGTCGAAACACGGGTCAT-3′
CEBPA	5′-CGGTGCGTCTAAGATGAGGG-3′	5′-AGGCACATATTTGCTCCCCC-3′
LGALS12	5′-GACCCGCTTCCTGACACCTT-3′	5′-CCCTCCACAAACGGGTTGAT-3′
TRIM55	5′-CGTAGGGCCTTCAGGTTCTG-3′	5′-GTTCCCTTACCACTCACCGC-3′
ACSL4	5′-TCTGTTCGCTGTTGCTGATTTG-3′	5′-GAGAGCCCGCCACACAAGT-3′
CCL2	5′-AAACGGAGACTTGGGCACAT-3′	5′-CTTGCAAGGACCCTTCCGTC-3′
DUOX2	5′-GCTCTGCATAAGACCAGAGGC-3′	5′-GTCAGTGAGAGTGCGTCCTG-3′
PTX3	5′-GCAGGTTGTGAAACAGCGAT-3′	5′-TTTGACCCAAATGCAGGCAC-3′
MMP25	5′-TTGTCCTGGCGTTTCTGTGT-3′	5′-GGCCATCAGCTTGGCTCATA-3′
LTF	5′-GTCACAGCCATCGCTAACCT-3′	5′-TTGCTCCTCCAAGCTTGACCT-3′
VNN2	5′-GATGTCCCTGAAAAGCCGGA-3′	5′-GTCACAGCAGGATCACGGAA-3′

2.6. Statistical Analysis

The data is presented as mean ± SD. GraphPad Prism 9 was used to analyze the RT-PCR data and the comparison between two groups for the fold change in gene expression was performed using One-way ANOVA with Bonferroni's posthoc correction and Students' t-test was used for statistical analysis. A probability (p) value of <0.05 was accepted as statistically significant. * p < 0.05, ** p < 0.01, *** p < 0.001 and **** p < 0.0001.

3. Results

3.1. Radiological Assessment and Histomorphology

Compared to the preoperative baseline assessment, post-surgical assessment of the FA involved in AVF showed partially blocked or stenosed FA with the presence of neointimal hyperplasia (NIH) and large plaques significantly obstructing the lumen of the vessel. The flow volume and velocities were decreased compared to baseline and contralateral FA. Angiography was performed for the FA in which guidewire can be advanced to the site of AVF. Angiography showed open FA in some swine while others, mainly in scrambled and vehicle groups, showed thrombosed and stenosed arteries. In FA showing no flow in doppler ultrasound, a guidewire could not be advanced, and no angiography was done. Similarly, OCT was performed only in the arteries which were open and not for the blocked arteries. Open arteries showed normal OCT while OCT was not done in thrombosed arteries due to technical difficulty in advancing the guidewire. The H&E staining and Movat-Pentachrome staining also showed blocked FA and extensive fibrosis around vessels at the AVF site (Figure 1).

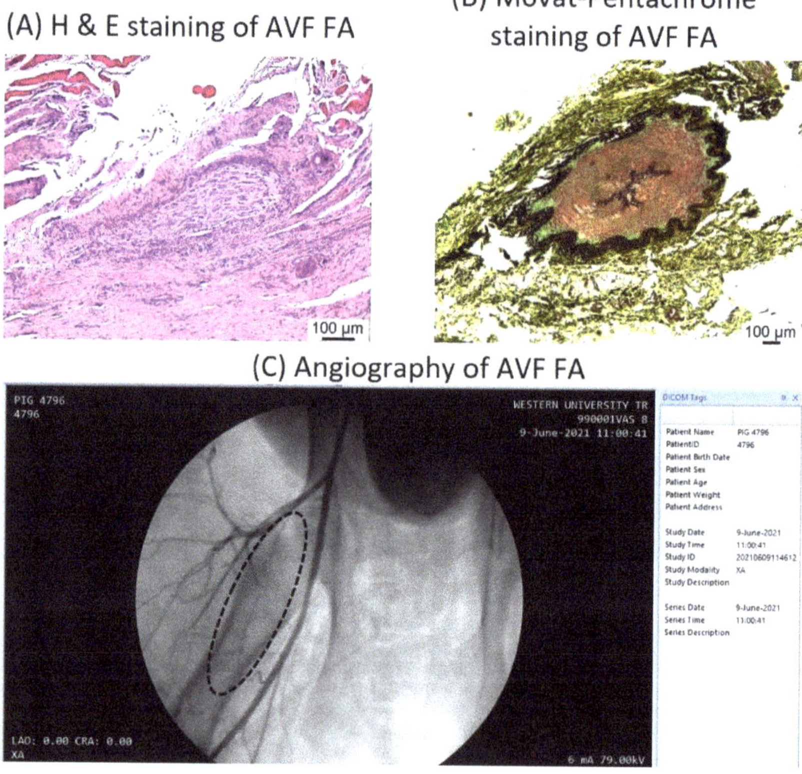

Figure 1. Histomorphological and radiological analysis of the femoral artery involved in the arteriovenous fistula. Hematoxylin and eosin staining (**panel A**) showed a thrombosed artery with inflammation in the adventitia. Movat-pentachrome staining (**panel B**) revealed a thrombosed artery with elastin degradation in the adventitia and increased collagen deposition in media and intima. Angiography (**panel C**) revealed a stenosed superficial femoral artery (the dotted circle) involved in AVF. These images are representative of the stenosed arteries.

3.2. DEGs Related to Inflammation

Bulk RNA sequencing of the tissues revealed a total of 14,554 genes, of which 415 were significantly expressed genes ($p < 0.05$ and $\log_2$ fold > 1). From the list of differentially expressed genes (DEGs) with $p < 0.05$ and $\log_2$ fold > 1, genes with p-value < 0.05 and $\log_2$ fold > 2 were sorted out (Supplementary Material 1). An extensive literature search was done to check all DEGs for their role in inflammation in any organ system of the body; migration, proliferation, and activation of immune cells including macrophages, neutrophils, natural killer (NK) cells, B- and T-lymphocytes, T-regulatory (Treg) cells, and dendritic cells (DCs); activation and regulation of innate and adaptive immune response; and regulation of cytoplasmic kinases including nuclear factor-kappa beta (NF-κB), phosphoinositide 3-kinases (PI3K), protein kinase B (PKB/Akt), and mitogen-activated protein kinase (MAPK). These parameters were included in the literature search because of their crucial role in regulating inflammation and inflammatory signaling. We found various DEGs (Tables 2–5) playing a crucial role in inflammation, regulation of inflammatory pathways, and immune regulation. Among the DEGs involved in inflammation, some of the genes which were differentially expressed (positive $\log_2$ fold value) in scrambled peptide treated FA were found downregulated (negative $\log_2$ fold values) after treatment with TREM-1 inhibitor (LR12) + TLR-4 inhibitor (TAK-242) suggesting the effect of attenuating inflammation by inhibiting TREM-1 and TLR-4. Additionally, the gene occurring in different comparison groups was included only in one table and not in others (the details of all DEGs involved in inflammation and immune regulation in each group can be found in Supplementary Material 1).

Table 2. Differentially expressed genes (DEGs) while comparing contralateral femoral artery with AVF femoral artery combining all tissues (+value = higher expression in AVF FA; −value = higher expression in contralateral FA).

Gene ID	$\log_2$ FoldChange	p-Value	Gene Name
ENSSSCG00000006216	8.11	0.00000000772	TRIM55
ENSSSCG00000007436	5.58	0.0000000953	MMP-9
ENSSSCG00000006590	4.48	0.00000979	S100A8
ENSSSCG00000009645	3.96	0.0000292	ADAMDEC1
ENSSSCG00000031053	3.69	0.000000542	S100A1
ENSSSCG00000022512	3.20	0.0000546	TRDC
ENSSSCG00000023842	3.05	0.000389967	TRAT1
ENSSSCG00000028331	3.01	0.000191823	IL1R2
ENSSSCG00000015037	2.84	0.0000120	IL-18
ENSSSCG00000006309	2.78	0.000659109	CD247
ENSSSCG00000006025	2.71	0.000426117	PKHD1L1
ENSSSCG00000014310	2.58	0.000507304	CXCL14
ENSSSCG00000006452	2.31	0.000310425	CD1D
ENSSSCG00000003113	2.08	0.000494243	C5AR2
ENSSSCG00000009051	2.04	0.000959894	IL-15
ENSSSCG00000008606	−2.48	0.000000136	OSR1

Table 3. Differentially expressed genes (DEGs) while comparing contralateral FA with LR-12 + TAK-242 treated AVF FA (+value = higher expression in AVF FA; −value = higher expression in contralateral FA).

Gene ID	$\log_2$ FoldChange	p-Value	Gene Name
ENSSSCG00000022490	7.12	0.037585934	GPR83
ENSSSCG00000016688	7.06	0.040958471	CPVL
ENSSSCG00000013056	5.41	0.000202896	LGALS12
ENSSSCG00000010478	4.80	0.000919971	FFAR4
ENSSSCG00000016878	4.17	0.011959826	FGF10
ENSSSCG00000015015	4.16	0.008270436	ARHGAP20
ENSSSCG00000002866	4.05	0.000885129	CEBPA

Table 3. *Cont.*

Gene ID	log$_2$ FoldChange	*p*-Value	Gene Name
ENSSSCG00000011579	4.04	0.00094779	PPARG
ENSSSCG00000015332	3.77	0.041851053	PON1
ENSSSCG00000026297	3.76	0.00389716	KLB
ENSSSCG00000008237	3.50	0.002121156	RETSAT
ENSSSCG00000007710	3.37	0.005462755	MLXIPL
ENSSSCG00000011831	3.21	0.003915618	APOD
ENSSSCG00000015267	2.77	0.010298314	FMO2
ENSSSCG00000017705	2.61	0.014997339	CCL5
ENSSSCG00000014880	2.23	0.045711441	AQP11
ENSSSCG00000005122	2.22	0.020695275	TEK
ENSSSCG00000008953	2.21	0.031773067	CXCL8
ENSSSCG00000025578	2.19	0.026455299	ALDH1A2
ENSSSCG00000003578	2.17	0.032743557	FGR
ENSSSCG00000008624	2.16	0.024212176	LPIN1
ENSSSCG00000029813	2.10	0.025936426	TSPAN5
ENSSSCG00000013886	2.07	0.038238135	B3GNT3
ENSSSCG00000009789	2.00	0.044237292	HCAR1
ENSSSCG00000000211	−3.42	0.017397621	AQP5
ENSSSCG00000015707	−3.07	0.192328785	GPR39
ENSSSCG00000008835	−2.84	0.007206502	RASL11B
ENSSSCG00000001834	−2.53	0.011049962	MFGE8

Table 4. Differentially expressed genes (DEGs) while comparing contralateral FA with scrambled peptide treated AVF FA (+value = higher expression in AVF FA treated with scrambled peptide; −value = higher expression in contralateral FA).

Gene ID	log$_2$ FoldChange	*p*-Value	Gene Name
ENSSSCG00000005951	9.28	0.002454177	TMEM71
ENSSSCG00000011862	8.97	0.004185163	MUC13
ENSSSCG00000029879	8.17	0.016038726	LTF
ENSSSCG00000015748	7.92	0.023568215	CSMD1
ENSSSCG00000022473	7.78	0.029138305	A4GNT
ENSSSCG00000008972	7.45	0.046820249	PPEF2
ENSSSCG00000027568	6.64	0.000784677	BLK
ENSSSCG00000025042	6.02	0.007096705	ICOS
ENSSSCG00000013378	5.78	0.041425329	ABCC8
ENSSSCG00000011131	5.58	0.016369542	PRKCQ
ENSSSCG00000017466	5.30	0.002770819	CCR7
ENSSSCG00000001613	5.11	0.001355889	TREML1
ENSSSCG00000006266	5.01	0.022187352	ST18
ENSSSCG00000002821	4.88	0.008741294	CCL22
ENSSSCG00000006734	4.68	0.005857569	CD101
ENSSSCG00000029668	4.52	0.011344171	IL2RB
ENSSSCG00000015093	4.50	0.004868833	CD3D
ENSSSCG00000021569	4.14	0.002023038	MMP25
ENSSSCG00000013115	4.01	0.00956326	CD5
ENSSSCG00000000257	3.85	0.004632273	ITGB7
ENSSSCG00000004195	3.84	0.004380652	ARG1
ENSSSCG00000030042	3.76	0.002440454	SBNO2
ENSSSCG00000006359	3.57	0.003335966	ADAMTS4
ENSSSCG00000017962	3.49	0.004592719	KDM6B
ENSSSCG00000000705	3.49	0.027859902	CD27
ENSSSCG00000009630	3.48	0.008741686	EGR3
ENSSSCG00000000605	3.46	0.036481122	ERP27
ENSSSCG00000013649	3.40	0.01209607	ICAM3

Table 4. *Cont.*

Gene ID	log$_2$ FoldChange	*p*-Value	Gene Name
ENSSSCG00000013839	3.36	0.043788189	RASAL3
ENSSSCG00000015550	3.15	0.045746825	RGS16
ENSSSCG00000000136	3.10	0.009086041	CSF2RB
ENSSSCG00000000688	3.07	0.02332297	LAG3
ENSSSCG00000014825	3.06	0.016131597	RELT
ENSSSCG00000004678	3.05	0.009097692	DUOX2
ENSSSCG00000011727	2.93	0.011434284	PTX3
ENSSSCG00000006588	2.91	0.044677952	S100A9
ENSSSCG00000003805	2.83	0.013758246	PDE4B
ENSSSCG00000000521	2.83	0.012411864	PHLDA1
ENSSSCG00000015299	2.79	0.028194104	STEAP4
ENSSSCG00000004179	2.69	0.028039502	VNN2
ENSSSCG00000006379	2.67	0.029518927	CD48
ENSSSCG00000004779	2.66	0.037184423	PLCB2
ENSSSCG00000008388	2.64	0.044215498	REL
ENSSSCG00000021944	2.63	0.025402199	RAC2
ENSSSCG00000000223	2.63	0.043774003	BIN2
ENSSSCG00000011443	2.62	0.018907549	STAB1
ENSSSCG00000006800	2.41	0.02415221	CD53
ENSSSCG00000013655	2.29	0.034717888	ICAM1
ENSSSCG00000017330	2.17	0.034875529	MAP3K14
ENSSSCG00000010219	2.15	0.033546298	ARID5B
ENSSSCG00000012583	2.10	0.038092469	ACSL4
ENSSSCG00000017723	2.09	0.038906102	CCL2
ENSSSCG00000006002	−2.12	0.035313216	NOV
ENSSSCG00000024259	−3.50	0.024778737	PPP1R11
ENSSSCG00000004948	−2.33	0.030111699	SMAD6

Table 5. Differentially expressed genes (DEGs) while comparing scrambled peptide treated FA with LR-12 + TAK-242 treated AVF FA (+value = higher expression in scrambled peptide treated AVF FA; −value = higher expression in LR-12 + TAK-242 treated AVF femoral artery).

Gene ID	log$_2$ FoldChange	*p*-Value	Gene Name
ENSSSCG00000007717	3.57	0.022045439	METTL27
ENSSSCG00000013056	3.53	0.007938274	LGALS12
ENSSSCG00000010184	3.33	0.007715639	AGT
ENSSSCG00000003399	3.27	0.021031658	RBP7
ENSSSCG00000012667	3.12	0.014533603	IGSF1
ENSSSCG00000028996	3.05	0.013185638	ALDH1A1
ENSSSCG00000029813	2.74	0.019674447	TSPAN5
ENSSSCG00000000576	2.69	0.019345908	LDHB
ENSSSCG00000020657	2.55	0.024205833	BCAM
ENSSSCG00000013260	2.45	0.031013151	MDK
ENSSSCG00000015249	2.40	0.036414276	ADAMTS8
ENSSSCG00000009492	2.34	0.040360908	GPR180
ENSSSCG00000017605	2.34	0.036679735	MMD
ENSSSCG00000016331	2.34	0.035784764	RAMP1
ENSSSCG00000022301	2.19	0.047966521	EIF4EBP1
ENSSSCG00000006764	−7.32	0.032418272	PTPN22
ENSSSCG00000005236	−7.12	0.043128193	DMRT1
ENSSSCG00000007956	−5.37	0.047045343	NLRC3
ENSSSCG00000007523	−5.15	0.000483159	TUBB1
ENSSSCG00000015656	−4.58	0.028406123	FCMR
ENSSSCG00000010772	−4.16	0.002898421	ADAM8
ENSSSCG00000009237	−3.74	0.00903231	HPSE
ENSSSCG00000006378	−3.73	0.007424014	SLAMF7

Table 5. *Cont.*

Gene ID	log$_2$ FoldChange	*p*-Value	Gene Name
ENSSSCG00000000653	−3.68	0.007197867	CD69
ENSSSCG00000017908	−3.66	0.036598715	GP1BA
ENSSSCG00000013853	−3.52	0.007956036	HSH2D
ENSSSCG00000010575	−3.20	0.009548539	PPRC1
ENSSSCG00000004369	−2.86	0.034230241	PRDM1
ENSSSCG00000013041	−2.38	0.034963931	FERMT3
ENSSSCG00000009761	−2.30	0.036755714	NCOR2
ENSSSCG00000017333	−2.27	0.041076061	FMNL1

3.3. Quantitative Real-Time PCR

Quantitative RT-PCR revealed increased expression of IL-8, IL-18, MYOD1, CEBPA, LGALS12, TRIM55, ACSL4, CCL2, DUOX2, PTX3, MMP25, LTF, and VNN2 in FA involved in AVF compared to contralateral FA (Figure 2). Significantly increased expression of these genes in AVF FA suggests the presence of inflammation and supports the findings of RNA sequencing data with higher expression of these DEGs in AVF FA compared to contralateral FA.

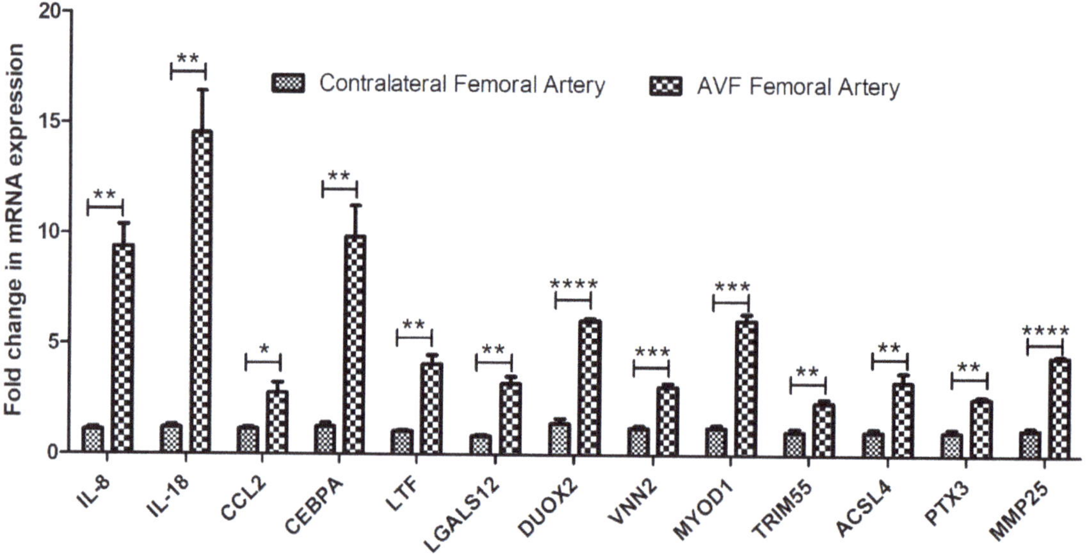

Figure 2. RT-PCR for mRNA expression of DEGs in femoral artery involved in arteriovenous fistula compared to the contralateral femoral artery. All data are presented as mean ± standard deviation (SD). A *p*-value < 0.05 was considered significant. * $p < 0.05$, ** $p < 0.01$, *** $p < 0.001$, **** $p < 0.0001$. Interleukin (IL), C-C Motif Chemokine Ligand 2 (CCL2), CCAAT Enhancer Binding Protein Alpha (CEBPA), Lactotransferrin (LTF), Galectin 12 (LGALS12), Dual Oxidase 2 (DUOX2), Vanin 2 (VNN2), Myoblast Determination Protein 1 (MYOD1), Tripartite Motif Containing 55 (TRIM55), Pentraxin 3 (PTX3), Matrix metalloproteinase 25 (MMP25).

3.4. Network Analysis

The network analysis (Signor for regulatory network analysis and STRING for protein-protein interaction networkanalyst.ca) using the selected DEGs as input showed the regulatory interaction of these genes with each other and the role of these DEGs in inflammation, angiogenesis, and fibrosis, in addition to the three contributing factors in vessel thrombosis (Figures 3 and 4). The network analysis also revealed the role of these DEGs in M1 and M2 macrophages polarization, T-cell activation, macrophage activation, basophils, granulocytes, and monocytes differentiation. The increased expression of these DEGs in FA involved in AVF and their association with inflammation, regulation of inflammation, immune cell differentiation, and activation suggest the role of these DEGs in chronic inflammation which precipitates vascular thrombosis.

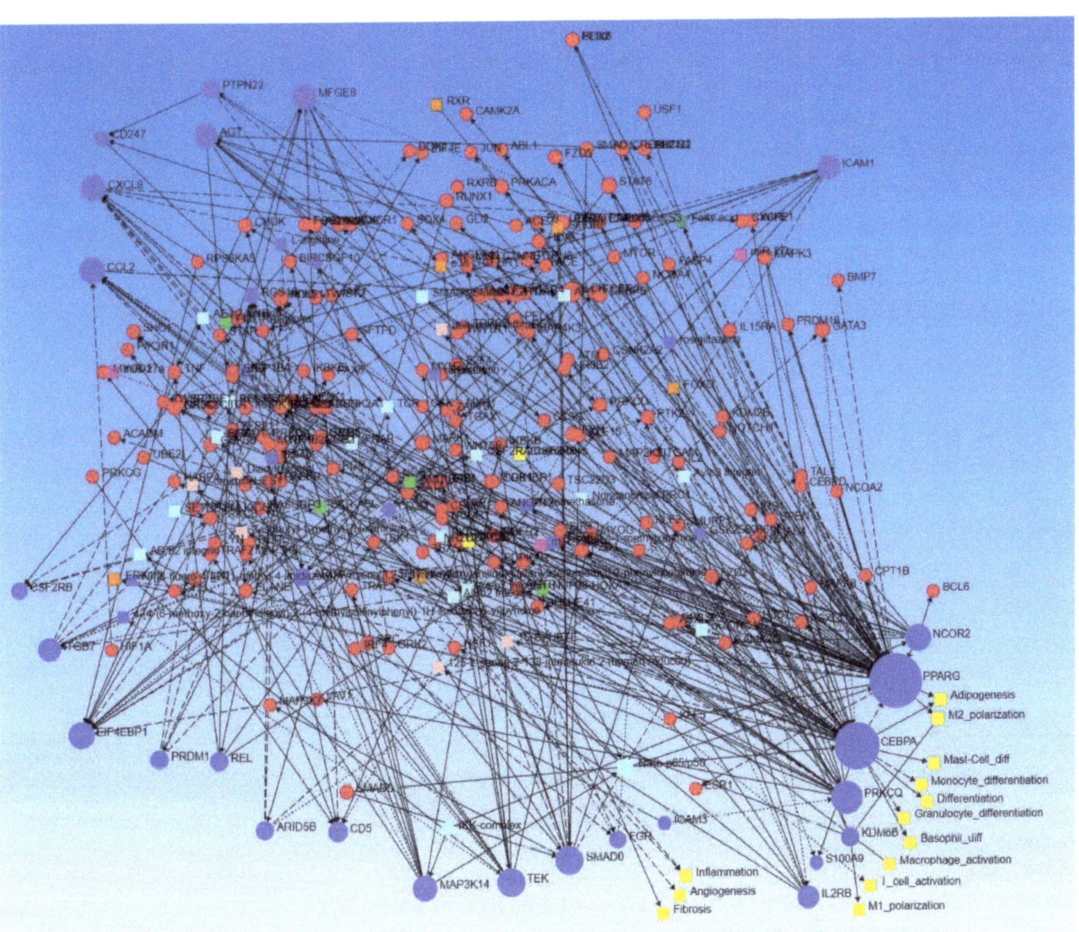

Figure 3. Signor network analysis of the DEGs. The network analysis showed the correlation of DEGs with each other. Blue circles show the DEGs listed in Tables 2–5 and involved in inflammation, immune response, and remodeling (CXCL8, CCL2, ICAM1, ICAM3, ITGB7, PPARG, REL, IL2, S100A9, etc.), yellow circles show cellular mechanisms involved in chronic inflammation and plaque formation including macrophage polarization, fibrosis, inflammation, angiogenesis, and lymphocytes activation.

Figure 4. STRING analysis for protein-protein interaction. The analysis showed that various DEGs are related to each other and regulate each other.

4. Discussion

Chronic inflammation and inflammatory immune cells play a critical role in the development and progression of atheromatous plaque formation contributing to thrombosis and vessel stenosis. The development and progression of atherosclerosis are characterized by deposition of low-density lipoproteins in vessel intima, formation of a fatty streak, foam cell formation, plaque formation, and immune cell infiltration followed by thrombosis and vessel stenosis. Persistent infiltration of pro-inflammatory immune cells causing increased secretion of pro-inflammatory cytokines and increased expression of inflammatory mediators leads to chronic inflammation contributing to thrombosis [11,15–19]. Thus, targeting inflammation seems a promising therapeutic strategy to attenuate the progression of atheromatous plaque and thrombosis. Thrombosis of the vessels participating in AVF is a common cause to precipitate early AVF failure [20,21]. Since inflammation plays a crucial role in vessel thrombosis and studies have shown the crucial role of inflammation-intimal hyperplasia-plaque-thrombosis, we focused on investigating the factors mediating adventitial inflammation and inflammation of the surrounding structure possibly playing a role in early vessel thrombosis and AVF failure. In other words, the focus of this study was to focus on the role of inflammation in the vicinity of AVF. The inflammation may either be due to surgical injury or the immune response of the body. Inflammation of the vessel adventitia, around a blood vessel and vicinity tissue, causes perivascular cuffing due to the accumulation of immune cells [22,23] and contributes to vessel stenosis [10].

A significantly increased TRIM55 (Murf2) expression in AVF FA tissue with RNA seq ($\log_2$ fold = 8.1, Table 2) suggests that increased expression of TRIM55 might have caused perivascular cuffing of the FA and chronic inflammatory milieu precipitating vessel thrombosis [24]. Increased TRIM55 might be in response to muscle injury while creating AVF or post-surgical muscle atrophy [25,26]. An association of increased perivascular cuffing with increased TRIM55 expression and decreased infiltration of immune cells and perivascular cuffing with TRIM55 knock out as reported previously [27,28] suggests TRIM55 as a potential therapeutic target to attenuate perivascular cuffing and an increased TRIM55 expression in AVF FA compared to contralateral FA suggests a chronic inflammatory state around the AVF FA mediating vessel thrombosis and stenosis. Another DEG that was significantly increased in AVF FA compared to control FA was myogenic differentiation 1 (MYOD1) ($\log_2$ fold = 7.29). MYOD1 is a transcription factor that regulates muscle differentiation [29], and its expression varies with the presence and absence of inflammation [30,31]. Although MYOD1 is associated with muscle differentiation, its association with inflammation coerced us to investigate the regulatory network of MYOD1. The gene regulatory network using all 425 DEGs revealed MYOD1 association with nuclear factor kappa beta (NF-κB), interferon-gamma (IFN-γ), integrins, interleukin (IL)-18, caspases, matrix metalloproteinases (MMPs), sirtuins, vascular endothelial growth factor A (VEGFA), angiopoietin 1 (ANGPT1), and immune cell activation (Figure 5). The association of MYOD1 with the genes involved in inflammation, angiogenesis, arteriogenesis, and immune cell activation suggests its probable role in the pathologies involved in vessel thrombosis (inflammation, immune cell activation, and angiogenesis). Increased expression of MYOD1 in association with other genes involved in inflammation namely IL-8 (CXCL8, $\log_2$ fold = 2.21), IL-18 ($\log_2$ fold = 2.84), CCR7 ($\log_2$ fold = 5.30), ITGB7 ($\log_2$ fold = 3.85), MMP-9 ($\log_2$ fold = 5.58), MMP-25 ($\log_2$ fold = 4.14), and TREML1 ($\log_2$ fold = 5.11) suggest the role of MYOD1 in perivascular cuffing. Furthermore, an association of these DEGs with thrombosed AVF FA suggests the role of these DEGs in perivascular cuffing and vessel thrombosis.

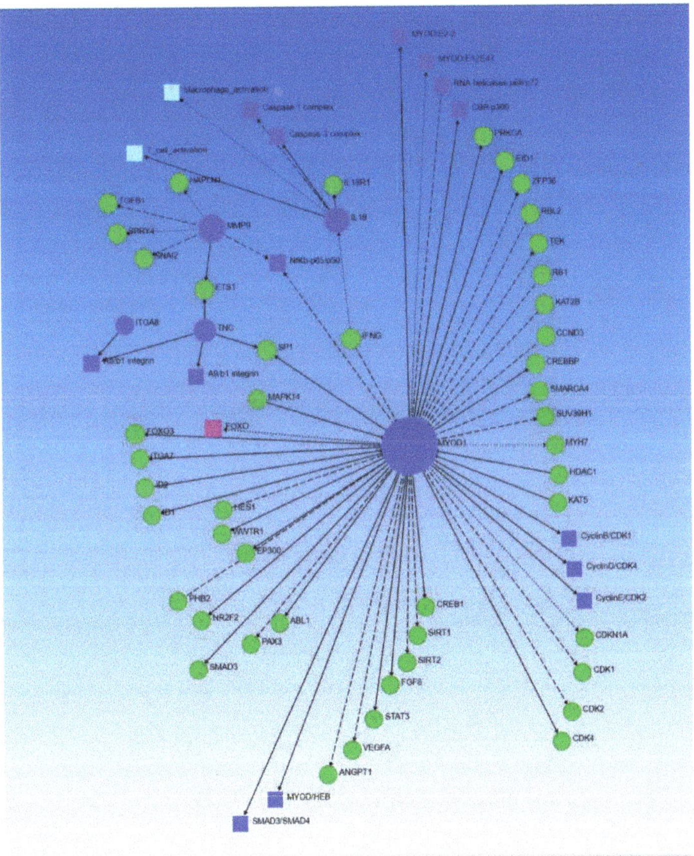

Figure 5. MYOD1 regulates the expression of various DEGs involved in chronic inflammation, plaque formation, and thrombosis.

Tissue injury is associated with increased secretion of S100 proteins. The S100 proteins S100A8, S100A9, and S100A12 play a crucial role in inflammation and atherosclerosis [16,32], while S100A1 plays a role in post-ischemic angiogenesis [33]. S100 proteins are potential therapeutic targets in atherosclerosis [32,34]. An increased expression of DEGs S100A1 ($\log_2$ fold = 3.69), S100A8 ($\log_2$ fold = 4.48), and S100A9 ($\log_2$ fold = 2.91) in RNA seq analysis of the FA involved in AVF compared to contralateral FA (Tables 2–5) suggest the pathologic role of calgranulins in vessel thrombosis and early AVF failure. Increased expression of S100 proteins is associated with inflammation, increased secretion of proinflammatory cytokines (IL-6, IL-8, and IL-18), and immune cell infiltration (macrophages) [35,36]. Increased expression of IL-8 (CXCL8, $\log_2$ fold = 2.21), IL-18 ($\log_2$ fold = 2.84), and CCR7 ($\log_2$ fold = 5.30) in association with S100 proteins suggest that IL-8 and IL-18 play a critical role in vessel thrombosis. This notion is supported by the role of IL-8 secreted from macrophages in the pathogenesis of atherosclerosis [37–39]. Similarly, the increased expression of IL-18 is associated with atherosclerosis, and it enhances atherosclerosis in association with IFN-γ [40–42]. An increased expression of CCR7 ($\log_2$ fold = 5.30, M1 macrophage marker) and association of MYOD1 with IFN-γ suggest a possible role of IL-8 and IL-18 in vascular thrombosis, an underlying pathology for early AVF failure. Another DEG related to inflammation and involved in angiogenesis was TEK [43–45] whose expression was increased in AVF FA ($\log_2$ fold = 2.22). The TEK gene is also known as Tie2 and is an angiopoietin receptor and is involved in neoangiogenesis, which contributes to the progression and stabilization of the plaques. Tie-2 plays a key role in vessel stabilization and destabilization in association with Ang-I mediated Tie-2 activation and Ang-II mediated inhibition of Tie-2 activation [46]. Increased TEK expression in AVF FA might be due to the locally active renin-angiotensin system in the vessel intima [47], as TEK expression is regulated by Ang-I and Ang-II. The involvement of TEK in angiogenesis and inflammation supports the hypothesis of its involvement in chronic inflammation and a probable role in vessel thrombosis. This is also supported by the fact that inhibition of TEK alleviates the release of inflammatory cytokines [44]. Another DEG with increased expression was bridging integrator 2 (BIN2) ($\log_2$ fold = 2.63). BIN2 regulates platelet activation in thrombosis, thrombo-inflammation, and atherosclerosis, and depletion of BIN2 is associated with protection from arterial thrombosis [48,49]. Increased expression of BIN2 in AVF FA samples in this study suggests a possible role of BIN2 in early vessel thrombosis and its contribution to early AVF failure. Another DEG phospholipase C-β2 (PLCB2) was found increased ($\log_2$ fold = 2.66) in AVF FA tissue. PLCB2 expression is regulated by NF-κB and is involved in platelet activation, inflammation, and atherosclerosis [50]. The involvement of PLCB2 in inflammation and atherosclerosis and its increased expression in AVF FA indicates the role of PLCB2 in vessel thrombosis and probably early AVF failure. Another DEG, ABL2 ($\log_2$ fold = 4.69) regulates vascular leakage during inflammation, and depletion of Arg/Abl2 associates with improvement in endothelial cell adhesion and prevents vascular leakage during inflammation [51]. LGALS12 (galectin-12, $\log_2$ fold = 5.41) enhances inflammation by promoting M1 macrophage polarization and negatively regulates M2 macrophage polarization [52]. Increased expression of these and multiple other DEGs (Tables 2–5) in AVF FA suggest the probable critical role of these DEGs in vessel thrombosis; however, this warrants future detailed mechanistic studies.

Along with the various DEGs involved and inducing inflammation, we also found various DEGs with an anti-inflammatory and antiatherosclerosis function. The DEGs were OSR1 ($\log_2$ fold= −2.48), FFAR4 ($\log_2$ fold = 4.80), CEBPA ($\log_2$ fold = 4.06), PON1 ($\log_2$ fold = 3.77), MLXIPL ($\log_2$ fold = 3.37), HCAR1 ($\log_2$ fold = 2.00), GPR39 ($\log_2$ fold = −3.07), MFGE8 ($\log_2$ fold = −2.53), A4GNT ($\log_2$ fold = 7.78), ABCC8 ($\log_2$ fold = 5.78), CD5 ($\log_2$ fold = 4.01), and ARID5B ($\log_2$ fold = 2.15). Odd-skipped related transcription factor 1 (OSR1), also known as oxidative stress-responsive kinase 1 (OSXR1), inhibits NF-κB [53] and regulates hepatic inflammation [54]. Free Fatty Acid Receptor 4 (FFAR4, GPR120) has anti-atherosclerotic potential and attenuates M1 macrophage activity, thus providing anti-inflammatory activity [55]. CCAAT/enhancer-binding protein alpha (CEBPA), gene encoding C/EBPα, plays a crucial role in myeloid lineage maturation and is expressed during the late phase of inflammatory responses. A decreased secretion of inflammatory cytokines TNF-α, IL-6, IL-1β, and IFN-γ with MTL-CEBPA, a small activating RNA targeting for upregulation of C/EBPα, suggests an anti-inflammatory role of CEBPA [56]. Paraoxonase 1 (PON1) protects against lipid oxidation and has an antioxidant and anti-inflammatory role in atherosclerosis [57]. The increased expression of MLXIPL by c-Jun inhibits inflammation in spinal cord nerve injury [58]. The increased expression of hydroxycarboxylic acid receptor 1 (HCAR1) is associated with anti-inflammatory response in glaucoma [59]. Under inflammatory conditions, G protein-coupled receptor 39 (GPR39) plays an anti-inflammatory role by enhancing IL-10 production from macrophages [60]. The increased expression of milk fat globule epidermal growth factor VIII (MFGE8) is associated with aging, atherosclerosis, hypertension, and diabetic arterial walls, and plays a crucial role in remodeling [61]. MFGE8 also has an anti-inflammatory response, and treatment with recombinant MFGE8 suppresses inflammation in mouse [62]. A4GNT may protect against inflammation-associated gastric adenocarcinoma (https://www.uniprot.org/uniprot/Q14BT6; accessed on 8 January 2022) and knocking out A4GNT is associated with gastric mucosal hyperplasia [63]. ATP-binding cassette transporter sub-family C member (ABCC) 8 encodes for sulfonylurea receptor 1 (Sur1) and silencing of Abcc8 or inhibition of Sur1-Trpm4 attenuate inflammation and disease progression in experimental autoimmune encephalomyelitis. This suggests the anti-inflammatory effect of silencing ABCC8 [64]. IL-10 is an anti-inflammatory cytokine secreted by CD5+ B cells [65], and increased expression of CD5 in AVF FA samples suggests the immune response of the body to increase IL-10 secretion. IL-10 is important as it protects against atherosclerosis by regulating atherogenic macrophage function [66] and can mitigate atherosclerosis [67,68]. Since neointimal hyperplasia and progressive plaque formation contribute to vessel thrombosis, increased CD5 expression in these samples suggests a protective mechanism. The AT-Rich Interaction Domain 5B (ARID5B) gene encodes a member of the AT-rich interaction domain (ARID) family of DNA binding proteins and methylation of ARID5B prevents inflammation and progression and development of atherosclerosis by inhibiting the ox-LDL/PI3K/Akt/NF-κB pathway [69].

The presence of DEGs contributing to pro-and anti-inflammatory pathogenesis in AVF FA samples compared to contralateral FA indicate the presence of chronic inflammation in AVF tissue as well as the anti-inflammatory immune response of the body. The presence of both pro-and anti-inflammatory DEGs suggests an immune response of the body to attenuate chronic inflammation but the outnumbering of the number of pro-inflammatory DEGs compared to the number of anti-inflammatory DEGs suggests the presence of persistent chronic inflammation. The presence of inflammatory DEGs and persistent inflammation might be the cause of early vessel thrombosis and early AVF failure. Furthermore, the presence of DEGs involved in perivascular cuffing suggests that for AVF patency, targeting inflammation around the vessel along with the pathologies inside the lumen (neointimal hyperplasia, intimal inflammation, plaque formation) should be considered. Targeting inflammation and immune cells to attenuate atherosclerosis supports the notion of targeting perivascular cuffing and luminal inflammation to enhance AVF maturation [19]. Targeting T-cells and macrophage accumulation to regulate adaptive venous remodeling increase AVF maturation [70], and increased expression of CCR7 ($\log_2$ fold = 5.30), a marker for the pro-inflammatory M1 macrophage, indicate the significance of targeting inflammatory macrophages for AVF maturation. Adaptive vascular extracellular matrix (ECM) remodeling favors outward remodeling of the vessel and contributes to AVF maturation. Thus, not only adaptive intimal ECM remodeling but adventitial remodeling also contributes to AVF maturation. Collagen deposition and elastin degradation play a crucial role in ECM remodeling, which is an attractive target for AVF maturation [71]. The presence of various DEGs associated with ECM remodeling including ELN, ITGA8, PRELP, FMOD, ITGA3, ADAMTSL4, HSPE, OSMR, CHST2, ECM1, and matrix metalloproteinases including MMP9, MMP7, MMP25, MMP17, and MMP8, and in our data indicate the probability of targeting vascular adventitial and ECM remodeling for AVF maturation. Targeting MMPs and ECM remodeling in association with inflammation is important because MMP activity is regulated by inflammatory cytokines and MMPs regulate inflammatory processes [72–74]. Targeting inflammation to enhance outward remodeling is also supported by favorable remodeling in mice deficient in the TLR-4 homolog RP105 [75].

5. Targeting DEGs and Translational Aspect

Chronic inflammation critically contributes to vessel thrombosis and stenosis. The presence of inflammatory DEGs and their correlation with inflammatory processes in this study support the notion of targeting these DEGs to attenuate chronic inflammation. Furthermore, measuring the serum expression of these DEGs may be used as biomarkers of the ongoing thrombosis and stenosis of the femoral vessels and AVF, as increased expression of these DEGs indicate thrombosis and an ongoing AVF failure. Thus, a timely intervention may be taken. Further, the presence of significantly increased inflammatory DEGs is also indicative of the ongoing inflammation and the need for anti-inflammatory treatment targeting these genes after AVF creation. The presence of thrombosed arteries also suggests the possibility of anti-platelet or anti-thrombotic treatment before and after AVF creation to attenuate plaque formation and early thrombosis [76,77].

Additionally, various DEGs delineated in this analysis involved in chronic inflammation also regulate other cellular mechanisms including fibrosis, ECM remodeling, vascular smooth muscle cell proliferation, and phenotype change (Figure 5, MYOD1 regulates many genes and TFs). These changes can be investigated using high-resolution scanning of the AVF, such as high-resolution 3D imaging [78]. High-resolution imaging will help in assessing AVF at various time points and an evolving thrombus or vessel stenosis can be evaluated early before complete occlusion. This will enhance the clinical outcome and may also help maintain AVF patency for a longer time. However, these assumptions warrant well-organized large-scale clinical trials.

6. Conclusions

Overall, the RNA sequencing results of AVF FA and contralateral control FA revealed multiple DEGs involved in inflammation, inflammatory pathways, immune cell migration, and proliferation and regulation of cytoplasmic kinases. Additionally, the increased expression of genes related to skeletal muscle injury and playing a role in vascular cuffing support the notion of targeting chronic inflammation in the vicinity of AVF along with the luminal pathologies. The tissues in this study were treated with the inhibitors of TREM-1 and TLR-4 to attenuate plaque formation by decreasing inflammation; however, the presence of various DEGs related to inflammation indicates a more aggressive approach to attenuate inflammation due to skeletal muscle injury and target perivascular cuffing, adventitial inflammation, and remodeling.

7. Limitations of the Study

This study elucidated several significantly expressed DEGs involved in inflammation, inflammatory pathogenesis, and the regulation of innate and adaptive immune response and proposed targeting chronic inflammation in the perivascular space due to vessel and skeletal muscle injury which might have therapeutic potential in AVF maturation. The limited number of samples and treatment with TREM-1 and TLR-4 inhibitors might have confounded the outcome. Furthermore, the involvement of several DEGs in inflammation as discussed here has not been shown in the perspective of atherosclerosis, and warrants future in vitro and in vivo studies. A comparison of AVF tissue without any treatment with treated tissues may reveal more DEGs involved in inflammation and as a therapeutic target. Despite these limitations, this study elucidated novel DEGs that can be targeted to attenuate or inhibit early vascular thrombosis to hasten AVF maturation. The focus on vascular adventitia remodeling and attenuating perivascular cuffing by targeting DEGs involved in inflammation is a key point of this manuscript.

Supplementary Materials: The following supporting information can be downloaded at: https://www.mdpi.com/article/10.3390/biomedicines10020433/s1, Table S1: Differentially Expressed Genes (DEGs) in various groups.

Author Contributions: Conceptualization, V.R. and D.K.A.; methodology, V.R.; software, V.R.; validation, D.K.A.; formal analysis, V.R.; investigation, V.R.; resources, D.K.A.; data curation, V.R.; writing—original draft preparation, V.R.; writing—review and editing, D.K.A.; supervision, D.K.A.; project administration, D.K.A.; funding acquisition, D.K.A. All authors have read and agreed to the published version of the manuscript.

Funding: V.R. is supported by an intramural grant IMR Rai 12397B from the Western University of Health Sciences, Pomona, California. The research work of D.K.A. is supported by the R01 HL144125 and R01 HL147662 grants from the National Institutes of Health, USA. The content of this critical review is solely the responsibility of the authors and does not necessarily represent the official views of the National Institutes of Health.

Institutional Review Board Statement: Not applicable. The Institutional Animal Care and Use Committee (IACUC) at Western University of Health Sciences approved protocol No. R20IACUC038 for this study.

Informed Consent Statement: Not applicable.

Data Availability Statement: All data supporting the results of this manuscript has been included in this manuscript along with a Supplementary File. Bulk RNA seq (Fastq. Files) can be provided from the corresponding authors on request.

Acknowledgments: The authors acknowledge Mohamed M Radwan who performed the surgery and radiology of the AVF in pigs.

Conflicts of Interest: As the corresponding author, I declare that this manuscript is original; that the article does not infringe upon any copyright or other proprietary rights of any third party; that neither the text nor the data have been reported or published previously. All the authors have no conflicts of interest and have read the journal's authorship statement.

References

1. Segal, M.; Qaja, E. Types of Arteriovenous Fistulas. In *StatPearls*; StatPearls Publishing: Treasure Island, FL, USA, 2022.
2. Kaygin, M.A.; Halici, U.; Aydin, A.; Dag, O.; Binici, D.N.; Limandal, H.K.; Arslan, U.; Kiymaz, A.; Kahraman, N.; Calik, E.S.; et al. The relationship between arteriovenous fistula success and inflammation. *Ren. Fail.* **2013**, *35*, 1085–1088. [CrossRef]
3. Chang, C.J.; Ko, Y.S.; Ko, P.J.; Hsu, L.A.; Chen, C.F.; Yang, C.W.; Hsu, T.S.; Pang, J.H. Thrombosed arteriovenous fistula for hemodialysis access is characterized by a marked inflammatory activity. *Kidney Int.* **2005**, *68*, 1312–1319. [CrossRef]
4. Stirbu, O.; Gadalean, F.; Pitea, I.V.; Ciobanu, G.; Schiller, A.; Grosu, I.; Nes, A.; Bratescu, R.; Olariu, N.; Timar, B.; et al. C-reactive protein as a prognostic risk factor for loss of arteriovenous fistula patency in hemodialyzed patients. *J. Vasc. Surg.* **2019**, *70*, 208–215. [CrossRef] [PubMed]
5. Vazquez-Padron, R.I.; Duque, J.C.; Tabbara, M.; Salman, L.H.; Martinez, L. Intimal Hyperplasia and Arteriovenous Fistula Failure: Looking Beyond Size Differences. *Kidney360* **2021**, *2*, 1360–1372. [CrossRef]
6. Siddiqui, M.A.; Ashraff, S.; Santos, D.; Carline, T. An overview of AVF maturation and endothelial dysfunction in an advanced renal failure. *Ren. Replace. Ther.* **2017**, *3*, 42. [CrossRef]
7. Smith, C.; Kruger, M.J.; Smith, R.M.; Myburgh, K.H. The inflammatory response to skeletal muscle injury: Illuminating complexities. *Sports Med.* **2008**, *38*, 947–969. [CrossRef] [PubMed]
8. Cholok, D.; Lee, E.; Lisiecki, J.; Agarwal, S.; Loder, S.; Ranganathan, K.; Qureshi, A.T.; Davis, T.A.; Levi, B. Traumatic muscle fibrosis: From pathway to prevention. *J. Trauma Acute Care Surg.* **2017**, *82*, 174–184. [CrossRef]
9. Howard, E.E.; Pasiakos, S.M.; Blesso, C.N.; Fussell, M.A.; Rodriguez, N.R. Divergent Roles of Inflammation in Skeletal Muscle Recovery From Injury. *Front. Physiol.* **2020**, *11*, 87. [CrossRef]
10. Gulati, A.; Bagga, A. Large vessel vasculitis. *Pediatr. Nephrol.* **2010**, *25*, 1037–1048. [CrossRef]
11. Rai, V.; Rao, V.H.; Shao, Z.; Agrawal, D.K. Dendritic Cells Expressing Triggering Receptor Expressed on Myeloid Cells-1 Correlate with Plaque Stability in Symptomatic and Asymptomatic Patients with Carotid Stenosis. *PLoS ONE* **2016**, *11*, e0154802. [CrossRef]
12. Rao, V.H.; Rai, V.; Stoupa, S.; Subramanian, S.; Agrawal, D.K. Tumor necrosis factor-alpha regulates triggering receptor expressed on myeloid cells-1-dependent matrix metalloproteinases in the carotid plaques of symptomatic patients with carotid stenosis. *Atherosclerosis* **2016**, *248*, 160–169. [CrossRef] [PubMed]
13. Rao, V.H.; Rai, V.; Stoupa, S.; Subramanian, S.; Agrawal, D.K. Data on TREM-1 activation destabilizing carotid plaques. *Data Brief* **2016**, *8*, 230–234. [CrossRef] [PubMed]

14. Singh, R.K.; Haka, A.S.; Asmal, A.; Barbosa-Lorenzi, V.C.; Grosheva, I.; Chin, H.F.; Xiong, Y.; Hla, T.; Maxfield, F.R. TLR4 (Toll-Like Receptor 4)-Dependent Signaling Drives Extracellular Catabolism of LDL (Low-Density Lipoprotein) Aggregates. *Arterioscler. Thromb. Vasc. Biol.* **2020**, *40*, 86–102. [CrossRef] [PubMed]

15. Rai, V.; Agrawal, D.K. Immunomodulation of IL-33 and IL-37 with Vitamin D in the Neointima of Coronary Artery: A Comparative Study between Balloon Angioplasty and Stent in Hyperlipidemic Microswine. *Int. J. Mol. Sci.* **2021**, *22*, 8824. [CrossRef]

16. Rai, V.; Agrawal, D.K. The role of damage- and pathogen-associated molecular patterns in inflammation-mediated vulnerability of atherosclerotic plaques. *Can. J. Physiol. Pharmacol.* **2017**, *95*, 1245–1253. [CrossRef]

17. Conti, P.; Shaik-Dasthagirisaeb, Y. Atherosclerosis: A chronic inflammatory disease mediated by mast cells. *Cent. Eur. J. Immunol.* **2015**, *40*, 380–386. [CrossRef]

18. Wolf, D.; Ley, K. Immunity and Inflammation in Atherosclerosis. *Circ. Res.* **2019**, *124*, 315–327. [CrossRef]

19. Soehnlein, O.; Libby, P. Targeting inflammation in atherosclerosis—From experimental insights to the clinic. *Nat. Rev. Drug. Discov.* **2021**, *20*, 589–610. [CrossRef]

20. MacRae, J.M.; Dipchand, C.; Oliver, M.; Moist, L.; Lok, C.; Clark, E.; Hiremath, S.; Kappel, J.; Kiaii, M.; Luscombe, R.; et al. Arteriovenous Access Failure, Stenosis, and Thrombosis. *Can. J. Kidney Health Dis.* **2016**, *3*, 2054358116669126. [CrossRef]

21. Yap, Y.-S.; Chi, W.-C.; Lin, C.-H.; Liu, Y.-C.; Wu, Y.-W. Association of early failure of arteriovenous fistula with mortality in hemodialysis patients. *Sci. Rep.* **2021**, *11*, 5699. [CrossRef]

22. Weyand, C.; Goronzy, J. Vasculitides: Giant Cell Arteritis, Polymyalgia Rheumatica, and Takayasu's Arteritis. In *Primer on the Rheumatic Diseases*; Klippel, J.H., Stone, J.H., Crofford, L.J., White, P.H., Eds.; Springer: New York, NY, USA, 2008; pp. 397–405. [CrossRef]

23. Agrawal, S.M.; Williamson, J.; Sharma, R.; Kebir, H.; Patel, K.; Prat, A.; Yong, V.W. Extracellular matrix metalloproteinase inducer shows active perivascular cuffs in multiple sclerosis. *Brain* **2013**, *136*, 1760–1777. [CrossRef] [PubMed]

24. Gralinski, L.E.; Ferris, M.T.; Aylor, D.L.; Whitmore, A.C.; Green, R.; Frieman, M.B.; Deming, D.; Menachery, V.D.; Miller, D.R.; Buus, R.J.; et al. Genome Wide Identification of SARS-CoV Susceptibility Loci Using the Collaborative Cross. *PLoS Genet.* **2015**, *11*, e1005504. [CrossRef] [PubMed]

25. Nguyen, T.; Bowen, T.S.; Augstein, A.; Schauer, A.; Gasch, A.; Linke, A.; Labeit, S.; Adams, V. Expression of MuRF1 or MuRF2 is essential for the induction of skeletal muscle atrophy and dysfunction in a murine pulmonary hypertension model. *Skelet Muscle* **2020**, *10*, 12. [CrossRef]

26. Moriscot, A.S.; Baptista, I.L.; Silva, W.J.; Silvestre, J.G.; Adams, V.; Gasch, A.; Bogomolovas, J.; Labeit, S. MuRF1 and MuRF2 are key players in skeletal muscle regeneration involving myogenic deficit and deregulation of the chromatin-remodeling complex. *JCSM Rapid Commun.* **2019**, *2*, 1–25. [CrossRef]

27. Kane, M.; Golovkina, T.V. Mapping viral susceptibility loci in mice. *Annu. Rev. Virol.* **2019**, *6*, 525–546. [CrossRef]

28. De Wit, E.; Van Doremalen, N.; Falzarano, D.; Munster, V.J. SARS and MERS: Recent insights into emerging coronaviruses. *Nat. Rev. Microbiol.* **2016**, *14*, 523–534. [CrossRef] [PubMed]

29. Hodge, B.A.; Zhang, X.; Gutierrez-Monreal, M.A.; Cao, Y.; Hammers, D.W.; Yao, Z.; Wolff, C.A.; Du, P.; Kemler, D.; Judge, A.R. MYOD1 functions as a clock amplifier as well as a critical co-factor for downstream circadian gene expression in muscle. *Elife* **2019**, *8*, e43017. [CrossRef]

30. Nakamura, S.; Yonekura, S.; Shimosato, T.; Takaya, T. Myogenetic Oligodeoxynucleotide (myoDN) Recovers the Differentiation of Skeletal Muscle Myoblasts Deteriorated by Diabetes Mellitus. *Front. Physiol.* **2021**, *12*, 679152. [CrossRef]

31. Chandran, R.; Knobloch, T.J.; Anghelina, M.; Agarwal, S. Biomechanical signals upregulate myogenic gene induction in the presence or absence of inflammation. *Am. J. Physiol. Cell. Physiol.* **2007**, *293*, C267–C276. [CrossRef]

32. Xiao, X.; Yang, C.; Qu, S.L.; Shao, Y.D.; Zhou, C.Y.; Chao, R.; Huang, L.; Zhang, C. S100 proteins in atherosclerosis. *Clin. Chim. Acta* **2020**, *502*, 293–304. [CrossRef]

33. Most, P.; Lerchenmuller, C.; Rengo, G.; Mahlmann, A.; Ritterhoff, J.; Rohde, D.; Goodman, C.; Busch, C.J.; Laube, F.; Heissenberg, J.; et al. S100A1 deficiency impairs postischemic angiogenesis via compromised proangiogenic endothelial cell function and nitric oxide synthase regulation. *Circ. Res.* **2013**, *112*, 66–78. [CrossRef] [PubMed]

34. Singh, H.; Rai, V.; Agrawal, D.K. Discerning the promising binding sites of S100/calgranulins and their therapeutic potential in atherosclerosis. *Expert Opin. Ther. Pat.* **2021**, *31*, 1045–1057. [CrossRef] [PubMed]

35. Xia, C.; Braunstein, Z.; Toomey, A.C.; Zhong, J.; Rao, X. S100 Proteins As an Important Regulator of Macrophage Inflammation. *Front. Immunol.* **2017**, *8*, 1908. [CrossRef] [PubMed]

36. Wang, S.; Song, R.; Wang, Z.; Jing, Z.; Wang, S.; Ma, J. S100A8/A9 in Inflammation. *Front. Immunol.* **2018**, *9*, 1298. [CrossRef]

37. Boisvert, W.A.; Curtiss, L.K.; Terkeltaub, R.A. Interleukin-8 and its receptor CXCR2 in atherosclerosis. *Immunol. Res.* **2000**, *21*, 129–137. [CrossRef]

38. Apostolopoulos, J.; Davenport, P.; Tipping, P.G. Interleukin-8 production by macrophages from atheromatous plaques. *Arterioscler. Thromb. Vasc. Biol.* **1996**, *16*, 1007–1012. [CrossRef]

39. Cavusoglu, E.; Marmur, J.D.; Yanamadala, S.; Chopra, V.; Hegde, S.; Nazli, A.; Singh, K.P.; Zhang, M.; Eng, C. Elevated baseline plasma IL-8 levels are an independent predictor of long-term all-cause mortality in patients with acute coronary syndrome. *Atherosclerosis* **2015**, *242*, 589–594. [CrossRef]

40. Whitman, S.C.; Ravisankar, P.; Daugherty, A. Interleukin-18 enhances atherosclerosis in apolipoprotein E($^{-/-}$) mice through release of interferon-gamma. *Circ. Res.* **2002**, *90*, E34–E38. [CrossRef]

41. Yamaoka-Tojo, M.; Tojo, T.; Wakaume, K.; Kameda, R.; Nemoto, S.; Takahira, N.; Masuda, T.; Izumi, T. Circulating interleukin-18: A specific biomarker for atherosclerosis-prone patients with metabolic syndrome. *Nutr. Metab.* **2011**, *8*, 3. [CrossRef]

42. Mallat, Z.; Corbaz, A.; Scoazec, A.; Besnard, S.; Leseche, G.; Chvatchko, Y.; Tedgui, A. Expression of interleukin-18 in human atherosclerotic plaques and relation to plaque instability. *Circulation* **2001**, *104*, 1598–1603. [CrossRef]

43. Horwood, N.J.; Urbaniak, A.M.; Danks, L. Tec family kinases in inflammation and disease. *Int. Rev. Immunol.* **2012**, *31*, 87–103. [CrossRef] [PubMed]

44. Meng, Y.; Sha, S.; Yang, J.; Ren, H. Effects of Tec Tyrosine Kinase Inhibition on the Inflammatory Response of Severe Acute Pancreatitis-Associated Acute Lung Injury in Mice. *Dig. Dis. Sci.* **2019**, *64*, 2167–2176. [CrossRef] [PubMed]

45. Camare, C.; Pucelle, M.; Negre-Salvayre, A.; Salvayre, R. Angiogenesis in the atherosclerotic plaque. *Redox Biol.* **2017**, *12*, 18–34. [CrossRef] [PubMed]

46. Peters, K.G.; Kontos, C.D.; Lin, P.C.; Wong, A.L.; Rao, P.; Huang, L.; Dewhirst, M.W.; Sankar, S. Functional significance of Tie2 signaling in the adult vasculature. *Recent Prog. Horm. Res.* **2004**, *59*, 51–71. [CrossRef] [PubMed]

47. Pacurari, M.; Kafoury, R.; Tchounwou, P.B.; Ndebele, K. The Renin-Angiotensin-aldosterone system in vascular inflammation and remodeling. *Int. J. Inflamm.* **2014**, *2014*, 689360. [CrossRef]

48. Volz, J.; Kusch, C.; Beck, S.; Popp, M.; Vogtle, T.; Meub, M.; Scheller, I.; Heil, H.S.; Preu, J.; Schuhmann, M.K.; et al. BIN2 orchestrates platelet calcium signaling in thrombosis and thrombo-inflammation. *J. Clin. Investig.* **2020**, *130*, 6064–6079. [CrossRef]

49. Fenning, R.S.; Burgert, M.E.; Hamamdzic, D.; Peyster, E.G.; Mohler, E.R.; Kangovi, S.; Jucker, B.M.; Lenhard, S.C.; Macphee, C.H.; Wilensky, R.L. Atherosclerotic plaque inflammation varies between vascular sites and correlates with response to inhibition of lipoprotein-associated phospholipase A2. *J. Am. Heart Assoc.* **2015**, *4*, 1477. [CrossRef]

50. Mao, G.; Jin, J.; Kunapuli, S.P.; Rao, A.K. Nuclear factor-kappaB regulates expression of platelet phospholipase C-beta2 (PLCB2). *Thromb. Haemost.* **2016**, *116*, 931–940. [CrossRef]

51. Amado-Azevedo, J.; van Stalborch, A.D.; Valent, E.T.; Nawaz, K.; van Bezu, J.; Eringa, E.C.; Hoevenaars, F.P.M.; De Cuyper, I.M.; Hordijk, P.L.; van Hinsbergh, V.W.M.; et al. Depletion of Arg/Abl2 improves endothelial cell adhesion and prevents vascular leak during inflammation. *Angiogenesis* **2021**, *24*, 677–693. [CrossRef]

52. Wan, L.; Lin, H.J.; Huang, C.C.; Chen, Y.C.; Hsu, Y.A.; Lin, C.H.; Lin, H.C.; Chang, C.Y.; Huang, S.H.; Lin, J.M.; et al. Galectin-12 enhances inflammation by promoting M1 polarization of macrophages and reduces insulin sensitivity in adipocytes. *Glycobiology* **2016**, *26*, 732–744. [CrossRef]

53. Chen, W.; Wu, K.; Zhang, H.; Fu, X.; Yao, F.; Yang, A. Odd-skipped related transcription factor 1 (OSR1) suppresses tongue squamous cell carcinoma migration and invasion through inhibiting NF-kappaB pathway. *Eur. J. Pharmacol.* **2018**, *839*, 33–39. [CrossRef] [PubMed]

54. Zhou, Y.; Liu, Z.; Lynch, E.C.; He, L.; Cheng, H.; Liu, L.; Li, Z.; Li, J.; Lawless, L.; Zhang, K.K.; et al. Osr1 regulates hepatic inflammation and cell survival in the progression of non-alcoholic fatty liver disease. *Lab. Investig.* **2021**, *101*, 477–489. [CrossRef] [PubMed]

55. Kiepura, A.; Stachyra, K.; Olszanecki, R. Anti-Atherosclerotic Potential of Free Fatty Acid Receptor 4 (FFAR4). *Biomedicines* **2021**, *9*, 467. [CrossRef]

56. Zhou, J.; Li, H.; Xia, X.; Herrera, A.; Pollock, N.; Reebye, V.; Sodergren, M.H.; Dorman, S.; Littman, B.H.; Doogan, D.; et al. Anti-inflammatory Activity of MTL-CEBPA, a Small Activating RNA Drug, in LPS-Stimulated Monocytes and Humanized Mice. *Mol. Ther.* **2019**, *27*, 999–1016. [CrossRef]

57. Litvinov, D.; Mahini, H.; Garelnabi, M. Antioxidant and anti-inflammatory role of paraoxonase 1: Implication in arteriosclerosis diseases. *N. Am. J. Med. Sci.* **2012**, *4*, 523–532. [CrossRef] [PubMed]

58. Zhan, H.; Wang, Y.; Yu, S.; Cai, G.; Zeng, Y.; Ma, J.; Liu, W.; Wu, W. Upregulation of Mlxipl induced by cJun in the spinal dorsal horn after peripheral nerve injury counteracts mechanical allodynia by inhibiting neuroinflammation. *Aging* **2020**, *12*, 11004–11024. [CrossRef]

59. Harun-Or-Rashid, M.; Inman, D.M. Reduced AMPK activation and increased HCAR activation drive anti-inflammatory response and neuroprotection in glaucoma. *J. Neuroinflamm.* **2018**, *15*, 313. [CrossRef]

60. Muneoka, S.; Goto, M.; Kadoshima-Yamaoka, K.; Kamei, R.; Terakawa, M.; Tomimori, Y. G protein-coupled receptor 39 plays an anti-inflammatory role by enhancing IL-10 production from macrophages under inflammatory conditions. *Eur. J. Pharmacol.* **2018**, *834*, 240–245. [CrossRef]

61. Wang, M.; Wang, H.H.; Lakatta, E.G. Milk fat globule epidermal growth factor VIII signaling in arterial wall remodeling. *Curr. Vasc. Pharmacol.* **2013**, *11*, 768–776. [CrossRef]

62. Hansen, L.W.; Yang, W.L.; Bolognese, A.C.; Jacob, A.; Chen, T.; Prince, J.M.; Nicastro, J.M.; Coppa, G.F.; Wang, P. Treatment with milk fat globule epidermal growth factor-factor 8 (MFG-E8) reduces inflammation and lung injury in neonatal sepsis. *Surgery* **2017**, *162*, 349–357. [CrossRef]

63. Kawakubo, M.; Komura, H.; Goso, Y.; Okumura, M.; Sato, Y.; Fujii, C.; Miyashita, M.; Arisaka, N.; Harumiya, S.; Yamanoi, K. Analysis of A4gnt Knockout Mice Reveals an Essential Role for Gastric Sulfomucins in Preventing Gastritis Cystica Profunda. *J. Histochem. Cytochem.* **2019**, *67*, 759–770. [CrossRef] [PubMed]

64. Makar, T.K.; Gerzanich, V.; Nimmagadda, V.K.; Jain, R.; Lam, K.; Mubariz, F.; Trisler, D.; Ivanova, S.; Woo, S.K.; Kwon, M.S.; et al. Silencing of Abcc8 or inhibition of newly upregulated Sur1-Trpm4 reduce inflammation and disease progression in experimental autoimmune encephalomyelitis. *J. Neuroinflamm.* **2015**, *12*, 210. [CrossRef] [PubMed]

65. Qin, J.; Zhao, N.; Wang, S.; Liu, S.; Liu, Y.; Cui, X.; Wang, S.; Xiang, Y.; Fan, C.; Li, Y.; et al. Roles of Endogenous IL-10 and IL-10-Competent and CD5+ B Cells in Autoimmune Thyroiditis in NOD.H-2h4 Mice. *Endocrinology* **2020**, *161*, bqaa033. [CrossRef] [PubMed]

66. Han, X.; Boisvert, W.A. Interleukin-10 protects against atherosclerosis by modulating multiple atherogenic macrophage function. *Thromb. Haemost.* **2015**, *113*, 505–512. [CrossRef]

67. Fourman, L.T.; Saylor, C.F.; Cheru, L.; Fitch, K.; Looby, S.; Keller, K.; Robinson, J.A.; Hoffmann, U.; Lu, M.T.; Burdo, T.; et al. Anti-Inflammatory Interleukin 10 Inversely Relates to Coronary Atherosclerosis in Persons With Human Immunodeficiency Virus. *J. Infect. Dis.* **2020**, *221*, 510–515. [CrossRef]

68. Mallat, Z.; Besnard, S.; Duriez, M.; Deleuze, V.; Emmanuel, F.; Bureau, M.F.; Soubrier, F.; Esposito, B.; Duez, H.; Fievet, C.; et al. Protective role of interleukin-10 in atherosclerosis. *Circ. Res.* **1999**, *85*, e17–e24. [CrossRef]

69. Jackson, A.O.; Regine, M.A.; Subrata, C.; Long, S. Molecular mechanisms and genetic regulation in atherosclerosis. *Int. J. Cardiol. Heart Vasc.* **2018**, *21*, 36–44. [CrossRef]

70. Matsubara, Y.; Kiwan, G.; Liu, J.; Gonzalez, L.; Langford, J.; Gao, M.; Gao, X.; Taniguchi, R.; Yatsula, B.; Furuyama, T.; et al. Inhibition of T-Cells by Cyclosporine A Reduces Macrophage Accumulation to Regulate Venous Adaptive Remodeling and Increase Arteriovenous Fistula Maturation. *Arterioscler. Thromb. Vasc. Biol.* **2021**, *41*, e160–e174. [CrossRef]

71. Shiu, Y.T.; He, Y.; Tey, J.C.S.; Knysheva, M.; Anderson, B.; Kauser, K. Natural Vascular Scaffolding Treatment Promotes Outward Remodeling During Arteriovenous Fistula Development in Rats. *Front. Bioeng. Biotechnol.* **2021**, *9*, 622617. [CrossRef]

72. Fingleton, B. Matrix metalloproteinases as regulators of inflammatory processes. *Biochim. Biophys. Acta Mol. Cell Res.* **2017**, *1864*, 2036–2042. [CrossRef]

73. Ries, C.; Petrides, P.E. Cytokine regulation of matrix metalloproteinase activity and its regulatory dysfunction in disease. *Biol. Chem. Hoppe Seyler* **1995**, *376*, 345–355. [PubMed]

74. Brown, R.D.; Jones, G.M.; Laird, R.E.; Hudson, P.; Long, C.S. Cytokines regulate matrix metalloproteinases and migration in cardiac fibroblasts. *Biochem. Biophys. Res. Commun.* **2007**, *362*, 200–205. [CrossRef] [PubMed]

75. Bezhaeva, T.; Wong, C.; de Vries, M.R.; van der Veer, E.P.; van Alem, C.M.A.; Que, I.; Lalai, R.A.; van Zonneveld, A.J.; Rotmans, J.I.; Quax, P.H.A. Deficiency of TLR4 homologue RP105 aggravates outward remodeling in a murine model of arteriovenous fistula failure. *Sci. Rep.* **2017**, *7*, 10269. [CrossRef] [PubMed]

76. Sharathkumar, A.; Hirschl, R.; Pipe, S.; Crandell, C.; Adams, B.; Lin, J.J. Primary thromboprophylaxis with heparins for arteriovenous fistula failure in pediatric patients. *J. Vasc. Access.* **2007**, *8*, 235–244. [CrossRef] [PubMed]

77. Chen, L.; Ling, Y.S.; Lin, C.H.; Guan, T.J. Combined use of heparin and anisodamine reduces the risk of early thrombosis in native arteriovenous fistula. *Vascular* **2013**, *21*, 369–374. [CrossRef] [PubMed]

78. Chen, J.; Sivan, U.; Tan, S.L.; Lippo, L.; De Angelis, J.; Labella, R.; Singh, A.; Chatzis, A.; Cheuk, S.; Medhghalchi, M.; et al. High-resolution 3D imaging uncovers organ-specific vascular control of tissue aging. *Sci. Adv.* **2021**, *7*, eabd7819. [CrossRef]

Article

The Inflammatory Pattern of Chronic Limb-Threatening Ischemia in Muscles: The TNF-α Hypothesis

Diego Caicedo [1,*], Clara V. Alvarez [2], Sihara Perez-Romero [2] and Jesús Devesa [3]

[1] Angiology and Vascular Surgery Department, Complejo Hospitalario de Santiago de Compostela, 15706 Santiago de Compostela, Spain

[2] Neoplasia & Endocrine Differentiation P0L5, Centro de Investigación en Medicina Molecular y Enferme-dades Crónicas (CIMUS), University of Santiago de Compostela (USC), 15783 Santiago de Compostela, Spain; clara.alvarez@usc.es (C.V.A.); siara.perez@usc.es (S.P.-R.)

[3] The Medical Center Foltra, 15886 Teo, Spain; devesa.jesus@gmail.com

* Correspondence: diego.caicedo.valdes@sergas.es; Tel.: +34-981-950-043

Abstract: Background: Vascular inflammation plays a crucial role in peripheral arterial disease (PAD), although the role of the mediators involved has not yet been properly defined. The aim of this work is to investigate gene expression and plasma biomarkers in chronic limb-threating ischemia (CLTI). **Methods:** Using patients from the GHAS trial, both blood and ischemic muscle samples were obtained to analyze plasma markers and mRNA expression, respectively. Statistical analysis was performed by using univariate (Spearman, t-Student, and X^2) and multivariate (multiple logistic regression) tests. **Results:** A total of 35 patients were available at baseline (29 for mRNA expression). Baseline characteristics (mean): Age: 71.4 ± 12.4 years (79.4% male); TNF-α: 10.7 ± 4.9 pg/mL; hsCRP:1.6 ± 2.2 mg/dL; and neutrophil-to-lymphocyte ratio (NLR): 3.5 ± 2.8. Plasma TNF-α was found elevated (≥8.1) in 68.6% of patients, while high hsCRP (≥0.5) was found in 60.5%. Diabetic patients with a high level of inflammation showed significantly higher levels of NOX4 expression at baseline (p = 0.0346). Plasma TNF-α had a negative correlation with *NOS3* (*eNOS*) expression (−0.5, p = 0.015) and plasma hsCRP with *VEGFA* (−0.63, p = 0.005). The expression of *NOX4* was parallel to that of plasma TNF-α (0.305, p = 0.037), especially in DM. Cumulative mortality at 12 months was related to NLR ≥ 3 (p = 0.019) and TNF-α ≥ 8.1 (p = 0.048). The best cutoff point for NLR to predict mortality was 3.4. **Conclusions:** *NOX4* and TNF-α are crucial for the development and complications of lower limb ischemia, especially in DM. hsCRP could have a negative influence on angiogenesis too. NLR and TNF-α represent suitable markers of mortality in CLTI. These results are novel because they connect muscle gene expression and plasma information in patients with advanced PAD, deepening the search for new and accurate targets for this condition.

Keywords: vascular inflammation; peripheral arterial disease (PAD); chronic limb-threatening ischemia (CLTI); GHAS trial; TNF-α; hsCRP; neutrophil-to-lymphocyte ratio (NLR); *NOX4*; *NOS3* (*eNOS*); *VEGFA*

Citation: Caicedo, D.; Alvarez, C.V.; Perez-Romero, S.; Devesa, J. The Inflammatory Pattern of Chronic Limb-Threatening Ischemia in Muscles: The TNF-α Hypothesis. *Biomedicines* **2022**, *10*, 489. https://doi.org/10.3390/biomedicines10020489

Academic Editor: Byeong Hwa Jeon

Received: 30 November 2021
Accepted: 14 February 2022
Published: 18 February 2022

Publisher's Note: MDPI stays neutral with regard to jurisdictional claims in published maps and institutional affiliations.

1. Introduction

Peripheral arterial disease (PAD) is an occlusive arterial disease, mainly of atherosclerotic origin, that affects lower limbs. It represents an independent risk factor for cardiovascular (CV) morbidity and mortality [1–3]. In fact, these patients have a CV event rate similar to those with established coronary or cerebral vascular disease [4–6]. Considering that there are few tools to identify which patients with PAD are at risk for an acute event, risk handling in this setting represents a major health challenge [4,7]. Inflammation plays a key role in the development of PAD, but the mediators involved in this condition have not yet been fully defined [8,9]. The isolated use of clinical staging of PAD (Rutherford class) ignores the importance of other factors that may also be crucial for long-term survival in these patients, such as biomarkers of systemic inflammation or endothelial dysfunction [9,10].

Both experimental and large cohort studies in humans have evaluated the role of inflammatory cytokines in atherosclerosis [3,11–14]. IL-6, TNF-α, and C-reactive protein of high sensitivity (hsCRP) have been found predictive of future CV events in healthy and diseased populations [15,16]. However, studies using Mendelian randomization have shown contradictory results depending on the marker, justifying the need for further study [9,15,17–19]. Moreover, it seems that different cytokines could influence the diverse vascular beds in different ways. In fact, murine and human data do not always coincide, as we will see below.

Inflammation has a clear deleterious effect on vessel endothelial function. Indeed, the CANTOS trial provided the first compelling evidence that inhibition of cytokine function, by decreasing the activity of IL-1ß and IL-6 signaling pathways, can reduce CV risk regardless of blood pressure or lipid level [20]. However, this breakthrough has only been demonstrated so far for coronary heart disease. In PAD, several cytokines have been associated with disease progression in one cohort study, finding a significant increase in the levels of IL-6, TNF-α, selectins, neopterin, CAMs, MMP-2, and MMP-9 [21]. However, a preceding study found no significant difference for IL-6 and IL-1ß in lower limb ischemia [22]. While in the CANTOS study the relevant role is for IL-6, in PAD the levels of TNF-α and IL-8 were clearly increased, which supports the statement of different inflammatory patterns in both atherosclerotic conditions. Further, IL-6 secretion is highly dependent on TNF-α and appears to have different behavior when secreted by monocytes (proinflammatory) or skeletal muscle (anti-inflammatory) [23]. In addition, nonrandomized observational studies have suggested a reduction in the rate of atherosclerotic events in patients treated with TNF-α inhibitors [24,25]. However, the causal association between these or other biomarkers and PAD has not yet been established.

The neutrophil-to-lymphocyte ratio (NLR), defined as the ratio between absolute count of neutrophils and lymphocytes, has been gaining relevance as a marker of CV disease. Many researchers have deeply evaluated NLR as a potential prognostic biomarker, predicting pathological and survival outcomes in patients with atherosclerosis and, in particular, in PAD, in which a strong relationship has been identified for systemic (long-term mortality) and local (lower limb) complications [26].

It is also well known that oxidative stress plays an important role in endothelial dysfunction in the inflammatory context of CV disease. Many studies have highlighted how oxidative distress triggers and impairs this condition, facilitating adverse CV events [27,28]. Although many studies currently address inflammation and redox balance in the vascular system, few of them have cross-linked gene and plasma information in a human trial.

Here we present new insights from an interventional study, the GHAS trial, in which information about muscle tissue mRNA expression and plasma biomarkers was combined in patients suffering from chronic limb-threatening ischemia (CLTI). Our results are novel because they were obtained in this PAD special subset of patients presenting CLTI, who represent those with the highest level of inflammation and CV risk.

2. Materials and Methods

From January 2016 to December 2018, all patients with diagnosis of CLTI who met inclusion criteria were enrolled in the Growth Hormone Angiogenic Study (GHAS) trial, registered in the Spanish Registry of Clinical Trials (REEC) with the number 2012-002228-34.

The GHAS study, a phase III randomized controlled trial, was designed to test the benefit of low dose of GH (0.4 mg per day, 5 days a week, during 2 months) for wound healing and rest pain relief in CLTI patients compared to placebo or control group, in which an injection of serum was administered with the same protocol instead of GH [28–30]. Blood samples were extracted for the determination of plasma biomarkers at the beginning of the study (basal) and after two months of treatment initiation (final). Skeletal muscle samples from the ischemic limb were also obtained with the same time interval.

2.1. Informed Consent and Recruitment

Written informed consent was obtained from each participant at the beginning of the investigation. The recruitment of the subjects for this study was made by vascular surgeons in the Angiology and Vascular Surgery Department of the Clinical Hospital of Pontevedra, Spain.

2.2. Medical Screening through Medical History and Physical Examination

At the beginning and at the end of this study, demographic information, CV risk factors, comorbid conditions, Rutherford class of ischemia, ankle–brachial index (ABI), ankle pressure (AK) and photoplethysmography (PPG), blood and muscle samples, and a list of current medications were collected from the medical history of each patient.

2.3. Inclusion/Exclusion/Withdrawal Criteria

2.3.1. Inclusion Criteria:

- Age > 18 years
- Diagnosis of CLTI: presence of trophic lesions and/or rest pain plus ABI less than 0.4 and/or AP < 50 mmHg or plain or damped PPG curves or toe pressure (TP) < 30 mmHg [31].
- Failure of a previous attempt of revascularization; patients considered at high risk of failure or at high risk of surgical complications during the procedure or in poor condition for surgery.
- High risk of limb loss.

2.3.2. Exclusion Criteria

- Pregnancy
- Legally incapacitated.
- Current cancer or during the last 5 years before the study.
- Current pneumonia or sepsis or severe foot infection.
- Untreated hypothyroidism and/or hypocortisolism.

2.3.3. Withdrawal Criteria

- Patient's own request.
- Decision of the physician due to adverse reactions supposedly secondary to the drug.
- Pneumonia/sepsis during the period of treatment.
- Increase in levels of IGF-1 more than 2 standard deviations.
- Increase in tumor markers.

A total of 37 consecutive subjects met the inclusion criteria. There were 2 deaths after signing the informed consent and before starting any treatment. Therefore, 35 patients were finally eligible for the study. However, both baseline and final muscle samples were obtained in 29 patients who completed the trial without amputation in the referred treatment period (2 months). A total of 16 patients were treated with GH (GH group), and 13 received placebo (control group or placebo group).

2.4. Measurements

2.4.1. ABI and AP

The method used for determination of ABI and AP has been extensively described and recently reviewed [32,33]. A bidirectional continuous doppler with an 8 MHz probe and specific software (Hadeco, es-100V3 model, Quermed S.A, Madrid, Spain) was used for curve analysis.

2.4.2. Inflammatory and Vascular Circulating Biomarkers

A blood sample from a peripheral vein was obtained for determination of plasma biomarkers. The markers and their reference values were: tumor necrosis factor-alpha (TNF-α): <8.1 pg/mL; high-sensitivity C-reactive protein (hsCRP): <0.5 mg/dL or 5mg/L;

beta-2 microglobulin (ß-2M): <0.25 mg/L; cystatin C (CyC): 0.53–0.95 mg/L; fibrinogen: 200–430 mg/dL; glycosylated hemoglobin (HbA1C): <5.5%; insulin-like growth factor I (IGF-1) and IGF-1 binding protein 3 (IGF-1-BP3): age-standardized values according to our laboratory reference values. Quantifications were performed by using ELISA technique according to the manufacturer's protocols. TNF-α, IGF-1, and IGF-BP3 were measured using IMMULITE 2000 IMMUNOASSAY SYSTEM (Siemens); for ß-2M and CyC the DIMENSION VISTA 1500 (Siemens) was used; for hsCRP, ADVIA 2400 CHEMISTRY SYSTEM (Siemens); for HbA1C, HPLC ADAMS A1C HA-8180 (Arkray); cell count was performed using ADVIA 2120 HEMATOLOGY SYSTEM (Siemens); fibrinogen was measured using ACL TOP 550 (Werfen).

The reference value of NLR in general population is considered 2.15. In PAD patients, it ranges between 2.5 and 5.25 in different studies. A cutoff point for elevated NLR has not been properly defined for PAD, though the recommendation is to consider a value between 2.5 and 3 [26]. In our study, a value ≥ 2.5 was chosen as elevated.

2.4.3. Skeletal Muscle Samples

Samples were taken from the soleus muscle using a cutting trocar and local anesthesia with 2% lidocaine. In the internal aspect of the leg, a small skin incision of 2–3 mm was made. Then, the trocar was inserted until the desired level, and a muscle cylinder was removed. Once extracted, the samples were conserved in RNA-later (AM7021, Invitrogen, Vilnius, Lithuania) at 4 °C and finally stored at −80 °C until analysis.

2.4.4. Real-Time PCR (RT-qPCR)

RNA extraction was performed using TRIzol™ reagent (15596026, Invitrogen, Carlsbad, CA, USA), following manufacturer's instructions. RNA was incubated with 1 IU RNase free DNase I (EN0521, Thermo, Carlsbad, CA, USA), 5 µL 10X buffer with MgCl2, and water for a total volume of 50 µL at 37 °C for 30 min. The reaction was terminated by inactivating DNase, and then RNA was purified with an affinity column using the GeneJET RNA Cleanup and Concentration micro kit (K0842, Thermo Fisher, Vilnius, Lithuania). RNA was finally quantified by spectrophotometry (Nanodrop 2000, Thermo Fisher, Vilnius, Lithuania).

Total RNA pool obtained from commercial human skeletal muscle was used as reference (636534, Clontech, Mountain View, CA, USA).

A total of 1 µg of total, previous to cDNA synthesis, is incubation with 1 IU of RNase-free DNase I (EN0521, Thermo Fisher, Carlsbad, CA, USA), 1 µL of MgCl2 buffer, and water to a final volume of 10 µL for 30 min at 37 °C. DNase was then inactivated by adding 1 µL of EDTA and incubating for 10 min at 65 °C. cDNA was synthesized following the supplier's protocol, adding 1.5 µL of 300 IU MMLV (28025-013, Invitrogen, Carlsbad, CA, USA), 6 µL 5X First-Strand Buffer, 1.5 µL 10 mM dNTPs(10297-018, Invitrogen, Carlsbad, CA, USA), 0.1 µL Random Primers(48190011; Invitrogen, Freder-ick, MD, USA, 3 µL 0.1 M DTT, 1 µL RNaseOUT™ Recombinant Ribonuclease Inhibitor (40 units/µL) (10777-019, Carlsbad, CA, USA), and H2O for a total 30 µL reaction. For human samples, 50, 25, and 12.5 ng of Poly A+ mRNA was similarly treated.

Expression was detected by qPCR using 1 µL of the cDNA reaction plus 6 µL 2x TaqMan Gene Expression MasterMix (4369016 Applied Biosystems, Foster City, CA, USA) and 6 µL diluted primers in 96 well-plates in a 7500 Real-Time PCR System (4351105, Applied Biosystems, Foster City, CA, USA). As control for general gene ex-pression, we used human negative controls of the reverse-transcription step (all reagents and RNA samples but without reverse transcriptase) and the PCR step (all reagents but no reverse-transcribed samples) were included in each assay plate. We used the house-keeping gene (TBP) as an expression control gene since we have previous experience in human tissue samples [34,35]. The following genes were determined: *VEGFA*: Vascular endothelial growth factor A; *IGF1*: Insulin-like growth factor I; *NOS3* or *eNOS*: Nitric oxide synthase 3 or endothelial NOS; *MSTN*: Myostatin; *NOX4*: NADPH (Nicotinamide adenine dinucleotide phosphate oxidase) 4; *MYOG*: Myogenin; *KDR*: VEGFA receptor 1; *IL6*: Interleukin 6; and *TNF*: Tumor Necrosis Factor. All of them appear in italic in the text and Figures.

ΔCt relative values with respect to the commercial pool of human muscle were calculated for each gene related to TBP. When compared from time zero (control time) to two months of treatment/placebo, $\Delta\Delta$Ct values were obtained. Primer sets and TaqMan assays (Applied Biosystems) used are shown in Table 1.

Table 1. Real-time PCR (RT-qPCR). Primer sets, TaqMan assays and conditions used for RT-qPCR in the GHAS trial for muscle gene expression analysis. Genes are written in italic.

Gene	Sequence	Amplification Size	Annealing T^a
TBP	Fw: 5′-GCCCGAAACGCCGAATAT-3′ Rv: 5′-TTCGTGGCTCTCTTATCCTCATG-3′ Pb: 5′-TCCCAAGCGGTTTGCTGCGGTA-3′	67 bp	60 °C
VEGFA	Applied Biosystems: Hs00900055_m1	67 bp	60 °C
IGF1	Applied Biosystems: Hs01547656_m1	68 bp	60 °C
NOS3	Applied Biosystems: Hs01574665_m1	86 bp	60 °C
MSTN	Applied Biosystems: Hs00976237_m1	69 bp	60 °C
NOX4	Applied Biosystems: Hs01379108_m1	64 bp	60 °C
MYOG	Applied Biosystems: Hs01072232_m1	76 bp	60 °C
KDR	Applied Biosystems: Hs00911700_m1	83 bp	60 °C
IL6	Applied Biosystems: Hs00174131_m1	95 bp	60 °C
TNF	Applied Biosystems: Hs00174128_m1	80 bp	60 °C

2.5. Statistical Analysis

Mean, median, and standard deviation and standard error values were calculated for each group. Serum biomarkers and molecular data were analyzed at baseline, and for differences between basal and final time points in the two treated groups. Results were also stratified by patients with or without DM, but not by type of DM.

Data were processed with GraphPad prism® v8 software (San Diego, CA, USA). The initial step was to check if data within a group followed a normal distribution and if their variances were equal or not. For normality distribution, we used the Shapiro–Wilk test (as the sample size was less than 50) and Fisher's test for variances. Based on the results, if the groups were normal and homoscedastic, we used t-test; if the groups were normal and not homoscedastic, we used *t*-test with Welch's correction; finally, if the groups followed a non-normal distribution, we used the test of Wilcoxon–Mann–Whitney. To compare among qualitative, or quantitative to qualitative variables, X^2 test was performed with Fisher's test when expected frequency was less than 5.

Correlations (r) were measured using the SPSS® software v27 (Armonk, NY, USA), mainly the Spearman correlation coefficient for non-normal qualitative or quantitative variables. The graphs were constructed by using the mentioned Graphpad prism® software. In addition, Kaplan–Meier and AUC–ROC curves (AUC: Area Under the Curve; ROC: Receiver Operating Characteristics) were carried out by using R statistics software (version 4.0.3, free software). Fisher and Mantel–Haenszel tests were used for odds ratios analysis.

In addition to the univariate study, a multivariate study was addressed by using multiple logistic regression analysis.

3. Results

3.1. General Characteristics

The baseline characteristics between both groups of the study were similar in terms of age, sex, hypertension (HT), DM with or without neuropathy, chronic kidney disease (CKD), dialysis, heart disease, or Rutherford class. However, on the one hand, some differences were detected between groups on tobacco consumption, with a higher number of nonsmokers in the intervention group (GH: 88.89% vs. placebo: 50%, $p = 0.0107$). On the other hand, the presence of trophic lesions in the foot was more frequent in the GH group than in the control group (GH: 83.3% vs. placebo: 50%, $p = 0.0381$). In CLTI studies, a higher proportion of men than women is usually found, as in our study (men/women: 79.4% vs. 20.6%), since this disease has a higher incidence in men, without any difference between GH-treated patients and controls ($p = 0.2715$). Our group previously published a table with all these characteristics (see Table 2 in [28]). In the sample analyzed for this study, 59% of patients suffered from DM, 57.6% of them with established neuropathy. All diabetic patients were type II, and all except one suffered from DM of more than 10 years of evolution. DM affected 53.3% of the men and 100% of the women in this study. Age was stratified into three groups (<65 years; 65–80 years; and >80 years). The percentage of diabetic patients in each age group was as follows: <65: 29.2%; 65–80: 33.3%; >80: 37.5% ($p = 0.2531$).

Table 2. Mean and median results of inflammatory parameters in CLTI patients from the GHAS trial. TG: Triglycerides; HDLc: High density lipoprotein cholesterol; LDLc: Low density lipoprotein cholesterol; ABI: Ankle–brachial index; AP: Ankle pressure; IGF-1: Insulin-like growth factor 1; IGFBP3: IGF-1 binding protein 3; TNF-α: Tumor necrosis factor alpha; hsCRP: C-reactive protein of high sensitivity; B2M: Beta-2 microglobulin; CyC: Cystatin C; HbA1C: Glycosylated hemoglobin; NLR: Neutrophil-to-lymphocyte ratio; Fibrin.: Fibrinogen; SD: Standard deviation; Min: Minimum; and Max: Maximum.

	TG	HDLc	LDLc	ABI	AP	Age	IGF-I	IGFBP3	TNF-α	hsCRP	B2-M	CyC	HbA1C	NLR	Fibrin.
Mean	164.1	45.5	97.9	0.23	38.6	71.5	134.9	3.06	10.66	1.6	0.4	1.45	6.5	3.5	560.21
Median	132.5	41	100	0.2	31.5	72	125	2.9	10	0.95	0.3	1.2	6.3	2.6	505
SD	88.1	15.1	32.1	0.23	37.02	12.4	53.2	1.1	4.9	2.2	0.3	0.9	1.02	2.8	128.51
Mín.	46	29	36	0	0	49	38	0.5	4	0.1	0.14	0.55	5.1	0.4	403
Max.	412	97	161	0.93	140	93	275	5.2	27	9.1	1.1	3.8	8.9	16.3	853

3.2. Hemodynamic Parameters and Plasma Biomarkers

The baseline characteristics (mean and median) of the hemodynamic and plasma parameters are schematized in Table 2: Age: 71.5 ± 12.4 years; ankle–brachial index (ABI): 0.23 ± 0.23; ankle pressure (AP): 38.6 ± 37.02; TNF-α: 10.66 ± 4.9 pg/mL; hsCRP: 1.6 ± 2.2 mg/dL; CyC: 1.45 ± 0.9 mg/L; B2M: 0.4 ± 0.3 mg/L; fibrinogen: 560.21 ± 128.51 mg/dL; HbA1C: 6.5 ± 1.02%; neutrophil-to-lymphocyte ratio (NLR): 3.5 ± 2.8; IGF-1: 134.9 ± 53.2 ng/ml; and IGF-1BP3: 3.06 ± 1.1 μg/mL.

An elevated level of plasma TNF-α (≥8.1 pg/mL) was seen in 68.57% of patients (GH: 83.3%; Placebo: 52.9%), while a high level of hsCRP (≥0.5 mg/dL) was detected in 60.5% (GH: 70%; Placebo: 50%). An elevated NLR (≥2.5) appeared in 52.6% of patients, and CyC was also increased in the GH group compared with the placebo. The plasma levels of IGF-1 increased in 8.11% of patients, mainly in the GH group, and the plasma levels of IGFBP3 were found low in 44.1% of patients: 45.5% vs. 42.9%, GH and placebo, respectively ($p = 0.634$).

Table 3 depicts relevant information about the groups. As can be seen, the level of plasma biomarkers of inflammation such as TNF-α, hsCRP, and CyC was significantly higher in the GH group at baseline, which implies a higher inflammatory state in this group. Even basal levels of fibrinogen were higher in the GH group (601.25 vs. 489.85,

$p = 0.0425$), which supports this statement. At the end of the study, only plasma TNF-α was significantly decreased compared to the placebo, as previously reported by our group [28].

Table 3. Biomarker distribution in the GHAS study. Basal: Baseline or pretreatment values; Final: Final values after 2 months of treatment. Statistical significance is highlighted in colors. GH: Growth hormone; Obs: Observed; SD: Standard deviation; TNF-α: Tumor necrosis factor; hsCRP: C-reactive protein of high sensitivity; B2M: Beta-2 microglobulin; and CyC: Cystatin C. * $p < 0.05$.

Plasma Marker	GH			Placebo			
	Obs.	Mean	SD	Obs.	Mean	SD	*p*-Value
TNF-α (Basal)	16	12.35	5.2	16	8.78	3.9	0.0184 *
TNF-α (Final)	15	10.93	5.12	14	8.04	3.6	0.0464 *
hsCRP (Basal)	18	2.07	2.86	16	0.79	0.70	0.0454 *
hsCRP (Final)	17	1.1	1.38	14	3.42	7.51	0.2188
B2M (Basal)	7	0.47	0.27	16	0.22	0.08	0.1269
B2M (Final)	8	0.56	0.51	14	0.21	0.12	0.3894
CyC (Basal)	7	1.75	0.93	4	0.76	0.17	0.035 *
CyC (Final)	8	1.71	1.12	2	0.8	0.14	0.3054

3.3. Basal mRNA Expression

After analyzing the basal mRNA expression of different genes in the skeletal ischemic muscle (Figure 1A–H), it was found that only *NOX4* had a consistently higher expression in the group with DM and GH compared to the group without DM and placebo ($p = 0.0346$) (Figure 1A).

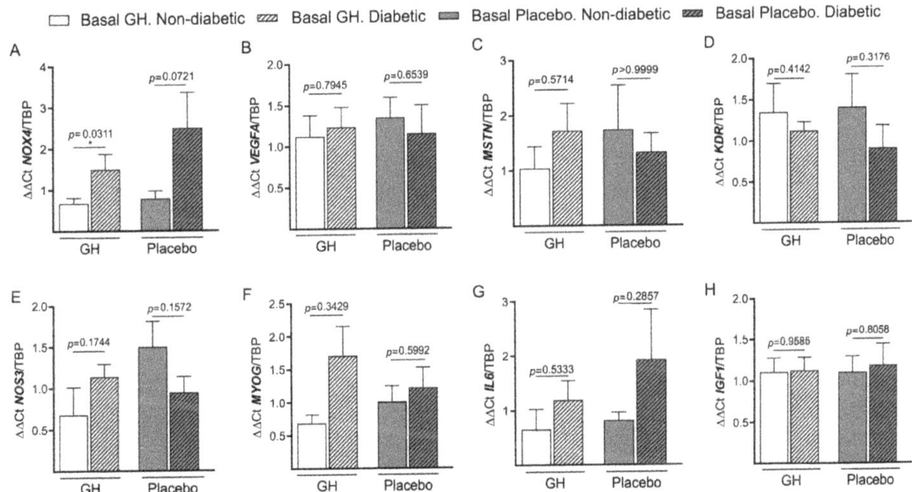

Figure 1. **mRNA expression of different genes at baseline in both groups, GH, and placebo) in DM and non-DM patients.** (**A**). *NOX4*: NADPH (Nicotinamide adenine dinucleotide phosphate oxidase) 4; (**B**). *VEGFA*: Vascular endothelial growth factor A; (**C**). *MSTN*: Myostatin; (**D**). *KDR*: VEGFA receptor 1. (**E**). *NOS3* or *eNOS*: Nitric oxide synthase 3 or endothelial NOS; (**F**). *MYOG*: Myogenin; (**G**). *IL6*: Interleukin 6; (**H**). *IGF1*: Insulin-like growth factor I; GH: Growth hormone; Placebo: control group; Basal: Baseline. * $p < 0.05$.

3.4. Basal and Final mRNA Expression

NOX4 mRNA expression underwent a real decrease in DM ischemic patients treated with GH compared to non-DM ($p = 0.0348$) (Figure 2A). *TNF* mRNA expression (*TNF-α*) was not reduced in the GH group, neither in patients with DM nor in those without DM

(Figure 2B). The angiogenesis marker *KDR* underwent a real increase in its expression during GH treatment, which was only significant in patients with DM ($p = 0.036$) (Figure 2C).

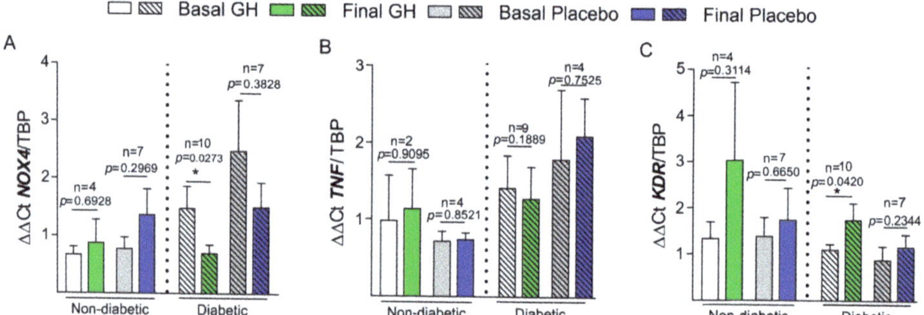

Figure 2. Basal to final mRNA expression of different genes. *NOX4* (**A**), *TNF* (*TNF-α*) (**B**), and *KDR* (**C**). Findings in both groups, GH, and placebo, with DM and without DM. * $p < 0.05$.

3.5. Plasma Biomarkers and mRNA Expression

Figure 3 depicts the relationship between basal plasma TNF-α and gene expression in ischemic skeletal muscle for DM patients in both groups of treatment (GH and placebo). First, a positive correlation was found between plasma TNF-α level and mRNA expression of *TNF-α* (r = 0.588, $p = 0.035$) (Figure 3A); this was especially relevant for those patients with TNF-α $\geq$ 8.1 (r = 0.802 $p = 0.001$) (see also Tables 4 and 5). Second, there was an inverse correlation between the expression of *TNF-α* and *NOS3* (r = −0.4999, $p = 0.0151$), more evident in the GH group (r = −0.78, $p = 0.0064$) (Figure 3B). Third, an inverse correlation was found between plasma TNF-α and the angiogenic factor *VEGFA* (r = −0.4321, $p = 0.0395$), also stronger in the GH group (r = −0.7727, $p = 0.0074$) (Figure 3C). Fourth, the high level of inflammation, expressed by plasma TNF-α $\geq$ 8.1, was related to the expression of *IGF-1* ($p = 0.0010$), showing a moderate but positive correlation between both (r = 0.5192, $p = 0.0111$) (Figure 3D). All these relationships, as mentioned above, appeared exclusively or were stronger in patients with DM, independent of the treatment received.

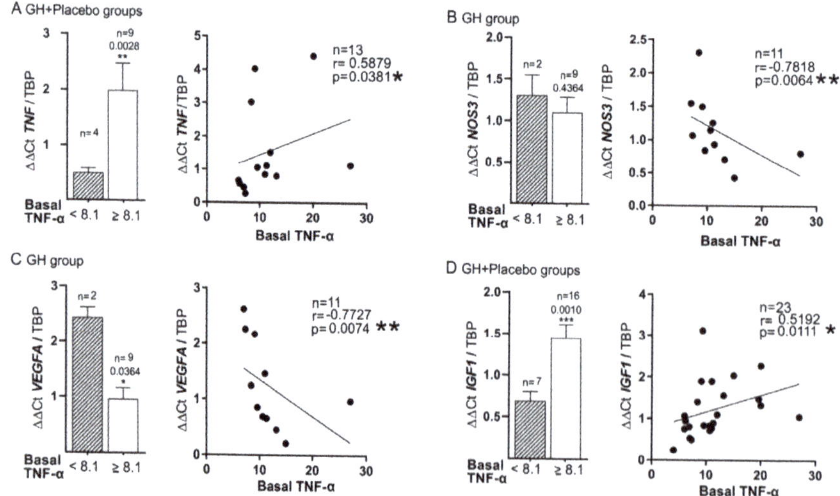

Figure 3. Baseline plasma levels of TNF-α related to different muscle gene expressions in DM patients. Both groups: GH and placebo. Correlation between plasma TNF-α and muscle *TNF* (**A**), *NOS3* (**B**), *VEGFA* (**C**), and *IGF-1* (**D**) mRNA expressions. TBP: housekeeping gene used as an expression control gene. * $p < 0.05$; and ** $p < 0.001$.

Table 4. Summary of correlations between plasma biomarkers and gene expression of *NOS3*, *VEGFA*, and *TNF*. * Indicates that this condition was also found significantly associated in non-DM population of the GHAS trial.

Group	Plasma Marker/Gene	NOS3 (Basal) Both	NOS3 (Basal) Placebo	NOS3 (Basal) GH	NOS3 (Final) GH	VEGFA (Basal) Both	VEGFA (Basal) GH	VEGFA (Final) Placebo	TNF (Basal) Both	TNF (Basal) GH
DM	TNF-α (Basal)	r = −0.49, p = 0.015		r = −0.78, p = 0.0064		r = −0.432, p = 0.039	r = −0.773, p = 0.005		r = 0.588, p = 0.035 *	
DM	TNF-α (Final)		r = 0.866, p = 0.012							
DM	TNF-α > 8.1 (Basal)								r = 0.802, p = 0.001 *	
DM	hsCRP (Basal)			r = 0.74, p = 0.009		r = −0.632, p = 0.005		r = −0.775, p = 0.041		
DM	hsCRP > 0.5 (Basal)			r = −0.693, p = 0.018		r = −0.546, p = 0.019		r = −0.866, p = 0.012		
DM	hsCRP > 0.5 (Final)				r = −0.711, p = 0.021 *			r = −0.866, p = 0.012		
DM	NLR > 3 (Basal)			r = −0.645, p = 0.032						
DM	NLR > 5 (Basal)			r = −0.69, p = 0.018						
DM	HbA1C (Basal)									r = 0.728, p = 0.026
DM	HbA1C (Final)							r = −0.975, p = 0.005		
Non-DM	hsCRP (Basal)				r = −0.9, p = 0.037					
Non-DM	LDLc					r = −0.827, p = 0.002				

Table 5. Summary of correlations between plasma biomarkers and gene expression of *IGF-1*, *MSTN*, *MYOG*, *KDR*, and *NOX4*. * Indicates that this condition was also found significantly associated in non-DM population of the GHAS trial.

Group	Plasma Marker/Gene	IGF-I (Basal) Both	IGF-I (Basal) Placebo	IGF-I (Final) GH	MSTN (Basal) Both	MSTN (Final) GH	MYOG (Final) GH	MYOG (Final) Placebo	KDR (Final) Placebo	NOX4 (Basal) Both	NOX4 (Final) GH
DM	TNF-α (Basal)	r = 0.519, p = 0.011	r = 0.821, p = 0.0341					r = 0.8, p = 0.023 *		r = 0.305, p = 0.037	
DM	TNF-α > 8.1 (Basal)	r = 0.598, p = 0.003			r = 0.586, p = 0.011						
DM	hsCRP (Basal)			r = −0.648, p = 0.031							
DM	hsCRP (Final)					r = −0.691, p = 0.027					r = 0.709, p = 0.003 *
DM	hsCRP > 0.5 (Final)										r = 0.773, p = 0.001 *
DM	NLR > 3 (Basal)			r = −0.662, p = 0.019							
DM	NLR > 5 (Basal)								r = −0.857, p = 0.014 *		
DM	HbA1C (Basal)	r = 0.597, p = 0.019									
Non-DM	TNF-α (Basal)									r = 0.645, p = 0.032	
Non-DM	TNF-α > 8.1 (Basal)									r = 0.717, p = 0.009	
Non-DM	hsCRP (Basal)							r = 0.9, p = 0.037		r = −0.847, p = 0.016	

In addition, those patients with the highest levels of plasma TNF-α at baseline had higher levels of redox stress ($p = 0.0475$), with a weak but significant correlation between plasma TNF-α and *NOX4* (r = 0.35, $p = 0.0375$) (Figure 4A). Interestingly, the latter relationship shows different behavior depending on the levels of plasma TNF-α ($\geq$8.1 or <8.1) (Figure 4B). It should be noted that patients without DM showed the best correlation between both markers (0.645, $p = 0.032$), especially when plasma TNF-α reached $\geq$ 8.1 pg/mL (r = 0.717, $p = 0.009$) (see Table 5).

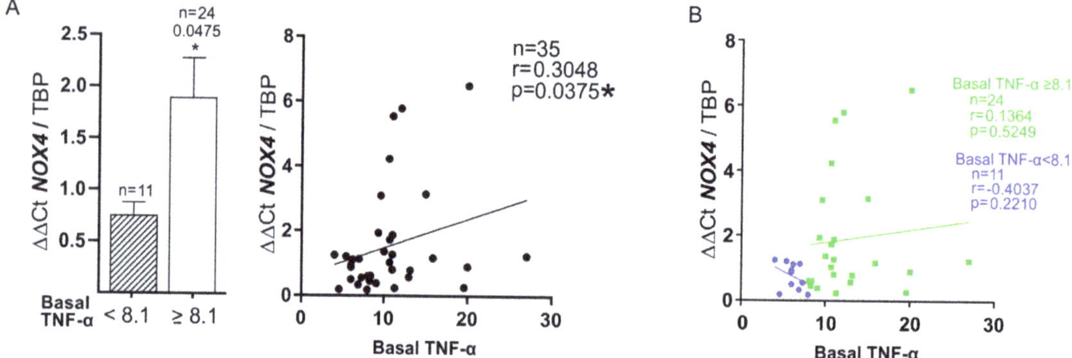

Figure 4. Relationship between high levels of plasma TNF-α (≥8.1 pg/mL) and *NOX4* expression in all patients of the study. (**A**). Left: expression of muscle *NOX4* mRNA in relation to plasma TNF-α. Striped bar: TNF-α < 8.1 pg/mL. White bar: TNF-α ≥ 8.1 pg/mL. Right: Spearman correlation between both markers. (**B**). Trend of this correlation differentiating by TNF-α values: ≥8.1 (green color) and <8.1 (blue color). * $p < 0.05$.

High levels of hsCRP were related to *NOS3* expression at baseline (r = −0.74, p = 0.009), also in DM patients treated with GH (Figure 5A), while at the end of the study (final time point) a value of hsCRP ≥ 0.5 was negatively correlated to muscle *NOS3* expression, both in patients with DM and without DM treated with GH (r = −0.711, p = 0.021) (see Table 4)

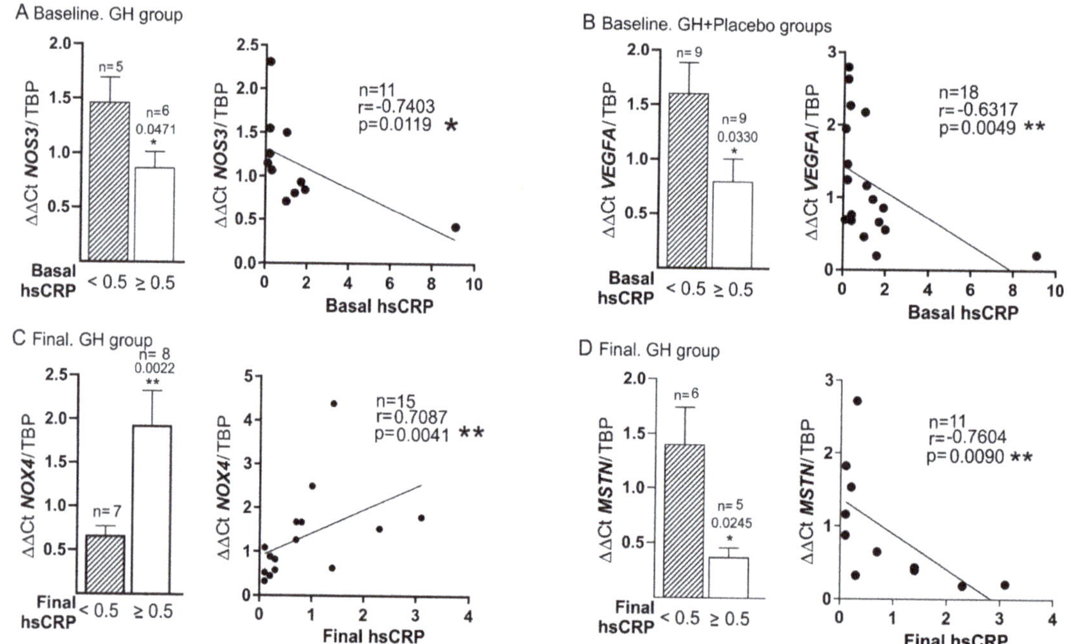

Figure 5. Correlation between plasma hsCRP and muscle gene expression in diabetic patients. (**A**). Baseline expression levels of hsCRP and *NOS3* in the GH group. (**B**). Baseline hsCRP and *VEGFA* levels in both groups. (**C**). Final hsCRP and redox stress levels measured by final *NOX4* in the GH-treated patients. (**D**). Final hsCRP levels were related to final *MSTN* mRNA expression in GH group. Striped bars: hsCRP < 0.5; White bars: hsCRP ≥ 0.5. * $p < 0.05$; and ** $p < 0.001$.

Basal levels of hsCRP were negatively associated with basal *VEGFA* expression too in both groups of treatment, although more specifically in DM (r = −0.632, *p* = 0.005) and for those patients with the highest levels of inflammation measured by hsCRP ≥ 0.5 (*p* = 0.0330) (Figure 5B). In the GH group, the final hsCRP was also related to the final expression of *NOX4* (r = 0.7, *p* = 0.0041) (Figure 5C), while its relationship with *MSTN* expression was negative (r = −0.7604, *p* = 0.009) (Figure 5D).

Of interest are also the findings of plasma HbA1C correlations at baseline. Indeed, HbA1C was highly correlated to *TNF-α* mRNA expression in the GH group with DM (r = 0.7280, *p* = 0.0321) (Figure 6A). The same occurred between HbA1C and *IGF-1* mRNA expression in patients with DM, but, in this case, in both groups of patients (GH and placebo, r = 0.7, *p* = 0.0039) (Figure 6B). At the end of the treatment, the final levels of HbA1C were also inversely related to the final muscle expression of *VEGFA* in DM, but only in the placebo group (r = −0.9747, *p* = 0.0333) (Figure 6C).

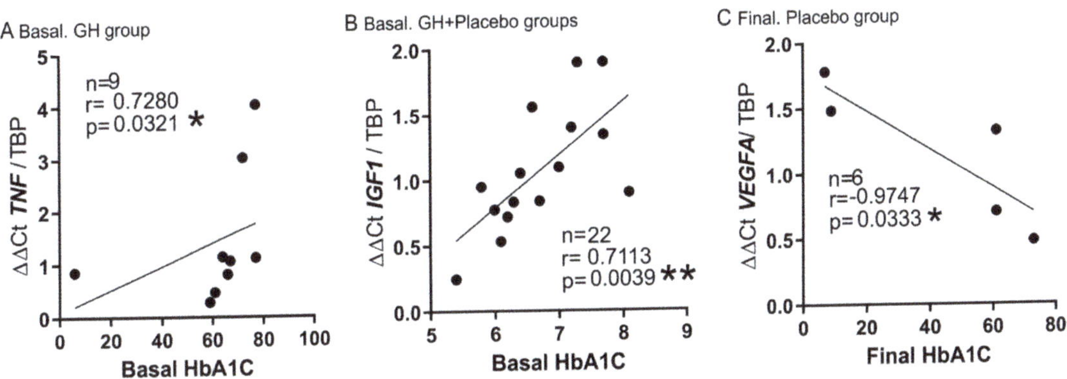

Figure 6. Correlation of plasma HbA1C and muscle mRNA expression in diabetic patients. (**A**). *TNF*. (**B**). *IGF-1*. **C**. *VEGFA*. (**A**,**B**) represent correlations at baseline, while (**C**) represents final time point. * *p* < 0.05; ** *p* < 0.001.

As our group previously published, *NOX4* expression was specifically reduced in the GH-treated group (*p* = 0.025), while it tended to increase in the placebo group (see Figure 5 in [28]).

All correlations between plasma markers and mRNA expression of different genes are summarized in Table 4 (*NOS3*, *VEGFA*, and *TNF-α*) and 5 (*IGF-1*, *MSTN*, *MYOG*, *KDR*, and *NOX4*).

3.6. Mortality in the GHAS Trial

Special attention should be paid to mortality in the GHAS trial. The cumulative mortality at 12 months reached 47.4% of CLTI patients, being higher in GH-treated patients than in patients receiving placebo, both at 2 months (5.5% vs. 0%, *p* = 0.42) and at 12 months (29.4% vs. 12.5%, *p* = 0.23) (Table 6).

Table 6. Mortality in the GHAS trial. Short-term mortality: 0–2 months (period of treatment). Long-term mortality: 2–12 months (observation period).

	0–2 Months	Mortality	*p*-Value	2–12 Months	Mortality	*p*-Value
Placebo	0/16	0%		2/16	12.5%	
GH	1/18	5.5%	0.42	5/17	29.4%	0.23
Cumulative		5.5%			47.4%	

In order to identify some possible predictors associated with mortality in the GHAS trial, we performed a careful analysis of variables that could be responsible for it. We found that a value of plasma TNF-$\alpha \geq 8.1$ pg/mL was significantly associated with cumulative mortality at 12 months or long-term mortality ($p = 0.0487$, OR = 2.5, CI (95%): 0.23–136.6, Fisher test: $p = 0.63$) (Figure 7A). In addition, a value of NLR ≥ 3 was also associated with long-term mortality ($p = 0.019$, OR = 6.9, CI (95%): 0.71–353.7, X^2 with McNemar correction: $p = 0.01946$, Fisher test: $p = 0.093$) (Figure 7B). For NLR ≥ 3, the Mantel–Haenszel test showed no influence of other variables such as smoking, DM, hsCRP, and TNF-α on the results.

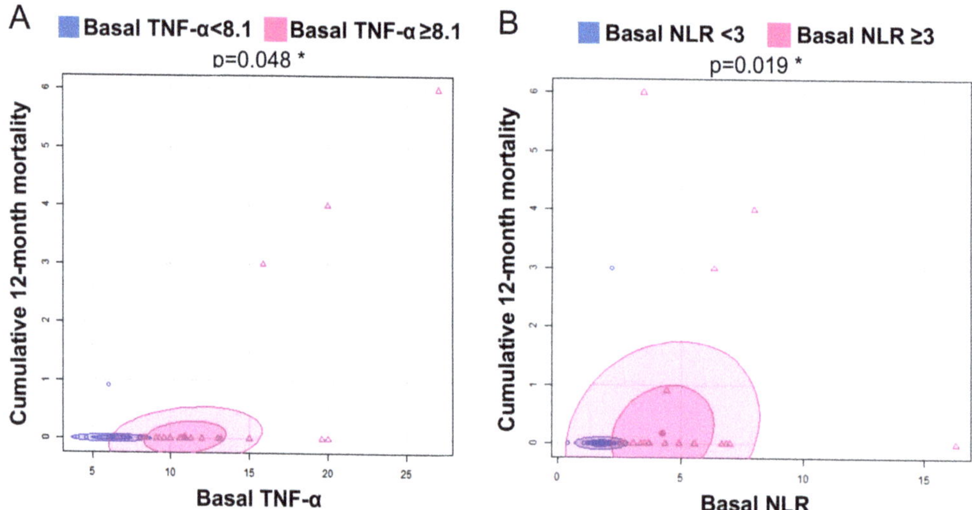

Figure 7. Cumulative 12-month mortality and plasma markers. (**A**). Basal TNF-α; (**B**). Basal NLR (Neutrophil-to-lymphocyte ratio). * $p < 0.05$.

The Kaplan–Meier survival analysis indicated that patients with NLR ≥ 3 and NLR ≥ 5 had a higher mortality rate (Figure 8A). AUC–ROC curves identified the best cutoff point for NLR to predict mortality in 3.4 (74.2%, 85.7%), with an AUC value of 0.811 (0.733–0.857) and a power of 0.77 (Figure 8B). The best cutoff point for plasma TNF-α was 15.4 (96.6%, 66.7%), with an AUC value of 0.7989 (52.4%–100%) and a power of 0.67 (Figure 8C). After comparing both biomarkers in terms of their predictive ability, we observed that there were no statistical differences between them ($p = 0.9264$) (Figure 8D).

Some clinical predictors of long-term mortality were also found, such as chronic obstructive pulmonary disease (COPD) ($p = 0.042$, OR = 5.8, CI (95%): 0.84–40.7) and an American Society of Anesthesiologists class 4 (ASA 4) compared to a class 3 (ASA 3) ($p = 0.0119$, OR = 15.75, CI (95%): 0.87–284.9). (Table 7).

Table 7. Plasma and clinical predictors of mortality in the GHAS trial. COPD: Chronic Obstructive Pulmonary Disease; ASA: The American Society of Anesthesiologists (ASA) class is a system for assessing the risk of patients before surgery. ASA 3 and 4 include patients in poor (ASA 3) and extremely poor (ASA 4) physical condition for surgery. CI: Confidence Interval. ($p < 0.05$ is considered to be significant).

	Predictors of Mortality in the GHAS Trial		
	p-Value	OR	CI (95%)
NLR ≥ 3 (Basal)	0.019	6.9	0.71–353.7
TNF-$\alpha \geq 8.1$ (Basal)	0.0487	2.5	0.23–136.6
COPD	0.042	5.8	0.84–40.7
ASA4-ASA3	0.0119	15.7	0.87–284.9

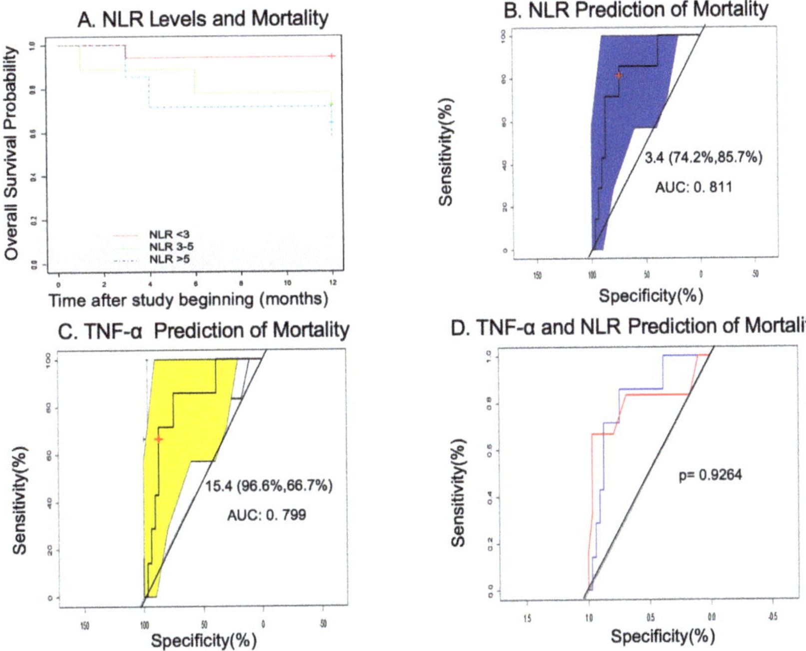

Figure 8. Relationship between NLR and plasma TNF-α with mortality. (**A**). Kaplan-Meier analysis of NLR and mortality. (**B**). AUC-ROC (Area Under the Curve-Receiver Operating Characteristics) curve of NLR related to mortality. (**C**). AUC-ROC curve of plasma TNF-α related to mortality. Color blue and yellow represent confidence intervals. Values in brackets represents sensitivity and specificity, respectively (**D**). Comparation between plasma TNF-α and NLR as predictors of mortality showed no significant differences ($p = 0.9264$).

It is also noteworthy that the basal level of hsCRP was related to short-term mortality (2 months) ($p = 0.008$). In addition, the level of NLR at baseline was closely related to the level of hsCRP (r = 0.5335, $p = 0.0006$) (Figure 9A), especially for the values of hsCRP ≥ 0.5 and NLR > 2.5 ($p = 0.013$) (Figure 9B).

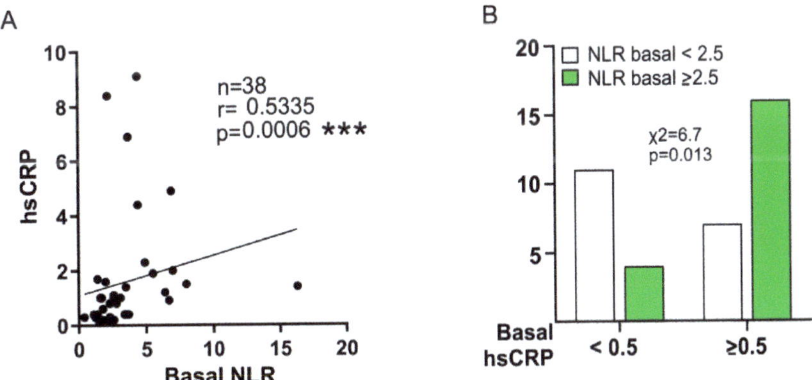

Figure 9. Relationship between basal levels of NLR and plasma hsCRP. (**A**). Spearman correlation. (**B**). Histogram comparing hsCRP ≥ 0.5 and NLR ≥ 2.5. Green bars: NLR ≥ 2.5; White bars: NLR < 2.5. *** $p < 0.0001$.

3.7. Statistical Study: Measures of Association of Variables

3.7.1. Univariate Analysis

The linear regression model showed that values of hsCRP $\geq$ 0.5 mg/dL at baseline tended to be related to the cumulative amputation ratio at 12 months (p = 0.056). The analysis of the COR curves indicated that the best cutoff point was 0.7mg/dL (AUC: 66.4%, 49.3–83.5%). However, the statistical power of this association was low (0.2579), therefore a larger sample size is needed to confirm this possible relation. In the placebo group, final levels of *NOX4* were significantly related to the Rutherford class of patients (r = 0.67, p = 0.011), while final *IGF-1* was related to ABI (r = −0.63, p = 0.011). In the GH group, the final level of *KDR* was affected by age (r = −0.56, p = 0.016), while the expression of *NOS3* correlated positively to wound healing (r = 0.55, p = 0.031). At baseline, *VEGF-A* expression was related to CKD (r = −0.41, p = 0.013) (Figure 10, top). The specific relationship between *NOX4* expression and Rutherford class is represented in Figure 10, bottom.

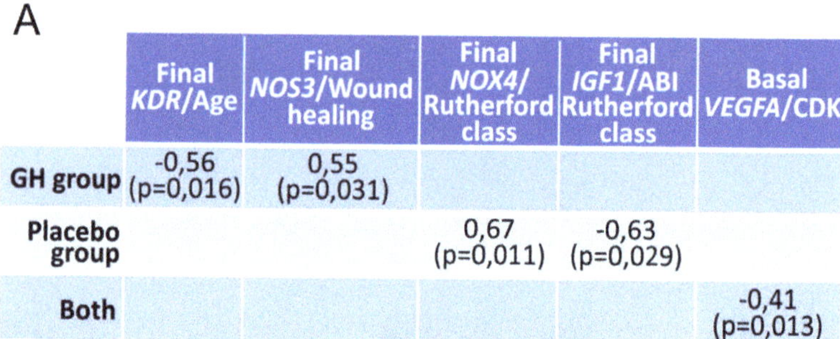

A

	Final KDR/Age	Final NOS3/Wound healing	Final NOX4/ Rutherford class	Final IGF1/ABI Rutherford class	Basal VEGFA/CDK
GH group	-0,56 (p=0,016)	0,55 (p=0,031)			
Placebo group			0,67 (p=0,011)	-0,63 (p=0,029)	
Both					-0,41 (p=0,013)

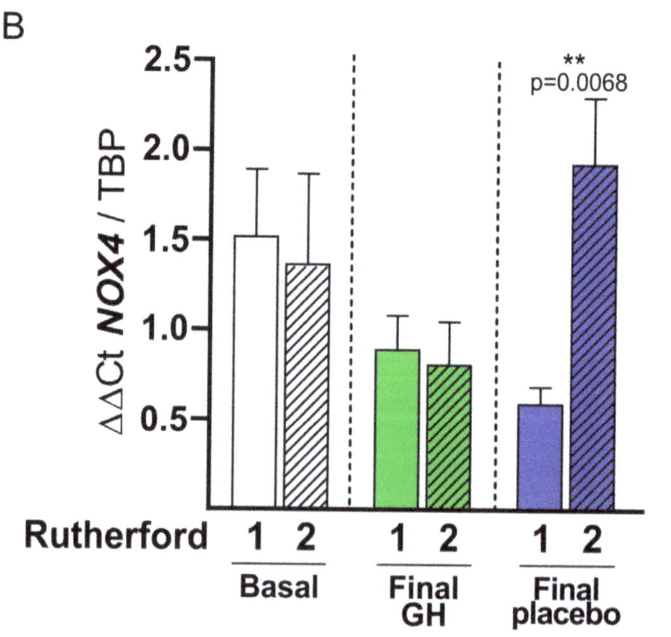

Figure 10. (**A**): Relation between different muscle genes and clinical variables in both groups. (**B**): Graph showing the link between *NOX4* mRNA expression and Rutherford class, grouped in 1: less severity of ischemia, and 2: more severity of ischemia. White box: basal level of *NOX4*; Green bars: final level of *NOX4* in the GH group; and Blue bars: final level of *NOX4* in the placebo group. Striped bars: group with severe ischemia. ** p < 0.001.

3.7.2. Multivariate Analysis

The multiple logistic regression model also demonstrated the relationship between basal NLR $\geq$ 2.5 and hsCRP $\geq$ 0.5 ($p = 0.013$), which was confirmed by using Kendall's Tau ($0.41, p = 0.010$).

However, high levels of plasma TNF-α ($\geq$8.1) at baseline were not related to anything using the multivariate analysis, although there was a trend to be related to basal hsCRP $\geq$ 0.5 ($p = 0.055$), NLR ($p = 0.063$), and CKD ($p = 0.059$). Basal hsCRP $\geq$ 0.5 was also not related to anything with this model, although there was a tendency to be associated with the Rutherford class ($p = 0.055$).

4. Discussion

In this study we analyzed the role of some plasma biomarkers and the expression of some muscle genes in the ischemic muscle of PAD patients, focusing on the data from the GHAS trial carried out in the special group of patients suffering from CLTI, which shows a high level of inflammation and, in parallel, a high rate of morbidity and mortality [32]. Most studies on this field have investigated plasma samples from patients with intermittent claudication, a less severe form of PAD, without information on the status of the ischemic muscle. Therefore, the real role of inflammatory cytokines in advanced PAD needs to be better defined. In the GHAS trial, we cross-linked plasma information with that obtained from gene expression at the level of ischemic skeletal muscle of the lower extremities.

Our data confirm the high level of inflammation in CLTI patients (Table 2) and the high mortality rate of this special population (Table 6). Interestingly, according to plasma biomarker data, patients in the GH group were significantly more affected by inflammation than those in the placebo group (Table 3). The high level of inflammation in GH-treated patients configured a severely ill population with higher levels of redox imbalance, biomarkers, genetic alterations, and mortality, as shown in the results, especially if they suffered from DM. This is also consistent with the fact that patients in the GH group had significant higher levels of fibrinogen and, as published by our group [28], more trophic lesions in the foot than those observed in the placebo group, indicating a more severe stage of ischemia. Nevertheless, at the end of the study, only plasma TNF-α experienced a significant reduction in GH-treated patients [28], which is consistent with previous studies [36,37]. This supports both the anti-inflammatory action of GH and the key role of TNF-α in PAD. The effect of GH on plasma hsCRP did not reach a significant reduction, probably as a consequence of the relatively small sample size studied. However, the MESA trial or Mendelian randomization studies have shown neutral results of plasma hsCRP in PAD [38], considering CRP an acute phase protein rather than a causal factor [20]. As we can see from our results, this statement might not be completely true for CLTI patients, as we found how hsCRP could influence on the gene expression of both processes: angiogenesis (*NOS3*, *VEGFA*) and redox balance (*NOX4*) (Figure 5A–C).

Again, we demonstrated here that oxidative stress plays a crucial role in CLTI, as *NOX4* was the only gene that our group found significantly increased in patients with high levels of inflammation and DM (Figure 1A), which supports the link between inflammation and DM, as it was advanced in a previous publication [28].

Despite the fact that *NOX4* was significantly reduced in all GH-treated patients, our data show that patients with DM benefited more compared to non-DM patients ($p = 0.0348$) (Figure 2A).

Although GH decreased plasma TNF-α, a parallel significant reduction in mRNA expression of this marker (*TNF*) was not detected in the ischemic muscle, neither in patients with DM nor in patients without DM (Figure 2B). The possible explanation could be related to the possibility that GH reduces the cleavage of that TNF-α joined to the cell membrane, which is the source of soluble TNF-α, hence lowering plasma TNF-α, but it would not decrease the expression of *TNF* at the cell membrane [39]. The binding of TNF-α to the cell membrane depends on cell–cell interactions, and, furthermore, there are many polymorphisms of the *TNF* gene, both functional and structural, that affect gene function, mRNA production, and its final expression [39]. The complex regulation of the *TNF* gene and the lack of information about the behavior of this gene in clinical studies with ischemia, among others, might explain this finding.

What is also of great relevance is the negative correlation between plasma levels of TNF-α and hsCRP with mRNA expression of *NOS3* and *VEGFA*, indicative of the deleterious effect of these cytokines on tissue regeneration related to angiogenesis (Figures 3 and 5). On the one hand, the role of plasma TNF-α in this setting was expected, as it was advanced in a previous experimental study [40]. Now our group first confirms that this important fact also occurs in humans with CLTI. Nevertheless, the possible role of CRP on angiogenesis was surprising. The data obtained in this regard confirm that this protein could be associated with homeostasis and angiogenesis in the state of high levels of inflammation, as indicated by an experimental study [41]. Therefore, it could be considered that this protein might have an active role rather than that of a passive acute phase marker. Both biomarkers, TNF-α and hsCRP, were also negatively correlated to *VEGFA* expression, supporting their impact on angiogenesis. Other authors have found a relationship between plasma levels of VEGF and hsCRP in experimental models of stroke [42]. Again, our group is the first to describe the connection between the expression of *VEFGA* in the skeletal muscle and plasma hsCRP in patients with lower limb ischemia.

The mentioned findings are parallel and consistent with the positive significant correlation between basal plasma TNF-α and *NOX4* in all patients of this study. The data obtained show that plasma TNF-α has a different behavior pattern depending on its levels related to *NOX4* expression (Figure 4B), which supports the fact that a high level of TNF-α, and hence, of inflammation, is responsible for its deleterious effect on redox balance. The fact that patients without DM seemed to have a greater correlation between TNF-α and *NOX4* needs confirmation and explanation.

In line with this finding, the use of TNF-α inhibitors in patients with a high level of inflammation, such as rheumatoid arthritis, decreased atherosclerotic CV events similarly to the use of IL-6 blockers [43], also improving the endothelial dysfunction [24]. These drugs have demonstrated an anti-inflammatory efficacy by decreasing the expression of several molecules [25,44]. However, the benefit of TNF-α inhibitors for PAD management seems to be determined by *TNF* gene polymorphisms, and pharmacogenetics could help identify which individuals benefit most from them [45].

One of the most important data obtained in our study is the link between the plasma levels of TNF-α and *NOS3* expression. As previously published, *NOS3* is one of the most important enzymes that maintains vascular homeostasis, and it is involved in vascular defense against chronic or excessive inflammation [27,28]. TNF-α can affect the activity of *NOS3* with an independent action on its gene promoter in a dose- and time-dependent manner [40]. This proinflammatory molecule facilitates the phosphorylation of *NOS3*, reducing the level of nitric oxide (NO), facilitating endothelial dysfunction, and altering the regeneration process [40,46]. This finding in humans suggests that TNF-α might be an accurate target for CLTI. To see the importance of the decrease in plasma TNF-α by GH, it is worth highlighting a meta-analysis of 54 prospective cohort studies on inflammation and PAD, in which plasma TNF-α was able to predict the risk of CV events with the same magnitude as the therapies for lowering blood pressure or lipids [47,48].

The observation that plasma TNF-α and *NOX4* expression run in parallel is also remarkable. NOX4 enzyme is the main isoform of NADPH oxidase responsible for TNF-α-induced oxidative distress and apoptosis of different cells in the body [49–51]. An important source of ROS in blood vessels comes from NOX catalytic enzymes, which are ubiquitously distributed in the three vessel layers. NOX4 enzyme, which maintains a physiological basal ROS generation, is highly expressed in cells under stress [27]. This protein could play a fundamental role in the regulation of angiogenic growth factors, such as VEGFA, since the inhibition of NOX and/or the production of mitochondrial ROS decreases the expression of this growth factor [52]. Thus, our finding of the link between TNF-α and *VEGFA* at baseline in those patients with a high level of inflammation (GH group) makes sense. Although TNF-α can favor vascular homeostasis when it is produced in small concentrations by endothelial cells (ECs), a chronically high production of this particle determines deleterious effects, overstimulating NOX4 enzyme, which eventually leads to an excess of ROS that ends up in the final inactivation of NO [22,53]. These circumstances warranted ECs activation, favoring the prothrombotic state and thrombotic complications associated with advanced PAD [29,54].

While in the CANTOS study the relevant role is for IL-6 and IL-1B in patients with ischemic heart disease [20], Gardner et al. demonstrated that in patients with PAD, TNF-α and IL-8 levels were increased but not IL-6 levels [22], which supports the fact that both diseases seem to show differences in the inflammatory pattern. However, Gardner's study was conducted in patients with intermittent claudication, a less severe form of ischemia with less inflammation, and the causal association between this or other biomarkers and PAD has not yet been established. In the GHAS trial, *IL-6* mRNA expression was not altered at baseline, and GH treatment did not affect this expression either. However, information on plasma IL-6 levels was not included in the GHAS study.

Baseline level of plasma HbA1C appears to be well correlated with TNF-α in the GH group with DM, supporting the role of this biomarker in patients with a high level of inflammation (Figure 6A). The final levels of HbA1C were inversely related to *VEGFA* mRNA expression in DM, but only in the placebo group (r = −0.9747, *p* = 0.0333) (Figure 6C), which means that untreated DM patients maintain a higher level of inflammation and, therefore, angiogenesis is negatively affected. Thus, HbA1C should be a possible target in these patients, highlighting the need to a good glycemic control.

On the other hand, mRNA expression of *IGF-1* seemed to follow an inverse behavior, increasing its expression in ischemic muscles with inflammation, probably as a defense mechanism. That is, in cultures of ECs with a high level of inflammation induced by the addition of CRP, this protein increases the phosphorylation of the enzyme eNOS, decreasing its activity. When IGF-1 protein is added to this medium rich in CRP, the activity of the enzyme eNOS increases [41], which means that IGF-1 protein exerts a negative feedback on the pernicious effect of CRP on ECs, and that these cells can produce IGF-1 in response to a high level of CRP. In our trial, the treatment with GH restored inflammatory and redox imbalances, and the association between CRP and IGF-1 reverted to inverse (Table 5). In the placebo group, this benefit did not appear, maintaining the positive association. Although CRP and IGF-1 were investigated in PAD–CLTI patients in a prior study [55], only the plasma level of these factors was studied and not the gene expression in the ischemic muscle.

Last but not least, we obtained relevant data on some plasma biomarkers and mortality. NLR has been proposed as a marker of adverse complications in CV disease, malignancy, and infection [56]. In PAD, NLR was incorporated into the ERICVA score for the prediction of poor prognosis in patients with CLTI [57]. In our study, we saw that NLR ≥ 3 at baseline was related to cumulative mortality at 12 months (Figure 8A) and was correlated with the level of plasma hsCRP (Figure 9). Although the appropriate cutoff point for NLR in CV disease has yet to be adequately defined [26], we found that a value of 3.4 was the best predictor for long-term mortality in CLTI patients (Figure 8B) with a good sensitivity and specificity and a light to moderate power of the test. This observed value was very close to the average NLR value in the series (3.5). Perhaps, in studies with acute limb ischemia, where NLR is normally higher, the threshold seems to be 5.4 [56], and in those studies performed in patients with intermittent claudication, the cutoff point was 5 [57]. It makes sense that the higher the level of inflammation, the lower the NLR cutoff point could be for predicting mortality, and that this cutoff point might be different depending on the morbid condition and the level of inflammation. Plasma TNF-α ≥ 8.1 was also found as a possible predictor of mortality. The calculation of the confidence interval for the odds ratio of both biomarkers did not find significant differences (see CI in Table 7), probably as a consequence of the small sample size. However, in the case of NLR, it can be seen a clear trend.

The main limitation of the GHAS trial is the relatively small sample size and the lack of confirmation of some of these data (with some exceptions) when the multivariate model was used for statistics, although the latter was only applied to cross-link clinical and plasma variables, not for gene expression. Furthermore, as this study represents the first clinical trial exploring the use of GH for angiogenesis, the dose and time of application of this hormone have not properly been established [28,29]. However, plasma and gene expression data at baseline are not affected by the intervention in this study. The main advantage of this study is that the ischemic muscle has been studied in depth, offering relevant data.

5. Conclusions

Vascular homeostasis and redox balance govern vascular health or disease states. Chronic limb-threatening ischemia is responsible for a high level of inflammation and mortality rate and seems to show differences in the inflammatory pattern compared to ischemic heart disease. The behavior of some biomarkers in patients with lower limb ischemia depends on the level of inflammation. Plasma TNF-α plays a crucial role in both inflammation-mediated redox distress and endothelial dysfunction and regeneration, affecting *NOS3* and *VEFGA* expression. CRP seems to play a more active role than previously assigned, affecting also redox balance and angiogenesis. Muscle NOX4 enzyme has been revealed as key for redox imbalance, and IGF-1 could be produced as a mechanism of defense against ischemia. Diabetic patients with ischemia of the lower limbs represent an especially targeted population for antioxidant and regenerative therapies as GH. NLR $\geq$ 3, together with plasma TNF-$\alpha \geq 8.1$, could be a good predictor of mortality in this morbid condition.

All these findings are clinically relevant and open new paths in the search for new drugs to decrease cardiovascular events and to relieve symptoms in patients with chronic limb-threatening ischemia. Our results are novel because they represent the link between plasma and muscle gene expression in humans with limb ischemia, but need to be corroborated in a larger clinical trial.

Author Contributions: Conceptualization, D.C. and J.D.; methodology, D.C., C.V.A. and S.P.-R.; software, S.P-R and D.C.; validation, D.C., C.V.A., and J.D.; formal analysis, D.C. and S.P.-R.; investigation, D.C., C.V.A. and S.P.-R.; resources, D.C. and C.V.A.; data curation, D.C. and S.P.-R.; writing—original draft preparation, D.C.; writing—review and editing, D.C., C.V.A., S.P.-R. and J.D.; visualization, D.C. and S.P.-R.; supervision, C.V.A. and J.D.; project administration, D.C.; funding acquisition, D.C. and C.V.A. All authors have read and agreed to the published version of the manuscript.

Funding: The GHAS trial was funded by the Carlos III Health Institute and the European Regional Development Fund (ISCIII-FEDER), Madrid, Spain. Grant number PI 13-00790. C.V.A. is funded by the Agencia Estatal de Investigación (AEI), grant number PID2019-110437RB-I00, and Xunta de Galicia -ERDF (Centro singular de investigación de Galicia accreditation 2019–2022).

Institutional Review Board Statement: The study was conducted according to the guidelines of the Declaration of Helsinki and approved by the Spanish Agency of Drugs and Health Products (AEMPs) with Eudract number 2012-002228-34 and the Autonomic Committee on Research Ethics in Galicia (CAEIG, 2012/378), Spain. The GHAS trial counts with the registration in the Spanish Registry of Clinical Trials (REEC): 2012-002228-34.

Informed Consent Statement: Informed consent was obtained from all subjects involved in the study.

Data Availability Statement: The data presented in this study are available on request from the corresponding author. The data are not publicly available due to data protection policy in Galicia, Spain.

Acknowledgments: We thank Santiago Pérez Cachafeiro for providing an important stimulus to develop this review and the GHAS trial. Without his help and expertise, this article would not have been possible. We equally thank Miguel Chenlo for his help in formatting the figures.

Conflicts of Interest: The authors declare no conflict of interest. The funders had no role in the design of the study; in the collection, analyses, or interpretation of data; in the writing of the manuscript, or in the decision to publish the results.

References

1. Heart Protection Study Collaborative Group. MRC/BHF Heart Protection Study of cholesterol lowering with simvastatin in 20,536 high-risk individuals: A randomised placebocontrolled trial. *Lancet* **2002**, *360*, 7–22. [CrossRef]
2. Kumbhani, D.J.; Steg, G.; Cannon, C.P.; Eagle, K.A.; Smith, S.C.; Goto, S.; Magnus Ohman, E.; Elbez, Y.; Sritara, P.; Baumgartner, I.; et al. Statin therapy and long-term adverse limb outcomes in patients with peripheral artery disease: Insights from the REACH registry. *Eur. Heart J.* **2014**, *35*, 2864–2872. [CrossRef] [PubMed]
3. Ohman, E.M.; Bhatt, D.L.; Steg, P.G.; Goto, S.; Hirsch, A.T.; Liau, C.S.; Mas, J.L.; Richard, A.J.; Röther, J.; Wilson, P.W.F. The REduction of Atherothrombosis for Continued Health (REACH) Registry: An international, prospective, observational investigation in subjects at risk for atherothrombotic events-study design. *Am. Heart J.* **2006**, *151*, 786.e1–786.e10. [CrossRef] [PubMed]
4. Haugen, S.; Casserly, I.P.; Regensteiner, J.G.; Hiatt, W.R. Risk assessment in the patient with established peripheral arterial disease. *Vasc. Med.* **2007**, *12*, 343–350. [CrossRef] [PubMed]
5. Diehm, C. Association of low ankle brachial index with high mortality in primary care. *Eur. Heart J.* **2006**, *27*, 1743–1749. [CrossRef]
6. Resnick, H.E.; Lindsay, R.S.; McDermott, M.M.; Devereux, R.B.; Jones, K.L.; Fabsitz, R.R.; Howard, B.V. Relationship of high and low ankle brachial index to all-cause and cardiovascular disease mortality: The Strong Heart Study. *Circulation* **2004**, *109*, 733–739. [CrossRef]
7. Kistorp, C. Risk Stratification in Secondary Prevention. *Circulation* **2006**, *114*, 184–186. [CrossRef] [PubMed]
8. Brevetti, G.; Giugliano, G.; Brevetti, L.; Hiatt, W.R. Inflammation in peripheral artery disease. *Circulation* **2010**, *122*, 1862–1875. [CrossRef]
9. Saenz-pipaon, G.; Martinez-aguilar, E.; Orbe, J.; Miqueo, A.G.; Fernandez-alonso, L.; Paramo, J.A.; Roncal, C. The role of circulating biomarkers in peripheral arterial disease. *Int. J. Mol. Sci.* **2021**, *22*, 3601. [CrossRef]
10. Aboyans, V.; Criqui, M.H. Can we improve cardiovascular risk prediction beyond risk equations in the physician's office? *J. Clin. Epidemiol.* **2006**, *59*, 547–558. [CrossRef]
11. Steering Committee of the Physicians' Health Study Research Group. Physicians' Health Study: Aspirin and Primary Prevention of Coronary Heart Disease. *N. Engl. J. Med.* **1989**, *321*, 1825–1828. [CrossRef] [PubMed]
12. Steering Committee of the Physicians' Health Study Research Group. Final Report on the Aspirin Component of the Ongoing Physicians' Health Study. *N. Engl. J. Med.* **1989**, *321*, 129–135. [CrossRef] [PubMed]
13. Ridker, P.M.; Hennekens, C.H.; Buring, J.E.; Rifai, N. C-Reactive Protein and Other Markers of Inflammation in the Prediction of Cardiovascular Disease in Women. *N. Engl. J. Med.* **2000**, *342*, 836–843. [CrossRef] [PubMed]
14. Kleemann, R.; Zadelaar, S.; Kooistra, T. Cytokines and atherosclerosis: A comprehensive review of studies in mice. *Cardiovasc. Res.* **2008**, *79*, 360–376. [CrossRef]
15. Ridker, P.M. Anticytokine Agents Targeting Interleukin Signaling Pathways for the Treatment of Atherothrombosis. *Circ. Res.* **2019**, *124*, 437–450. [CrossRef]
16. Kaptoge, S.; Seshasai, S.R.K.; Gao, P.; Freitag, D.F.; Butterworth, A.S.; Borglykke, A.; Di Angelantonio, E.; Gudnason, V.; Rumley, A.; Lowe, G.D.O.; et al. Inflammatory cytokines and risk of coronary heart disease: New prospective study and updated meta-analysis. *Eur. Heart J.* **2014**, *35*, 578–589. [CrossRef]
17. Interleukin-6 Receptor Mendelian Randomisation Analysis (IL6R MR) Consortium. The interleukin-6 receptor as a target for prevention of coronary heart disease: A mendelian randomisation analysis. *Lancet* **2012**, *379*, 1205–1213. [CrossRef]
18. Swerdlow, D.I.; Holmes, M.V.; Kuchenbaecker, K.B.; Engmann, J.; Shah, T.; Sofat, R. The interleukin-6 receptor as a potential target for coronary heart disease prevention: Evaluation using Mendelian randomization. *Lancet* **2012**, *379*, 1214–1224. [CrossRef]
19. Qasim, A.N.; Reilly, M.P. Genetics of Atherosclerotic Cardiovascular Disease. In *Emery and Rimoin's Principles and Practice of Medical Genetics*; Elsevier: Amsterdam, The Netherlands, 2013; pp. 1–37.
20. Ridker, P.M.; Everett, B.M.; Thuren, T.; MacFadyen, J.G.; Chang, W.H.; Ballantyne, C.; Fonseca, F.; Nicolau, J.; Koenig, W.; Anker, S.D.; et al. Antiinflammatory therapy with canakinumab for atherosclerotic disease. *N. Engl. J. Med.* **2017**, *377*, 1119–1131. [CrossRef]
21. Signorelli, S.S.; Anzaldi, M.; Libra, M.; Navolanic, P.M.; Malaponte, G.; Mangano, K.; Quattrocchi, C.; Di Marco, R.; Fiore, V.; Neri, S. Plasma Levels of Inflammatory Biomarkers in Peripheral Arterial Disease. *Angiology* **2016**, *67*, 870–874. [CrossRef]
22. Gardner, A.W.; Parker, D.E.; Montgomery, P.S.; Sosnowska, D.; Casanegra, A.I.; Esponda, O.L.; Ungvari, Z.; Csiszar, A.; Sonntag, W.E. Impaired Vascular Endothelial Growth Factor A and Inflammation in Patients With Peripheral Artery Disease. *Angiology* **2014**, *165*, 683–690. [CrossRef] [PubMed]
23. Brandt, C.; Pedersen, B.K. The Role of Exercise-Induced Myokines in Muscle Homeostasis and the Defense against Chronic Diseases. *J. Biomed. Biotechnol.* **2010**, *2010*, 520258. [CrossRef] [PubMed]
24. Barnabe, C.; Martin, B.-J.; Ghali, W.A. Systematic review and meta-analysis: Anti-tumor necrosis factor α therapy and cardiovascular events in rheumatoid arthritis. *Arthritis Care Res.* **2011**, *63*, 522–529. [CrossRef] [PubMed]
25. Murdaca, G.; Spanò, F.; Cagnati, P.; Puppo, F. Free radicals and endothelial dysfunction: Potential positive effects of TNF-α inhibitors. *Redox Rep.* **2013**, *18*, 95–99. [CrossRef] [PubMed]
26. Paquissi, F. The role of inflammation in cardiovascular diseases: The predictive value of neutrophil–lymphocyte ratio as a marker in peripheral arterial disease. *Ther. Clin. Risk Manag.* **2016**, *12*, 851. [CrossRef]

27. Bir, S.C.; Kolluru, G.K.; Fang, K.; Kevil, C.G. Redox balance dynamically regulates vascular growth and remodeling. *Semin. Cell Dev. Biol.* **2012**, *23*, 745–757. [CrossRef]

28. Caicedo, D.; Devesa, P.; Alvarez, C.V.; Devesa, J. Why Should Growth Hormone (GH) Be Considered a Promising Therapeutic Agent for Arteriogenesis? Insights from the GHAS Trial. *Cells* **2020**, *9*, 807. [CrossRef]

29. Caicedo, D.; Díaz, O.; Devesa, P.; Devesa, J. Growth Hormone (GH) and Cardiovascular System. *Int. J. Mol. Sci.* **2018**, *19*, 290. [CrossRef]

30. Caicedo, D.; Devesa, J. Growth Hormone (GH) and Wound Healing. In *Wound Healing-Current Perspectives*; IntechOpen: London, UK, 2019. [CrossRef]

31. Conte, M.S.; Bradbury, A.W.; Kolh, P.; White, J.V.; Dick, F.; Fitridge, R.; Mills, J.L.; Ricco, J.B.; Suresh, K.R.; Murad, M.H.; et al. Global Vascular Guidelines on the Management of Chronic Limb-Threatening Ischemia. *Eur. J. Vasc. Endovasc. Surg.* **2019**, *58*, S1–S109.e33. [CrossRef]

32. Aboyans, V.; Ricco, J.-B.; Bartelink, M.-L.E.L.; Björck, M.; Brodmann, M.; Cohnert, T.; Collet, J.-P.; Czerny, M.; De Carlo, M.; Debus, S.; et al. 2017 ESC Guidelines on the Diagnosis and Treatment of Peripheral Arterial Diseases, in collaboration with the European Society for Vascular Surgery (ESVS). *Eur. J. Vasc. Endovasc. Surg.* **2018**, *55*, 305–368. [CrossRef]

33. Suárez, C.; Lozano, F. *Guía Española de Consenso Multidisciplinar en Enfermedad Arterial Periférica de Extremidades Inferiores*, 1st ed.; Luzán, S.A., Ed.; FESEMI: Madrid, Spain, 2012; ISBN 978-84-7989-716-1.

34. Garcia-Rendueles, A.R.; Rodrigues, J.S.; Garcia-Rendueles, M.E.R.; Suarez-Fariña, M.; Perez-Romero, S.; Barreiro, F.; Bernabeu, I.; Rodriguez-Garcia, J.; Fugazzola, L.; Sakai, T.; et al. Rewiring of the apoptotic TGF-β-SMAD/NFκB pathway through an oncogenic function of p27 in human papillary thyroid cancer. *Oncogene* **2017**, *36*, 652–666. [CrossRef] [PubMed]

35. Chenlo, M.; Rodriguez-Gomez, I.A.; Serramito, R.; Garcia-Rendueles, A.R.; Villar-Taibo, R.; Fernandez-Rodriguez, E.; Perez-Romero, S.; Suarez-Fariña, M.; Garcia-Allut, A.; Cabezas-Agricola, J.M.; et al. Unmasking a new prognostic marker and therapeutic target from the GDNF-RET/PIT1/p14ARF/p53 pathway in acromegaly. *EBioMedicine* **2019**, *43*, 537–552. [CrossRef] [PubMed]

36. Carlson, B.M. Part.II. Developmento of the body systems. Cardiovascular System. In *Human Embryology and Developmental Biology*; Carlson, B., Ed.; Elsevier: St Louis, MO, USA, 2019; p. 496, ISBN 9780323523752.

37. Byrd, N.; Grabel, L. Hedgehog signaling in murine vasculogenesis and angiogenesis. *Trends Cardiovasc. Med.* **2004**, *14*, 308–313. [CrossRef] [PubMed]

38. Criqui, M.H.; McClelland, R.L.; McDermott, M.M.; Allison, M.A.; Blumenthal, R.S.; Aboyans, V.; Ix, J.H.; Burke, G.L.; Liu, K.; Shea, S. The ankle-brachial index and incident cardiovascular events in the MESA (Multi-Ethnic Study of Atherosclerosis). *J. Am. Coll. Cardiol.* **2010**, *56*, 1506–1512. [CrossRef]

39. Fragoso, J.; Sierra, M.; Vargas, G.; Barrios, A.; Ramírez, J. El factor de necrosis tumoral α (TNF-α) en las enfermedades cardiovasculares: Biología molecular y genética. *Gac. Med. Mex.* **2013**, *149*, 521–530.

40. Anderson, H.D.I.; Rahmutula, D.; Gardner, D.G. Tumor Necrosis Factor-α Inhibits Endothelial Nitric-oxide Synthase Gene Promoter Activity in Bovine Aortic Endothelial Cells. *J. Biol. Chem.* **2004**, *279*, 963–969. [CrossRef]

41. Liu, S.-J.; Zhong, Y.; You, X.-Y.; Liu, W.-H.; Li, A.-Q.; Liu, S.-M. Insulin-like growth factor 1 opposes the effects of C-reactive protein on endothelial cell activation. *Mol. Cell. Biochem.* **2014**, *385*, 199–205. [CrossRef]

42. Åberg, N.D.; Wall, A.; Anger, O.; Jood, K.; Andreasson, U.; Blennow, K.; Zetterberg, H.; Isgaard, J.; Jern, C.; Svensson, J. Circulating levels of vascular endothelial growth factor and post-stroke long-term functional outcome. *Acta Neurol. Scand.* **2020**, *141*, 405–414. [CrossRef]

43. Giles, J.T.; Sattar, N.; Gabriel, S.; Ridker, P.M.; Gay, S.; Warne, C.; Musselman, D.; Brockwell, L.; Shittu, E.; Klearman, M.; et al. Cardiovascular Safety of Tocilizumab Versus Etanercept in Rheumatoid Arthritis: A Randomized Controlled Trial. *Arthritis Rheumatol.* **2020**, *72*, 31–40. [CrossRef]

44. Murdaca, G.; Spanò, F.; Miglino, M.; Puppo, F. Effects of TNF-α inhibitors upon the mechanisms of action of VEGF. *Immunotherapy* **2013**, *5*, 113–115. [CrossRef]

45. De Simone, C.; Farina, M.; Maiorino, A.; Fanali, C.; Perino, F.; Flamini, A.; Caldarola, G.; Sgambato, A. TNF-alpha gene polymorphisms can help to predict response to etanercept in psoriatic patients. *J. Eur. Acad. Dermatol. Venereol.* **2015**, *29*, 1786–1790. [CrossRef] [PubMed]

46. Yan, G.; You, B.; Chen, S.-P.; Liao, J.K.; Sun, J. Tumor Necrosis Factor-α Downregulates Endothelial Nitric Oxide Synthase mRNA Stability via Translation Elongation Factor 1-α 1. *Circ. Res.* **2008**, *103*, 591–597. [CrossRef] [PubMed]

47. Kaptoge, S.; Di Angelantonio, E.; Lowe, G.; Pepys, M.B.; Thompson, S.G.; Collins, R.; Danesh, J.; Tipping, R.W.; Ford, C.E.; Pressel, S.L.; et al. C-reactive protein concentration and risk of coronary heart disease, stroke, and mortality: An individual participant meta-analysis. *Lancet* **2010**, *375*, 132–140. [CrossRef] [PubMed]

48. McDermott, M.M.; Guralnik, J.M.; Corsi, A.; Albay, M.; Macchi, C.; Bandinelli, S.; Ferrucci, L. Patterns of inflammation associated with peripheral arterial disease: The InCHIANTI study. *Am. Heart J.* **2005**, *150*, 276–281. [CrossRef] [PubMed]

49. Basuroy, S.; Tcheranova, D.; Bhattacharya, S.; Leffler, C.W.; Parfenova, H. Nox4 NADPH oxidase-derived reactive oxygen species, via endogenous carbon monoxide, promote survival of brain endothelial cells during TNF-α-induced apoptosis. *Am. J. Physiol. Cell Physiol.* **2011**, *300*, C256–C265. [CrossRef]

50. Moe, K.T.; Yin, N.O.; Naylynn, T.M.; Khairunnisa, K.; Wutyi, M.A.; Gu, Y.; Atan, M.S.M.; Wong, M.C.; Koh, T.H.; Wong, P. Nox2 and Nox4 mediate tumour necrosis factor-α-induced ventricular remodelling in mice. *J. Cell. Mol. Med.* **2011**, *15*, 2601–2613. [CrossRef]
51. Moe, K.T.; Aulia, S.; Jiang, F.; Chua, Y.L.; Koh, T.H.; Wong, M.C.; Dusting, G.J. Differential upregulation of Nox homologues of NADPH oxidase by tumor necrosis factor-alpha in human aortic smooth muscle and embryonic kidney cells. *J. Cell. Mol. Med.* **2006**, *10*, 231–239. [CrossRef]
52. Xia, C.; Meng, Q.; Liu, L.-Z.; Rojanasakul, Y.; Wang, X.-R.; Jiang, B.-H. Reactive Oxygen Species Regulate Angiogenesis and Tumor Growth through Vascular Endothelial Growth Factor. *Cancer Res.* **2007**, *67*, 10823–10830. [CrossRef]
53. Maekawa, Y.; Ishikawa, K.; Yasuda, O.; Oguro, R.; Hanasaki, H.; Kida, I.; Takemura, Y.; Ohishi, M.; Katsuya, T.; Rakugi, H. Klotho suppresses TNF-alpha-induced expression of adhesion molecules in the endothelium and attenuates NF-kappaB activation. *Endocrine* **2009**, *35*, 341–346. [CrossRef]
54. Signorelli, S.; Marino, E.; Scuto, S. Inflammation and Peripheral Arterial Disease. *J* **2019**, *2*, 142–151. [CrossRef]
55. Brevetti, G.; Colao, A.; Schiano, V.; Pivonello, R.; Laurenzano, E.; Di Somma, C.; Lombardi, G.; Chiariello, M. IGF system and peripheral arterial disease: Relationship with disease severity and inflammatory status of the affected limb. *Clin. Endocrinol.* **2008**, *69*, 894–900. [CrossRef] [PubMed]
56. Coelho, N.H.; Coelho, A.; Augusto, R.; Semião, C.; Peixoto, J.; Fernandes, L.; Martins, V.; Canedo, A.; Gregório, T. Pre-operative Neutrophil to Lymphocyte Ratio is Associated With 30 Day Death or Amputation After Revascularisation for Acute Limb Ischaemia. *Eur. J. Vasc. Endovasc. Surg.* **2021**, *62*, 74–80. [CrossRef] [PubMed]
57. Brizuela Sanz, J.A.; González Fajardo, J.A.; Taylor, J.H.; Río Solá, L.; Muñoz Moreno, M.F.; Vaquero Puerta, C. Design of a New Risk Score in Critical Limb Ischaemia: The ERICVA Model. *Eur. J. Vasc. Endovasc. Surg.* **2016**, *51*, 90–99. [CrossRef] [PubMed]

MDPI
St. Alban-Anlage 66
4052 Basel
Switzerland
www.mdpi.com

Biomedicines Editorial Office
E-mail: biomedicines@mdpi.com
www.mdpi.com/journal/biomedicines

www.ingramcontent.com/pod-product-compliance
Lightning Source LLC
LaVergne TN
LVHW080457180726
843515LV00007BA/1783